D1248456

Methods of Complex Analysis in Partial Differential Equations with Applications

CANADIAN MATHEMATICAL SOCIETY
SERIES OF MONOGRAPHS
AND ADVANCED TEXTS

Monographies et Études de la Société Mathématique du Canada

Frank H. Clarke **Optimization and Nonsmooth Analysis**
Erwin Klein and Anthony C. Thompson **Theory of Correspondences: Including Applications to Mathematical Economics**
I. Gohberg, P. Lancaster, and L. Rodman **Invariant Subspaces of Matrices with Applications**
Jonathan Borwein and Peter Borwein **PI and the AGM—A Study in Analytic Number Theory and Computational Complexity**
Subhashis Nag **The Complex Analytic Theory of Teichmüller Spaces**
Ernst J. Kani and Robert A. Smith **The Collected Papers of Hans Arnold Heilbronn**
Manfred Kracht and Erwin Kreyszig **Methods of Complex Analysis in Partial Differential Equations with Applications**

Methods of Complex Analysis in Partial Differential Equations with Applications

MANFRED KRACHT

University of Düsseldorf
Düsseldorf, Germany

ERWIN KREYSZIG

Carleton University
Ottawa, Ontario, Canada

A Wiley-Interscience Publication

JOHN WILEY & SONS

New York · Chichester · Brisbane · Toronto · Singapore

Library of Congress Cataloging in Publication Data:

Kracht, Manfred.
Methods of complex analysis in partial differential equations with applications.

(Canadian Mathematical Society series of monographs and advanced texts)
"A Wiley-Interscience publication."
Bibliography: pp. 349–378.
1. Functions of complex variables. 2. Differential equations, Partial. I. Kreyszig, Erwin. II. Title. III. Series.

QA331.K783 1987 515.3′53 87-33993
ISBN 0-471-83091-7

Printed in the United States of America

10 9 8 7 6 5 4 3 2 1

Dedicated to the Memory
of Our Dear Friend and Colleague
Stefan Bergman

Preface

This book is devoted to the development of complex function theoretic methods in partial differential equations and to the study of the analytic behavior of solutions. The main purpose of the book is the presentation of some basic facts and recent results, emphasizing the method of integral operators as initiated by S. Bergman, M. Eichler, and I. N. Vekua and developed further by D. L. Colton, R. P. Gilbert, E. Lanckau, J. Mitchell, M. M. Schiffer, and many others, and the method of differential operators as used by K. W. Bauer, H. Florian, G. Jank, E. Peschl, S. Ruscheweyh, and their schools.

These methods of applying concepts and results of complex analysis in the study of partial differential equations has become a large field of its own. Indeed, the number of papers in this field and its application to fluid flow, elasticity problems, quantum mechanics, and other areas is growing rapidly, and these results are widely scattered throughout many journals. Special results have been reported in a number of research monographs (e.g., Bauer–Ruscheweyh [1980],[1] Bergman [1971], Colton [1976c, 1976d, 1980], Gilbert [1969b, 1974], Gilbert–Buchanan [1983], Krzywoblocki [1960], Tutschke [1977], and Wendland [1979]) and of symposium proceedings and collections (e.g., Gilbert–Weinacht [1976], Lanckau–Tutschke [1983], Meister–Weck–Wendland [1976], Rassias [1986], Ruscheweyh [1973b], and Tutschke [1980]). Since, for the most part, these reports concentrate only on certain aspects of the general theory, and many important new results have not yet been included in any book at all, we shall pay particular attention to that progress during the last years as well as to the unification of various approaches and results, thereby exhibiting common features and comparing advantages and limitations of different methods.

Roughly speaking, the book may be divided into three parts as follows.

First, we investigate the existence and representation theory of integral operators and differential operators that transform analytic functions of one complex variable into solutions of linear second-order partial differential

[1]See the bibliography at the end of the book.

equations (Chaps. 2–5). Here, the main tools will be Bergman integral operators and—as far as they exist—Laplace–Darboux–Bauer differential operators. In view of applications, special emphasis is placed upon constructive methods for obtaining a variety of explicit and simple representations of solutions. Furthermore, relations to the theories of similar operators are discussed, e.g., for the Bergman–Gilbert operator, Eichler operator, Lanckau operator, complex Le Roux operator, Riemann–Vekua operator, and the transmutation theory. In order to enrich the instrumental apparatus for the investigation of solutions, we study the intrinsic relations between Bergman's method of integral operators of the first kind and the Riemann–Vekua approach, and we include methods and recent results in the determination of the complex Riemann function (Chaps. 6–7).

Second, the results of the preceding chapters, particularly on the existence and representation of solutions by suitable integral or differential operators, are applied to exploit the highly developed theory of analytic functions of one complex variable in the study of solutions of partial differential equations. Here, those operators work as a "translation principle." Instead of attempting to give a survey of the entire area, we shall emphasize ideas and techniques that demonstrate the power of this constructive operator method and give guidelines for further investigations in this field. Accordingly, we concentrate on operators that preserve various function theoretical properties of analytic functions and on applications of general interest. This includes results on the coefficient problem and the location and type of singularities (Chap. 8) and on the approximation of solutions (Chap. 9). Moreover, we establish a value distribution theory of solutions, with analogs and generalizations of Nevanlinna's Fundamental Theorems and the Great Picard Theorem (Chap. 10).

Third, we discuss applications of the theory in the case of three classes of equations that are of special significance in mathematical physics, beginning with the so-called Bauer–Peschl equation (Chap. 11). By certain transformations or separation processes we may relate this equation, which is a generalization of Laplace's equation, to the wave equation, the equation of generalized axially symmetric potential theory (GASPT), and certain second-order ordinary differential equations of special functions in mathematical physics. Here, using mainly the aforementioned differential operators, we establish analogs of theorems of geometric function theory, for solutions of partial differential equations. Moreover, we develop a special function theory of solutions that may be interpreted as a function theory of generalized (higher-order) Cauchy–Riemann equations. As a second class, in Chap. 12, we shall be concerned with basic equations of compressible fluid flow and shall discuss three methods for obtaining solutions. Finally, in Chap. 13 we shall present an integral operator approach to transonic flow. The basic tool is a modified Bergman integral operator of the second kind. This operator is used to solve Cauchy problems for the Tricomi equation with data given on the sonic line. Special expansions of solutions are investigated in detail. Relations to a special Eichler integral operator are utilized to obtain families of famous solutions from a unique source.

The bibliography at the end of the book contains about 600 entries. It is intended to give the reader not only the references cited in the book, but also background material and possible extensions with respect to special topics. At the end of every chapter we list those references from the bibliography that are closely related to the problems of the chapter. These short lists will serve as a guide to the bibliography.

This book evolved from a joint seminar at the Technical University of Karlsruhe, from seminars by the first author at the University of Düsseldorf, and from seminars by the second author at Stanford (jointly with S. Bergman) and at Ohio State University. We have made every effort to provide a presentation that is self-contained. The only prerequisite is a modest amount of complex analysis, as it is usually covered in a first course in that area. Special results needed in some applications are presented in full, with references to standard monographs.

We wish to express our thanks to the Natural Sciences and Engineering Research Council of Canada for support of this work under grant A9097, and to John Wiley and Sons, in particular, to Mrs. Maria Taylor, Editor, and Mr. Bob Hilbert, Editorial Supervisor, for efficient and pleasant cooperation.

MANFRED KRACHT
ERWIN KREYSZIG

Düsseldorf, Germany
Ottawa, Canada
January 1988

Contents

Chapter 1 ***Introduction*** *1*

1.1 Motivation and General Ideas, 1
1.2 General Orientation, 3

Chapter 2 ***Bergman Operators: General Theory*** *8*

2.1 Equations to Be Considered, 8
2.2 Integral Operators of Bergman Type, 12
2.3 Conditions for Kernels, 15
2.4 Integral Operators of the First Kind, 20
2.5 Further Representations of Operators of the First Kind, 26
2.6 The Inversion Problem, 31
2.7 Related Ideas and Operators, 34
Additional References, 44

Chapter 3 ***Integral Operators with Polynomial Kernels and Differential Operators*** *45*

3.1 Operators of Class P, 45
3.2 Integral-Free Representations of Solutions, 51
3.3 Differential Operators, 53
3.4 Inversion Problem for Class P Operators, 57
Additional References, 62

Chapter 4 ***Polynomial Kernels: Existence and Construction*** *63*

4.1 Existence of Polynomial Kernels in Special Cases, 63
4.2 Existence of Polynomial Kernels for Type I Representations, 68
4.3 Existence of Polynomial Kernels for Type II Representations, 77
4.4 Construction Principle for Class P Operators and Equations, 81

4.5 Second Construction Principle, 88
4.6 Applications, 94
4.7 Self-Adjoint Equations, 98
Additional References, 107

Chapter 5 ***Further Closed-Form Kernels*** *108*

5.1 Exponential Operators and Kernels, 109
5.2 Properties of Exponential Operators, 113
5.3 Further Results on Exponential Operators, 116
5.4 General Remarks. Rational Kernels, 118
Additional References, 123

Chapter 6 ***Riemann–Vekua Representation and Further Methods Related to Bergman Kernels*** *124*

6.1 Fundamental Domains and Le Roux Operators, 125
6.2 Complex Riemann Function: Existence and Uniqueness, 128
6.3 Properties of the Complex Riemann Function, 134
6.4 Riemann–Vekua Representation and Bergman Representation, 137
6.5 Bergman–Gilbert Operator, 147
6.6 Relations to Carroll's Theory of Transmutations, 158
Additional References, 165

Chapter 7 ***Determination of Riemann Functions*** *166*

7.1 Classical Methods and Generalizations, 167
7.2 Addition of Riemann Functions, 176
7.3 Further Results, 180
Additional References, 196

Chapter 8 ***Coefficient Problem and Singularities of Solutions*** *197*

8.1 General Setting and Approach, 198
8.2 Use of Operators of the First Kind, 199
8.3 Further Coefficient Theorems, 201
8.4 Coefficient Theorems for Solutions, 203
8.5 Singularities of Solutions, 207
Additional References, 208

Chapter 9 ***Approximation of Solutions*** *209*

9.1 Analogs of Runge's and Mergelyan's Theorems, 210
9.2 Approximation Theorem of Walsh Type, 213

9.3 Further Approximation Theorems for Solutions, 216
Additional References, 222

Chapter 10 ***Value Distribution Theory of Solutions*** **223**

10.1 General Idea and Setting, 224
10.2 Generalized Nevanlinna's Second Theorem, 236
10.3 Analog of the Little Picard Theorem, 246
10.4 Generalization of the Great Picard Theorem, 249
Additional References, 255

Chapter 11 ***Applications of Class P Operators. Function Theory of the Bauer–Peschl Equation*** **256**

11.1 Mathematical and Physical Importance of the Equation, 257
11.2 Representations of Solutions, 260
11.3 Properties of Solutions Satisfying Additional Conditions, 272
11.4 Analogs of Theorems from Geometric Function Theory, 289
11.5 A Function Theory of Solutions Satisfying Generalized Cauchy–Riemann Equations, 298
Additional References, 305

Chapter 12 ***Application to Compressible Fluid Flow*** **306**

12.1 Basic Concepts and Equations, 306
12.2 Bernoulli's Law, Chaplygin's Equation, 309
12.3 Method of Pseudoanalytic Functions, 312
12.4 Chaplygin's Approach, 314
12.5 Bergman's Integral Operator Method, 317
Additional References, 321

Chapter 13 ***Integral Operators Applied to Transonic Flow. Tricomi Equation*** **322**

13.1 Equations for the Stream Function, 323
13.2 Integral Operator for the Tricomi Equation, 325
13.3 Cauchy Problems for the Tricomi Equation, 329
13.4 Entire Cauchy Data for the Tricomi Equation. Borel Transform, 333
13.5 Polynomial Expansions of Solutions of the Tricomi Equation, 335
13.6 A Special Cauchy Problem, 339

13.7 Eichler's Integral Operator for the Tricomi Equation, 340
13.8 Families of Solutions of the Tricomi Equation, 344
Additional References, 348

Bibliography ***349***

Symbol Index ***379***

Author Index ***381***

Subject Index ***387***

Methods of Complex Analysis in Partial Differential Equations with Applications

Chapter One

Introduction

1.1 MOTIVATION AND GENERAL IDEAS

Two-dimensional potential theory—the so-called theory of the *logarithmic potential*—is very closely related to the theory of complex analytic functions of one variable. This well-known fact has important consequences. For studying and solving *two-dimensional potential problems*, one can utilize various methods from complex analysis, such as techniques of conformal mapping, ideas related to the concept of complex potential, the Cauchy and Poisson formulas, maximum principles, or Harnack's convergence theorem. In this way, results on complex analytic functions can easily be *translated* into theorems on harmonic functions in two real variables.

This close relation to potential theory is one of the main reasons for the basic importance of complex analysis in physics and engineering, another being the more satisfactory *complex* theory of families of special functions arising as particular solutions in the separation of the wave, heat, potential, and other equations in different coordinate systems.

The idea of relating the *two-dimensional Laplace equation* to complex function theory appeared early and led to an interrelated evolution of the two areas, in the works of Riemann, H. A. Schwarz, Dirichlet, C. Neumann, Poincaré, and others.

The wealth of detailed and specific deep results obtained in complex analysis over a long period of time would make it very desirable to establish relations to more general partial differential equations, so that these results could be used in developing and applying a comprehensive unified theory of solutions of these equations, for which at first only very few analogs of methods and theorems of complex analytic function theory are known. That this is not an easy task becomes obvious if one recalls the extent to which the situation changes in the transition from the Laplace to other equations.

However, in such a case, one may still obtain relations of that type by introducing suitable *integral operators* which map complex analytic functions onto solutions of given linear partial differential equations. In this way, from methods and theorems of complex analysis, these operators yield results on various properties of solutions. This approach has lately become known as the ***function theoretical method*** in partial differential equations.

Integral operators of various kinds have been used for a long time. What distinguishes this modern Bergman–Eichler–Vekua approach from older work is the idea of creating classes of integral operators on algebras of analytic functions with the explicit purpose that these operators act as a *translation principle* for methods and results in complex function theory in order to obtain detailed information on solutions of given partial differential equations in a systematic fashion.

Such operators were introduced by S. Bergman [1937b][1] and independently by I. N. Vekua [1937] and M. Eichler [1942], but a general, almost explosive development in the area started only much later, a few years after the 1961 first edition of Bergman's monograph [1971] made these ideas known in wide circles of mathematicians and physicists.

We emphasize strongly that this integral operator method is ***constructive.*** This fact is basic in the solution of boundary and initial value problems. Beyond that, a main goal of the method is the characterization of general properties of solutions of given equations, using a unified procedure and emphasizing properties that are independent of the special form of the coefficients of the equation or depend on it in a simple way. The development up to the present has produced an abundance of basic results of this kind, for instance, on the growth of solutions, their domain of regularity, the location and type of singularities, on coefficient problems for various representations, on the approximation by "polynomial-type" solutions and other classes of functions (generalized Runge-, Mergelyan-, and Walsh-type theorems), on the behavior near the boundary of the domain of regularity, on continuation and global representation problems, and so on.

The specific character of results available from complex analysis entails the advantage that one can gain deep and detailed insight into the nature of solutions and a complete characterization of basic properties of large classes of solutions. Also, the whole approach, being quite different from classical methods, yields results not easily obtainable from those methods or other modern ones.

Such investigations of properties of solutions can be kept independent of boundary or other side conditions, an aspect which has recently become of interest in quantum field theory and in applications of inverse methods. This also yields a new access to problems of elasticity, hydrodynamics, and other areas of physics.

Furthermore, these operators also provide a practical method for obtaining complete families of solutions, as they are needed in *boundary value problems*. Here, the discussion can be carried up to the very threshold of numerical

[1]See the bibliography at the end of the book.

computation, as Bergman–Herriot [1965], Gilbert–Atkinson [1970], and others have shown. This includes various reductions to most suitable forms for specific applications.

In this connection, these operators will in general produce new families of particular solutions different from those in classical work, and thus provide for applications of the theory of special functions or, conversely, will enrich the latter by new aspects and results. An early illustration of this circle of ideas was given by P. Henrici [1953] in connection with the wave equation; and Bergman, Vekua, and others have treated the Helmholtz equation in a similar spirit. Various other relations to the theory of special functions are outlined in the monographs by R. P. Gilbert [1969b] and D. L. Colton [1980].

In the light of all these facts, it should not be surprising that the present methods have found significant applications in compressible fluid flow—Bergman's early work, aimed at improving Chaplygin's famous method, was in this field—and various other boundary value and initial value problems, to shell theory and more general problems in elasticity, to surface bending and other differential geometric mapping problems, to approximation theory, and so on.

The present integral operator approach is particularly powerful in connection with ***ill-posed problems.*** Classically a curiosity of marginal interest, these problems are becoming of increasing practical importance. For simple examples, see Colton [1980] and for more sophisticated ones (related to aerodynamical transition problems), see our Chap. 13. For computer applications in the design of airfoils, see Bauer et al. [1972, 1975]. Together with *inverse problems*, these problems are developing into an interdisciplinary area of joint research of mathematicians, computer scientists, geophysicists and engineers, relating to inverse scattering, atmospheric studies, remote sensing, profile inversion, seismology, biomedical tomography, and system identification, to mention just a few areas. For a general (nontechnical) introduction to motivations and ideas in this field, see also Payne [1975].

1.2 GENERAL ORIENTATION

The integral operator method characterized in general terms in the previous section applies to any linear partial differential equation with analytic coefficients and to systems of such equations, regardless of order or type.

For a given equation there will in general exist various such integral operators. Since these operators are supposed to act as a translation principle (cf. Sec. 1.1), a main problem usually is to introduce operators which are most suitable in that respect, and to represent them in different forms, depending on the specific purpose. Much work and experience has been incorporated in this task of extending the theory beyond its original general frame, as given in the fundamental papers by Bergman and others.

In this book, our emphasis will be on recent developments, giving preference to topics of general interest and unifying aspects in theory and application, and we shall also include unpublished work of our own. The presentation will be self-contained.

Over the past two decades, the whole field has grown very large and diversified, and the amount of literature is truly overwhelming. This made it mandatory to concentrate on central areas in the field in order to obtain a balanced account of the general ideas. Accordingly, we have selected second-order equations in two complex variables for consideration. These equations may be considered (but not necessarily) as obtained from real hyperbolic or elliptic equations by extension to complex values and suitable transformation. (In the elliptic case, the transition to real-valued solutions is quite simple.) In many respects, these equations constitute a model case of the whole theory and its applications, in the sense that they exhibit various features which have counterparts in connection with equations of higher order or in more than two variables. None of the latter will be considered in this book. We shall also leave aside systems of equations; for these we refer to the recent monographs by Gilbert–Buchanan [1983] and Wendland [1979]. Our choice was influenced by the observation that second-order equations in two variables are probably at present the most active area of research in the rapidly developing integral operator method.

We shall also leave aside other basic tools in the function theoretic approach to partial differential equations, namely, *pseudoanalytic* (generalized analytic) and *generalized hyperanalytic functions*. For these functions we refer to the excellent survey articles and monographs by Bers [1956], Vekua [1963], Položiĭ [1965b], and Tutschke [1977] in the case of pseudoanalytic functions, and by Gilbert [1983] and Gilbert–Buchanan [1983] in the case of generalized hyperanalytic functions; see also our introductory remarks to Sec. 11.5.

Let us now turn to a brief orientation about the content of the following chapters.

Chapter 2 contains the basic concepts and fundamental theorems on Bergman and related integral operators, in particular, Representation Theorem 2.3-1 involving the *kernel equation* (2.3-2a). This includes a thorough introduction to the so-called *operators of the first kind*, whose importance results from the property that they have a very simple inverse. This is basic in most applications. The reader will notice that we have made great effort in closing gaps existing in certain monographs and papers, and we have also added material which should prove helpful for practical purposes. Section 2.4 is concerned with the *existence* of operators of the first kind, and in Sec. 2.5 we consider further representations of solutions by these operators whose practical value becomes apparent in applications, e.g., in Chap. 9. Section 2.6 gives the solution of the inversion problem which yields the solution of *Goursat initial value problems*. In order to show the model character of the approach just described, we consider two natural kinds of generalization to higher-dimensional equations, also of parabolic type, as proposed by Colton [1976c, d, 1980] and Lanckau [1983b, 1986].

Chapter 3 is the first one devoted to the idea of operators whose kernels are of *closed form* so that they are suitable for obtaining ***global representations** of solutions*. In this chapter, these kernels are polynomials in the variable of integration t with coefficient functions depending on the independent variables z_1, z_2. They apply to a large class of equations which includes equations related to the wave equation, Delassus equation, Euler–Poisson–Darboux equation, and other equations of practical interest. The corresponding operators with those kernels, we have termed ***class P operators.*** They have recently attracted great attention and have been studied in a large number of papers that appeared after the authors succeeded in showing that certain classes of *differential operators* are actually class P operators that have been converted into an integral-free form, so that those operators can be elegantly treated within the framework of the present integral operator theory, as we show in Chap. 3. The *inversion problem* for those operators is solved in Sec. 3.4.

Chapter 4 is concerned with the *existence problem* of class P operators and exhibits necessary and sufficient conditions for equations for which these operators exist. Special attention is attached to several ***construction principles*** that simultaneously admit the construction of equations and corresponding operators. Some typical examples are given for illustration.

Chapter 5 concerns further closed-form kernels, which are of interest since they permit a global study of solutions, as has been mentioned before. The main topic in the chapter is ***exponential operators*** and kernels. Although these kernels are more complicated than class P kernels, one can establish satisfactory necessary and sufficient conditions for the coefficients of equations in order that the latter admit such operators. (For class P operators, the problem of handling those criteria depends severely on the degree of the operators.) The Helmholtz and other equations are of this type, and our study exemplifies, in terms of Bessel functions, the interrelation between the present operator approach and the theory of special functions. The chapter concludes with suggestions of how to introduce further classes of closed-form kernels. First results concerning two promising ideas in this direction are discussed.

Chapter 6 establishes relations of the integral representations of solutions obtained by the present operators to other *integral representations by Riemann, Le Roux, and Vekua*. We show how to obtain such relations in a systematic and relatively simple fashion. This includes the technical details for explicitly determining the complex Riemann function from the Bergman kernel of the first kind (see Theorem 6.4-1) and vice versa (Theorem 6.4-4). Thus, the Bergman approach or the Riemann–Vekua approach may be optionally applied depending on the problem or on the purpose. Furthermore, we consider Gilbert's G-function, the ***Bergman–Gilbert operator*** as well as its interpretation within Carroll's theory of *transmutations*.

Chapter 7 is devoted to the problem of the actual *determination of the **Riemann function*** in specific cases. Because of the great practical importance of its explicit knowledge, not least because of the relations to the present operator approach, we give a concise overview on methods and results from classical up to the most recent development.

In this book, we have tried to strike a balance between theory and applications. Accordingly, after having devoted the first seven chapters to the basic ideas, methods, and techniques in the theory of Bergman-type operators, in the remaining six chapters (8–13) we turn to typical *practical tasks* that show how the "*translation principle*" works in practice, that is, how integral operators lead from results in complex analysis to theorems on general properties of large classes of solutions.

Chapter 8, the first chapter on applications, mainly discusses the ***coefficient problem.*** This problem has been extensively studied in complex analysis, so that there are many coefficient theorems for analytic functions. We shall derive from them similar theorems for representations of solutions of partial differential equations. This includes theorems on location and type of *singularities of solutions*.

Chapter 9 concerns ***approximation theory***, an area in which one has numerous results. Earlier work in the complex domain was surveyed in the standard monographs by Walsh [1969] and Gaier [1980]. We explain methods for establishing from those and other theorems an approximation theory for solutions of partial differential equations, thus, for instance, obtaining analogs of the approximation theorems of Runge, Mergelyan, Sewell, and Walsh.

Chapter 10 extends the theory of ***value distribution*** (Picard–Nevanlinna's circle of ideas) to partial differential equations and contains analogs of the fundamental theorems in this area. In connection with the extension of the Great Picard Theorem, the theory of multianalytic functions in the sense of Krajkiewicz is applied.

The last three chapters consider applications of the theory to three classes of equations which are important in mathematical physics.

Chapter 11 investigates the so-called ***Bauer–Peschl equation*** and the corresponding function theory of its solutions. We have shown that this theory can be incorporated into the Bergman-type operator theory, and this has subsequently led to a very rapid development of this area, which is also of practical interest. For the physical importance of the equation see Sec. 11.1. Here, in particular, analogs of theorems of *geometric function theory* are proved for solutions of this equation (cf. Sec. 11.4 and also part of Sec. 11.3). Moreover, we consider solutions that satisfy a higher-order Cauchy–Riemann differential equation.

Chapter 12 on ***compressible fluid flow*** is self-contained inasmuch as it first surveys the physical foundations and their mathematical formulation. We then discuss the basic idea of the *Bergman operator method* in this field and compare this approach with the classical *Chaplygin method*.

Chapter 13 is the second chapter on fluid flow and deals with *transition problems* (involving both subsonic and supersonic flow) and the ***Tricomi equation*** as the standard model, the best known and most frequently applied equation in this area. We show that the operator method is superior in obtaining and discussing solutions of this equation and their application to practical problems. Here we use the so-called *Bergman operator of the second* *kind* to solve Cauchy problems with data given on the sonic line. Furthermore, a theorem by Bergman and Bojanić is applied to obtain a corresponding *Eichler integral operator* which is specially adapted for the generation of famous particular solutions of the Tricomi equation. This chapter represents a remarkable example for the unifying character of the integral operator approach.

Various other applications of the operator approach have been adequately covered in existing monographs, so that we need not again discuss them here. This, for instance, includes scattering problems in quantum mechanics and quantum field theory (see Gilbert [1969b]), problems in elasticity theory (Vekua [1967]), and the inverse Stefan problem (Colton [1980]).

Chapter Two

Bergman Operators: General Theory

This chapter includes concepts which will be used throughout this book. In Sec. 2.1 we introduce the kind of equations to be considered in most chapters, namely, second-order elliptic or hyperbolic linear partial differential equations in two real or complex variables. In Sec. 2.2 we state the definitions which are basic in connection with Bergman operators, and we also indicate the motivations for these operators and their historical roots. In the next section, representation theorem 2.3-1 contains the important ***kernel equation*** (2.3-2a). This is the main condition which a Bergman kernel must satisfy. It will occur in almost all of the subsequent chapters. The remaining sections of the present chapter will be devoted to a particularly important class of operators, which are called *operators of the first kind*. (Operators of the *second kind* will be considered in Chap. 13.) The existence of operators of the first kind is proved in Sec. 2.4. Several practically useful representations of these operators are derived in Sec. 2.5. Operators of the first kind owe their significance mainly to the fact that they have a relatively simple inverse. This is of interest, for instance, in the solution of initial value problems, as we shall see in Sec. 2.6. Finally, in Sec. 2.7, we shall discuss the problem of extending the idea of such operators to parabolic equations and to higher dimension, a problem that was recently investigated by Colton [1976c, d] and Lanckau [1983b].

2.1 EQUATIONS TO BE CONSIDERED

NOTATIONS

The following notations will be used throughout.[1]

j always assumes the values 1 and 2.

$j^* = 3 - j$

[1] For a more extensive list of symbols used, see at the end of the book (pp. 379, 380).

$\mathbb{N} = \{1, 2, \cdots\}$, $\mathbb{N}_0 = \{0\} \cup \mathbb{N}$

$z = (z_1, z_2)$, $s = (s_1, s_2)$, z_j, s_j complex

$\Omega_j \subset \mathbb{C}$ a simply connected domain in the z_j-plane

$\Omega = \Omega_1 \times \Omega_2$

$B_j(s_j, r_j) = \{z_j \in \mathbb{C} \mid |z_j - s_j| < r_j\}$, $r_j > 0$

$B_0 = \{t \in \mathbb{C} \mid |t| < 1\}$

$\overline{B}_0 = B_0 \cup \partial B_0$

$D_j = \partial/\partial z_j$, $D_t = \partial/\partial t$

$L = D_1 D_2 + a_1(z) D_1 + a_2(z) D_2 + a_0(z)$.

EQUATIONS TO BE CONSIDERED

We shall be concerned with differential equations of the form

$$Lw = D_1 D_2 w + a_1(z) D_1 w + a_2(z) D_2 w + a_0(z) w = 0. \tag{2.1-1}$$

We assume that $a_0, a_1, a_2 \in C^\omega(\Omega)$. Here, as usual, $C^\omega(\Omega)$ denotes the set of functions holomorphic on $\Omega \subset \mathbb{C}^2$. Slightly abusing notation, we shall write $w(z)$ as well as $w(z_1, z_2)$, whichever will be more convenient in a specific case; similarly for the coefficients.

In addition to (2.1-1) we shall also employ one or the other of the ***reduced equations***

$$L_j u_j = 0, \qquad L_j = D_1 D_2 + b_j(z) D_{j^*} + c_j(z), \qquad j = 1, 2, \tag{2.1-2}$$

where

$$j^* = 3 - j.$$

This is no restriction of generality because

$$u_j = w/e_j$$

where

$$e_j(z) = \exp\left[\eta_j(z_j) - \int_{s_{j^*}}^{z_{j^*}} a_j(\tilde{z})\Big|_{\tilde{z}_j = z_j} d\tilde{z}_{j^*}\right], \tag{2.1-3}$$

$$s_{j^*} \in \Omega_{j^*}; \quad \eta_j \in \mathbb{C}^\omega(\Omega_j) \text{ arbitrary},$$

and the coefficients are given by

$$\begin{aligned} b_j(z) &= D_j \eta_j(z_j) + a_{j^*}(z) - \int_{s_{j^*}}^{z_{j^*}} D_j\big(a_j(\tilde{z})|_{\tilde{z}_j = z_j}\big)\, d\tilde{z}_{j^*} \\ c_j(z) &= a_0(z) - a_1(z) a_2(z) - D_j a_j(z). \end{aligned} \tag{2.1-4}$$

We have thus obtained the following simple relations, which we note for later use.

2.1-1 Remark

With the notations just defined we have for solutions in a domain Ω:

$$L_1u_1 = 0 \quad \text{implies} \quad L(u_1e_1) = 0, \quad L_2(u_1e_1/e_2) = 0$$

$$L_2u_2 = 0 \quad \text{implies} \quad L(u_2e_2) = 0, \quad L_1(u_2e_2/e_1) = 0$$

$$Lw = 0 \quad \text{implies} \quad L_1(w/e_1) = 0, \quad L_2(w/e_2) = 0.$$

RELATION TO REAL ELLIPTIC EQUATIONS

We further observe that (2.1-1) can be derived from a real *elliptic equation*

$$\text{(2.1-5)} \quad \hat{L}\hat{w} = \Delta\hat{w} + \hat{a}_1(x_1, x_2)\hat{w}_{x_1} + \hat{a}_2(x_1, x_2)\hat{w}_{x_2} + \hat{a}_0(x_1, x_2)\hat{w} = 0$$

involving two independent real variables x_1, x_2 and real-analytic coefficients $\hat{a}_0$, $\hat{a}_1$, $\hat{a}_2$ in a domain of the x_1x_2-plane. This may be readily accomplished by continuing the coefficients analytically to complex x_1 and x_2, introducing

$$\text{(2.1-6)} \qquad z_1 = x_1 + ix_2, \qquad z_2 = x_1 - ix_2$$

as new independent variables and using the ***Wirtinger operators***

$$\text{(2.1-7)} \qquad \frac{\partial}{\partial z_j} = \frac{1}{2}\left[\frac{\partial}{\partial x_1} + (-1)^j i \frac{\partial}{\partial x_2}\right].$$

Note that $z_2 = \bar{z}_1$ if and only if x_1 and x_2 are real.

If (2.1-1) is obtained from (2.1-5) in this way, the coefficients are related by

$$a_0(z) = \frac{1}{4}\hat{a}_0\left(\frac{z_1 + z_2}{2}, \frac{z_1 - z_2}{2i}\right)$$

(2.1-8)

$$a_j(z) = \frac{1}{4}\left[\hat{a}_1\left(\frac{z_1 + z_2}{2}, \frac{z_1 - z_2}{2i}\right) + (-1)^{j+1} i\hat{a}_2\left(\frac{z_1 + z_2}{2}, \frac{z_1 - z_2}{2i}\right)\right],$$

$$j = 1, 2.$$

"*REAL*" *SOLUTIONS*

2.1-2 *Remark*

Obviously, for real x_1, x_2 we have from (2.1-8) the formulas

$$a_0(z_1, \bar{z}_1) = \overline{a_0(z_1, \bar{z}_1)} \tag{2.1-9}$$

$$a_1(z_1, \bar{z}_1) = \overline{a_2(z_1, \bar{z}_1)}\,.$$

For an analytic function a_2 we shall denote $\overline{a_2(z_1, \bar{z}_1)}$ simply by $\bar{a}_2(\bar{z}_1, z_1)$. Thus, if

$$a_2(z_1, z_2) = \sum_{m=0}^{\infty} \sum_{n=0}^{\infty} \alpha_{mn} z_1^m z_2^n$$

then

$$\bar{a}_2(\bar{z}_1, z_1) = \sum_{m=0}^{\infty} \sum_{n=0}^{\infty} \bar{\alpha}_{mn} \bar{z}_1^m z_1^n$$

as is customary. Accordingly, if we now continue x_1, x_2 in $z_1, \bar{z}_1$ to *complex* values, we obtain a function a_2^* defined by

$$a_2^*(z_1, z_2) = \sum_{m=0}^{\infty} \sum_{n=0}^{\infty} \bar{\alpha}_{mn} z_1^n z_2^m$$

and satisfying

$$a_2^*(z_1, \bar{z}_1) = \bar{a}_2(\bar{z}_1, z_1) = \overline{a_2(z_1, \bar{z}_1)}\,.$$

For a function of a single variable we may proceed similarly; that is, we denote $\overline{f(z_1)}$ by $\bar{f}(\bar{z}_1)$, and that process of continuation yields a function f^* satisfying

$$f^*(z_2)\big|_{z_2=\bar{z}_1} = \overline{f(z_1)}\,.$$

Hence if the coefficients of (2.1-1) satisfy (2.1-9) and w is a solution of (2.1-1) on a domain $\Omega = B_1(0, r) \times B_2(0, r)$, $r > 0$, by using the corresponding function w^* we obtain another solution

$$W = \tfrac{1}{2}(w + w^*) \tag{2.1-10}$$

of (2.1-1) on Ω. This function is called a ***real solution*** of (2.1-1) (even for complex values of the arguments) because it satisfies

$$W(z_1, \bar{z}_1) = \tfrac{1}{2}[w(z_1, \bar{z}_1) + \bar{w}(\bar{z}_1, z_1)] = \operatorname{Re} w(z_1, \bar{z}_1).$$

Consequently, the present process generalizes the operation of taking the real part, and we note the following result.

2.1-3 *Lemma*

If equation (2.1-1) *is obtained from an equation* (2.1-5) *by means of* (2.1-6, 7) *and* $w \in C^{\omega}(\Omega)$ *is a solution of* (2.1-1) *on* $\Omega = B_1(0, r) \times B_2(0, r)$, $r > 0$, *then* (2.1-10) *is a solution of* (2.1-1) *on* Ω *which is real-analytic for* $z_2 = \bar{z}_1$. *(This also holds for domains* Ω *given by Def.* 6.1-1.)

In the present connection, the idea of a real solution defined in this way was suggested by S. Bergman and has proved very useful in boundary value problems, approximation theory, and other applications.

2.2 *INTEGRAL OPERATORS OF BERGMAN TYPE*

We shall now introduce integral operators for representing solutions of equations of the form (2.1-1). The purpose of these operators has been described in Chap. 1, and typical applications will be considered later, starting in Chap. 8. The general framework of this approach was created by S. Bergman [1937b, 1971] and developed further by numerous mathematicians whom we shall mention as we proceed. Applications require many further details of this theory which we shall provide, along with simplifications of Bergman's original work and subsequent extensions of it.

Historically, Bergman operators arose first in connection with the ***Helmholtz equation***

$$\Delta \hat{w} + c^2 \hat{w} = 0, \qquad c = \mathit{const} \neq 0. \tag{2.2-1}$$

In our notation (2.1-1), the Helmholtz equation takes the form

$$Lw = D_1 D_2 w + \tfrac{1}{4} c^2 w = 0. \tag{2.2-2}$$

According to a classical result [which we shall derive later, cf. (6.4-6, 17, 18)] for equation (2.2-2) the solution of the ***Goursat problem*** (characteristic initial value problem) with analytic initial data

$$\begin{aligned} w(z_1, s_2) &= \sum_{n=0}^{\infty} a_{n0}(z_1 - s_1)^n \\ w(s_1, z_2) &= \sum_{n=0}^{\infty} a_{0n}(z_2 - s_2)^n \end{aligned} \tag{2.2-3}$$

can be represented in terms of Bessel functions of the first kind, namely, by a series

$$w(z) = a_{00}w_{10}(z) + \sum_{n=1}^{\infty} \left[a_{n0}w_{1n}(z) + a_{0n}w_{2n}(z)\right].$$

Here

$$w_{1n}(z) = (z_1 - s_1)^n \sum_{m=0}^{\infty} \frac{(-1)^m n!}{m!(n+m)!} \left[\tfrac{1}{4}c^2(z_1 - s_1)(z_2 - s_2)\right]^m$$

(2.2-4a)

$$= n!\left(\frac{2}{c}\right)^n \left(\frac{z_1 - s_1}{z_2 - s_2}\right)^{n/2} J_n\left(c\sqrt{(z_1 - s_1)(z_2 - s_2)}\right), \qquad n \in \mathbb{N}_0,$$

(2.2-4b) $$w_{2n}(z_1, z_2) = w_{1n}(z_2 + s_1 - s_2, z_1 + s_2 - s_1), \qquad n \in \mathbb{N}$$

with J_n denoting the Bessel function of the first kind of order n. Now from the defining integral of the beta function B we have

$$\int_{-1}^{1} t^{2m}(1 - t^2)^{n-1/2}\, dt = B\left(n + \tfrac{1}{2}, m + \tfrac{1}{2}\right)$$

(2.2-5)

$$= \frac{\pi(2n)!(2m)!}{4^{n+m}n!m!(n+m)!}.$$

Hence by interchanging the order of summation and integration we obtain from (2.2-4a)

$$w_{1n}(z) = \int_{-1}^{1} \cos\left(c\sqrt{(z_1 - s_1)(z_2 - s_2)}\, t\right) \frac{(n!)^2 8^n}{\pi(2n)!} \left[\tfrac{1}{2}(z_1 - s_1)\tau\right]^n \tau^{-1/2}\, dt$$

where $\tau = 1 - t^2$, and a corresponding integral representation of w_{2n} from (2.2-4b). Consequently, w takes the form

$$w = w_1 + w_2$$

(2.2-6) $$w_j(z) = \int_{-1}^{1} \cos\left(c\sqrt{(z_1 - s_1)(z_2 - s_2)}\, t\right) f_j(\zeta_j) \tau^{-1/2}\, dt, \quad j = 1, 2,$$

$$\zeta_j = s_j + \tfrac{1}{2}(z_j - s_j)\tau, \qquad \tau = 1 - t^2$$

with the f_j's given by the series developments

$$f_j(\zeta_j) = \sum_{n=0}^{\infty} \gamma_{jn}(\zeta_j - s_j)^n$$

(2.2-7)
$$\gamma_{jn} = \frac{(n!)^2 8^n}{\pi(2n)!}(a_{n0}\delta_{1j} + a_{0n}\delta_{2j}), \qquad n \in \mathbb{N},$$

$$\gamma_{10} + \gamma_{20} = \frac{1}{\pi} a_{00}$$

where δ_{ij} is the Kronecker symbol. Our derivation shows that w_1 and w_2 are solutions of the Helmholtz equation (2.2-2) which are holomorphic for all $z \in \mathbb{C}^2$ such that ζ_j with $\tau \in [-1, 1]$ is in the domain of holomorphy of f_j, and $w = w_1 + w_2$ solves the Goursat problem (2.2-2, 3).

BERGMAN OPERATORS, BERGMAN KERNELS

We now utilize the result (2.2-6), in a modified form, as a model for introducing Bergman operators for the general equation (2.1-1), as follows.

Let Ω_0 denote the domain Ω or some simply connected subdomain of it. Let $C_j \subset \overline{B}_0$, $j = 1, 2$, be rectifiable simple curves beginning at $t = -1$ and ending at $t = 1$, where B_0 is the open unit disk in the complex t-plane. Then, taking two functions $\tilde{k}_j \in C^\omega(\Omega_0 \times N(C_j, \delta))$ with

$$N(C_j, \delta) = \{t \in \mathbb{C} \mid \exists\, \tilde{t} \in C_j\colon |t - \tilde{t}| < \delta\}$$

($\delta > 0$ arbitrarily small, fixed), we define a pair of integral operators

$$T_j\colon\ V(\tilde{\Omega}_j) \ \to\ V(\Omega_0), \qquad j = 1, 2,$$

by setting

(2.2-8)
$$T_j f_j(z) = \int_{C_j} \tilde{k}_j(z, t) f_j(\zeta_j) \tau^{-1/2}\, dt.$$

Here $V(\tilde{\Omega}_j)$ and $V(\Omega_0)$ are the vector spaces of functions $f_j \in C^\omega(\tilde{\Omega}_j)$ and $w \in C^\omega(\Omega_0)$, respectively, and $\tilde{\Omega}_j \subset \mathbb{C}$ is a domain containing

$$\zeta_j = \rho_j + \tfrac{1}{2}(z_j - \sigma_j)\tau, \quad \rho_j, \sigma_j \in \mathbb{C} \text{ arbitrarily fixed}; \quad \tau = 1 - t^2$$

for all $z_j \in \Omega_j$ and $t \in N(C_j, \delta)$.

2.2-1 *Definition*

The operator T_j in (2.2-8) is called a ***Bergman operator*** for equation (2.1-1) (on Ω_0) if $T_j \neq 0$, and if for every $f_j \in C^\omega(\tilde{\Omega}_j)$ the image $w_j = T_j f_j$ is a solution of (2.1-1) on $\Omega_0 \subset \Omega$. The kernel $\tilde{k}_j$ of a Bergman operator T_j is called a ***Bergman kernel***. The pre-image f_j of w_j is known as an ***associated function*** of w_j (with respect to T_j). If $\rho_j = \sigma_j = s_j$, then $s = (s_1, s_2)$ is called the ***point of reference*** of T_j.

One can often choose $\rho_j = \sigma_j = s_j = 0$, $j = 1, 2$ (as is frequently done in the literature, mainly in order to simplify formulas), but we shall later meet cases in which another choice is preferable or even mandatory.

The reason for introducing *pairs* of operators T_1, T_2 results from initial value problems, as illustrated at the beginning of this section, and will be discussed further in Sec. 2.6 and later on.

2.3 *CONDITIONS FOR KERNELS*

Given an equation (2.1-1), one faces the problem of how to obtain a corresponding Bergman operator. Sufficient conditions for $\tilde{k}_j$ to be a Bergman kernel are stated in the following basic theorem by Bergman [1937b, 1971]. We give it under weaker assumptions and in a modified form which will be needed in later extensions.

2.3-1 *Representation Theorem*

Let $s \in \Omega$ and Ω_0 be a simply connected subdomain of Ω. Let e_j and L_j be given by (2.1-2, 3, 4). *Suppose that*

$$k_j = \tilde{k}_j / e_j \tag{2.3-1}$$

is a not identically vanishing $C^\omega(\Omega_0 \times N(C_j, \delta))$-solution of the equation

$$M_j k_j = \left[\tau D_{j*} D_t - t^{-1} D_{j*} + 2(z_j - \sigma_j) t L_j\right] k_j = 0 \tag{2.3-2a}$$

satisfying the condition

$$(z_j - \sigma_j)^{-1} t^{-1} D_{j*} k_j \in C^0\left(\Omega_0 \times N(C_j, \delta)\right). \tag{2.3-2b}$$

Then $\tilde{k}_j$ is a Bergman kernel for (2.1-1) *on Ω_0. Thus, if $a_0, a_1, a_2 \in C^\omega(\Omega)$, then for every $f_j \in C^\omega(\tilde{\Omega}_j)$ the function*

$$w_j = T_j f_j \tag{2.3-3a}$$

is a $C^{\omega}(\Omega_0)$*-solution of* (2.1-1). *Hence the same is true for*

$$w = \alpha_1 w_1 + \alpha_2 w_2, \qquad \alpha_1, \alpha_2 \in \mathbb{C} \tag{2.3-3b}$$

whenever the assumptions are satisfied for both $j = 1$ *and* $j = 2$.

Proof. Obviously, it suffices to consider $w = w_1$ and, because of Remark 2.1-1, to show that $L_1 u = 0$, where $u = u_1 = w_1/e_1$.

We first show that, without loss of generality, under our assumptions on k_1 and f_1, we may assume that C_1 in (2.2-8) is replaced by a rectifiable simple curve $\tilde{C}_1$ in $N(C_1, \delta)$ obtained from C_1 by a suitable deformation and containing the intervals $[-1, -1 + \varepsilon_0]$ and $[1 - \varepsilon_0, 1]$ of the real axis for some sufficiently small $\varepsilon_0 \in (0, \delta)$, where $0 < \delta < 1$.

Indeed, let $B_t(-1, \varepsilon_0)$, $B_t(-1, \varepsilon)$, etc., be the disks of radius ε_0, ε, etc., about $t = -1$ in the t-plane. Further, let $\tilde{C}_{12}(\varepsilon_0)$ and $\tilde{C}_{12}(\varepsilon)$ denote the arcs from $t = -1 + \varepsilon_0$ and $t = -1 + \varepsilon$ along $\partial B_t(-1, \varepsilon_0)$ and $\partial B_t(-1, \varepsilon)$, $\varepsilon \in (0, \varepsilon_0)$, to the first intersection points (call them t_{ε_0} and t_ε), respectively, with the curve C_1. Furthermore, by $C_{11}(\varepsilon_0)$ and $C_{11}(\varepsilon)$ we denote the parts of C_1 from $t = -1$ to $t = t_{\varepsilon_0}$ and to $t = t_\varepsilon$, respectively; similarly, by $\tilde{C}_{11}(\varepsilon_0)$ and $\tilde{C}_{11}(\varepsilon)$ we denote the parts of the real axis from $t = -1$ to $t = -1 + \varepsilon_0$ and to $t = -1 + \varepsilon$, respectively. Now the integrand in (2.2-8) is holomorphic in t in the region

$$\left\{ t \in N(C_1, \delta) \mid \exists\, \tilde{t} \in \bar{B}_0 \colon |t - \tilde{t}| < \frac{\varepsilon}{2} \right\} \Big\backslash \left[B_t\left(-1, \frac{\varepsilon}{2}\right) \cup B_t\left(1, \frac{\varepsilon}{2}\right) \right].$$

Consequently, by Cauchy's integral theorem we may replace the contour $C_{11}(\varepsilon_0) - C_{11}(\varepsilon)$ with $-\tilde{C}_{12}(\varepsilon) + [\tilde{C}_{11}(\varepsilon_0) - \tilde{C}_{11}(\varepsilon)] + \tilde{C}_{12}(\varepsilon_0)$. Since

$$\lim_{\varepsilon \to +0} \int_{-\tilde{C}_{12}(\varepsilon)} k_1(z, t) f_1(\zeta_1) \tau^{-1/2}\, dt$$

$$= \lim_{\varepsilon \to +0} \left[\varepsilon^{1/2} \int_{\arg t_\varepsilon}^{0} k_1(z, -1 + \varepsilon e^{i\varphi}) f_1\big(\rho_1 + \tfrac{1}{2}(z_1 - \sigma_1)\varepsilon e^{i\varphi}(2 - \varepsilon e^{i\varphi})\big) \times \right.$$

$$\left. \times\, i e^{i\varphi/2} (2 - \varepsilon e^{i\varphi})^{-1/2}\, d\varphi \right] = 0$$

and

$$\lim_{\varepsilon \to +0} \int_{-\tilde{C}_{11}(\varepsilon)} k_1(z, t) f_1(\zeta_1) \tau^{-1/2}\, dt$$

$$= \lim_{\varepsilon \to +0} \lim_{\tilde{\varepsilon} \to +0} \int_{\arcsin(-1+\tilde{\varepsilon})}^{\arcsin(-1+\varepsilon)} k_1(z, \sin\varphi) f_1\big(\rho_1 + \tfrac{1}{2}(z_1 - \sigma_1)\cos^2\varphi\big)\, d\varphi = 0$$

we conclude that the integral over $C_{11}(\varepsilon_0)$ equals the limit as $\varepsilon \to 0$ of the integral over $C_{11}(\varepsilon_0) - C_{11}(\varepsilon)$, which contour can be replaced as just mentioned, so that that limit equals the integral over $\tilde{C}_{11}(\varepsilon_0) + \tilde{C}_{12}(\varepsilon_0)$. Near $t = 1$ the conclusions are similar. Hence in (2.2-8) we can replace C_1 by the aforementioned contour and have

$$u(z) = \int_{\tilde{C}_1} k_1(z,t) f_1(\zeta_1) \tau^{-1/2}\, dt \tag{2.3-4}$$

with

$$\tilde{C}_1 = \tilde{C}_{11}(\varepsilon_0) + \tilde{C}_{12}(\varepsilon_0) + [C_1 - C_{11}(\varepsilon_0) - C_{13}(\varepsilon_0)] + \tilde{C}_{13}(\varepsilon_0) + \tilde{C}_{14}(\varepsilon_0),$$

where $C_{13}(\varepsilon_0)$, $\tilde{C}_{13}(\varepsilon_0)$, and $\tilde{C}_{14}(\varepsilon_0)$ mean the analogous parts of C_1 and $\tilde{C}_1$ corresponding to $C_{11}(\varepsilon_0)$, $\tilde{C}_{11}(\varepsilon_0)$, and $\tilde{C}_{12}(\varepsilon_0)$, in a neighborhood of $t = 1$. A well-known result of complex analysis (cf. Behnke–Sommer [1972], p. 104, Theorem I.75) asserts that, under our assumptions on k_1 and f_1 [which imply that $F = k_1 f_1 \in C^{\omega}(\Omega_0 \times \tilde{C}_1)$], the integral

$$\int_{-\pi/2}^{\arcsin(-1+\varepsilon_0)} F(z, \sin\varphi)\, d\varphi$$

is complex differentiable with respect to z_j in Ω_0 and its derivative equals the integral of the derivative of the integrand; call the latter integral I. Now this derivative is equal to

$$D_j \lim_{\varepsilon \to 0} \int_{-1+\varepsilon}^{-1+\varepsilon_0} F(z,t)\tau^{-1/2}\, dt = D_j \int_{\tilde{C}_{11}(\varepsilon_0)} F(z,t)\tau^{-1/2}\, dt$$

and I is simply

$$\lim_{\varepsilon \to 0} \int_{-1+\varepsilon}^{-1+\varepsilon_0} D_j F(z,t)\tau^{-1/2}\, dt = \int_{\tilde{C}_{11}(\varepsilon_0)} D_j F(z,t)\tau^{-1/2}\, dt.$$

The same argument holds for the integral along $\tilde{C}_{13}(\varepsilon_0)$ instead of $\tilde{C}_{11}(\varepsilon_0)$, and another application of the preceding theorem yields

$$D_j \int_{\tilde{C}_1 - [\tilde{C}_{11}(\varepsilon_0) + \tilde{C}_{13}(\varepsilon_0)]} F(z,t)\tau^{-1/2}\, dt = \int_{\tilde{C}_1 - [\tilde{C}_{11}(\varepsilon_0) + \tilde{C}_{13}(\varepsilon_0)]} D_j F(z,t)\tau^{-1/2}\, dt.$$

Summarizing these results, we conclude that the derivative $D_j u$ exists in Ω_0 and is given by

$$D_j u(z) = \int_{\tilde{C}_1} D_j [k_1(z,t) f_1(\zeta_1)] \tau^{-1/2}\, dt, \qquad j = 1,2. \tag{2.3-5a}$$

Applying the same reasoning to $D_2 u$, we also obtain

$$D_1 D_2 u(z) = \int_{\tilde{C}_1} D_1 D_2 [k_1(z,t) f_1(\zeta_1)] \tau^{-1/2} \, dt. \tag{2.3-5b}$$

Thus, applying L_1 to u, we may perform the differentiations under the integral sign.

We now observe that

$$D_1[f_1(\zeta_1)] = \tfrac{1}{2}\tau f_1'(\zeta_1), \qquad D_t[f_1(\zeta_1)] = -(z_1 - \sigma_1) t f_1'(\zeta_1)$$

whence

$$(z_1 - \sigma_1) t D_1 [f_1(\zeta_1)] = -\tfrac{1}{2}\tau D_t [f_1(\zeta_1)]. \tag{2.3-6}$$

Assumption (2.3-2b) yields

$$(z_1 - \sigma_1)^{-1} t^{-1} D_2 k_1 \in C^0(\Omega_0 \times \tilde{C}_1).$$

Hence, from (2.3-5b, 6) it follows that

$$\begin{aligned} D_1 D_2 u(z) &= \int_{\tilde{C}_1} [D_1 D_2 k_1(z,t)] f_1(\zeta_1) \tau^{-1/2} \, dt \\ &\quad - \frac{1}{2} \int_{\tilde{C}_1} \left[(z_1 - \sigma_1)^{-1} t^{-1} D_2 k_1(z,t) \tau^{1/2} \right] D_t [f_1(\zeta_1)] \, dt. \end{aligned} \tag{2.3-7}$$

From (2.3-2b) and $D_2 k_1 \in C^\omega(\Omega_0 \times N(C_1, \delta))$ we conclude that

$$D_t \left[(z_1 - \sigma_1)^{-1} t^{-1} D_2 k_1(z,t) \tau^{1/2} \right]$$

exists for all (z, t) in some neighborhood of

$$\Omega_0 \times \left(\tilde{C}_1 - \left[\tilde{C}_{11}(\varepsilon) + \tilde{C}_{13}(\tilde{\varepsilon}) \right] \right),$$

$\tilde{C}_{13}(\tilde{\varepsilon})$ being the curve from $t = 1 - \tilde{\varepsilon}$ to $t = 1$ along the real axis with arbitrary $\tilde{\varepsilon} \in (0, \varepsilon_0)$. We also conclude that it is of class C^ω in this neighborhood, and that it can be rewritten as

$$(z_1 - \sigma_1)^{-1} t^{-1} \tau^{-1/2} \left[\tau D_t D_2 - t^{-1} D_2 \right] k_1(z,t).$$

Therefore, by integration by parts, the second term on the right-hand side of

(2.3-7) takes the form

$$-\frac{1}{2}\int_{\tilde{C}_1}\left[(z_1-\sigma_1)^{-1}t^{-1}D_2k_1(z,t)\tau^{1/2}\right]D_t\left[f_1(\zeta_1)\right]dt$$

$$= \lim_{\varepsilon\to 0}\lim_{\tilde{\varepsilon}\to 0}\left\{\left[-\frac{1}{2}(z_1-\sigma_1)^{-1}t^{-1}D_2k_1(z,t)\tau^{1/2}f_1(\zeta_1)\right]_{t=-1+\varepsilon}^{t=1-\tilde{\varepsilon}}\right.$$

$$+\int_{\tilde{C}_1-[\tilde{C}_{11}(\varepsilon)+\tilde{C}_{13}(\tilde{\varepsilon})]}\left[2(z_1-\sigma_1)t\right]^{-1}\times$$

$$\left.\times\left[\left(\tau D_tD_2-t^{-1}D_2\right)k_1(z,t)\right]f_1(\zeta_1)\tau^{-1/2}\,dt\right\}.$$

Since, as we have shown, the left-hand side exists, and since the double limit in the first line on the right obviously exists and is zero, the double limit in the second line on the right must also exist. Inserting the result into (2.3-7), we get

$$\begin{aligned} D_1D_2u(z) &= \int_{\tilde{C}_1}\left\{\left[2(z_1-\sigma_1)t\right]^{-1}\left[\tau D_tD_2-t^{-1}D_2\right.\right. \\ &\qquad \left.\left.+\,2(z_1-\sigma_1)tD_1D_2\right]k_1(z,t)\right\}f_1(\zeta_1)\tau^{-1/2}\,dt. \end{aligned} \tag{2.3-8}$$

We mention that u is of class $C^\omega(\Omega_0)$; this can be seen by repeating the proof which gave (2.3-5). Finally, from (2.3-4, 5a, 8) we obtain

$$\begin{aligned} L_1u(z) &= \left[D_1D_2+b_1(z)D_2+c_1(z)\right]u(z) \\ &= \int_{\tilde{C}_1}\left[2(z_1-\sigma_1)t\right]^{-1}\left[M_1k_1(z,t)\right]f_1(\zeta_1)\tau^{-1/2}\,dt. \end{aligned}$$

Thus, assumption (2.3-2a) implies $L_1u=0$ on Ω_0. This proves the theorem. ■

TYPE I AND TYPE II REPRESENTATIONS

2.3-2 Definition

We call (2.3-3b) with $\alpha_1=0$ or $\alpha_2=0$ [as well as (2.3-3a)] a ***Type I representation***, and with $\alpha_1\alpha_2\neq 0$ a ***Type II representation***.

2.3-3 Remark

Condition (2.3-2b) can easily be met in practice, perhaps by a suitable choice of σ_j and C_j. The essential condition is the "*kernel equation*" (2.3-2a), and we shall discuss various methods of obtaining solutions of this equation. For the time being, in order to beware of trivial misunderstandings, we should emphasize the following fact. Since $M_j k_j = 0$ contains *three* independent variables, it is certainly more complicated than the given equation (2.1-1). Hence at first glance it seems that we have merely increased the difficulties. However, this is not the case, because if we generate a *single* not identically vanishing solution of (2.3-2a), we can represent and study *whole classes* of solutions of (2.1-1). Actually, under suitable assumptions on $\tilde{k}_j$ we may even be able to obtain *all* $C^\omega(\Omega_0)$-solutions of (2.1-1) on a given domain $\Omega_0 \subset \Omega$. Indeed, the existence of such kernels will be proved as a basic result in the next section.

2.4 INTEGRAL OPERATORS OF THE FIRST KIND

The previous discussion suggests our next task: the solution of the kernel equation (2.3-2a). This may mean:

(i) the development of general methods applicable in the case of any equation (2.1-1) with analytic coefficients, or

(ii) the development of special methods which yield "better" results, but are restricted to certain large classes of equations (2.1-1).

We shall see that for a given equation (2.1-1) in general there exist various Bergman operators whose kernels may have quite different properties, and the construction or choice of one of them depends on the purpose. At present, since we also want to make sure that the assumptions in Theorem 2.3-1 can be met for *every* equation (2.1-1), it is clear that we should begin with methods of the kind described under (i). This section is devoted to such a method which yields operators of great practical importance.

OPERATORS OF THE FIRST KIND

2.4-1 Definition

Let s be the point of reference of an operator T_j, i.e., $\rho_j = \sigma_j = s_j$ (cf. Def. 2.2-1). The operator T_j and its kernel $\tilde{k}_j$ are said to be ***of the first kind*** if $\tilde{k}_j$ is a Bergman kernel such that $k_j = \tilde{k}_j / e_j$ satisfies

$$k_j(z^{(m)}, t) = 1, \qquad m = 1, 2, \tag{2.4-1}$$

where

$$z^{(1)} = (z_1, s_2), \qquad z^{(2)} = (s_1, z_2).$$

We shall see that an operator of the first kind has a particularly simple inverse, a property which will be shown later and is of considerable interest in connection with the coefficient problem (cf. Chap. 8), approximation of solutions (cf. Chap. 9), and other applications.

EXISTENCE AND REPRESENTATION OF OPERATORS OF THE FIRST KIND

In order to settle the problem of the (local!) existence of operators of the first kind, we prove the following result by Bergman [1937b, 1971].

2.4-2 Existence and Representation Theorem

In equation (2.1-1), *suppose that* $a_0, a_1, a_2 \in C^{\omega}(\Omega)$. *Let* $s \in \Omega$ *and* $r_m > 0$ *such that*

$$B_m(s_m, r_m) \cup \partial B_m(s_m, r_m) \subset \Omega_m, \qquad m = 1, 2.$$

Finally, let $\eta_m \in C^{\omega}(\Omega_m)$, $m = 1, 2$. *Then assertions* $(a), (b), (c)$ *hold*:

(a) *Let* $p_{j0}, p_{j1}, \cdots$ *be defined by the recursion*

$$p_{j0}(z) = 1, \qquad p_{j1}(z) = -2c_j(z),$$

$$p_{j,m+1}(z) = -\frac{2}{2m+1}\left[D_j p_{jm}(z) + b_j(z)p_{jm}(z) + c_j(z)\int_{s_{j^*}}^{z_{j^*}} p_{jm}(\tilde{z})\Big|_{\tilde{z}_j = z_j} d\tilde{z}_{j^*}\right], \qquad m \in \mathbb{N} \tag{2.4-2}$$

with b_j *and* c_j *as in* (2.1-4), *and let*

$$\tilde{q}_{j0}(z) = p_{j0}(z) = 1,$$

$$\tilde{q}_{jm}(z) = \int_{s_{j^*}}^{z_{j^*}} p_{jm}(\tilde{z})\Big|_{\tilde{z}_j = z_j} d\tilde{z}_{j^*}. \tag{2.4-3}$$

Then the series

$$\sum_{m=0}^{\infty} \tilde{q}_{jm}(z)(z_j - s_j)^m t^{2m} \tag{2.4-4}$$

converges uniformly in a domain containing $\overline{G}_j \times \overline{B}_0$, *where* $\overline{G}_j$ *is the closure of*

$$G_j = \left[B_1(s_1, \hat{r}_1) \times B_2(s_2, \hat{r}_2)\right]_{\hat{r}_j = r_j/2,\ \hat{r}_{j^*} = r_{j^*}}.$$

(b) *The series* (2.4-4) *represents a function* $k_j \in C^\omega(G_j \times \overline{B}_0)$ *such that* $\tilde{k}_j = e_j k_j$ *is a kernel of a Bergman operator of the first kind for* (2.1-1).

(c) *If* $f_j \in C^\omega(B_j(s_j, \frac{1}{4}r_j))$ *and* $C_j = [-1, 1]$, *then*

$$w_j(z) = T_j f_j(z) = \int_{-1}^{1} \tilde{k}_j(z,t) f_j(\xi_j) \tau^{-1/2}\, dt, \qquad j = 1, 2, \tag{2.4-5}$$

$$\xi_j = s_j + \tfrac{1}{2}(z_j - s_j)\tau,$$

are C^ω*-solutions of* (2.1-1) *on* G_j; *hence* (*cf. Theorem* 2.3-1)

$$w = w_1 + w_2$$

is a C^ω*-solution of* (2.1-1) *on*

$$G_1 \cap G_2 = B_1(s_1, \tfrac{1}{2}r_1) \times B_2(s_2, \tfrac{1}{2}r_2).$$

Proof. Suppose first that part (a) has already been shown. Accordingly, there exists a (sufficiently small) $\tilde{\varepsilon} > 0$ such that

$$\overline{B}_j(\tilde{\varepsilon}) = B_j(\tilde{\varepsilon}) \cup \partial B_j(\tilde{\varepsilon}) \subset \Omega_j, \qquad j = 1, 2, \tag{2.4-6}$$

where $B_j(\tilde{\varepsilon}) = B_j(s_j, r_j + \tilde{\varepsilon})$, and, furthermore, the series (2.4-4) converges uniformly on the closure of $G_j(\tilde{\varepsilon}) \times B_0(\tilde{\varepsilon})$, where

$$\begin{aligned} G_j(\tilde{\varepsilon}) &= \left[B_1(s_1, \hat{r}_1 + \tilde{\varepsilon}) \times B_2(s_2, \hat{r}_2 + \tilde{\varepsilon})\right]_{\hat{r}_j = r_j/2,\, \hat{r}_{j*} = r_{j*}} \\ B_0(\tilde{\varepsilon}) &= \{t \in \mathbb{C} \mid |t| < 1 + \tilde{\varepsilon}\}. \end{aligned} \tag{2.4-7}$$

Since $a_0, a_1, a_2 \in C^\omega(\Omega)$, also $b_j, c_j \in C^\omega(\Omega)$. Hence from (2.4-2, 3) we see that the series (2.4-4) defines a function $k_j \in C^\omega(G_j(\tilde{\varepsilon}) \times B_0(\tilde{\varepsilon}))$. It follows that

$$\begin{aligned} &\text{(a)} \qquad k_j \in C^\omega(\overline{G}_j \times \overline{B}_0) \\ &\text{(b)} \quad D_{j*}k_j \in C^\omega(\overline{G}_j \times \overline{B}_0). \end{aligned} \tag{2.4-8}$$

Substituting that series into the kernel equation (2.3-2a) and determining the coefficient function of each power of t, we see that $M_j k_j = 0$ together with (2.4-1) is equivalent to the system (2.4-2). Hence, by (2.4-8a), the function k_j satisfies the first of the conditions in Theorem 2.3-1. Furthermore, condition (2.3-2b) follows from formula (2.4-8b) and

$$D_{j*}k_j(z,t) = \sum_{m=0}^{\infty} p_{j,m+1}(z)(z_j - s_j)^{m+1} t^{2m+2}.$$

Thus $\tilde{k}_j = e_j k_j$ is a Bergman kernel for (2.1-1) on G_j.

Statement (b) is obvious from (2.4-3, 4) and the definition of a kernel of the first kind given in Def. 2.4-1.

Statement (c) follows from Theorem 2.3-1 by noting that, for $\Omega_j = B_j(s_j, \frac{1}{2}r_j)$ and $C_j = [-1, 1]$, we can replace $\tilde{\Omega}_j$ by the disk $B_j(s_j, \frac{1}{4}r_j)$.

We prove (a). Under the assumptions on a_0, a_1, a_2, η_j, r_1, r_2, and $\tilde{\varepsilon}$ there exists an $\varepsilon > \tilde{\varepsilon}$ such that (2.4-6) with $\tilde{\varepsilon}$ replaced by ε holds and, furthermore, the generalized Cauchy inequalities yield on

$$B_1(\varepsilon) \times B_2(\varepsilon) = B_1(s_1, r_1 + \varepsilon) \times B_2(s_2, r_2 + \varepsilon)$$

the following dominants (here and later, $\ll$ means that the series on the right is a *dominant* of that on the left):

$$\begin{aligned} b_j(z) &= \sum_{\rho=0}^{\infty} \sum_{\sigma=0}^{\infty} b_{j\rho\sigma}(z_1 - s_1)^{\rho}(z_2 - s_2)^{\sigma} \\ &\ll \sum_{\rho=0}^{\infty} \sum_{\sigma=0}^{\infty} \alpha \left(\frac{z_1 - s_1}{r_1 + \varepsilon} \right)^{\rho} \left(\frac{z_2 - s_2}{r_2 + \varepsilon} \right)^{\sigma} = \alpha A_1 A_2 \end{aligned}$$

where

$$A_m = \left(1 - \frac{z_m - s_m}{r_m + \varepsilon}\right)^{-1}, \qquad m = 1, 2$$

and similarly

$$c_j(z) \ll \alpha A_1 A_2;$$

here, $\alpha \geq 1$ is an upper bound of $|b_j(z)|$, $|c_j(z)|$ on $\overline{B}_1(\varepsilon) \times \overline{B}_2(\varepsilon)$.

We now need

2.4-3 Lemma

Under the assumptions of Theorem 2.4-2 we have for the series (2.4-4)

$$\sum_{m=0}^{\infty} \tilde{q}_{jm}(z)(z_j - s_j)^m t^{2m} \ll 1 + \beta_j \kappa_j t^2 {}_2F_1(\gamma_j + 1, 1; 1; \kappa_j t^2)$$

on $\overline{G}_j(\tilde{\varepsilon})$ [defined in (2.4-7)], where β_j, γ_j and κ_j are suitable positive constants (whose values are given in the proof) and ${}_2F_1$ denotes the hypergeometric series.

Proof. From (2.4-2) we see that $p_{jm} \in C^{\omega}(B_1(\varepsilon) \times B_2(\varepsilon))$, so that the series

$$p_{jm}(z) = \sum_{\rho=0}^{\infty} \sum_{\sigma=0}^{\infty} p_{jm\rho\sigma}(z_1 - s_1)^{\rho}(z_2 - s_2)^{\sigma}$$

converges on that dicylinder. We modify (2.4-2) by dropping the two minus signs and replacing b_j and c_j by their dominants $\alpha A_1 A_2$. The resulting modified system defines functions which we denote by $\tilde{p}_{j0}, \tilde{p}_{j1}, \cdots$. Clearly, $\tilde{p}_{jm} \in C^{\omega}(B_1(\varepsilon) \times B_2(\varepsilon))$ for all $m \in \mathbb{N}$. Furthermore, the coefficients of the power series of $\tilde{p}_{jm}$ with center $s = (s_1, s_2)$ are real and nonnegative. Also, each such coefficient $\tilde{p}_{jm\rho\sigma}$ is an upper bound of the corresponding $|p_{jm\rho\sigma}|$ in the preceding series expansion of p_{jm}. Hence

$$p_{jm}(z) \ll \tilde{p}_{jm}(z), \qquad m \in \mathbb{N}_0. \tag{2.4-9}$$

We now prove uniform convergence of the dominants $\tilde{p}_{jm}$. Define $\hat{p}_{jm}$ by

$$\hat{p}_{jm}(y) = \frac{(2m)!}{2^{2m-1}m!}\left(1 - \frac{z_j - s_j}{r_j + \varepsilon}\right)^m \tilde{p}_{jm}(z), \qquad y = z_{j*} - s_{j*}. \tag{2.4-10}$$

From the recursive system for the $\tilde{p}_{jm}$ we see that $\hat{p}_{jm}$, $m \in \mathbb{N}_0$, is indeed independent of z_j and satisfies

$$\hat{p}_{j0} = 1, \qquad \hat{p}_{j1}(y) = 2\alpha A_{j*},$$

$$\hat{p}_{j,m+1}(y) = \left[\frac{m}{r_j + \varepsilon} + \alpha A_{j*}\right]\hat{p}_{jm}(y) + \alpha A_{j*}\int_0^y \hat{p}_{jm}(\tilde{y})\, d\tilde{y}.$$

Since the power series of $\tilde{p}_{jm}$ about s has real and nonnegative coefficients, the same holds for the power series of $\hat{p}_{jm}$ about 0. We thus have

$$|\hat{p}_{jm}(y)| \leq \hat{p}_{jm}(|y|).$$

Also

$$\left|\int_0^y \hat{p}_{jm}(\tilde{y})\, d\tilde{y}\right| \leq (r_{j*} + \varepsilon)\hat{p}_{jm}(|y|)$$

and

$$|A_{j*}| \leq R_{j*}, \qquad \text{where} \qquad R_{j*} = \frac{r_{j*} + \varepsilon}{\varepsilon - \tilde{\varepsilon}}.$$

From that recursion for the $\hat{p}_{jm}$ we thus obtain the inequalities

$$|\hat{p}_{j1}(y)| \leq 2\alpha R_{j*}$$

$$|\hat{p}_{j,m+1}(y)| \leq \frac{m + \gamma_j}{r_j + \varepsilon}\hat{p}_{jm}(|y|), \qquad m \in \mathbb{N},$$

where

$$\gamma_j = \alpha(r_j + \varepsilon)(1 + r_{j*} + \varepsilon)R_{j*}.$$

Hence

$$|\hat{p}_{jm}(y)| \le 2\alpha \frac{\Gamma(\gamma_j + m)}{\Gamma(\gamma_j + 1)} \frac{R_{j*}}{(r_j + \varepsilon)^{m-1}}$$

or, by (2.4-10), for $z_j \in B_j(\varepsilon)$ and $z_{j*} \in \overline{B}_{j*}(\tilde{\varepsilon})$,

$$|\tilde{p}_{jm}(z)| \le \alpha \frac{2^{2m} m! \Gamma(\gamma_j + m)}{(2m)! \Gamma(\gamma_j + 1)} \frac{(r_j + \varepsilon) R_{j*}}{|r_j + \varepsilon - z_j + s_j|^m}, \qquad m \in \mathbb{N}.$$

Here, $2^{2m}m!/(2m)! \le 2/(m-1)!$. Using (2.4-9), for $|z_j - s_j| \le \frac{1}{2} r_j + \tilde{\varepsilon}$, we thus obtain

$$|(z_j - s_j)^m \tilde{q}_{jm}(z)| \le \frac{\beta_j}{(m-1)!} \frac{\Gamma(\gamma_j + m)}{\Gamma(\gamma_j + 1)} \kappa_j^m$$

where

$$\beta_j = 2\alpha(r_j + \varepsilon)(r_{j*} + \varepsilon) R_{j*},$$

$$\kappa_j = \frac{r_j + 2\tilde{\varepsilon}}{r_j + 2(\varepsilon - \tilde{\varepsilon})}.$$

From this, the assertion of Lemma 2.4-3 follows. ■

Proof of Theorem 2.4-2 (Continued). For $\tilde{\varepsilon} \in [0, \frac{1}{2}\varepsilon]$ define χ_j by

$$\chi_j(\tilde{\varepsilon}) = \frac{(r_j + 2\tilde{\varepsilon})(1 + \tilde{\varepsilon})^2}{r_j + 2(\varepsilon - \tilde{\varepsilon})}.$$

χ_j is a positive and strictly monotone increasing continuous function such that

$$\chi_j(0) < 1 - \frac{\varepsilon}{r_j + 2\varepsilon} < \chi_j\left(\tfrac{1}{2}\varepsilon\right).$$

Hence there is an $\tilde{\varepsilon}_0 \in (0, \frac{1}{2}\varepsilon)$ such that

$$\chi_j(\tilde{\varepsilon}) < 1 - \frac{\varepsilon}{r_j + 2\varepsilon} < 1 \qquad \text{for all } \tilde{\varepsilon} \in (0, \tilde{\varepsilon}_0).$$

Consequently, for $t \in \overline{B}_0(\tilde{\varepsilon})$ with $\tilde{\varepsilon} \in (0, \tilde{\varepsilon}_0)$ we have

$$|\kappa_j t^2| \le \chi_j(\tilde{\varepsilon}_0) \le 1 - \frac{\varepsilon}{r_j + 2\varepsilon} < 1.$$

This proves uniform convergence of the hypergeometric series in Lemma 2.4-3. By this lemma, it implies uniform convergence of the series (2.4-4) on $G_j(\tilde{\varepsilon}) \times B_0(\tilde{\varepsilon})$, where $\tilde{\varepsilon} \in (0, \tilde{\varepsilon}_0)$; this domain clearly contains $\overline{G}_j \times \overline{B}_0$ occurring in part (a) of Theorem 2.4-2. The proof of Theorem 2.4-2 is now complete. ■

2.5 FURTHER REPRESENTATIONS OF OPERATORS OF THE FIRST KIND

In general, to a given equation (2.1-1) there correspond various integral operators for representing solutions, and the choice of such an operator depends on the purpose. But even after one has decided to use a certain operator, one has the possibility of selecting one or another of the different representations that operator may have. To demonstrate the latter fact and to prepare for applications, for instance in connection with the approximation of solutions, let us continue our study of operators of the first kind. A first representation of a kernel of the first kind is given in Theorem 2.4-2a. The corresponding representation of solutions is obtained by substituting (2.4-4) into (2.2-7), as indicated in Theorem 2.4-2c. Other representations can be derived as follows.

2.5-1 Theorem

Suppose that the assumptions of Theorem 2.4-2 hold. Then the solution w_j in Theorem 2.4.2c can be represented in the form

$$(2.5\text{-}1) \qquad w_j(z) = e_j(z)\left[h_j(z_j) + \sum_{m=1}^{\infty} \tilde{q}_{jm}(z)\frac{(2m)!}{2^{2m}m!}H_{jm}(z_j)\right],$$

where

$$(2.5\text{-}2) \qquad h_j(z_j) = \int_{-1}^{1} f_j(\xi_j)\tau^{-1/2}\,dt, \qquad \xi_j = s_j + \tfrac{1}{2}(z_j - s_j)\tau$$

and

$$(2.5\text{-}3) \qquad H_{jm}(z_j) = \int_{s_j}^{z_j}\int_{s_j}^{z_{jm}} \cdots \int_{s_j}^{z_{j2}} h_j(z_{j1})\,dz_{j1} \cdots dz_{j,m-1}\,dz_{jm}$$

which can also be written as a single integral:

$$(2.5\text{-}4) \qquad H_{jm}(z_j) = \frac{1}{(m-1)!}\int_{s_j}^{z_j}(z_j - \tilde{z}_j)^{m-1}h_j(\tilde{z}_j)\,d\tilde{z}_j, \qquad m \in \mathbb{N}.$$

Proof. Let $f_j \in C^\omega(B_j(s_j, \frac{1}{4}r_j))$ have the Taylor series

$$f_j(\zeta_j) = \sum_{n=0}^{\infty} \gamma_{jn}(\zeta_j - s_j)^n. \tag{2.5-5}$$

Then, by (2.4-4, 5),

$$w_j(z) = e_j(z)\left[h_j(z_j) + \sum_{m=1}^{\infty} \tilde{q}_{jm}(z)(z_j - s_j)^m \phi_{jm}(z_j)\right] \tag{2.5-6}$$

where by (2.2-5)

$$\begin{aligned} \phi_{jm}(z_j) &= \int_{-1}^{1} t^{2m} f_j(\xi_j)\tau^{-1/2}\,dt \\ &= \sum_{n=0}^{\infty} \gamma_{jn} 2^{-n}(z_j - s_j)^n \int_{-1}^{1} t^{2m}\tau^{n-1/2}\,dt \\ &= \frac{(2m)!}{2^{2m}m!} \sum_{n=0}^{\infty} \frac{\pi(2n)!}{8^n n!(m+n)!}\gamma_{jn}(z_j - s_j)^n. \end{aligned} \tag{2.5-7}$$

On the other hand, using (2.5-3, 2, 5) and (2.2-5), in this order, we obtain

$$H_{jm}(z_j) = \sum_{n=0}^{\infty} \frac{\pi(2n)!}{8^n(n!)^2}\gamma_{jn} \int_{s_j}^{z_j}\int_{s_j}^{z_{jm}} \cdots \int_{s_j}^{z_{j2}} (z_{j1} - s_j)^n\,dz_{j1} \cdots dz_{jm}.$$

Together with (2.5-7) this yields

$$(z_j - s_j)^m \phi_{jm}(z_j) = \frac{(2m)!}{2^{2m}m!} H_{jm}(z_j).$$

Substitution into (2.5-6) yields (2.5-1) with H_{jm} given by (2.5-3). Finally, (2.5-4) follows from (2.5-3) by Liouville's formula. This completes the proof. ■

PRESERVATION OF BERGMAN SPACE PROPERTY

Theorem 2.5-1 is often used in applications. For example, Marzuq [1984] utilizes representation (2.5-1) with (2.5-2, 4) in order to prove that the operator T_1 of the first kind preserves the Bergman space property. This means that for h_1 defined in terms of f_1 by (2.5-2) and for equations (2.1-1) with entire coefficients, the following holds.

$$\text{If} \quad h_1 \in A^p(B_1(0,1)), \quad \text{then} \quad w_1 = T_1 f_1 \in A^p(B_1(0,1) \times B_2(0,1)), \quad 1 \le p < \infty.$$

Here, the Bergman space $A^p(\tilde{B}(n))$, where

$$\tilde{B}(1) = B_1(0,1), \qquad \tilde{B}(2) = B_1(0,1) \times B_2(0,1), \qquad \text{etc.},$$

is defined as the space of all functions $f \in C^\omega(\tilde{B}(n))$ with norm

$$\|f\| = \left[\int_0^1 \cdots \int_0^1 M_p^p(r,f) r_1\, dr_1 \cdots r_n\, dr_n\right]^{1/p} < \infty$$

where

$$M_p^p(r,f) = (2\pi)^{-n} \int_{-\pi}^{\pi} \cdots \int_{-\pi}^{\pi} \left|f\left(r_1 e^{i\theta_1}, \dots, r_n e^{i\theta_n}\right)\right|^p d\theta_1 \cdots d\theta_n .$$

Theorem 2.5-1 will also play an essential role in some of our proofs, for instance in Chap. 9.

REPRESENTATION OF THE KERNEL. UNIQUENESS

We shall now derive a formal expression for k_j and establish its validity by means of Theorem 2.4-2. Substitution of

$$k_j(z,t) = \sum_{m=0}^{\infty} \tilde{q}_{jm}(z)(z_j - s_j)^m t^{2m} \tag{2.5-8}$$

[cf. (2.4-4)] into the kernel equation (2.3-2a) yields the system

$$\begin{aligned} D_{j*}\tilde{q}_{j0} &= 0 \\ D_{j*}\tilde{q}_{jm} &= -\frac{2}{2m-1} L_j \tilde{q}_{j,m-1}, \qquad m \in \mathbb{N} . \end{aligned} \tag{2.5-9}$$

For our present k_j this is equivalent to (2.3-2a). In order that k_j be holomorphic on $\Omega \times N(C_j, \delta)$, we must require that

$$(z_j - s_j)^m \tilde{q}_{jm} \in C^\omega(\Omega) .$$

Then (2.5-9) is equivalent to

$$\tilde{q}_{jm} = \frac{(-4)^m m!}{(2m)!} S_{jm} \cdots S_{j0} 0$$

where

$$S_{j\mu} g(z) = \int_{s_{j*}}^{z_{j*}} L_j g(\tilde{z})\Big|_{\tilde{z}_j = z_j} d\tilde{z}_{j*} + h_{j\mu}(z_j) \tag{2.5-10}$$

and $h_{j\mu}$ is such that

$$(z_j - s_j)^\mu h_{j\mu} \in C^\omega(\Omega_j)$$

but otherwise arbitrary. This proves:

2.5-2 *Theorem*

Suppose that $k_j \in C^\omega(\Omega \times N(C_j, \delta))$ *is an* even *function of* t. *Then* $M_j k_j = 0$ [*cf*. (2.3-2*a*)] *if and only if*

$$(2.5\text{-}11)\quad k_j(z,t) = \sum_{m=0}^{\infty} \frac{(-4)^m m!}{(2m)!} S_{jm} \cdots S_{j0} 0(z)(z_j - s_j)^m t^{2m}$$

with $S_{j\mu}$ *defined in* (2.5-10).

Theorem 2.5-2 implies a *uniqueness statement* for kernels of the first kind, as follows:

2.5-3 *Corollary*

Let $\eta_j \in C^\omega(\Omega_j)$ *be arbitrary and fixed. Let* $\tilde{k}_j$ *be a kernel of the first kind* (*with point of reference* s) *and* $\tilde{k}_{je} = e_j k_{je}$ *its even part in* t. *Then* $\tilde{k}_{je}$ *is uniquely determined and* k_{je} *has the representation*

$$(2.5\text{-}12)\qquad k_{je}(z,t) = 1 + \sum_{m=1}^{\infty} \frac{(-4)^m m!}{(2m)!} S_j^m 1(z)(z_j - s_j)^m t^{2m}$$

on $\overline{G}_j \times \overline{B}_0$ (*see Theorem* 2.4-2 *for notation*), *where* S_j *denotes* $S_{j\mu}$ *defined in* (2.5-10), *with* $h_{j\mu} = 0$.

Note that in Corollary 2.5-3, uniqueness follows readily from (2.5-11) by taking into account the conditions (2.4-1) for a kernel of the first kind. The domain of validity of (2.5-12) is given by Theorem 2.4-2a.

SOME APPLICATIONS

2.5-4 *Example*

In the case of the ***Helmholtz equation***

$$\Delta\hat{w} + c^2\hat{w} = 0, \qquad c = const \neq 0,$$

or

$$D_1 D_2 w + \tfrac{1}{4}c^2 w = 0$$

choosing $\eta_j = 0$, so that $e_j = 1$, we have from (2.5-10)

$$S_j^m 1(z) = \frac{1}{m!}\left(\tfrac{1}{4}c^2\right)^m (z_{j*} - s_{j*})^m, \qquad m \in \mathbb{N}.$$

Hence (2.5-12) yields

$$\tilde{k}_j(z,t) = \tilde{k}_{je}(z,t) = 1 + \sum_{m=1}^{\infty} \frac{(-1)^m}{(2m)!} c^{2m} (z_{j*} - s_{j*})^m (z_j - s_j)^m t^{2m}$$

$$= \cos\left(c\sqrt{(z_1 - s_1)(z_2 - s_2)}\, t\right), \qquad j = 1,2.$$

Note that this agrees with the kernels in (2.2-6).

2.5-5 *Example*

The equation

$$\text{(2.5-13)} \qquad Lw = D_1 D_2 w + \frac{\lambda n(n+1)}{(1 + \lambda z_1 z_2)^2} w = 0, \qquad \lambda \in \mathbb{C}, \quad n \in \mathbb{N}$$

is of importance, for instance in the case $\lambda = -1$, because of its relation to the wave equation, from which it can be obtained by separating variables in spherical coordinates and introducing suitable new variables in the resulting equation for the angular part. It has been investigated by many authors (references will be given in Chap. 11) and has various interesting properties, as we shall see later. For this equation we obtain from (2.5-10)

$$S_j^m 1(z) = \frac{(2m)!}{m!} \binom{n+m}{2m} \left(\frac{\lambda z_{j*}}{1 + \lambda z_1 z_2} \right)^m$$

as can be shown by induction. Corollary 2.5-3 thus yields

$$\text{(2.5-14)} \qquad \tilde{k}_j(z,t) = \sum_{m=0}^{n} \binom{n+m}{2m} \left(\frac{-4\lambda z_1 z_2}{1 + \lambda z_1 z_2} \right)^m t^{2m}, \qquad j = 1,2.$$

It is remarkable that for the present equation the series in Corollary 2.5-3 reduces to a *polynomial* in t. This is a great advantage in applying integral operators to practical problems. As we proceed, we shall see that there is a large class of equations (2.1-1) which admit Bergman operators with such "polynomial kernels."

2.6 THE INVERSION PROBLEM

In various applications of the integral operator method it is of basic importance that one has a procedure for determining associated functions of a given solution of (2.1-1). This amounts to the problem of inverting operators. Whereas for certain operators this may be quite involved, for operators of the *first kind* this inversion problem is relatively simple. Indeed, it is one of the basic properties of operators of the first kind that they have a simple inverse. To pose and solve that problem for an operator of the first kind, we may proceed as follows.

GOURSAT PROBLEM

Our consideration in Sec. 2.4 suggests that we begin with a Goursat problem consisting of (2.1-1) and data

$$w(z_1, s_2) = \varphi_1(z_1), \qquad w(s_1, z_2) = \varphi_2(z_2) \tag{2.6-1}$$

where $\varphi_j \in C^\omega(\Omega_j)$ and $\varphi_1(s_1) = \varphi_2(s_2)$. Our task is the determination of a pair of associated functions f_1, f_2 of the solution w of (2.1-1), (2.6-1) with respect to a pair of operators of the first kind.

2.6-1 Remark

In general, in terms of operators of the first kind, the solution w of the problem (2.1-1), (2.6-1) has a Type II representation (cf. Def. 2.3-2). This solution can be reduced to a Type I representation with $\alpha_j \neq 0$ (j fixed) if φ_{j*} is restricted by the condition

$$\varphi_{j*}(z_{j*}) = \varphi_j(s_j) \exp\left[-\int_{s_{j*}}^{z_{j*}} a_j(\tilde{z}) \Big|_{\tilde{z}_j = s_j} dz_{j*} \right] \tag{2.6-2}$$

relating it to φ_j. This follows from (2.2-8) together with

$$\varphi_j(s_j) = \left(\exp \eta_j(s_j)\right) f_j(s_j) \pi$$

and the definition of an operator of the first kind. In this case we need only a single associated function, which will turn out to be unique. For a Type II representation, however, the pair f_1, f_2 is not unique. This is not surprising because already in the case of the Laplace equation, to a real harmonic function $\hat{w}$ there corresponds a complex analytic function w, with $\hat{w} = \operatorname{Re} w$, whose imaginary part is determined only up to a constant, as is well known.

INVERSION OF AN OPERATOR OF THE FIRST KIND

From the definition of a kernel of the first kind $\tilde{k}_j$ it follows that the corresponding k_j must have a representation of the form

$$k_j(z, t) = 1 + (z_1 - s_1)(z_2 - s_2)t\check{k}_j(z, t)$$

with a suitable function $\check{k}_j \in C^\omega(\Omega_0 \times N(C_j, \delta))$. Consequently, if we choose $\check{k}_j \in C^\omega(\Omega_0 \times \bar{B}_0)$, condition (2.3-2b) is satisfied for all $C_j \subset \bar{B}_0$. This suggests that we take $C_j \subset \mathbb{R}$, thus, $C_j = [-1, 1]$.

2.6-2 *Theorem*

Let $\Omega = \Omega_1 \times \Omega_2$, where the domains Ω_1 and Ω_2 are stars with respect to s_1 and s_2, respectively. Let w be the solution of the Goursat problem (2.1-1), (2.6-1). Let T_1, T_2 be Bergman operators of the first kind for a Type II representation, where $C_1 = C_2 = [-1, 1]$. Then there exist pairs of associated functions (f_1, f_2) with $f_j \in C^\omega(\tilde{\Omega}_j)$ (cf. Def. 2.2-1) such that w can be represented in the form

$$w = T_1 f_1 + T_2 f_2 \,. \tag{2.6-3}$$

If we prescribe the value of one of these functions at s, for instance,

$$e_1(s) f_1(s_1) = e_2(s) f_2(s_2)$$

or

$$f_{j^*}(s_{j^*}) = 0 \,, \qquad j \text{ fixed},$$

then these functions are uniquely determined. In terms of the initial data,

$$f_j(\xi_j) = \frac{1}{\pi}\tilde{\varphi}_j(s_j) + \frac{4}{\pi}(\xi_j - s_j)\int_0^{\pi/2} \left.\frac{d\tilde{\varphi}_j(\tilde{z}_j)}{d\tilde{z}_j}\right|_{\tilde{z}_j = \tilde{\psi}_j(\xi_j)} \sin\theta \, d\theta \tag{2.6-4}$$

where

$$\begin{aligned} \tilde{\varphi}_j(\tilde{z}_j) &= \left[\varphi_j(\tilde{z}_j) - e_{j^*}(\tilde{z})\big|_{\tilde{z}_{j^*} = s_{j^*}} \pi f_{j^*}(s_{j^*})\right] \Big/ e_j(\tilde{z})\Big|_{\tilde{z}_{j^*} = s_{j^*}} \\ \tilde{\psi}_j(\xi_j) &= 2(\xi_j - s_j)\sin^2\theta + s_j \end{aligned} \tag{2.6-5}$$

and, as before, $\xi_j = s_j + \frac{1}{2}(z_j - s_j)\tau$, $\tau = 1 - t^2$, $j = 1, 2$.

Proof. From Theorem 2.3-1 we infer that (2.6-3) with f_1, f_2 given by (2.6-4, 5) is a $C^\omega(\Omega)$-solution of (2.1-1). We show that w also satisfies the initial conditions (2.6-1). From (2.6-3) and the definition of an operator of the first kind (Def. 2.4-1), writing

$$W(z_j) = w(z)|_{z_{j^*}=s_{j^*}}$$

for simplicity, we obtain

$$W(z_j) = e_j(z)\Big|_{z_{j^*}=s_{j^*}} \int_{-1}^{1} f_j(\xi_j)\tau^{-1/2}\,dt + e_{j^*}(z)\Big|_{z_{j^*}=s_{j^*}} \pi f_{j^*}(s_{j^*}). \tag{2.6-6}$$

Substitution of (2.6-4) yields

$$W(z_j) = e_{j^*}(z)|_{z_{j^*}=s_{j^*}}\pi f_{j^*}(s_{j^*}) + e_j(z)|_{z_{j^*}=s_{j^*}} \times$$

$$\times\left[\tilde{\varphi}_j(s_j) + \frac{2}{\pi}\int_{-1}^{1}(z_j - s_j)\tau\int_0^{\pi/2}\frac{d\tilde{\varphi}_j(\tilde{z}_j)}{d\tilde{z}_j}\Bigg|_{\tilde{z}_j=\psi_j}\sin\theta\,d\theta\,\tau^{-1/2}\,dt\right] \tag{2.6-7}$$

where $\psi_j = \tilde{\psi}_j(\xi_j) = (z_j - s_j)\tau\sin^2\theta + s_j$. We introduce ψ_j as a new variable of integration in the inner integral, finding for the integrals on the right,

$$\frac{2}{\pi}\int_0^1\int_{s_j}^{\omega_j}\frac{d\tilde{\varphi}_j(\psi_j)}{d\psi_j}\frac{(z_j - s_j)^{1/2}}{\left[(z_j - s_j)\tau - \psi_j + s_j\right]^{1/2}}\,d\psi_j\,dt$$

where $\omega_j = (z_j - s_j)\tau + s_j$. Introducing ω_j as a new variable of integration in the integral from 0 to 1, and interchanging the order of integration (by applying Dirichlet's formula), we obtain from (2.6-7)

$$W(z_j) = e_{j^*}(z)\Big|_{z_{j^*}=s_{j^*}}\pi f_{j^*}(s_{j^*}) + e_j(z)\Big|_{z_{j^*}=s_{j^*}}\Bigg[\tilde{\varphi}_j(s_j)$$

$$+\int_{s_j}^{z_j}\left(\int_{\psi_j}^{z_j}\frac{d\omega_j}{\pi(\omega_j - \psi_j)^{1/2}(z_j - \omega_j)^{1/2}}\right)\frac{d\tilde{\varphi}_j(\psi_j)}{d\psi_j}\,d\psi_j\Bigg].$$

Since the inner integral equals 1, by integrating with respect to ψ_j and inserting the definition (2.6-5) of $\tilde{\varphi}_j$, we finally have

$$\begin{aligned}W(z_j) &= e_{j^*}(z)|_{z_{j^*}=s_{j^*}}\pi f_{j^*}(s_{j^*}) + e_j(z)|_{z_{j^*}=s_{j^*}}\tilde{\varphi}_j(s_j)\\ &\quad + \left[\varphi_j(z_j) - e_{j^*}(z)|_{z_{j^*}=s_{j^*}}\pi f_{j^*}(s_{j^*})\right] - e_j(z)|_{z_{j^*}=s_{j^*}}\tilde{\varphi}_j(s_j)\\ &= \varphi_j(z_j).\end{aligned}$$

That is, (2.6-1) is satisfied.

We prove the asserted uniqueness statement. We write the Taylor series of f_j with center s_j in the form

$$f_j(\xi_j) = \sum_{n=0}^{\infty} \gamma_{jn}(\xi_j - s_j)^n .$$

Hence, the initial conditions and formula (2.6-6) yield

$$\sum_{n=0}^{\infty} \frac{\pi(2n)!}{8^n(n!)^2}\gamma_{jn}(z_j - s_j)^n$$

$$= \exp\left[-\eta_j(z_j)\right]\left(\varphi_j(z_j) - \pi f_{j^*}(s_{j^*})e_{j^*}(z)\big|_{z_{j^*}=s_{j^*}}\right).$$

If for fixed j the value $f_{j^*}(s_{j^*})$ is given, the right-hand side is uniquely determined and so are the coefficients γ_{jn} on the left. By analytic continuation the assertion follows in $\tilde{\Omega}_j$. This completes the proof. ■

Theorem 2.6-2 readily implies the assertion of uniqueness regarding Type I representations; cf. Remark 2.6-1.

2.6-3 Corollary

If the problem in Theorem 2.6-2 *satisfies* (2.6-2) *with fixed* j *and if* T_j *is an operator of the first kind with* $C_j = [-1, 1]$, *then there is a unique associated function* $f_j \in C^{\omega}(\tilde{\Omega}_j)$ *of the solution* w *such that* $w = T_j f_j$. *This function* f_j *is given by* (2.6-4, 5) *with* $f_{j^*} = 0$.

2.7 RELATED IDEAS AND OPERATORS

We shall primarily concentrate on integral operators for formally elliptic and hyperbolic equations, which have been studied and developed most extensively. However, it is interesting that the corresponding ideas can be modified to suit *parabolic* equations. That is, for these equations, one can obtain integral operators of a similar nature defined on sets of complex analytic functions (or pairs of such functions) with range in the space of solutions of a given partial differential equation.

TWO OPERATORS BY COLTON

In the case of a single space variable, for instance, without loss of generality, one may consider

$$D_x^2 w + c(x, t)w - c_0(x, t)D_t w = 0$$

($D_x = \partial/\partial x$). For this equation, Colton [1976c] introduced such an operator defined by

$$w(x,t) = \underset{\sim}{P}_1[f_1, f_2](x,t)$$

$$= -\frac{1}{2\pi i}\sum_{j=1}^{2}\int_{|\sigma - t| = \delta} k^{(j)}(x,t,\sigma) f_j(\sigma)\, d\sigma .$$

For the kernels he assumed series expansions

$$k^{(1)}(x,t,\sigma) = \frac{1}{t-\sigma} + \sum_{n=2}^{\infty} x^n p_n(x,t,\sigma)$$

and

$$k^{(2)}(x,t,\sigma) = \frac{x}{t-\sigma} + \sum_{n=3}^{\infty} x^n q_n(x,t,\sigma)$$

p_n and q_n being suitable functions to be determined and the convergence of the series being proved by the majorant method. See also the references in the research notes by Colton [1976c, d]. Note that, in contrast to Bergman kernels, these kernels have singularities. These operators yield analytic solutions. But it is well known (cf. Friedman [1964]) that for parabolic equations, analyticity of the coefficients may not entail analyticity of solutions. Colton [1976c] shows that one can define more satisfactory operators on solutions of the heat equation. (Thereby he loses contact with complex analysis. See also Widder [1975].)

Colton also introduced an operator for parabolic equations in two space variables:

$$D_1 D_2 w + b(z,t) D_2 w + c(z,t) w - c_0(z,t) D_t w = 0 \tag{2.7-1}$$

where $z = (z_1, z_2)$. This operator $\underset{\sim}{P}_2$ is defined on functions of two complex variables and is given by the formula

$$w(z,t) = \underset{\sim}{P}_2 f(z,t)$$

$$= -\frac{1}{2\pi i}\int_{|\sigma_2 - t| = \delta}\int_{-1}^{1} k(z,t,\sigma_1,\sigma_2) f\left(\tfrac{1}{2}z_1\left(1-\sigma_1^2\right), \sigma_2\right) \times \tag{2.7-2}$$

$$\times \left(1-\sigma_1^2\right)^{-1/2} d\sigma_1\, d\sigma_2 .$$

$\underset{\sim}{P}_2$ is such that for equations and solutions independent of t it reduces to the Bergman operator of the first kind [with $j = 1$, $\eta_1 = 0$, and point of reference

$s = (0,0)$]. For more details we refer to the above literature; see also Colton's monograph [1980].

GENERATING OPERATORS IN THE SENSE OF LANCKAU

A further useful generalization of the Bergman operator of the first kind was recently introduced by Lanckau [1983b]. Lanckau's method is quite remarkable since it provides a unified treatment of time-dependent processes in the plane, for instance, processes described by hyperbolic, parabolic, and pseudoparabolic equations; and we shall discuss this method in some detail.

For simplicity, we consider equations of the form

$$D_1 D_2 w + Sw = 0. \tag{2.7-3}$$

Here, w is a function of the independent variables $z = (z_1, z_2)$ and (real or complex) t, $(z, t) \in \Omega \times \Omega_3$, $\Omega = \Omega_1 \times \Omega_2$, and Ω_1 is a star with respect to $z_1 = 0$; also

$$\Omega_2 = \Omega_1^* = \{ z_2 = x_1 - ix_2 \in \mathbb{C} \mid z_1 = x_1 + ix_2 \in \Omega_1 \}$$

and Ω_3 is an open interval or a domain in the plane. S denotes a linear operator depending on t, but independent of z and such that (2.7-3) is self-adjoint with respect to z_1, z_2.

We remember the Type I representation of solutions for the Helmholtz equation obtained in Example 2.5-4. Letting $s = (0,0)$ and substituting $t = \sin\alpha$, we obtain, by (2.2-6),

$$\begin{aligned} w(z) &= \int_{-1}^{1} \cos\left(c\sqrt{z_1 z_2}\, t\right) f_1\left(\tfrac{1}{2} z_1 (1 - t^2)\right)(1 - t^2)^{-1/2}\, dt \\ &= \int_{-\pi/2}^{\pi/2} \cos\left(c\sqrt{z_1 z_2}\, \sin\alpha\right) f_1\left(\tfrac{1}{2} z_1 \cos^2\alpha\right) d\alpha \qquad (2.7\text{-}4) \\ &= \int_{-\pi/2}^{\pi/2} \sum_{m=0}^{\infty} \frac{1}{(2m)!} \left(-4 z_1 z_2 \sin^2\alpha\right)^m \left(\frac{c^2}{4}\right)^m \tilde{f}(\tilde{\varphi}(z_1, \alpha))\, d\alpha \end{aligned}$$

with $\tilde{f} \circ \tilde{\varphi} = f_1 \circ \left(\frac{1}{2}\tilde{\varphi}\right)$ and $\tilde{\varphi}(z_1, \alpha) = z_1 \cos^2\alpha$. Here, in the case of the Helmholtz equation, we simply have $S = \tilde{S}$, where $\tilde{S}$ denotes multiplication with $c^2/4$. Therefore, w can be rewritten as

$$w(z) = \int_{-\pi/2}^{\pi/2} \tilde{E}\tilde{g}(z, \alpha)\, d\alpha, \qquad \tilde{g} = \tilde{f} \circ \tilde{\varphi} \tag{2.7-5a}$$

where the "generating operator" $\tilde{\tilde{E}}$ is given by

$$\tilde{\tilde{E}}\tilde{g} = \sum_{m=0}^{\infty} \frac{1}{(2m)!}(-4z_1z_2 \sin^2 \alpha)^m \tilde{S}^m \tilde{g}. \tag{2.7-5b}$$

This operator $\tilde{\tilde{E}}$ suggests defining an analogous operator $\tilde{E}$ for the general case of (2.7-3), where S depends on time t. We now introduce the following concept.

2.7-1 *Definition*

(a) A function f is said to be ***of class F*** if $f(\cdot, t)$ is holomorphic in Ω_1 and continuous on $\Omega_1 \times \partial\Omega_1$ $(t \in \Omega_3)$, and if there are constants $\beta, \gamma > 0$ such that for all $m \in \mathbb{N}_0$,

$$|S^m f(z_1, t)| \leq \beta\gamma^m (2m)! \qquad \text{in } \Omega_1 \times \check{\Omega}_3, \qquad \check{\Omega}_3 \subset \Omega_3.$$

(b) $\tilde{E}$ is called a ***generating operator*** if for $f \in F$ [F as defined in (a), thus depending on S in (2.7-3)],

$$\tilde{E}f = \sum_{m=0}^{\infty} \frac{1}{(2m)!}(-4z_1z_2 \sin^2 \alpha)^m S^m f \qquad \text{[cf. (2.7-5b)]}.$$

The assumptions on F guarantee the existence of $\tilde{E}$ in a neighborhood of $(0,0) \in \Omega$ and for all $t \in \check{\Omega}_3$ and $\alpha \in [-\frac{1}{2}\pi, \frac{1}{2}\pi]$. Indeed,

$$\begin{aligned} |\tilde{E}f(z,t,\alpha)| &\leq \sum_{m=0}^{\infty} \frac{4^m}{(2m)!}|z_1z_2|^m |S^m f(z_1,t)| \\ &\leq \beta \sum_{m=0}^{\infty} [4\gamma|z_1z_2|]^m = \beta/[1 - 4\gamma|z_1z_2|]. \end{aligned} \tag{2.7-6}$$

REPRESENTATION OF SOLUTIONS

In analogy to the representation (2.7-5), the operator $\tilde{E}$ can also be used for generating solutions in the time-dependent case of equation (2.7-3).

2.7-2 *Theorem*

For all $f \in F$, a solution of (2.7-3) is given by

$$w(z,t) = \int_{-\pi/2}^{\pi/2} \tilde{E}(f \circ \varphi)(z,t,\alpha)\, d\alpha; \tag{2.7-7}$$

here, $\tilde{E}$ is a generating operator according to Def. 2.7-1b, φ is defined by

$$\varphi(z_1, t, \alpha) = \left(z_1 \cos^2 \alpha, t\right)$$

and $z \in \Omega$ is contained in a neighborhood of $(0,0)$, $t \in \breve{\Omega}_3$.

Proof. Keeping in mind the preceding idea of the introduction of the operator $\tilde{E}$ and the representation of solutions, we see that the present theorem can be proved in a manner similar to that of the proof of Theorem 2.3-1. For the verification of (2.7-7) we only mention the following steps. Inserting (2.7-7) into the left-hand side of (2.7-3), because of (2.7-6), we can use term-by-term differentiation. We next observe that S is independent of z_1; hence we have

$$D_1 S^m(f \circ \varphi) = S^m D_1(f \circ \varphi), \qquad m \in \mathbb{N}_0.$$

Furthermore, we use

$$D_1(f \circ \varphi) = -\frac{1}{2z_1} \cot \alpha \frac{\partial}{\partial \alpha}(f \circ \varphi).$$

Interchanging the order of summation and integration, we thus obtain the result

$$\begin{aligned} D_1 D_2 w(z, t) = \sum_{m=1}^{\infty} \frac{1}{(2m-1)!}(-4z_1 z_2)^{m-1} \times \\ \times S^m \left[\int_{-\pi/2}^{\pi/2} \sin^{2m-1} \alpha \cos \alpha \frac{\partial}{\partial \alpha}(f \circ \varphi)(z_1, t, \alpha)\, d\alpha \right. \\ \left. -2m \int_{-\pi/2}^{\pi/2} \sin^{2m} \alpha\, (f \circ \varphi)(z_1, t, \alpha)\, d\alpha \right]. \end{aligned}$$

Applying integration by parts to the first integral, we get for the term in the brackets [$\cdots$] the expression

$$-(2m-1) \int_{-\pi/2}^{\pi/2} \sin^{2(m-1)} \alpha\, (f \circ \varphi)(z_1, t, \alpha)\, d\alpha.$$

Hence, the right-hand side of the last equation turns out to be equal to $-Sw(z, t)$, and the proof of the theorem is complete. ■

CLOSED-FORM SOLUTIONS

As applications, we consider some equations for which Theorem 2.7-2 yields solutions in *closed form*.

2.7-3 *Example*

Let

$$S = -(a + bD_t)^2, \qquad a, b \text{ constant}.$$

Then equation (2.7-3) becomes

$$D_1 D_2 w - b^2 D_t^2 w - 2abD_t w - a^2 w = 0. \tag{2.7-8}$$

This is a three-dimensional equation which is hyperbolic for real $b \neq 0$ and elliptic (with real coefficients) for pure imaginary $a, b \neq 0$. From Theorem 2.7-2 we obtain the following representation of solutions of (2.7-8):

$$w(z, t) = \int_{-\pi/2}^{\pi/2} \sum_{m=0}^{\infty} \frac{1}{(2m)!} (4 z_1 z_2 \sin^2 \alpha)^m (a + bD_t)^{2m} f(z_1 \cos^2 \alpha, t)\, d\alpha.$$

This expression, however, can be simplified considerably. First, we observe that we may add a zero term of the form

$$\int_{-\pi/2}^{\pi/2} \sum_{m=0}^{\infty} \frac{1}{(2m+1)!} (4 z_1 z_2 \sin^2 \alpha)^{m+1/2} \times$$

$$\times (-1)^{2m+1} (a + bD_t)^{2m+1} f(z_1 \cos^2 \alpha, t)\, d\alpha.$$

Thus, symbolically, we can rewrite w in the form

$$w(z, t) = \int_{-\pi/2}^{\pi/2} \exp\left[-2a\sqrt{z_1 z_2} \sin \alpha\right] \times$$

$$\times \exp\left[-2b\sqrt{z_1 z_2} \sin \alpha D_t\right] f(z_1 \cos^2 \alpha, t)\, d\alpha.$$

If we now choose $f \in F$ in a suitable way, for instance, as an entire function in t, we can introduce the Cauchy integral formulas for the derivatives of f with respect to t, then use the Maclaurin series of the exponential function, and finally obtain

$$\sum_{m=0}^{\infty} \frac{1}{m!} \left(-2b\sqrt{z_1 z_2} \sin \alpha\right)^m D_t^m f(z_1 \cos^2 \alpha, t)$$

$$= f\left(z_1 \cos^2 \alpha, t - 2b\sqrt{z_1 z_2} \sin \alpha\right).$$

Substituting this into the last equation for w yields the simple representation

$$w(z,t) = \int_{-\pi/2}^{\pi/2} \exp\left[-2a\sqrt{z_1 z_2}\sin\alpha\right] \times \tag{2.7-9}$$
$$\times f\left(z_1\cos^2\alpha,\, t - 2b\sqrt{z_1 z_2}\sin\alpha\right) d\alpha .$$

APPLICATION OF THEOREM 2.7-2 TO COLTON'S OPERATOR $\underset{\sim}{P}_2$

The following example demonstrates the remarkable fact that, for an equation (2.7-1) with $b = 0$ and constant c and c_0, Theorem 2.7-2 gives an *explicit representation* [cf. (2.7-14)] for the function k in Colton's formula (2.7-2).

2.7-4 *Example*

Consider

$$S = c - c_0 D_t, \qquad c,\, c_0 \text{ constant},$$

that is, the parabolic equation in two space variables in complex form

$$D_1 D_2 w + cw - c_0 D_t w = 0. \tag{2.7-10}$$

Substituting this S into the definition of $\tilde{E}$ (Def. 2.7-1b) and interchanging the order of summation, we obtain $\tilde{E}$ in the form

$$\tilde{E} = \sum_{\mu=0}^{\infty}\sum_{m=0}^{\infty}\binom{m+\mu}{\mu}\frac{(-c)^m c_0^\mu}{(2m+2\mu)!}\left(4z_1 z_2\sin^2\alpha\right)^{m+\mu} D_t^\mu . \tag{2.7-11}$$

Introducing (2.7-11) and Cauchy's representation

$$f(z_1,t) = \frac{1}{2\pi i}\int_{|\sigma_2 - t| = \delta} f(z_1,\sigma_2)(\sigma_2 - t)^{-1}\, d\sigma_2$$

($\delta > 0$ sufficiently small, such that the circle lies in $\check{\Omega}_3$) into (2.7-7), we get

$$w(z,t) = \frac{1}{2\pi i}\int_{-\pi/2}^{\pi/2}\int_{|\sigma_2 - t| = \delta}\tilde{E}\left(\frac{1}{\sigma_2 - t}\right)(z,t,\alpha) \times \tag{2.7-12}$$
$$\times f\left(z_1\cos^2\alpha,\, \sigma_2\right) d\sigma_2\, d\alpha .$$

Here, by (2.7-11), the kernel is explicitly given and can be expressed as a *hypergeometric function of Horn–Birkeland type* of two auxiliary variables U, V:

$$
\begin{aligned}
&(\sigma_2 - t)\tilde{E}\left(\frac{1}{\sigma_2 - t}\right)(z, t, \alpha) \\
(2.7\text{-}13)\qquad &= \sum_{\mu=0}^{\infty} \sum_{m=0}^{\infty} \frac{4^{\mu+m}(\mu + m)!\mu!}{[2(\mu + m)]!} \frac{1}{\mu! m!} U^{\mu} V^{m} \\
&= \Phi_3\left(1; \tfrac{1}{2}; U, V\right)
\end{aligned}
$$

[cf. Horn's list in Erdélyi et al. [1953–1955], Vol. I, p. 225 especially formula 5.7.1.(22)]. Here

$$
U = c_0 z_1 z_2 \sin^2\alpha / (\sigma_2 - t), \qquad V = -c z_1 z_2 \sin^2\alpha
$$

the series being convergent for all $(U, V) \in \mathbb{C}^2$. Formula (2.7-12), however, restates Colton's representation (2.7-2). For, substitute $\sigma_1 = \sin\alpha$ into (2.7-12), set

$$
f_1\left(\tfrac{1}{2}z_1(1 - \sigma_1^2), \sigma_2\right) \equiv f\left(z_1(1 - \sigma_1^2), \sigma_2\right)
$$

and, according to (2.7-12, 13), write

$$
(2.7\text{-}14)\quad k(z, t, \sigma_1, \sigma_2) = -\frac{1}{\sigma_2 - t}\Phi_3\left(1; \frac{1}{2}; \frac{c_0 z_1 z_2 \sigma_1^2}{\sigma_2 - t}, -c z_1 z_2 \sigma_1^2\right).
$$

Then (2.7-12) takes the form

$$
\begin{aligned}
&w(z, t) = -\frac{1}{2\pi i}\int_{|\sigma_2 - t| = \delta}\int_{-1}^{1} k(z, t, \sigma_1, \sigma_2) \times \\
(2.7\text{-}15)\qquad &\times f_1\left(\tfrac{1}{2}z_1(1 - \sigma_1^2), \sigma_2\right)\left(1 - \sigma_1^2\right)^{-1/2} d\sigma_1\, d\sigma_2 .
\end{aligned}
$$

This is the representation (2.7-2) of solutions to the special equation (2.7-10).
In the special case $c_0 = 0$, $c = -\tilde{c}^2/4$, formula (2.7-14) reduces to

$$
(2.7\text{-}16)\qquad k(z, t, \sigma_1, \sigma_2) = -(\sigma_2 - t)^{-1}\cos\left(\tilde{c}\sqrt{z_1 z_2}\,\sigma_1\right).
$$

Hence when f_1 is independent of σ_2, then (2.7-15) again gives (2.7-4).

In the special case $c = 0$, formula (2.7-14) yields

$$k(z, t, \sigma_1, \sigma_2) = -(\sigma_2 - t)^{-1} {}_1F_1\left(1; \tfrac{1}{2}; \frac{c_0 z_1 z_2 \sigma_1^2}{\sigma_2 - t}\right) \tag{2.7-17}$$

where ${}_1F_1$ denotes ***Kummer's function*** (***confluent hypergeometric function***; cf. Magnus et al. [1966], Chap. VI).

A PSEUDOPARABOLIC EQUATION OF THIRD ORDER

2.7-5 ***Example***

Assuming $0 \in \check{\Omega}_3$, let

$$S = -c_0 + c\int_0^t \cdot \, d\tau, \qquad c, c_0 \text{ constant}.$$

Then (2.7-3) becomes an *integro-differential equation*. From this we obtain the following *pseudoparabolic equation of third order*:

$$D_1 D_2 D_t w + cw - c_0 D_t w = 0. \tag{2.7-18}$$

We now apply the binomial formula and write the μ-fold integrals in S^m by Liouville's formula as single integrals. Then from Def. 2.7-1b we get

$$\tilde{E}g(z, t, \alpha) = \sum_{m=0}^{\infty} \frac{1}{(2m)!}(4z_1 z_2 \sin^2\alpha)^m \times$$

$$\times \sum_{\mu=0}^{m} \binom{m}{\mu} c_0^{m-\mu}(-c)^\mu D_t \frac{1}{\mu!}\int_0^t (t - \sigma_2)^\mu g(z_1, \alpha, \sigma_2)\, d\sigma_2$$

$$g = f \circ \varphi.$$

Interchanging the order of summation and suitably rearranging terms, we can represent $\tilde{E}g$ in the form [cf. (2.7-13)]

$$\tilde{E}g(z, t, \alpha) = D_t \int_0^t \sum_{\mu=0}^{\infty} \sum_{m=0}^{\infty} \frac{4^{\mu+m}(\mu + m)!\mu!}{[2(\mu + m)]!} \times$$

(2.7-19)

$$\times \frac{1}{\mu! m!} \frac{1}{(\mu!)^2}\left[-(t - \sigma_2)^2 U\right]^\mu V^m g(z_1, \alpha, \sigma_2)\, d\sigma_2.$$

Here U and V are as in (2.7-13). The double series converges for all $(U, V) \in \mathbb{C}^2$ and $t, \sigma_2 \in (\check{\Omega}_3 \subset)\mathbb{C}$, as can be seen from (2.7-13). Furthermore, from (2.7-19) it follows that the generating operator may be represented by a Duhamel product, namely,

$$\tilde{E}g(z,t,\alpha) = D_t\int_0^t \tilde{E}(1)(z,t-\sigma_2,\alpha)g(z_1,\alpha,\sigma_2)\,d\sigma_2.$$

Thus, Theorem 2.7-2 now gives the solution

$$\begin{aligned} w(z,t) &= \int_{-\pi/2}^{\pi/2} D_t\int_0^t \tilde{E}(1)(z,t-\sigma_2,\alpha)f(z_1\cos^2\alpha,\sigma_2)\,d\sigma_2\,d\alpha \\ &= D_t\int_0^t\int_{-1}^1 k(z,t,\sigma_1,\sigma_2)f_1\left(\tfrac{1}{2}z_1(1-\sigma_1^2),\sigma_2\right)\left(1-\sigma_1^2\right)^{-1/2}d\sigma_1\,d\sigma_2. \end{aligned} \tag{2.7-20}$$

Here k has the form [cf. (2.7-19)]

$$\begin{aligned} k(z,t,\sigma_1,\sigma_2) &= \sum_{\mu=0}^{\infty}\sum_{m=0}^{\infty}\frac{4^{\mu+m}(\mu+m)!}{[2(\mu+m)]!m!(\mu!)^2}\times \\ &\quad\times c_0^m(-c)^\mu(z_1z_2)^{\mu+m}(t-\sigma_2)^\mu\sigma_1^{2(\mu+m)}. \end{aligned} \tag{2.7-21}$$

This series converges for all $(z,t,\sigma_1,\sigma_2)\in\mathbb{C}^5$.

Considerable simplifications are obtained if $c_0=0$ or $c=0$. In the first case, (2.7-21) reduces to a generalized hypergeometric function of type ${}_0F_2$, namely,

$$\begin{aligned} k(z,t,\sigma_1,\sigma_2) &= \sum_{\mu=0}^{\infty}\frac{4^\mu}{\mu!(2\mu)!}\left[cz_1z_2(\sigma_2-t)\sigma_1^2\right]^\mu \\ &= {}_0F_2\left(1,\tfrac{1}{2};\,cz_1z_2(\sigma_2-t)\sigma_1^2\right). \end{aligned} \tag{2.7-22}$$

In the second case, by setting $c_0=-\tilde{c}^2/4$, it reduces to

$$k(z,t,\sigma_1,\sigma_2) = \cos\left(\tilde{c}\sqrt{z_1z_2}\,\sigma_1\right). \tag{2.7-23}$$

Hence if f_1 is independent of σ_2, then (2.7-20) with (2.7-23) again yields (2.7-4).

A GENERALIZATION OF THE RADIAL CASE OF EQUATION (2.7-3)

If the solution w depending on $z=(z_1,z_2)$ and t is assumed to be a function $\hat{w}$ of the variables $r=(z_1z_2)^{1/2}$ and t, then equation (2.7-3) may equivalently be written as

$$\hat{w}_{rr} + \frac{1}{r}\hat{w}_r + \hat{S}\hat{w} = 0 \tag{2.7-24}$$

where $\hat{S} = 4S$ is a linear operator depending only on t, as before. Thus, for $\hat{S} = \partial^2/\partial t^2$ and $z_2 = \bar{z}_1$, this equation contains the equation of axially symmetric potential theory. We wish to mention here that in a subsequent paper, Lanckau modifies the idea of the approach described for (2.7-3) and (2.7-24) to an extension of the equation of generalized axially symmetric potential theory (GASPT) [cf. (11.1-7)] of the form

$$\hat{w}_{rr} + \frac{2\lambda + 1}{r}\hat{w}_r + \hat{S}\hat{w} = 0, \qquad \lambda = const \tag{2.7-25}$$

see Lanckau [1986].

For further analogous investigations, but with respect to applications of a generalized Riemann–Vekua operator (cf. Chap. 6), also see Lanckau [1980b, 1983a].

Additional References

Bergman [1937b, 1943b, 1971]
Bhatnagar–Gilbert [1977]
Colton [1976c, 1976d, 1980]
Colton–Gilbert [1974]
Gilbert [1969b, 1974]
Gilbert–Newton [1968]
Gilbert–Weinacht [1976]
Heersink [1976a]
Heins–MacCamy [1963]
Henrici [1952]
Jank [1971]
Kracht [1970, 1974a]
Kracht–Kreyszig–Schröder [1981, 1982]
Kreyszig [1960, 1962, 1972]
Kürcz [1978]
Lanckau [1981, 1983b, 1986]
Lanckau–Tutschke [1983]
Marzuq [1984]
Meister–Weck–Wendland [1976]
Stecher [1975]
Tomantschger [1983b, 1985]
Wallner [1983]
Watzlawek [1971a]

Chapter Three

Integral Operators with Polynomial Kernels and Differential Operators

One of the main reasons for the importance of the integral operator method in applications is that this method is *constructive*. Moreover, from a practical point of view, a particularly valuable property of the method is as follows. We have shown in the previous chapter that for every equation (2.1-1) with analytic coefficients there exist corresponding Bergman operators, and the kernel of such an operator can usually be obtained in the form of an infinite series. However, it is quite remarkable that several equations of practical importance belong to large classes for which that kernel can be derived in *finite form*. Among other things, this property then permits *global* representations and study of solutions.

The present chapter is devoted to a class of such operators which has attracted particular interest and attention. These are the Bergman operators T_j with "*polynomial kernels*," that is, kernels which are polynomials in the variable of integration t, with coefficient functions depending on z_1, z_2.

In Sec. 3.1 we give the definitions and some simpler relations. In Sec. 3.2 we discuss one of their most remarkable properties, namely, their conversion to differential operators and vice versa. The latter will be studied in more detail in Sec. 3.3. We then turn to the inversion problem in Sec. 3.4.

3.1 OPERATORS OF CLASS P

In this section we define a large and remarkable class of Bergman operators and discuss some of their properties. These operators have kernels that are polynomials in the variable of integration with coefficients depending on z_1, z_2. We also include simple illustrative examples.

3.1-1 Definition

An operator T_j is called an ***operator of class P*** if its kernel $\tilde{k}_j$ is a *polynomial kernel*, that is, if the function $k_j = \tilde{k}_j/e_j$ has the form

$$(3.1\text{-}1) \qquad k_j(z,t) = \sum_{m=0}^{n} q_{jm}(z) t^{2m}, \qquad q_{jn} \neq 0, \qquad n \in \mathbb{N}_0 .$$

n is called the ***degree*** of the kernel.[1]

Furthermore, $\tilde{k}_j$ is said to be *minimal* for a given equation (2.1-1) on Ω if for that equation there does not exist a polynomial kernel on Ω of degree less than n.

The concept of minimality is suggested by the fact that the translation principle effected by operators of the present kind becomes more powerful the more simply we can choose kernels and their representations.

Clearly, we can impose on a Bergman kernel both of the conditions (2.4-1) and (3.1-1), so that we then obtain a *class P kernel of the first kind* (presupposed $\rho_j = \sigma_j = s_j$; cf. Defs. 2.2-1 and 2.4-1).

Class P operators were introduced by Kreyszig [1968]. In general they have the following three advantageous properties.

(i) They lead to global results on solutions in a simple way. This includes that they also permit a study of solutions at infinity.

(ii) There exist constructive principles for obtaining explicit representations of polynomial kernels.

(iii) Solutions obtained by class P operators can be cast into an integral-free form. This form also results by applying certain differential operators. The coefficient functions of the latter are related to those of the kernel of the original integral operators by simple formulas.

CLASSES OF DIFFERENTIAL EQUATIONS CORRESPONDING TO CLASS P OPERATORS

Clearly, the assumption of a special form of a Bergman kernel (possibly involving undetermined coefficients) imposes additional conditions on the coefficients of L in equation (2.1-1). In the case of (3.1-1) the situation is as follows. If we start from a finite sum [cf. (2.5-8)]

$$(3.1\text{-}2) \qquad k_j(z,t) = \sum_{m=0}^{n} \tilde{q}_{jm}(z)(z_j - \sigma_j)^m t^{2m}$$

[1] The use of the degree in t^2 rather than in t will later [in (3.2-2)] turn out to be more practical. Also, the integral over odd powers, taken along a path obtainable from $[-1,1]$ by the deformation principle, would be zero. For e_j, see (2.1-3).

we obtain a system analogous to (2.5-9), namely,

$$D_{j*}\tilde{q}_{jm} = -\frac{2}{2m-1}L_j\tilde{q}_{j,m-1}, \quad \tilde{q}_{j,-1} = 0; \quad m = 0,\cdots,n+1 \tag{3.1-3a}$$

and the additional condition

$$\tilde{q}_{j,n+1} = 0. \tag{3.1-3b}$$

Thus, this system is overdetermined and restrictions on L_j (hence on L) are inevitable. We shall discuss this question in later sections, where we shall obtain necessary and sufficient conditions for L to admit polynomial kernels.

Classes of operators L for which one can use class P operators with kernels as defined in Def. 3.1-1 will be denoted as follows.

3.1-2 Notations

For $n \in \mathbb{N}_0$ and sets of operators L in (2.1-1) we write

$$P_{jn} = \{L \mid L \text{ admits a polynomial kernel } \tilde{k}_j \text{ of degree } n\},$$

$$P_{jn}^0 = \{L \mid L \in P_{jn} \text{ with } \tilde{k}_j \text{ minimal}\},$$

$$FP_{jn} = \{L \mid L \in P_{jn} \text{ with } \tilde{k}_j \text{ of the first kind}\},$$

$$FP_{jn}^0 = \{L \mid L \in FP_{jn} \text{ with } \tilde{k}_j \text{ minimal}\}.$$

3.1-3 Remark

For $n \in \mathbb{N}$,

$$FP_{jn}^0 \subset FP_{jn} \subset P_{jn} \qquad \text{and} \qquad FP_{jn}^0 \subset P_{jn}^0 \subset P_{jn}$$

all these inclusions being proper.

The validity of those inclusions follows from 3.1-1 and 3.1-2, and equality is excluded, as can be seen from counterexamples, for instance from examples in 3.1-6 and 3.1-7, below.

Theoretically, class FP_{jn} seems to be particularly useful, mainly because of the properties of operators of the first kind. However, concentrating exclusively on FP_{jn} would entail a further restriction on k_j in (3.1-2) and thus on the class of equations (2.1-1); cf. Example 3.1-6, below.

We now mention two simple results on polynomial kernels. In the first place, by straightforward calculation one can verify the following.

3.1-4 Remark

For all ν, $n \in \mathbb{N}_0$ and $j = 1, 2$, the relation $P_{jn} \subset P_{j,n+\nu}$ holds. Furthermore, from a given polynomial kernel $\tilde{k}_j$ of degree n we can obtain a polynomial kernel $\tilde{k}_j^*$ of degree $n + \nu$ with $k_j^* = \tilde{k}_j^*/e_j$ in the form

$$k_j^*(z,t) = \sum_{m=0}^{n+\nu} \sum_{k=\chi}^{\lambda} \alpha_k \frac{[2(m-k)]!m!}{(m-k)!(2m)!} (z_j - \sigma_j)^k q_{j,m-k}(z) t^{2m}$$

where $\chi = \max(0, m - n)$, $\lambda = \min(\nu, m)$, $m = 0, \cdots, n + \nu$; α_k are arbitrary complex constants, with $\alpha_\nu \neq 0$, and $q_{j0}, \cdots, q_{jn}$ are the coefficients of $\tilde{k}_j$; cf. (3.1-1).

The second result is obtained by combining Defs. 2.4-1 and 3.1-1 and Corollary 2.5-3:

3.1-5 Corollary

If a given L admits a polynomial kernel $\tilde{k}_j$ of the first kind, then $\tilde{k}_j$ is unique, except for the choice of s and η_j in (2.1-3).

CLASS P OPERATORS FOR SPECIAL EQUATIONS

3.1-6 Example

The kernel (2.5-14) is a minimal polynomial kernel for the equation (2.5-13) in Example 2.5-5. This example also demonstrates that for all $n \in \mathbb{N}$ and $j = 1, 2$ (cf. 3.1-2),

$$\text{(i)} \quad FP_{jn}^0 \neq \varnothing$$

$$\text{(ii)} \quad P_{jn}^0 \neq P_{jn}$$

$$\text{(iii)} \quad FP_{jn} \neq P_{jn}.$$

Here, (i) also holds for $n = 0$. Assertion (ii) follows by observing that Remark 3.1-4 yields polynomial kernels for (2.5-13) from (2.5-14) which are not minimal. For proving (iii), we note that the Laplace operator is an element of FP_{j0}^0 $(j = 1, 2)$ as well as P_{jn}. The latter fact is a consequence of 3.1-4, but, by Corollary 3.1-5, that operator is not in FP_{jn} for $n \in \mathbb{N}$.

3.1-7 Example

Let

$$L = D_1 D_2 + 4z_j \left(1 + z_j^2 z_{j^*}\right)^{-2}, \qquad j \text{ fixed}.$$

Then $L \in FP_{j2} \cap P_{j1}^0$. Indeed, a polynomial kernel $\tilde{k}_j$ of the first kind and second degree (in t^2, with point of reference $s = 0$) is given by

$$\tilde{k}_j(z, t) = 1 + 8z_j^2 z_{j*}\left(1 + z_j^2 z_{j*}\right)^{-1}\left(-t^2 + \tfrac{2}{3}t^4\right).$$

Another polynomial kernel $\tilde{\tilde{k}}_j$, of first degree (in t^2), is given by

$$\tilde{\tilde{k}}_j(z, t) = 1 + \left(2 - 8z_j^2 z_{j*}\left(1 + z_j z_{j*}\right)^{-1}\right)t^2;$$

this kernel is minimal, as can be seen directly from (2.3-2a). Because of Corollary 3.1-5 we further have $L \notin FP_{j1}^0$ and, directly from the definition in 3.1-2, also $L \notin FP_{j2}^0$. Consequently, we may conclude

$$\text{(i)} \quad FP_{j2}^0 \neq FP_{j2}$$

$$\text{(ii)} \quad FP_{j1}^0 \neq P_{j1}^0$$

$$\text{(iii)} \quad FP_{j2} \not\subset P_{j2}^0 .$$

The next example illustrates the following remarkable fact: If for an L there exists a polynomial kernel $\tilde{k}_j$ of a certain kind for a Type I representation of solutions (cf. Def. 2.3-2), it does generally *not* follow that a polynomial kernel $\tilde{k}_{j*}$ of the same kind for a Type II representation exists.

3.1-8 Example

We consider

$$L = D_1 D_2 - \alpha z_j^{n-1} z_{j*}^n D_{j*} + n\alpha(z_1 z_2)^{n-1}$$

for $n \in \mathbb{N}$, fixed j, and complex $\alpha \neq 0$. We choose $s = 0$ as well as $\eta_1 = \eta_2 = 0$. Then the polynomial kernel $\tilde{k}_j$ of the first kind is given by

$$\tilde{k}_j(z, t) = 1 + \alpha z_1^n z_2^n \sum_{m=1}^{n} (-4)^m \binom{n}{m} \left[\binom{2m}{m}\right]^{-1} t^{2m}.$$

Furthermore, the uniquely determined kernel $\tilde{k}_{j*}$ of the first kind which is even in t [cf. Corollary 2.5-3 and (3.1-1)] is given by

$$\tilde{k}_{j*}(z, t) = \exp\left(\frac{1}{n}\alpha z_1^n z_2^n\right) k_{j*}(z, t)$$

$$k_{j*}(z, t) = 1 + \sum_{m=1}^{\infty} \sum_{\mu=1}^{m} \beta_\mu(n, m)(\alpha z_1^n z_2^n)^\mu t^{2m}.$$

Here the constants β_μ are known and, whereas they have a complicated form in general, for $n = 1$ they simply are

$$\beta_\mu(1, m) = \delta_{\mu m}(-4)^m (m+1)!/(2m)!.$$

Hence in that case the series representing k_{j*} becomes the Maclaurin expansion of a ***Kummer function*** (cf. Sec. 2.7), namely,

$$\begin{aligned} k_{j*}(z,t) &= \sum_{m=0}^{\infty} \frac{(m+1)!}{(2m)!}\left(-4\alpha z_1 z_2 t^2\right)^m \\ &= {}_1F_1\left(2; \tfrac{1}{2}; -\alpha z_1 z_2 t^2\right). \end{aligned}$$

This shows that there does not exist a polynomial kernel $\tilde{k}_{j*}$ of the first kind (although the preceding $\tilde{k}_j$ is of that form). It follows that $L \notin FP_{j*p}$ for all $p \in \mathbb{N}_0$.

The subsequent example gives a very simple necessary condition for the existence of polynomial kernels in a special case.

3.1-9 Example

We consider the self-adjoint operator

$$L = D_1 D_2 + a_0(z) \tag{3.1-4}$$

and require $L \in FP_{jn}$ with n ($\in \mathbb{N}$) and j fixed. Then, by Corollary 3.1-5, for $\eta_j = 0$ and fixed s, the class P kernel $\tilde{k}_j$ of the first kind of degree n and point of reference s is unique, $\tilde{k}_j = k_{je}$ [cf. (2.5-12)]. By the use of the system (3.1-3), Corollary 2.5-3, and $\eta_j = 0$ (hence $L_j = L$), we thus obtain

$$LS_j^n 1 = 0 \tag{3.1-5}$$

where

$$S_j g(z) = \int_{s_{j*}}^{z_{j*}} Lg(\tilde{z})\bigg|_{\tilde{z}_j = z_j} d\tilde{z}_{j*}.$$

Carrying out the differentiations with respect to z_{j*} in (3.1-5), we can rewrite the left-hand side of (3.1-5) as a sum of $D_j^n a_0$ and terms containing at least one integral from s_{j*} to z_{j*} (but no operator D_{j*}) and, consequently,

$$D_j^n a_0(z)\big|_{z_{j*} = s_{j*}} = 0. \tag{3.1-6}$$

Hence, on the plane $z_{j*} = s_{j*}$, the coefficient a_0 of L must necessarily reduce to a polynomial in z_j of degree not exceeding $n - 1$.

For further investigations of self-adjoint operators, in general not of class FP_{jn}, we refer to Sec. 4.7.

3.2 INTEGRAL-FREE REPRESENTATIONS OF SOLUTIONS

We shall now prove the important fact that solutions obtained by Bergman integral operators with polynomial kernels can be converted to an *integral-free form*. This property establishes a relation of those integral operators to certain *differential operators* which also yield those solutions and have been used by various authors, as we shall mention in more detail below. The basic result can be formulated as follows.

3.2-1 Theorem

Let (2.1-1) with $z \in \Omega$ be such that there exists an operator T_j of class P with kernel $\tilde{k}_j$ and with point of reference s, $s \in \Omega$. Furthermore, let $f_j \in C^{\omega}(\tilde{\Omega}_j)$ and $w_j = T_j f_j$, so that

$$(3.2\text{-}1) \qquad w_j(z) = e_j(z) \sum_{m=0}^{n_j} \tilde{q}_{jm}(z)(z_j - s_j)^m \int_{-1}^{1} t^{2m} f_j(\zeta_j) \tau^{-1/2}\, dt$$

where, as in (2.2-8),

$$\zeta_j = s_j + \tfrac{1}{2}(z_j - s_j)\tau, \qquad \tau = 1 - t^2.$$

Then we have $w_j = \tilde{w}_j$, where

$$\tilde{w}_j(z) = \tilde{T}_j \tilde{f}_j(z)$$

$$(3.2\text{-}2) \qquad \tilde{T}_j = e_j(z) \sum_{m=0}^{n_j} \frac{(2m)!}{2^{2m} m!} \tilde{q}_{jm}(z) D_j^{n_j - m}$$

and the function $\tilde{f}_j$ is the analytic continuation into Ω_j of the series

$$(3.2\text{-}3) \qquad \sum_{\nu=n_j}^{\infty} \tilde{\gamma}_{j\nu}(z_j - s_j)^{\nu}, \qquad \tilde{\gamma}_{j\nu} = \frac{[2(\nu - n_j)]!\pi}{8^{\nu - n_j}(\nu - n_j)!\nu!} \gamma_{j, \nu - n_j}.$$

Here, the $\gamma_{j\kappa}$'s are the coefficients of the Taylor series of f_j with center s_j in $B_j(s_j, \tilde{r}_j) \subset \tilde{\Omega}_j$, where $\tilde{r}_j > 0$. The analytic continuation $\tilde{f}_j$ of (3.2-3) is given by H_{jn_j} in the case $n_j \in \mathbb{N}$, and by $H_{j0} = h_j$ in the case $n_j = 0$, respectively, with H_{jn_j} and h_j defined as in Theorem 2.5-1. [In (3.2-1) we can take the path $C_j = [-1, 1] \subset \mathbb{R}$ because k_j has the special form (3.1-1), and formula (3.1-3a) implies that $D_{j*}\tilde{q}_{j0} = 0$, so that $0 \in C_j$ is admissible.]

Proof. Since the condition $\tilde{q}_{j0} = 1$ is not essential to the proof of Theorem 2.5-1, we may proceed exactly as in that proof, obtaining

$$w_j(z) = e_j(z) \sum_{m=0}^{n_j} \frac{(2m)!}{2^{2m} m!} \tilde{q}_{jm}(z) H_{jm}(z_j); \tag{3.2-4}$$

cf. (2.5-1). Now, by (2.5-7), the function H_{jm} has a series expansion about s_j of the form

$$H_{jm}(z_j) = \sum_{\nu=0}^{\infty} \frac{(2\nu)!\pi}{8^{\nu}\nu!(m+\nu)!} \gamma_{j\nu}(z_j - s_j)^{m+\nu}.$$

Hence by using $\tilde{f}_j$ as defined in the statement involving (3.2-3) and performing $n_j - m$ partial differentiations with respect to z_j, we see that

$$D_j^{n_j - m} \tilde{f}_j(z_j) = D_j^{n_j - m} H_{jn_j}(z_j) = H_{jm}(z_j).$$

By substituting this into (3.2-4) we obtain the assertion $w_j = \tilde{w}_j$, and the proof is complete. ■

This theorem was first obtained by the authors [1969]. As an application, we also proved that for equation (2.5-13), representations $\tilde{w}_j = \tilde{T}_j \tilde{f}_j$ of solutions can be derived from Bergman integral representations.

For equation (2.5-13) with $\lambda = \pm 1$, those differential operators have been used extensively by Bauer, Peschl, Ruscheweyh, and others beginning around 1965; see, for instance, Bauer [1965, 1966a, b, 1967] and Bauer and Peschl [1966, 1967]. During the past decade, that equation has often been referred to as the *Bauer–Peschl equation*; cf. Bauer and Ruscheweyh [1980], Part II. The purpose of those and other papers was the development of a function theory of solutions of those equations and a detailed investigation of various general properties of solutions. We shall present an outline of the more significant of those and related results in Chap. 11.

Theorem 3.2-1 is not only a representation theorem for solutions of an equation (2.1-1) admitting a polynomial kernel, but it also implies the existence of a differential operator for representing solutions of that equation; cf. also Theorem 3.3-1. Hence Bauer's theory can be incorporated in Bergman's theory. On the other hand, that connection also suggests an extension of Bauer's work to solutions of more general equations. Indeed, our results [1969] have sparked a large number of publications by Bauer, Florian, Heersink, Jank, Ruscheweyh, Schröder, Watzlawek, the authors, and others, as can be seen from the list of references at the end of this book. The main topics in those papers are

(i) existence, uniqueness and representation theorems for solutions, with application to initial value problems, and

(ii) the study of function theoretic properties of solutions.

We shall consider (i) here and in the following chapter, and present a characteristic selection from (ii) in Chap. 11.

3.3 DIFFERENTIAL OPERATORS

In Theorem 3.2-1 we have introduced differential operators

$$\tilde{T}_j\colon\ V(\Omega_j) \quad\to\quad \tilde{V}(\Omega)$$

of order n_j, which are defined by (3.2-2). Here $V(\Omega_j)$ is the vector space of holomorphic functions $\tilde{f}_j$ on Ω_j and $\tilde{V}(\Omega)$ is the vector space of holomorphic functions $\tilde{w}$ on $\Omega \setminus S(\tilde{w})$ with $S(\tilde{w})$ denoting a (possibly empty) set of pole planes $z_j = s_j$ of $\tilde{w}$, of order not exceeding n_j. If for a given equation (2.1-1) a class P (integral) operator T_j exists and has been determined, there also exists a corresponding differential operator $\tilde{T}_j$ that can be readily obtained from T_j. From Theorem 3.2-1 we see that if (3.2-3) holds, then $\tilde{T}_j$ generates a $C^\omega(\Omega)$-solution $\tilde{w}_j = \tilde{T}_j\tilde{f}_j$ of (2.1-1). Note that, in particular, (3.2-3) implies

$$D_j^m\tilde{f}_j(s_j) = 0\,, \qquad m = 0,\cdots,n_j - 1\,; \quad n_j \in \mathbb{N}\,. \tag{3.3-1}$$

Now, for k_j to be holomorphic on $\Omega \times [-1,1]$ in (3.1-2) we only need to require that

$$(z_j - \sigma_j)^m \tilde{q}_{jm} \in C^\omega(\Omega)\,, \qquad m = 0,\cdots,n_j\,.$$

However, if we make the stronger assumption that $\tilde{q}_{jm} \in C^\omega(\Omega)$ for $m = 0,\cdots,n_j$, then we have

$$\tilde{T}_j\colon\ V(\Omega_j) \quad\to\quad V(\Omega)\,,$$

where $V(\Omega)$ is the vector space of functions $\tilde{w} \in C^\omega(\Omega)$, and we are able to extend the preceding result insofar as we may drop the restrictions (3.2-3) and (3.3-1); that is, for *arbitrary* $\tilde{f}_j \in C^\omega(\Omega_j)$ this $\tilde{T}_j$ generates $C^\omega(\Omega)$-solutions of equation (2.1-1).

SOLUTIONS GENERATED BY DIFFERENTIAL OPERATORS

3.3-1 Theorem

Let (2.1-1) *be such that on* Ω *there exists a polynomial kernel* $\tilde{k}_j$ *of degree* n_j *given by* k_j *in* (3.1-2) *with* $\tilde{q}_{jm}$ *in* $C^\omega(\Omega)$, $m = 0,\cdots,n_j$, *and with point of reference* s (*hence*, $\sigma_j = s_j$), $s \in \Omega$. *Then for every* $C^\omega(\Omega_j)$*-function* $\tilde{f}_j$ *the image*

$$\tilde{w}_j = \tilde{T}_j\tilde{f}_j \tag{3.3-2}$$

with $\tilde{T}_j$ *as defined in* (3.2-2), *is a* $C^\omega(\Omega)$*-solution of* (2.1-1).

Proof. By assumption, from (3.1-2) we have (3.1-3), in particular,

$$D_{j*}\tilde{q}_{j0} = 0\,, \qquad L_j\tilde{q}_{jn_j} = 0\,. \tag{3.3-3}$$

We now split $\tilde{f}_j$ into two parts,

$$\tilde{f}_j = \tilde{f}_{j1} + \tilde{f}_{j2}\,, \tag{3.3-4}$$

such that $\tilde{f}_{j1}$ satisfies conditions (3.3-1) and $\tilde{f}_{j2}$ is a polynomial (possibly 0) of degree not exceeding $n_j - 1$. From Theorem 3.2-1 and its proof it follows that

$$D_j^{n_j}\tilde{f}_{j1}(z_j) = D_j^{n_j}H_{jn_j}(z_j) = h_j(z_j) = \int_{-1}^{1} f_{j1}(\zeta_j)\tau^{-1/2}\,dt\,.$$

This yields the inversion formula [cf. (2.6-4) and the proof of Theorem 2.6-2]

$$\begin{aligned} f_{j1}(\zeta_j) &= \frac{1}{\pi}D_j^{n_j}\tilde{f}_{j1}(s_j) \\ &\quad + \frac{4}{\pi}(\zeta_j - s_j)\int_0^{\pi/2}\left(D_j^{n_j+1}\tilde{f}_{j1}\right)\Big|_{z_j=\tilde{\psi}_j(\zeta_j)} \sin\theta\,d\theta \end{aligned} \tag{3.3-5}$$

where $\tilde{\psi}_j(\zeta_j) = 2(\zeta_j - s_j)\sin^2\theta + s_j$. Now, by applying the Bergman operator T_j with the given kernel $\tilde{k}_j$ to the function f_{j1} in (3.3-5) we obtain a $C^\omega(\Omega)$-solution w_{j1} of (2.1-1). Hence, by Theorem 3.2-1,

$$w_{j1} = \tilde{w}_{j1}\,, \qquad \tilde{w}_{j1} = \tilde{T}_j\tilde{f}_{j1}\,.$$

Thus, it remains to be shown that $\tilde{w}_{j2} = \tilde{T}_j\tilde{f}_{j2}$ is also a solution. Equivalently, because of Remark 2.1-1, we may prove that

$$L_ju_j = 0 \quad \text{in } \Omega\,, \qquad \text{where} \qquad u_j = \tilde{w}_{j2}/e_j\,.$$

By the definition of $\tilde{f}_{j2}$ and $\tilde{w}_{j2}$ we have

$$\begin{aligned} u_j(z) &= \sum_{m=0}^{n_j} \tilde{q}_{jm}(z)\hat{f}_{jm}(z_j) \\ \hat{f}_{jm}(z_j) &= \frac{(2m)!}{2^{2m}m!}D_j^{n_j-m}\sum_{\nu=0}^{n_j-1}\tilde{\gamma}_{j\nu}(z_j - s_j)^\nu \end{aligned} \tag{3.3-6}$$

where the $\tilde{\gamma}_{j\nu}$ are the coefficients of the Taylor series of $\tilde{f}_j$, and thus of $\tilde{f}_{j2}$, with center s_j. Consequently, we obtain

$$L_ju_j = \sum_{m=0}^{n_j}\left[\hat{f}_{jm}L_j\tilde{q}_{jm} + \left(D_j\hat{f}_{jm}\right)D_{j*}\tilde{q}_{jm}\right].$$

By using (3.1-3) and (3.3-3, 6) we can reduce this to the form

$$L_j u_j = \sum_{m=1}^{n_j-1} \left(-m - \tfrac{1}{2}\right) \hat{f}_{jm} D_{j*} \tilde{q}_{j,m+1} + \sum_{m=2}^{n_j} \left(D_j \hat{f}_{jm}\right) D_{j*} \tilde{q}_{jm}.$$

Now from (3.3-6) we see that

$$D_j \hat{f}_{jm} = \left(m - \tfrac{1}{2}\right) \hat{f}_{j,m-1}.$$

From the last two equations, it follows that $L_j u_j = 0$ holds, and the theorem is proved. ■

CONVERSION OF DIFFERENTIAL OPERATORS TO INTEGRAL FORM

The *converse* of Theorems 3.2-1 and 3.3-1 also holds. Indeed, uisng the first part of the previous proof, from representation (3.3-2)—if it exists—we can derive an integral representation of the form (3.2-1). Thus, by Defs. 2.2-1 and 3.1-1, we have obtained a representation of solutions by a class P operator. We summarize this as follows.

3.3-2 Theorem

Let

$$\tilde{T}_j = \sum_{m=0}^{n_j} A_{jm}(z) D_j^m, \qquad (z_j - s_j)^{n_j - m} A_{jm} \in C^\omega(\Omega)$$

be a differential operator which maps every $\tilde{f}_j \in C^\omega(\Omega_j)$ *satisfying* (3.3-1) *onto a* $C^\omega(\Omega)$*-solution* $\tilde{w}_j = \tilde{T}_j \tilde{f}_j$ *of the equation* $Lw = 0$ *given by* (2.1-1). *Then we have* $L \in P_{jn_j}$, *a corresponding polynomial kernel* $\tilde{k}_j$ *with point of reference s being defined by*

$$\text{(3.3-7)} \qquad \begin{aligned} \tilde{k}_j(z,t) &= e_j(z) \sum_{m=0}^{n_j} \tilde{q}_{jm}(z)(z_j - s_j)^m t^{2m} \\ \tilde{q}_{jm}(z) &= \frac{2^{2m} m!}{(2m)!} A_{j,n_j-m}(z)/e_j(z), \qquad m = 0, \cdots, n_j. \end{aligned}$$

Furthermore, if $\tilde{f}_j$ *satisfies* (3.3-1), *then*

$$\tilde{w}_j = T_j f_j$$

with f_j *and* $\tilde{f}_j$ *related by* (3.3-5).

Hence in the case of equations (2.1-1) as characterized in the last two theorems, for investigating solutions one has available both a Bergman integral operator with a polynomial kernel and a differential operator; and in applications, the choice of one of the two operators will depend on the purpose. Actually, in each case we have at our disposal an even larger variety of operators, as follows.

3.3-3 *Remark*

Suppose that L is such that there exists a differential operator $\tilde{T}_j$ of order n_j as defined in Theorem 3.3-2, so that $\tilde{T}_j$ generates solutions of $Lw = 0$. Then for that L and all $\nu \in \mathbb{N}$, there exist operators of the same kind which are of order $n_j + \nu$.

To verify this, note first that Theorem 3.3-2 gives a polynomial kernel $\tilde{k}_j$ with holomorphic coefficient functions. Remark 3.1-4 then implies that $L \in P_{j, n_j+\nu}$ and yields a kernel $\tilde{k}_j^*$ of degree $n_j + \nu$ in terms of $\tilde{k}_j$, so that the existence of a differential operator of order $n_j + \nu$ now follows from Theorem 3.3-1.

For the time being, as an illustrative example let us again consider the particularly important equation (2.5-13), postponing further equations to Chap. 4.

3.3-4 *Example*

For the equation given by (2.5-13), that is,

$$Lw = D_1 D_2 w + \frac{\lambda n(n+1)}{(1+\lambda z_1 z_2)^2} w = 0, \quad \lambda \in C; \quad n \in \mathbb{N},$$

the situation is as follows. According to Example 3.1-6 and Theorem 3.3-1, this equation admits differential operators $\tilde{T}_1$ and $\tilde{T}_2$ for Type II representations of solutions

$$\text{(3.3-8)} \qquad \tilde{w} = \tilde{w}_1 + \tilde{w}_2, \qquad \tilde{w}_j = \tilde{T}_j \tilde{f}_j, \qquad \tilde{f}_j \in C^\omega(\Omega_j);$$

here, Ω_j $(j = 1, 2)$ is assumed to be such that $1 + \lambda z_1 z_2 \neq 0$ everywhere in Ω_j. Such operators can be obtained from (2.5-14) and (3.2-2) in the form

$$\text{(3.3-9)} \qquad \begin{aligned} \tilde{T}_j &= \sum_{m=0}^{n} A_{jm}(z) D_j^m, \qquad j = 1,2 \\ A_{jm}(z) &= \frac{(2n-m)!}{(n-m)!m!} \left(\frac{-\lambda z_{j*}}{1 + \lambda z_1 z_2} \right)^{n-m}. \end{aligned}$$

We emphasize that *all* $C^\omega(\Omega)$-solutions of the equation under consideration can be represented by (3.3-8, 9). This property follows immediately from the fact that (2.5-14) yields Bergman operators of the first kind, so that our claim now results from Theorems 2.6-2 and 3.2-1.

3.4 INVERSION PROBLEM FOR CLASS P OPERATORS

We shall now consider and solve the inversion problem for operators T_j of class P and corresponding differential operators $\tilde{T}_j$. Throughout this section we assume that $\rho_j = \sigma_j = s_j$, $j = 1, 2$, i.e., that s is the point of reference of the operators T_1 and T_2; $s \in \Omega$.

CASE OF CLASS P KERNELS OF THE FIRST KIND

A partial result is immediate. Indeed, if T_j is *of the first kind*, then, reasoning as in Example 3.3-4, we obtain the following corollary to Theorems 2.6-2 and 3.2-1.

3.4-1 Corollary

Consider the Goursat problem of Sec. 2.6 *for equation* (2.1-1) *in* Ω (*with* Ω *as in* Theorem 2.6-2), *that is,*

$$w(z_1, s_2) = \varphi_1(z_1), \qquad w(s_1, z_2) = \varphi_2(z_2)$$

where $\varphi_j \in C^\omega(\Omega_j)$ *and* $\varphi_1(s_1) = \varphi_2(s_2)$ [*cf.* (2.6-1)]. *Suppose that there exist Bergman polynomial kernels* $\tilde{k}_1$ *and* $\tilde{k}_2$ *of the first kind, of respective degrees* n_1 *and* n_2, *for Type II representations of solutions of* $Lw = 0$. *Let* $\tilde{T}_1$ *and* $\tilde{T}_2$ *denote the differential operators obtained from those kernels by means of* (3.2-2). *Then there exist pairs* $(\tilde{f}_1, \tilde{f}_2) \in C^\omega(\Omega_1) \times C^\omega(\Omega_2)$ *which satisfy the conditions* (3.3-1); *that is,*

$$\left(D_j^m \tilde{f}_j\right)(s_j) = 0, \qquad m = 0, \cdots, n_j - 1; \quad j = 1, 2$$

and are such that the solution w *of that Goursat problem is given by*

$$w = \tilde{T}_1 \tilde{f}_1 + \tilde{T}_2 \tilde{f}_2 .$$

The pair $(\tilde{f}_1, \tilde{f}_2)$ *is uniquely determined by* (3.3-1) *and*

$$e_1(s) D_1^{n_1} \tilde{f}_1(s_1) = e_2(s) D_2^{n_2} \tilde{f}_2(s_2) . \tag{3.4-1}$$

Here we have $\tilde{f}_j = H_{jn_j}$, *where* [*cf*. (2.5-3)]

$$H_{jn_j}(z_j) = \int_{s_j}^{z_j}\int_{s_j}^{z_{jn_j}} \cdots \int_{s_j}^{z_{j2}} h_j(z_{j1})\, dz_{j1} \cdots dz_{j,n_j-1}\, dz_{jn_j} \tag{3.4-2a}$$

and, in terms of the initial data, h_j *has the form*

$$h_j(z_j) = \left[\varphi_j(z_j) - \tfrac{1}{2}\varphi_j(s_j)e_{j*}(z)\big|_{z_{j*}=s_{j*}} \Big/ e_{j*}(s)\right] \Big/ e_j(z)\big|_{z_{j*}=s_{j*}}. \tag{3.4-2b}$$

Condition (3.4-1) is similar to the condition for f_1 and f_2 in Theorem 2.6-2. Our present result shows that the functions $\tilde{f}_1$ and $\tilde{f}_2$ can be obtained from (3.4-2), that is, essentially by n integrations of the initial data of w on the planes $z_1 = s_1$, $z_2 = s_2$. Note further that Corollary 3.4-1 could also be derived from Theorem 3.4-2; cf. our paper [1979].

THE GENERAL CASE

We now turn to the inversion problem for class P operators which are not necessarily of the first kind. For such an operator, the solution of the problem can be formulated as follows.

3.4-2 Theorem

Suppose (2.1-1) *to be such that there exist kernels* $\tilde{k}_1$ *and* $\tilde{k}_2$ *for a Type II representation in* Ω; *let* [*cf. formulas* (3.1-2) *and* (3.3-7)]

$$\tilde{k}_j(z,t) = e_j(z) \sum_{m=0}^{n_j} \tilde{q}_{jm}(z)(z_j - s_j)^m t^{2m} \tag{3.4-3a}$$

where we assume that $\tilde{q}_{j0}, \cdots, \tilde{q}_{jn_j} \in C^\omega(\Omega)$ *and*

$$\tilde{q}_{j0}(z) \neq 0 \quad \text{in } \Omega, \qquad j = 1,2. \tag{3.4-3b}$$

Then (a)–(c) *hold:*

(a) *For every solution* $w \in C^\omega(\Omega)$ *of equation* (2.1-1) *there exist two unique functions* $\tilde{f}_1$, $\tilde{f}_2$ *such that*

$$\tilde{f}_1 \in C^\omega(\Omega_1), \qquad \tilde{f}_2 \in C^\omega(\Omega_2) \tag{3.4-4a}$$

and [*cf*. (3.3-1) *and* (3.4-1)]

$$D_j^m \tilde{f}_j(s_j) = 0, \qquad m = 0, \cdots, n_j - 1; \quad j = 1,2, \tag{3.4-4b}$$

$$\tilde{q}_{10}(s)e_1(s)D_1^{n_1}\tilde{f}_1(s_1) = \tilde{q}_{20}(s)e_2(s)D_2^{n_2}\tilde{f}_2(s_2) \tag{3.4-4c}$$

and $w = \tilde{w}$, *where*

$$\tilde{w} = \tilde{w}_1 + \tilde{w}_2 \qquad \text{and} \qquad \tilde{w}_j = \tilde{T}_j \tilde{f}_j$$

with $\tilde{T}_j$ *given by* (3.2-2). *Furthermore, by the use of* T_j *with* $\tilde{k}_j$ *as in* (3.4-3a), $w = w_1 + w_2$ *also holds, with* $w_j = T_j f_j$ *of the form* (3.2-1). *Here* f_1, f_2 *are obtained from* $\tilde{f}_1, \tilde{f}_2$ *by means of* (3.2-3): f_j *is the analytic continuation into* $\tilde{\Omega}_j$ *of the series*

$$\text{(3.4-5)} \qquad \sum_{\nu=0}^{\infty} \gamma_{j\nu}(\zeta_j - s_j)^{\nu}, \qquad \gamma_{j\nu} = \frac{8^{\nu}\nu!(\nu + n_j)!}{(2\nu)!\pi}\tilde{\gamma}_{j,\nu+n_j}$$

which converges in $B_j(s_j, \frac{1}{2}r_j) \subset \tilde{\Omega}_j$, *and* $\tilde{\gamma}_{j\kappa}$ *are the coefficients of the Taylor series of* $\tilde{f}_j$ *in* $B_j(s_j, r_j) \subset \Omega_j$.

(b) *In the representation* $w = \tilde{w} = \tilde{w}_1 + \tilde{w}_2$ [*cf*. (3.2-2)] *of a given solution* $w \in C^{\omega}(\Omega)$ *of* (2.1-1) *we have*

$$\text{(3.4-6)} \qquad \tilde{f}_j(z_j) = \tilde{\gamma}_{jn_j}(z_j - s_j)^{n_j} + \tilde{\tilde{f}}_j(z_j)$$

where

$$\text{(3.4-7)} \qquad \tilde{\gamma}_{jn_j} = \frac{w(s)}{n_j!2A_{jn_j}(s)}, \qquad A_{jn_j}(s) = \tilde{q}_{j0}(s)e_j(s)$$

and $\tilde{\tilde{f}}_j \in C^{\omega}(\Omega_j)$ *is the unique solution of the ordinary differential equation*

$$\text{(3.4-8)} \qquad \begin{aligned} \sum_{m=0}^{n_j+1} M_{jm}(z_j)D_j^m\tilde{\tilde{f}}_j &= \left[D_j w(z) + a_{j*}(z)w(z)\right]\Big|_{z_{j*}=s_{j*}} \\ &\quad - \tilde{\gamma}_{jn_j}\sum_{m=0}^{n_j}\binom{n_j}{m}m!(z_j - s_j)^{n_j-m}M_{jm}(z_j) \end{aligned}$$

with

$$\text{(3.4-9)} \quad M_{jm}(z_j) = \left[D_j A_{jm}(z) + A_{jm}(z)a_{j*}(z) + A_{j,m-1}(z)\right]\Big|_{z_{j*}=s_{j*}}$$

satisfying the initial conditions

$$\text{(3.4-10)} \qquad D_j^m\tilde{\tilde{f}}_j(s_j) = 0, \qquad m = 0,\cdots,n_j.$$

In this representation, $A_{j,-1} = A_{j,n_j+1} = 0$, *and* $A_{j0},\cdots,A_{jn_j}$ *are the coeffi-*

cient functions in $\tilde{T}_j$ as given by (3.2-2) [*cf.* (3.3-7)]; *that is,*

$$A_{jm}(z) = \frac{[2(n_j - m)]!}{4^{n_j - m}(n_j - m)!}\tilde{q}_{j, n_j - m}(z) e_j(z). \tag{3.4-11}$$

(c) *The analytic continuation of the germ* (3.4-5) *of f_j into $\tilde{\Omega}_j$ is obtained by*

$$f_j(\zeta_j) = \frac{1}{\pi} n_j! \tilde{\gamma}_{jn_j} + \frac{4}{\pi}(\zeta_j - s_j)\int_0^{\pi/2}\left(D_j^{n_j+1}\tilde{\tilde{f}}_j\right)\bigg|_{z_j = \tilde{\psi}_j(\zeta_j)} \sin\theta \, d\theta \tag{3.4-12}$$

where $\tilde{\psi}_j(\zeta_j) = 2(\zeta_j - s_j)\sin^2\theta + s_j$, *and* $\tilde{\gamma}_{jn_j}$ *and* $\tilde{\tilde{f}}_j$ *are given by* (3.4-7)–(3.4-11).

Proof. We first prove (a) and (b). Suppose that w can be represented in the form

$$w(z) = \tilde{w}(z) = \sum_{j=1}^{2}\sum_{m=0}^{n_j} A_{jm}(z) D_j^m \tilde{f}_j(z_j) \tag{3.4-13}$$

with A_{jm} as in (3.4-11) and $\tilde{f}_j$ satisfying conditions (3.4-4). Because of (3.4-4b) the function $\tilde{f}_j$ has a Taylor series

$$\tilde{f}_j(z_j) = \sum_{\nu=n_j}^{\infty} \tilde{\gamma}_{j\nu}(z_j - s_j)^{\nu} = \frac{1}{n_j!} D_j^{n_j}\tilde{f}_j(s_j)(z_j - s_j)^{n_j} + \tilde{\tilde{f}}_j(z_j) \tag{3.4-14}$$

where $\tilde{\tilde{f}}_j$ satisfies (3.4-10). Condition (3.4-4c) with (3.4-3b, 11) and condition (3.4-13) with (3.4-4b) together imply that the derivatives

$$D_j^{n_j}\tilde{f}_j(s_j) = n_j! \tilde{\gamma}_{jn_j}$$

are uniquely determined by the system of equations

$$A_{1n_1}(s) D_1^{n_1}\tilde{f}_1(s_1) + A_{2n_2}(s) D_2^{n_2}\tilde{f}_2(s_2) = w(s)$$

$$A_{1n_1}(s) D_1^{n_1}\tilde{f}_1(s_1) - A_{2n_2}(s) D_2^{n_2}\tilde{f}_2(s_2) = 0.$$

Now (3.4-3b) implies that $A_{jn_j}(s) \neq 0$, $j = 1, 2$. Hence that system has a unique solution which yields (3.4-7). From this we conclude that $\tilde{f}_j$ must be of the form (3.4-6, 7). Furthermore, representation (3.4-5) follows immediately from Theorem 3.2-1.

To complete the proof of (a) and (b), we have to show that there exist unique functions $\tilde{\tilde{f}}_1$ and $\tilde{\tilde{f}}_2$ having the properties stated in (b), and that w can be represented in terms of these functions in the form (3.4-13) with (3.4-6).

For this purpose we first note that Theorem 2.6-2 implies that under our assumptions a Goursat problem (2.1-1), (2.6-1) has a unique solution. Hence it remains to be shown that $\tilde{\tilde{f}}_j$ can be uniquely determined from the initial values $w(z)$ with $z_j^* = s_j^*$ of the given solution w on the characteristic planes $z_1 = s_1$ and $z_2 = s_2$. Substituting (3.4-14) as well as (3.4-7) into (3.4-13) and noting (3.4-11), we have

(3.4-15)
$$\begin{aligned} w(z)\big|_{z_{j*}=s_{j*}} = \sum_{m=0}^{n_j} A_{jm}(z)\Big|_{z_{j*}=s_{j*}} \left[\tilde{\gamma}_{jn_j}\binom{n_j}{m} m!(z_j - s_j)^{n_j-m} + D_j^m\tilde{\tilde{f}}_j(z_j)\right] \\ + \tilde{q}_{j*0}(s) e_{j*}(z)\big|_{z_{j*}=s_{j*}} \tilde{\gamma}_{jn_j} n_{j*}!. \end{aligned}$$

Note that $\tilde{q}_{j*0}$ does not depend on z_j since $D_j\tilde{q}_{j*0} = 0$, as follows from (3.1-3a). By means of (2.1-3) and differentiation with respect to z_j we obtain

(3.4-16)
$$\begin{aligned} D_j w(z)\big|_{z_{j*}=s_{j*}} = \sum_{m=0}^{n_j+1} \left[D_j A_{jm}(z) + A_{j,m-1}(z)\right]\Big|_{z_{j*}=s_{j*}} \times \\ \times \left[\tilde{\gamma}_{jn_j}\binom{n_j}{m} m!(z_j - s_j)^{n_j-m} + D_j^m\tilde{\tilde{f}}_j(z_j)\right] \\ - a_{j*}(z)\big|_{z_{j*}=s_{j*}} \tilde{q}_{j*0}(s) e_{j*}(z)\big|_{z_{j*}=s_{j*}} \tilde{\gamma}_{jn_j} n_{j*}! \end{aligned}$$

where $A_{j,-1} = A_{j,n_j+1} = 0$. Multiplying (3.4-15) by $a_{j*}(z)|_{z_{j*}=s_{j*}}$ and adding (3.4-16) to the resulting equation, we obtain equation (3.4-8) for $\tilde{\tilde{f}}_j$. Now the function

$$M_{j,n_j+1} = A_{jn_j}\big|_{z_{j*}=s_{j*}}$$

does not vanish in Ω, as follows from (3.4-3b). Hence we may divide (3.4-8) by that function, the result being an ordinary differential equation for $\tilde{\tilde{f}}_j$ of the form

$$D_j^{n_j+1}\tilde{\tilde{f}}_j + \sum_{m=0}^{n_j} \delta_j(z_j) D_j^m \tilde{\tilde{f}}_j = \sigma_j(z_j) \tag{3.4-17}$$

where δ_j and σ_j are holomorphic on Ω_j. From the theory of ordinary differential equations it is known that this equation, which is of order $n_j + 1$, has a unique solution in $B_j(s_j, \rho_j) \subset \Omega_j$ ($\rho_j > 0$, suitable) satisfying the $n_j + 1$ initial conditions (3.4-10). The monodromy theorem implies that this solution can be continued analytically into Ω_j in a unique fashion. Hence there exists at most one solution corresponding to (3.4-15). On the other hand, a solution of (3.4-17) is also a solution of (3.4-8). If we replace w (with $z_j^* = s_j^*$) in (3.4-8) by a function φ_j (depending on z_j) and regard the resulting equation as an ordinary differential equation for φ_j, the solution satisfying the initial condition $\varphi_j(s_j) = w(s)$ is

$$\varphi_j(z_j) = \exp\left[-\int_{s_j}^{z_j} a_{j*}(\tilde{z})\Big|_{\tilde{z}_{j*}=z_{j*}} d\tilde{z}_j\right] \times$$

$$\times\left(w(s) + \int_{s_j}^{z_j} \Lambda(\tilde{z}_j) \exp\left[\int_{s_j}^{\tilde{z}_j} a_{j*}(\tilde{\tilde{z}})\Big|_{\tilde{\tilde{z}}_{j*}=z_j} d\tilde{\tilde{z}}_j\right] d\tilde{z}_j\right)$$

where

$$\Lambda(\tilde{z}_j) = \sum_{m=0}^{n_j+1} M_{jm}(\tilde{z}_j)\left[\tilde{\gamma}_{jn_j}\binom{n_j}{m} m!(\tilde{z}_j - s_j)^{n_j-m} + D_j^m \tilde{f}_j(\tilde{z}_j)\right].$$

Consequently, $\varphi_j = w|_{z_{j*}=s_{j*}}$ holds, and the unique solution of equation (3.4-17) is identical with the unique solution of equation (3.4-15). This completes the proof of parts (a) and (b).

The proof of (c) is immediate. Indeed, from (b) we conclude that $\tilde{f}_j$ is of the same form as $\tilde{f}_{j1}$ in (3.3-4) in the proof of Theorem 3.3-1. Consequently, (3.4-12) follows by substituting formula (3.4-6) into (3.3-5). Theorem 3.4-2 is proved. ■

Additional References

Bauer [1965, 1966a, 1966b]
Bauer–Florian [1976]
Bauer–Peschl [1966]
Bauer–Ruscheweyh [1980]
Florian–Heersink [1974]
Heersink [1976a, 1976b]
Jank [1971]
Kracht [1974b, 1976]
Kracht–Kreyszig [1969, 1975, 1979]
Kracht–Schröder [1973]
Kreyszig [1968, 1970, 1971]
Lo [1980]
Watzlawek [1971b, 1973, 1983b]

Chapter Four

Polynomial Kernels: Existence and Construction

Integral operators with polynomial kernels were introduced and motivated in the previous chapter. There we also discussed relations of these integral operators to certain *differential* operators. This constitutes a theory which provides a new unified approach to those differential operators which have been studied and applied for some time.

In the present chapter we shall develop a theory of polynomial kernels, beginning with the existence problem. In Sec. 4.1 we motivate such a theory and begin its development by presenting first results on certain equations which were obtained earlier, by Bauer, Peschl, and others, using a method quite different from the present one. In Sec. 4.2 we shall obtain an existence and representation theorem for these operators and in Sec. 4.3 the extension of this result to pairs of operators, as they are needed in Type II representations for initial value problems and other applications.

We emphasize that the whole method has the advantage of being *constructive*. Indeed, the concluding four sections of the chapter will be devoted to four construction principles for equations and corresponding polynomial kernels, and to applications.

4.1 EXISTENCE OF POLYNOMIAL KERNELS IN SPECIAL CASES

The use of differential operators in a systematic function theoretic investigation of solutions of partial differential equations was initiated in the mid 1960s by Bauer and Peschl in a series of articles; see, for instance, Bauer [1965, 1966a, b, 1967, 1968], Bauer–Peschl [1966, 1967], and Bauer–Ruscheweyh

[1980]. Those papers concern the equation

$$(4.1\text{-}1) \qquad Lw = D_1 D_2 w + \frac{\varepsilon n(n+1)}{(1+\varepsilon z_1 z_2)^2} w = 0, \qquad n \in \mathbb{N};\ \varepsilon = \pm 1$$

which is (2.5-13) with $\lambda = \varepsilon = \pm 1$ (cf. Examples 2.5-5, 3.1-6, and 3.3-4) and is often called the *Bauer–Peschl equation*, as mentioned in Sec. 3.2. Extensions to larger classes of equations were subsequently obtained between 1969 and 1974, after the authors [1969] had shown that solutions generated by Bauer's differential operators can be represented by special Bergman operators (as explained in the preceding chapter). Since that time, both approaches were used to determine important subclasses of P_{jn_j}. The authors [1975] characterized the classes P_{jn_j} and $P_{1n_1} \cap P_{2n_2}$ by necessary and sufficient conditions which are of a constructive type. Based on those results as well as our results [1979], in this chapter we develop four construction principles for equations and corresponding differential and integral operators. In Secs. 4.4, 4.6, and 4.7 we shall demonstrate that *all explicitly known equations and operators of that type can be derived in this constructive way*. Thus, at present it suffices to illustrate the situation by some special results of practical significance.

We use the notations of classes of operators L as given in 3.1-2; moreover, by attaching an asterisk to these notations we shall denote those subclasses whose operators L have coefficients $a_j = 0$ ($j = 1$ or $j = 2$, fixed) and $a_0 \neq 0$. As before we assume that $a_0 \neq 0$, in order to exclude the trivial case that $Lw = 0$ reduces to an ordinary differential equation (for $D_{j*}w$). Thus, for instance, in (4.1-1) we have

$$L \in \left(FP_{1n}^0\right)^* \cap \left(FP_{2n}^0\right)^*.$$

The case $L \in P_{j0}$ being explicitly solved (see Remark 4.2-5), we may assume that $n_j > 0$. An *implicit* characterization of $L \in P_{jn_j}$ ($n_j > 0$) can be obtained by means of (2.5-11) and (3.1-2, 3), namely,

$$(4.1\text{-}2) \qquad S_{j,n_j+1} \cdots S_{j1} S_{j0} 0 = 0,$$

where the operators $S_{j\mu}$ are defined in (2.5-10). (Cf. Kreyszig [1968], Kracht [1974a].) Equation (4.1-2) is an integro-differential equation for the coefficients b_j and c_j of L_j whose order increases with n_j.

Obviously, from this criterion or from (3.1-3), assuming that the kernel $\tilde{k}_j$ is given by

$$(4.1\text{-}3) \qquad \tilde{k}_j(z,t) = \tilde{q}_{j0}(z_j) + \tilde{q}_{j1}(z)(z_j - s_j)t^2$$

one obtains as a necessary and sufficient condition for $L \in P_{j1}^*$ that a_{j*} and

a_0 have the form

$$\begin{aligned} a_{j*}(z) &= b_j(z) = \tfrac{1}{2}\tilde{q}_{j1}(z)/\tilde{q}_{j0}(z_j) - \big(D_1 D_2 \tilde{q}_{j1}(z)\big)/D_{j*}\tilde{q}_{j1}(z) \\ a_0(z) &= c_j(z) = -\tfrac{1}{2}\big(D_{j*}\tilde{q}_{j1}(z)\big)/\tilde{q}_{j0}(z_j). \end{aligned} \tag{4.1-4}$$

Observe that we cannot have $\tilde{q}_{j0} = 0$ or $D_{j*}\tilde{q}_{j1} = 0$ because this would lead to $a_0 = 0$ or $\tilde{k}_j = 0$.

An illustration for arbitrary $n \in \mathbb{N}$ is given by (2.5-13) in Example 2.5-5 (cf. also Example 3.1-6) or by (4.1-1). The latter equation may easily be transformed into

$$Lw = D_1 D_2 w - \frac{n(n+1)}{(z_1 + z_2)^2} w = 0, \qquad n \in \mathbb{N}, \tag{4.1-5}$$

and a corresponding polynomial kernel is obtained from (2.5-14) in a similar manner.

We now extend our result by setting

$$z_j = \psi_j(\check{z}_j)$$

where $\psi_j \in C^{\omega}(\psi_j^{-1}(\Omega_j))$ and $\check{D}_j\psi_j(\check{z}_j) \neq 0$ when $\check{z}_j \in \psi_j^{-1}(\Omega_j)$; in this expression $\check{D}_j = \partial/\partial\check{z}_j$, $j = 1, 2$. We also write

$$\check{w}(\check{z}_1, \check{z}_2) = w(\psi_1(\check{z}_1), \psi_2(\check{z}_2)).$$

We consider

$$\check{L}\check{w} = \check{D}_1\check{D}_2\check{w} - n(n+1)\frac{\check{D}_1\psi_1(\check{z}_1)\check{D}_2\psi_2(\check{z}_2)}{[\psi_1(\check{z}_1) + \psi_2(\check{z}_2)]^2}\check{w} = 0. \tag{4.1-6}$$

Let M_j denote the operator given by (2.3-2a), (4.1-5), and $\eta_j = 0$, $s = 0$. Similarly, let $\check{M}_j$ denote the operator coresponding to (2.3-2a) and (4.1-6). Then the above transformation yields

$$\begin{aligned} L &= \big[\check{D}_1\psi_1(\check{z}_1)\check{D}_2\psi_2(\check{z}_2)\big]^{-1}\check{L} \\ M_j &= \big[\check{D}_{j*}\psi_{j*}(\check{z}_{j*})\big]^{-1}\check{M}_j - \big[\check{z}_j - \psi_j(\check{z}_j)/\check{D}_j\psi_j(\check{z}_j)\big]\big[2t/\check{D}_{j*}\psi_{j*}(\check{z}_{j*})\big]\check{L} \end{aligned}$$

with L as in (4.1-5). Hence a solution of (4.1-5) is transformed into a solution of (4.1-6), and from a Bergman kernel $\tilde{k}_j$ for (4.1-5) we obtain a Bergman kernel $\check{\tilde{k}}_j$ for (4.1-6) by setting

$$\check{\tilde{k}}_j(\check{z}_1, \check{z}_2, t) = \tilde{k}_j(\psi_1(\check{z}_1), \psi_2(\check{z}_2), t) \tag{4.1-7}$$

provided $\check{z}_j = \psi_j(\check{z}_j)/\check{D}_j\psi_j(\check{z}_j)$ or

$$\psi_j(\check{z}_j) = \lambda_j \check{z}_j, \qquad \lambda_j \in \mathbf{C}\setminus\{0\}, \quad j = 1 \text{ or } 2, \text{ fixed}. \tag{4.1-8}$$

We note that this result can also be obtained if we start from a kernel $\tilde{k}_j = k_j$ of the form (3.1-2) and set

$$\tilde{q}_{jm}(z) = \mu_{jm}\left[\lambda_j z_j + \psi_{j*}(z_{j*})\right]^{-1}, \qquad \mu_{jm} \in \mathbf{C}\setminus\{0\}. \tag{4.1-9}$$

This determines a Bergman kernel for (4.1-6) with (4.1-8), provided we take

$$\begin{aligned} \mu_{jm} &= \frac{(-4)^m (n+m)!}{(2m)!(n-m)!}\mu_{j0}\lambda_j^m \\ &= \frac{(-4)^m m!}{(2m)!}\binom{n}{m}\mu_{j0}\frac{(n+m)!}{n!}\lambda_j^m \end{aligned} \qquad \mu_{j0} \neq 0. \tag{4.1-10}$$

By slightly modifying (4.1-10), namely, by replacing the expression $(n+m)!\lambda_j^m/n!$ with

$$\prod_{\nu=1}^{m}\left[(n+\nu)\lambda_j - \kappa_j\right], \qquad \kappa_j \in \mathbf{C}; \quad \kappa_j \neq (n+\nu)\lambda_j; \quad \nu = 1, \cdots, n$$

we obtain the equation (see Kreyszig [1970, 1971]; cf. Kracht [1974a])

$$Lw = D_1 D_2 w + b_j(z) D_{j*} w + c_j(z) w = 0 \tag{4.1-11}$$

with fixed $j = 1$ or 2 and coefficients given by

$$\begin{aligned} b_j(z) &= \kappa_j\left[\lambda_j z_j + \psi_{j*}(z_{j*})\right]^{-1} + D_j \eta_j(z_j) \\ c_j(z) &= -n\left[(n+1)\lambda_j - \kappa_j\right]\left[\lambda_j z_j + \psi_{j*}(z_{j*})\right]^{-2} D_{j*}\psi_{j*}(z_{j*}) \end{aligned} \tag{4.1-12}$$

where $\eta_j \in C^\omega(\Omega_j)$ is arbitrary. Here $L \in P_{jn}^*$; indeed, a corresponding polynomial kernel is

$$\tilde{k}_j = \exp\left[-\eta_j + \ln \mu_{j0}\right] k_j \tag{4.1-13a}$$

where

$$\begin{aligned} k_j(z,t) &= \sum_{m=0}^{n} \tilde{\mu}_{jm}\left(\frac{z_j}{\lambda_j z_j + \psi_{j*}(z_{j*})}\right)^m t^{2m} \\ \tilde{\mu}_{jm} &= \frac{(-4)^m m!}{(2m)!}\binom{n}{m}\prod_{\nu=1}^{m}\left[(n+\nu)\lambda_j - \kappa_j\right]. \end{aligned} \tag{4.1-13b}$$

WEAKENING OF THE ASSUMPTIONS

From the results just obtained it is clear that the assumption of holomorphy of ψ_1 and ψ_2 may be weakened. In fact, it suffices to require ψ_1 and ψ_2 to be such that they generate coefficient functions in (4.1-6) and (4.1-11, 12) which are of class $C^\omega(\Omega)$, where $\Omega = \Omega_1 \times \Omega_2$. If, for instance, ψ_{j^*} in (4.1-12) has a pole at $z_{j^*} = 0$, then (4.1-13) defines a polynomial kernel of the first kind with point of reference $s = (0, 0)$. Hence for L in (4.1-11) with coefficients (4.1-12) we have $L \in (FP_{jn})^*$.

By similar arguments (cf. Kracht [1974a]) one can show that $L \in P_{jn}^*$ $(n \in \mathbb{N})$ when L is of the form (4.1-11) with coefficient functions given by

$$
\begin{aligned}
b_j(z) &= \left[n\lambda_j + \tfrac{1}{2}\tilde{\psi}_{j^*}(z_{j^*})/\tilde{\psi}_j(z_j)\right]\left[\lambda_j z_j + \tilde{\lambda}_j\right]^{-1} \\
c_j(z) &= \left[-\tfrac{1}{2}D_{j^*}\tilde{\psi}_{j^*}(z_{j^*})/\tilde{\psi}_j(z_j)\right]\left[\lambda_j z_j + \tilde{\lambda}_j\right]^{-1}
\end{aligned}
\tag{4.1-14}
$$

where $\tilde{\psi}_j \in C^\omega(\Omega_j)$, $\tilde{\psi}_{j^*} \in C^\omega(\Omega_{j^*})$, $\tilde{\psi}_j \neq 0$, $\tilde{\psi}_{j^*} \neq const$, $\lambda_j \in \mathbb{C} \setminus \{0\}$, and $\tilde{\lambda}_j \in \mathbb{C}$. A corresponding polynomial kernel for (4.1-11) with coefficients (4.1-14) is

$$
\tilde{k}_j(z, t) = \tilde{\psi}_j(z_j) + \tilde{\psi}_{j^*}(z_{j^*}) \sum_{m=1}^{n} \rho_{jm} \left(\frac{z_j}{\lambda_j z_j + \tilde{\lambda}_j} \right)^m t^{2m}
\tag{4.1-15a}
$$

where

$$
\rho_{jm} = \frac{2}{n} \binom{n}{m} (-4\lambda_j)^{m-1} \bigg/ \binom{2m}{m}.
\tag{4.1-15b}
$$

We should remark that condition (4.1-8) has only been introduced in order to obtain the relation (4.1-7) between those Bergman kernels. Likewise, in (4.1-12) and (4.1-14), the variable z_j may be replaced by $\psi_j(z_j)/\lambda_j$ without affecting the property that $L \in P_{jn}^*$ in (4.1-11). This seems plausible in the light of our present discussion and the experience resulting from it, and will be generally confirmed in the next section (by Theorem 4.2-4 and Lemma 4.2-7). [Observe, however, that after that replacement, polynomial kernels can in general no longer be obtained by imposing a similar transformation on representation (4.1-13) or (4.1-15), but may be derived by employing Theorem 3.3-2; cf. Theorem 4.4-2.]

Investigation in the direction of our present discussion includes works by Bauer [1971, 1976a], Bauer–Florian [1976], Florian–Jank [1971], Kracht–Schröder [1973], Kreyszig [1975], Watzlawek [1972], and others. We do not enter into a detailed discussion of these results, some of which are rather involved. Here it is sufficient—as well as essential—to state that all the equations and representations of solutions in those papers can be derived by utilizing the construction principles to be developed in the present chapter. This will be demonstrated by applications in Secs. 4.4, 4.6, and 4.7.

4.2 *EXISTENCE OF POLYNOMIAL KERNELS FOR TYPE I REPRESENTATIONS*

In this and the next section we shall investigate the general existence problem for operators L (and equations $Lw = 0$) that admit polynomial kernels $\tilde{k}_j$ for Type I representations and polynomial kernels $\tilde{k}_1, \tilde{k}_2$ for Type II representations of solutions. For this purpose we shall utilize the *method of Laplace invariants* (cf. Darboux [1972] or Tricomi [1968]). We shall obtain general existence theorems, for instance, of the following kind. If for fixed $j = 1$ or 2, the operator L admits a minimal polynomial kernel $\tilde{k}_j$ of degree n_j for Type I representations of solutions,

$$\tilde{w}_j = \tilde{T}_j \tilde{f}_j ,$$

it follows that there also exists a polynomial kernel $\tilde{k}_{j*}$ of some degree $n_{j*} \in \mathbb{N}_0$ for Type II representations

$$\tilde{w} = \tilde{w}_1 + \tilde{w}_2 = \tilde{T}_1 \tilde{f}_1 + \tilde{T}_2 \tilde{f}_2$$

if and only if a certain auxiliary function satisfies a homogeneous *ordinary* linear differential equation of order $n_1 + n_2 + 1$. From those theorems, one obtains two construction principles for operators L and corresponding polynomial kernels. We shall also prove that an application of those principles leads to the construction of the most general operators derived by Bauer [1976a]. Furthermore, in a relatively simple way we shall obtain a uniqueness theorem for the functions $\tilde{f}_1$ and $\tilde{f}_2$ associated with a solution $\tilde{w}$.

This section is devoted to Type I representations. Results about Type II representations could be obtained by two applications of the present ideas and constructions. We shall presently not pursue this matter since for Type II representations we shall provide a much better method of approach in the next five sections.

NOTATIONS IN THIS CHAPTER

In this and the subsequent sections of this chapter, the following notations will be convenient.

4.2-1 Notations

For $r \in \mathbb{N}_0$ let

(4.2-1) $$L^{(r)} = D_1 D_2 + a_{1r}(z) D_1 + a_{2r}(z) D_2 + a_{0r}(z) ;$$

here $a_{m0} = a_m$, $m = 0, 1, 2$ [so that $L^{(0)} = L$; cf. (2.1-1)] and the a_{mr} with $r > 0$ follow successively from the formulas given below. For $r \in \mathbb{N}_0$ and fixed $j = 1$ or 2 let

$$w_{jr} = D_j^{(r)} w \tag{4.2-2}$$

where, for $r \in \mathbb{N}$ and j as before,

$$\begin{aligned} D_j^{(0)} &= 1 \\ D_j^{(r)} &= D_{jr} \cdots D_{j1} \\ D_{jr} &= D_{j*} + a_{j,r-1}(z). \end{aligned} \tag{4.2-3}$$

Furthermore, for $r \in \mathbb{N}$ and j as before, let

$$\begin{aligned} &\text{(a)} \quad L^{(r)} D_{jr} = h_{j,r-1} D_{jr} \frac{1}{h_{j,r-1}} L^{(r-1)} && \text{if } h_{j,r-1} \neq 0 \\ &\text{(b)} \quad L^{(r)} D_{jr} = D_{jr} L^{(r-1)} && \text{if } h_{j,r-1} = 0 \end{aligned} \tag{4.2-4}$$

where

$$h_{jr} = D_j a_{jr} + a_{1r} a_{2r} - a_{0r}, \qquad r \in \mathbb{N}_0. \tag{4.2-5}$$

Finally, for $r \in \mathbb{N}_0$ let

$$e_{jr}(z) = \exp\left[\eta_{jr}(z_j) - \int_{s_{j*}}^{z_{j*}} a_{jr}(\tilde{z})\Big|_{\tilde{z}_j = z_j} d\tilde{z}_{j*}\right]. \tag{4.2-6}$$

This is similar to (2.1-3); indeed, $e_{j0} = e_j$.

The introduction of the operators (4.2-4) is suggested by the following idea. By n_j applications of the operator D_{jr} we seek to obtain from $Lw = 0$ [cf. (2.1-1)] an equation

$$L^{(n_j)} w_{jn_j} = 0$$

which we can solve explicitly for

$$w_{jn_j} = D_j^{(n_j)} w$$

and then obtain

$$w = \left(D_j^{(n_j)}\right)^{-1} w_{jn_j}.$$

LAPLACE INVARIANTS

Of basic importance in this method is the form of the functions h_{jr} in (4.2-5). These functions are called ***Laplace invariants.*** The name "invariants" results from their transformation behavior which will be characterized in Lemma 4.2-7 and its proof.

We first investigate the relations between the coefficients of the operators defined by (4.2-1, 4).

4.2-2 Lemma

If $h_{jr} \neq 0$ *for* $r = 0, \cdots, n_j - 1$, *then* $L^{(n_j)}$ *has the coefficients*

$$\begin{aligned} a_{jn_j} &= a_{j0} - D_{j*}\Lambda_{jn_j}, \qquad \Lambda_{jn_j} = \sum_{r=0}^{n_j-1} \log h_{jr} \\ a_{j*n_j} &= a_{j*0} \\ a_{0n_j} &= D_j a_{jn_j} + a_{1n_j} a_{2n_j} - h_{jn_j}. \end{aligned} \tag{4.2-7}$$

Furthermore,

$$\begin{aligned} h_{jn_j} &= (n_j + 1) h_{j0} - n_j h_{j*0} - D_1 D_2 \tilde{\Lambda}_{jn_j} \\ \tilde{\Lambda}_{jn_j} &= \sum_{r=0}^{n_j-1} (n_j - r) \log h_{jr} \\ h_{j*n_j} &= h_{j,n_j-1}. \end{aligned} \tag{4.2-8}$$

Conversely,

$$\begin{aligned} a_j &= a_{j0} = a_{jn_j} + D_{j*}\Lambda_{jn_j} \\ a_{j*} &= a_{j*0} = a_{j*n_j} \\ a_0 &= a_{00} = a_{1n_j} a_{2n_j} + D_1 D_2 \Lambda_{jn_j} + D_j a_{jn_j} - h_{j0} + a_{j*n_j} D_{j*} \Lambda_{jn_j}. \end{aligned} \tag{4.2-9}$$

Proof. Formula (4.2-9) follows from (4.2-5, 7). Now

$$L^{(0)} w = D_{j^*1} w_{j1} - h_{j0} w.$$

Hence by (4.2-4a) we obtain

$$L^{(1)} w_{j1} = h_{j0} \left(D_{j1} \frac{1}{h_{j0}} D_{j^*1} w_{j1} - w_{j1} \right).$$

From this we conclude that $L^{(1)}$ has the coefficients

$$a_{j1} = a_{j0} - D_{j*}(\log h_{j0})$$
$$a_{j*1} = a_{j*0}$$
$$a_{01} = a_{00} - D_j a_{j0} + D_{j*} a_{j*0} - a_{j*0} D_{j*}(\log h_{j*0}).$$

Also, by (4.2-5),

$$h_{j1} = 2h_{j0} - h_{j*0} - D_1 D_2(\log h_{j0})$$
$$h_{j*1} = h_{j0}.$$

This takes care of the case $n_j = 1$. For $n_j > 1$ there exist similar relations between the coefficients of $L^{(r)}$ and $L^{(r-1)}$, as well as between the h_{jr}'s. In particular,

$$h_{jr} - 2h_{j,r-1} + h_{j,r-2} = -D_1 D_2(\log h_{j,r-1}), \quad r = 1,\cdots,n_j$$

where $h_{j,-1} = h_{j*0}$. Adding the first $n_j - m$ $(0 \le m \le n_j - 1)$ of these equations, we obtain

$$h_{j,n_j-m} - h_{j,n_j-m-1} = h_{j0} - h_{j*0} - D_1 D_2 \Lambda_{j,n_j-m}.$$

The addition of these n_j equations yields (4.2-8). The expression for a_{0n_j} in (4.2-7) follows from (4.2-5). Finally, a_{1n_j} and a_{2n_j}, as given in (4.2-7), are obtained by n successive applications of the idea by which we derived the preceding formulas for the coefficients of $L^{(1)}$, at the beginning of this proof. This proves the lemma. ■

ORDINARY DIFFERENTIAL EQUATION FOR w_{jn_j}

It is of basic importance that under the assumptions of Lemma 4.2-2 the solutions w of $L^{(0)}w = Lw = 0$ can be expressed in terms of w_{jn_j}, and that w_{jn_j} can be determined as a solution of an *ordinary* differential equation. Indeed, since $a_{j*r} = a_{j*0}$ by (4.2-9), and $L^{(r-1)}w_{j,r-1} = 0$, it follows that

$$(D_j + a_{j*0})w_{jr} = h_{j,r-1}w_{j,r-1}, \qquad r = 1,\cdots,n_j.$$

Hence, from (2.1-3) and $h_{j,r-1} \ne 0$,

$$w_{j,r-1} = \frac{e_{j*}}{h_{j,r-1}} D_j \frac{w_{jr}}{e_{j*}}, \qquad r = 1,\cdots,n_j$$

and

$$(4.2\text{-}10) \qquad w = w_{j0} = e_{j*}\left[\prod_{\lambda=0}^{n_j-1}\left(\frac{1}{h_{j\lambda}}D_j\right)\right]\left(\frac{1}{e_{j*}}w_{jn_j}\right).$$

If $h_{jn_j} = 0$, then (4.2-1, 2, 3, 5) imply that

$$D_{j*,n_j+1}w_{j,n_j+1} = D_{j*,n_j+1}D_{j,n_j+1}w_{jn_j}$$
$$= L^{(n_j)}w_{jn_j} = 0.$$

This is a homogeneous linear ordinary differential equation for w_{j,n_j+1}, the independent variable being z_j. Solving it and using (4.2-2), we obtain a nonhomogeneous linear ordinary differential equation for w_{jn_j} with z_{j*} as the independent variable. Holomorphic solutions of this equation on $B_1(s_1, r_1) \times B_2(s_2, r_2)$ are given by

$$(4.2\text{-}11) \qquad w_{jn_j} = e_{jn_j}\left(F_j + \int_{s_{j*}}^{z_{j*}} \frac{e_{j*}}{e_{jn_j}} F_{j*}\, d\tilde{z}_{j*}\right)$$

where $F_j \in C^\omega(B_j(s_j, r_j))$ and $F_{j*} \in C^\omega(B_{j*}(s_{j*}, r_{j*}))$ are arbitrary, and e_{jn_j} is defined as in (4.2-6). This proves the following result.

4.2-3 Lemma

Let $L^{(r)}w_{jr} = 0$ with $r = 1, \cdots, n_j$ ($n_j \in \mathbb{N}$) be obtained from $L^{(0)}w_{j0} = Lw = 0$ by means of (4.2-2, 3) and by successive application of (4.2-4a) to w_{j0}; that is, suppose that $h_{jr} \neq 0$ ($r = 0, \cdots, n_j - 1$) in (4.2-5) and let

$$L^{(r)}w_{jr} = h_{j,r-1}D_{jr}\frac{1}{h_{j,r-1}} \cdots h_{j0}D_{j1}\frac{1}{h_{j0}}L^{(0)}w_{j0}$$

with $L^{(0)} = L$ and $w_{jr} = D_j^{(r)}w$; cf. (4.2-1, 2). Furthermore, assume that $h_{jn_j} = 0$. Then the coefficients of L can be represented by (4.2-9), and every holomorphic solution of $Lw = 0$ on $B_1(s_1, r_1) \times B_2(s_2, r_2)$ by (4.2-10) with w_{jn_j} given by (4.2-11).

This lemma will play a key role in the proof of the following existence and representation theorem for polynomial kernels (cf. Kracht–Kreyszig [1975]), which is the main result in this section.

EXISTENCE AND REPRESENTATION OF CLASS P OPERATORS

4.2-4 Existence and Representation Theorem

Let $n_j \in \mathbb{N}$ and $j \in \{1, 2\}$ fixed. Then $L \in P_{jn_j}^0$, that is, there exists a class P operator for L with a minimal polynomial kernel of degree n_j (cf. 3.1-2) if and *only if the application of* $D_j^{(n_j)}$ *to* w *produces from* $Lw = 0$ *an equation*

$$L^{(n_j)} w_{jn_j} = 0, \qquad \text{where} \qquad w_{jn_j} = D_j^{(n_j)} w$$

such that $h_{jn_j} = 0$, *whereas* $h_{jr} \neq 0$ *when* $r = 0, \cdots, n_j - 1$ (*cf. Lemma* 4.2-3). *If* $L \in P_{jn_j}^0$, *a corresponding minimal polynomial kernel* $\tilde{k}_j$ *can be represented by*

$$\text{(4.2-12)} \quad \tilde{k}_j(z, t) = e_{j*}(z) \sum_{m=0}^{n_j} \frac{2^{2m} m!}{(2m)!} (z_j - s_j)^m \vartheta_{jm}\left(\frac{e_{jn_j}}{e_{j*}}\right)(z) t^{2m}$$

where

$$\vartheta_{jm} = \sum_{|\mu|=m} \prod_{\lambda=0}^{n_j - 1} \frac{1}{h_{j\lambda}} D_j^{\mu_{\lambda+1}}$$

$$\mu = (\mu_1, \cdots, \mu_{n_j}), \qquad \mu_r = 0, 1, \qquad r = 1, \cdots, n_j,$$

$$|\mu| = \mu_1 + \cdots + \mu_{n_j}.$$

Proof. (a) Suppose that $h_{jn_j} = 0$ and $h_{jr} \neq 0$ for $r = 0, \cdots, n_j - 1$. Then Lemma 4.2-3 yields the representation (4.2-10) with w_{jn_j} given by (4.2-11). Taking $F_{j*} = 0$, we thus obtain a polynomial kernel $\tilde{k}_j$ of degree $\nu \leq n_j$; cf. Theorem 3.3-2. From this and Remark 3.1-4 we conclude that $L \in P_{jn_j}$. Now from part (b) of this proof it will follow that $L \in P_{jm}^0$ for some $m < n_j$ implies $h_{jm} = 0$. This would contradict that $h_{jr} \neq 0$ for $r = 0, \cdots, n_j - 1$. Hence we must have $L \in P_{jn_j}^0$.

(b) Suppose that $L \in P_{jn_j}^0$, so that, by definition, there exists a minimal polynomial kernel $\tilde{k}_j$ of degree n_j. By Theorem 3.2-1 we then obtain solutions w (holomorphic on some domain in $\mathbb{C}^2$) of the form

$$w(z) = \sum_{m=0}^{n_j} A_{jm}(z) D_j^m F_j(z_j)$$

where $F_j \in C^\omega(B_j(s_j, r_j))$ satisfies (3.3-1) and

$$A_{j0} = \frac{(2n_j)!}{2^{2n_j} n_j!} e_j \tilde{q}_{jn_j} \neq 0.$$

Substituting this w into $w_{jr} = D_j^{(r)} w$, we see that $L^{(r)} w_{jr} = 0$ yields the system

$$\left(L^{(r)} D_j^{(r)} A_{j0}\right) F_j + \sum_{m=1}^{n_j - r} \left(L^{(r)} D_j^{(r)} A_{jm} + D_j^{(r+1)} A_{j,m-1}\right) D_j^m F_j$$

$$+ \left(D_j^{(r+1)} A_{j,n_j - r}\right) D_j^{n_j + 1 - r} F_j = 0$$

where $r = 0, \cdots, n_j$. For fixed r we thus obtain $n_j + 2 - r$ equations of the form

$$L^{(r)}D_j^{(r)}A_{j0} = 0$$

$$L^{(r)}D_j^{(r)}A_{jm} + D_j^{(r+1)}A_{j,m-1} = 0, \qquad m = 1, \cdots, n_j - r,$$

$$D_j^{(r+1)}A_{j,n_j-r} = 0$$

where the second of these three lines has to be omitted when $r = n_j$. In particular,

$$D_j^{(r+1)}A_{j,n_j-p} = 0 \quad \text{for} \quad p = 0, \cdots, r.$$

Hence, in proceeding from $w_{j,r-1}$ to w_{jr} $(r \geq 1)$, the order of differentiation decreases by at least one unit. On the other hand, since $\tilde{k}_j$ is minimal, none of its coefficient functions can be identically zero. Hence that order can decrease by at most one unit. After n_j steps we thus obtain

$$w_{jn_j} = D_j^{(n_j)}w = \left(D_j^{(n_j)}A_{j0}\right)F_j$$

where

$$D_j^{(n_j)}A_{j0} \neq 0. \tag{4.2-13}$$

Accordingly, for $r = n_j$ the preceding system reduces to

$$\begin{aligned} L^{(n_j)}D_j^{(n_j)}A_{j0} &= 0 \\ D_j^{(n_j+1)}A_{j0} &= 0. \end{aligned} \tag{4.2-14}$$

Because of (4.2-1, 3, 5) this yields

$$\begin{aligned} 0 &= L^{(n_j)}D_j^{(n_j)}A_{j0} \\ &= \left(D_{j^*,n_j+1}D_{j,n_j+1} - h_{jn_j}\right)D_j^{(n_j)}A_{j0} \\ &= D_{j^*,n_j+1}D_j^{(n_j+1)}A_{j0} - h_{jn_j}D_j^{(n_j)}A_{j0} \\ &= -h_{jn_j}D_j^{(n_j)}A_{j0}. \end{aligned}$$

From this and (4.2-13) we thus obtain $h_{jn_j} = 0$. We now claim that $h_{jr} \neq 0$ when $r = 0, \cdots, n_j - 1$. Indeed, if this were not true, considering the smallest $r < n_j$ such that $h_{jr} = 0$, by part (a) of this proof we would have $L \in P_{jr}$, contradicting $L \in P_{jn_j}^0$. Finally, the representation given by (4.2-12) follows readily from Theorem 3.3-2 and Lemma 4.2-3 with $F_j = \tilde{f}_j$ and $F_{j*} = 0$. This completes the proof of Theorem 4.2-4. ■

4.2-5 *Remark*

It is only for technical reasons that in Theorem 4.2-4 we have omitted the case $n_j = 0$. Indeed, one also has $L \in P_{j0}^0$ if and only if $h_{j0} = 0$. The proof is similar to the preceding one (but much simpler). We thus see that $L \in P_{j0}$ is equivalent to

$$\begin{aligned} L &= D_1D_2 + a_1D_1 + a_2D_2 + (a_1a_2 + D_ja_j) \\ &= D_{j*1}D_{j1}. \end{aligned}$$

APPLICATION OF THE EXISTENCE AND REPRESENTATION THEOREM

As a simple application of Theorem 4.2-4, let us show how we may derive the kernels in Example 3.1-7.

4.2-6 *Example*

We consider the operator

$$L = D_1D_2 + \frac{4z_j}{(1 + z_j^2 z_{j*})^2}$$

with fixed j in 3.1-7. From (4.2-5, 7, 8) we have

$$h_{10}(z) = -a_{30}(z) = -4z_j/\lambda_j^2 = h_{20}(z)$$

where

$$\lambda_j = 1 + z_j^2 z_{j*}$$

and, furthermore,

$$\begin{aligned} a_{j1}(z) &= -D_{j*}\log h_{j0}(z) = 2z_j^2/\lambda_j \\ a_{j*1}(z) &= a_{j*0}(z) = 0 \\ h_{j1}(z) &= (2h_{j0} - h_{j*0} - D_1D_2\log h_{j0})(z) = 0 \\ e_{j1}(z) &= \chi_j(z_j)/\lambda_j^2, \qquad \chi_j \in C^\omega(B_j(s_j, r_j)) \end{aligned}$$

$$\begin{aligned} \left(e_{j*}D_j\frac{e_{j1}}{e_{j*}}\right)(z) &= D_je_{j1}(z) \\ &= (D_j\chi_j(z_j) - 4z_1z_2\chi_j(z_j)/\lambda_j)/\lambda_j^2. \end{aligned}$$

From Theorem 4.2-4 we see that $L \in P_{j1}^{0}$, and (4.2-12) yields

$$\tilde{k}_j(z,t) = -\frac{1}{4z_j}\chi_j(z_j)$$

(4.2-15)

$$+\frac{1}{2z_j}\left(-D_j\chi_j(z_j) + 4z_1z_2\chi_j(z_j)/\lambda_j\right)(z_j - s_j)t^2.$$

Setting $s = 0$ and $\chi_j(z_j) = -4z_j$, we obtain

$$\tilde{\tilde{k}}_j(z,t) = 1 + \left(2 - \frac{8z_j^2 z_{j*}}{\lambda_j}\right)t^2$$

in agreement with 3.1-7. From this and (3.2-2) we have

$$\tilde{T}_j = e_j(z)\left(\tfrac{1}{2}\tilde{q}_{j1}(z) + \tilde{q}_{j0}(z)D_j\right)$$

$$= e_j(z)\left(\frac{1 - 3z_j^2 z_{j*}}{z_j\lambda_j} + D_j\right).$$

Setting $\eta_j = 0$ and applying $\tilde{T}_j$ to $(D_j\tilde{f}_j - z_j^{-1}\tilde{f}_j)$, we finally get

$$\tilde{w}_j(z) = \tilde{T}_j\left(D_j\tilde{f}_j - z_j^{-1}\tilde{f}_j\right)(z) = \tilde{\tilde{T}}_j\tilde{f}_j(z)$$

where

$$\tilde{\tilde{T}}_j = \frac{4z_{j*}}{\lambda_j} - \frac{4z_1z_2}{\lambda_j}D_j + D_j^2, \qquad \lambda_j = 1 + z_j^2 z_{j*}.$$

Consequently, (3.3-7) in Theorem 3.3-2 now yields the polynomial kernel of the first kind given in 3.1-7 in a systematic fashion.

TRANSFORMATIONS FOR CLASS P KERNELS

It is of practical importance to note that from known polynomial kernels, one can derive further equations and corresponding polynomial kernels by means of certain transformations which do not affect the conditions on the functions h_{jr} in Theorem 4.2-4. Indeed, the following result is readily obtained.

4.2-7 *Lemma*

Consider the equation $L^{(r)}w = 0$ *with* $L^{(r)}$ *as in* (4.2-1) *and the three transformations*

(a) $z_1 = \check{z}_1, \qquad z_2 = \check{z}_2, \qquad w(z) = \mu(z)\check{w}(\check{z})$

(b) $z_1 = \psi_1(\check{z}_1), \qquad z_2 = \psi_2(\check{z}_2), \qquad w(z) = \check{w}(\check{z})$

(c) $z_1 = \psi_2(\check{z}_2), \qquad z_2 = \psi_1(\check{z}_1), \qquad w(z) = \check{w}(\check{z});$

$$z = (z_1, z_2); \quad \check{z} = (\check{z}_1, \check{z}_2).$$

Each of these yields a transformed equation of the same kind as the original equation, namely,

$$\check{L}^{(r)}\check{w} = \left[\check{D}_1\check{D}_2 + \check{a}_{1r}(\check{z})\check{D}_1 + \check{a}_{2r}(\check{z})\check{D}_2 + \check{a}_{0r}(\check{z})\right]\check{w} = 0$$

where $\check{D}_j = \partial/\partial \check{z}_j$, *and the coefficients, with* $j = 1, 2$, *are given by*

(a)
$$\check{a}_{jr} = a_{jr} + \frac{1}{\mu} D_{j*}\mu$$
$$\check{a}_{0r} = a_{0r} + \frac{1}{\mu}(a_{1r}D_1\mu + a_{2r}D_2\mu + D_1D_2\mu)$$

(b)
$$\check{a}_{jr} = a_{jr}\check{D}_{j*}\psi_{j*}$$
$$\check{a}_{0r} = a_{0r}\check{D}_1\psi_1\check{D}_2\psi_2$$

(c)
$$\check{a}_{jr} = a_{j*r}\check{D}_{j*}\psi_{j*}$$
$$\check{a}_{0r} = a_{0r}\check{D}_1\psi_1\check{D}_2\psi_2$$

and for the functions

$$\check{h}_{jr} = \check{D}_j\check{a}_{jr} + \check{a}_{1r}\check{a}_{2r} - \check{a}_{0r}$$

with $j = 1, 2$ *we have in cases* (a) *and* (b)

$$\check{h}_{jr} \neq 0 \qquad \text{if and only if} \qquad h_{jr} \neq 0$$

and in case (c)

$$\check{h}_{jr} \neq 0 \qquad \text{if and only if} \qquad h_{j*r} \neq 0 .$$

Proof. The formulas for the transformed equation and its coefficients follow by straightforward calculation. The assertion on the h_{jr}'s and $\check{h}_{jr}$'s is obtained from

$$\check{h}_{jr} = h_{jr} \qquad \text{in case (a)}$$
$$\check{h}_{jr} = h_{jr}\check{D}_1\psi_1\check{D}_2\psi_2 \qquad \text{in case (b)}$$
$$\check{h}_{jr} = h_{j*r}\check{D}_1\psi_1\check{D}_2\psi_2 \qquad \text{in case (c)}. \quad ■$$

4.3 EXISTENCE OF POLYNOMIAL KERNELS FOR TYPE II REPRESENTATIONS

We shall now turn to the problem of the existence of Type II representations of solutions by operators with polynomial kernels. Applying Theorem 4.2-4 twice, once for $j = 1$ and once for $j = 2$, we obtain a first criterion for that problem, as was mentioned in the preceding section. The following theorem provides another approach, which is constructive, as we shall see in the next section.

4.3-1 Theorem

(a) *If* $L^{(0)} \in P^0_{jn_j} \cap P_{j^*n_{j^*}}$, $n_1, n_2 \in \mathbb{N}_0$, *and if*

$$\beta_{jn_j} = e_{j^*}/e_{jn_j} \tag{4.3-1}$$

then β_{jn_j} *satisfies an* ***ordinary*** *differential equation of the form*

$$\sum_{r=0}^{n_1+n_2+1} (-1)^r D^r_{j^*}\left(\alpha_{jr}\beta_{jn_j}\right) = 0 \tag{4.3-2}$$

where $\alpha_{j,\,n_1+n_2+1} = 1$ *and* $\alpha_{j0}, \cdots, \alpha_{j,\,n_1+n_2}$ *are functions depending only on* z_{j^*}.

(b) *If* β_{jn_j} *is a solution of an equation of the form* (4.3-2) *with arbitrary* $C^\omega(\Omega_{j^*})$*-coefficients satisfying the preceding requirements, then there exists an equation*

$$L^{(n_j)} w_{jn_j} = 0$$

with

$$L^{(n_j)} \in P^0_{j0} \cap P_{j^*,\,n_1+n_2}; \tag{4.3-3}$$

that is, $L^{(n_j)}$ *then admits polynomial kernels* $\tilde{k}_j$ *of degree* 0 *and* $\tilde{k}_{j^*}$ *of degree* $n_1 + n_2$ *for Type II representations of solutions of the equation*

$$L^{(n_j)} w_{jn_j} = 0.$$

If, in addition,

$$w_{jn_j} = D_j^{(n_j)} w \qquad \text{[cf. (4.2-2)]}$$

is such that $h_{jr} \neq 0$ *for* $r = 0, \cdots, n_j - 1$, *then there exists an equation* $L^{(0)} w = 0$ *with*

$$L^{(0)} \in P^0_{jn_j} \cap P_{j^*n_{j^*}}; \tag{4.3-4}$$

that is, $L^{(0)}$ *admits polynomial kernels* $\tilde{k}_j$ *of degree* n_j *(being minimal) and* $\tilde{k}_{j^*}$ *of degree* n_{j^*} *for Type II representations of solutions of the equation* $L^{(0)} w = 0$, *with coefficients given by* (4.2-9).

Remark

With regard to Remark 3.1-4 the results (4.3-3) and (4.3-4) imply

$$L^{(n_j)} \in P_{j\mu} \cap P_{j^*,\, n_1+n_2+\nu} \tag{4.3-5}$$

and

$$L^{(0)} \in P_{j,\, n_j+\mu} \cap P_{j^*,\, n_{j^*}+\nu} \tag{4.3-6}$$

respectively; here, $\mu, \nu \in \mathbb{N}_0$ are arbitrary.

Proof of the Theorem. Let $L^{(0)} \in P^0_{jn_j} \cap P_{j^*n_{j^*}}$. As in part (b) of the proof of Theorem 4.2-4, let

$$w(z) = \sum_{m=0}^{n_j} A_{jm} D_j^m F_j(z_j) + \sum_{r=0}^{n_{j^*}} A_{j^*r} D_{j^*}^r F_{j^*}(z_{j^*}) \tag{4.3-7}$$

so that, as in that proof, we obtain

$$w_{jn_j} = D_j^{(n_j)} w = C_{j0} F_j + \sum_{r=0}^{n_1+n_2} C_{j^*r} D_{j^*}^r F_{j^*}\,. \tag{4.3-8}$$

Here, the coefficients C_{j0} and C_{j^*r} are uniquely determined, and, by Theorem 4.2-4 and Remark 4.2-5, w_{jn_j} satisfies an equation

$$L^{(n_j)} w_{jn_j} = 0 \qquad \text{with} \qquad h_{jn_j} = 0\,.$$

It follows that we obtain a representation [cf. (4.2-11)]

$$w_{jn_j} = e_{jn_j} \hat{F}_j + e_{jn_j} \int_{s_{j^*}}^{z_{j^*}} \beta_{jn_j} \hat{F}_{j^*}\, d\tilde{z}_{j^*} \tag{4.3-9}$$

where $\hat{F}_j$ depends only on z_j and $\hat{F}_{j^*}$ only on z_{j^*}. Taking (4.2-14) into account, we now obtain the ordinary differential equation

$$D_{j,\, n_j+1} C_{j0} = 0$$

which has the nontrivial solution

$$C_{j0} = e_{jn_j}\,. \tag{4.3-10}$$

From $h_{jn_j} = 0$ we have $D_{j^*,\, n_j+1} D_{j,\, n_j+1} w_{jn_j} = 0$. Using (4.3-8) and considering the coefficient of the expression $D_{j^*}^{n_1+n_2+1} F_{j^*}$, we get the equation

$$D_{j^*,\, n_j+1} C_{j^*,\, n_1+n_2} = 0\,.$$

Consequently,

$$C_{j^*, n_1+n_2} = e_{j^*}. \tag{4.3-11}$$

Comparing (4.3-8) and (4.3-9), and using (4.3-10, 11), we see that $F_j = \hat{F}_j$ and

$$\sum_{r=0}^{n_1+n_2+1} \frac{1}{e_{jn_j}} C_{j^*r} D_{j^*}^r F_{j^*} = \int_{s_{j^*}}^{z_{j^*}} \beta_{jn_j} \hat{F}_{j^*} \, d\tilde{z}_{j^*}. \tag{4.3-12}$$

Differentiation with respect to z_{j^*} yields

$$\hat{F}_{j^*} = \sum_{r=0}^{n_1+n_2+1} \alpha_{jr} D_{j^*}^r F_{j^*} \tag{4.3-13}$$

where

$$\begin{aligned} \alpha_{j0} &= \beta_{jn_j}^{-1} D_{j^*}\left(C_{j^*0}/e_{jn_j}\right) \\ \alpha_{jr} &= \beta_{jn_j}^{-1}\left[D_{j^*}\left(C_{j^*r}/e_{jn_j}\right) + C_{j^*, r-1}/e_{jn_j}\right] \\ \alpha_{j, n_1+n_2+1} &= 1. \end{aligned} \tag{4.3-14}$$

Eliminating C_{j^*r}/e_{jn_j} $(r = 0, \cdots, n_1 + n_2)$ from (4.3-14), we obtain an equation of the form (4.3-2). This proves part (a) of the theorem. Part (b) will be proved in the next section, where we shall derive it from a construction principle to be developed there. ■

FROBENIUS IRREDUCIBILITY

Following Frobenius, we say that an ordinary differential equation of the form

$$\sum_{r=0}^{n_1+n_2} \alpha_{jr} D_{j^*}^r F_{j^*} + D_{j^*}^{n_1+n_2+1} F_{j^*} = 0 \tag{4.3-15}$$

is ***irreducible*** if it does not admit any solution $F_{j^*} \neq 0$ satisfying a homogeneous linear differential equation of order less than $n_1 + n_2 + 1$. The irreducibility of (4.3-15) implies that there does not exist a transformation given by

$$\chi_j = \sum_{\lambda=0}^{\kappa} \tilde{\chi}_{j\lambda} D_{j^*}^{\lambda} F_{j^*}$$

with $\kappa \leq n_1 + n_2$ and the $\tilde{\chi}_{j\lambda}$'s depending on z_{j^*}, such that the left-hand side of (4.3-15) can be represented as a linear combination of χ_j and its derivatives of order not exceeding $n_1 + n_2$. Because of (4.3-12, 13) this property carries over to (4.3-9) with $n_1 + n_2$ instead of $n_1 + n_2 + 1$. From this and Theorem 3.2-1 it follows that there cannot exist a polynomial kernel $\tilde{k}_{j*}$ of degree less than $n_1 + n_2$. We thus have

$$L^{(n_j)} \in P^0_{j*,\, n_1+n_2}$$

in Theorem 4.3-1b. Similarly, one can prove that $L^{(0)} \in P^0_{j*n_2}$. This proves part (a) of the following Corollary, part (b) being a known result by Frobenius.

4.3-2 Corollary

(a) *Let the functions* α_{jr} *in* (4.3-2) *be such that* (4.3-15) *is irreducible in the sense of Frobenius. Then in Theorem* 4.3-1b, *in addition to* (4.3-3) *and* (4.3-4) *we also have*

$$L^{(n_j)} \in P^0_{j0} \cap P^0_{j*,\, n_1+n_2}$$

and

$$L^{(0)} \in P^0_{1n_1} \cap P^0_{2n_2}$$

respectively.

(b) *For all* $n_1, n_2 \in \mathbb{N}_0$ *there exist functions* α_{jr} $(r = 0, \cdots, n_1 + n_2)$ *such that* (4.3-15) *is irreducible.*

Since the condition on the h_{jr}'s in Theorem 4.3-1b can be met (see the next section; see also Example 3.1-6), we have

$$P^0_{1n_1} \cap P^0_{2n_2} \neq \emptyset \qquad \text{for all } n_1, n_2 \in \mathbb{N}_0 .$$

4.4 CONSTRUCTION PRINCIPLE FOR CLASS P OPERATORS AND EQUATIONS

In this section we continue and finish the proof of Theorem 4.3-1b by introducing Construction Principle 4.4-1 for equations (2.1-1) such that there exist corresponding polynomial kernels. Applying this principle, we shall also derive the most general equation considered by Bauer [1976a] as well as corresponding integral and differential operators.

4.4-1 Construction Principle

We start from an ordinary differential equation of the form (4.3-2), that is,

$$(4.4\text{-}1) \qquad \sum_{r=0}^{n_1+n_2} (-1)^r D^r_{j*}\left(\alpha_{jr}\beta_{jn_j}\right) + \left(-D_{j*}\right)^{n_1+n_2+1}\beta_{jn_j} = 0$$

where the α_{jr}'s are arbitrary $C^{\omega}(\Omega_{j*})$-functions, as assumed in part (b) of Theorem 4.3-1. We take a solution $\beta_{jn_j} \neq 0$ of (4.4-1) and set

$$\beta_{jn_j} = e_{j*}/e_{jn_j} \tag{4.4-2}$$

as in (4.3-1). This yields a_{jn_j} in terms of $a_{j^*n_j}$ and

$$a_{0n_j} = D_j a_{jn_j} + a_{1n_1} a_{2n_2}$$

as well as an equation [cf. (4.2-1)]

$$L^{(n_j)} w_{jn_j} = 0$$

for which there exist polynomial kernels $\tilde{k}_1$ and $\tilde{k}_2$ needed in Type II Bergman representations of solutions. Here, $\tilde{k}_j$ is of degree 0 and $\tilde{k}_{j*}$ of degree $n_1 + n_2$. As in (4.3-10, 11) we set

$$C_{j^*, n_1+n_2} = e_{j^*}, \qquad C_{j0} = e_{jn_j}. \tag{4.4-3}$$

We now determine $C_{j^*0}, \cdots, C_{j^*, n_1+n_2-1}$ recursively from (4.3-14). By using (4.3-8) we then obtain a representation of w_{jn_j}. Now, if there exist functions $h_{j0}, \cdots, h_{j, n_j-1}$, not identically zero, so that (4.2-4a) is solvable for $L^{(r-1)}$, $r = 1, \cdots, n_j$, then from Theorem 4.2-4 we conclude that there exist (i) an equation

$$Lw = L^{(0)} w_{j0} = 0$$

of the form (2.1-1) and (ii) Bergman kernels of the form (4.2-12). Furthermore, the integral-free form (4.3-7) of solutions according to

$$\tilde{w}(z) = \tilde{w}_{j0}(z) = \left(D_j^{(n_j)}\right)^{-1} w_{jn_j}(z) \tag{4.4-4}$$

is given by (3.2-2), and the coefficients of $L = L^{(0)}$ are obtained from (4.2-9) in Lemma 4.2-2.

As described in this construction principle, we have thus obtained

$$L^{(n_j)} \in P_{j0}^{0} \cap P_{j^*, n_1+n_2}$$

and, by virtue of Theorem 4.2-4, also $L^{(0)} \in P_{jn_j}^{0} \cap P_{j^*n_{j^*}}$. Remark 3.1-4 then guarantees that for all $\mu, \nu \in \mathbb{N}_0$,

$$L^{(n_j)} \in P_{j\mu} \cap P_{j^*, n_1+n_2+\nu}, \qquad L^{(0)} \in P_{j, n_j+\mu} \cap P_{j^*, n_{j^*}+\nu}.$$

This completes the proof of Theorem 4.3-1b and its subsequent remark. ■

APPLICATION OF THE CONSTRUCTION PRINCIPLE

We shall now apply Construction Principle 4.4-1, as announced. We start from (4.4-1) with all α_{jr}'s zero, that is,

$$D_{j*}^{n_1+n_2+1}\beta_{jn_j} = 0\,, \qquad j = 1 \text{ or } j = 2\,, \text{ fixed}.$$

A solution is

$$\beta_{jn_j}(z) = (z_1 + z_2)^{n_1+n_2}. \tag{4.4-5}$$

We choose

$$a_{j*}(z) = a_{j*n_j}(z) = \frac{n_j - n_{j*}}{z_1 + z_2} \tag{4.4-6}$$

and, according to (4.4-3) and (4.2-5),

$$e_{j*}(z) = (z_1 + z_2)^{n_{j*}-n_j} = C_{j*,\,n_1+n_2}.$$

Then, by (4.4-2, 3, 5) we have

$$e_{jn_j}(z) = (z_1 + z_2)^{-2n_j} = C_{j0}. \tag{4.4-7}$$

From this, by (4.2-6) with

$$\eta_{jn_j}(z_j) = -2n_j \log(z_j + s_{j*})$$

we obtain

$$a_{jn_j}(z) = \frac{2n_j}{z_1 + z_2} \tag{4.4-8}$$

and, furthermore,

$$a_{0n_j}(z) = \left(D_j a_{jn_j} + a_{1n_j}a_{2n_j}\right)(z) = \frac{2n_j(n_j - n_{j*} - 1)}{(z_1 + z_2)^2}.$$

Hence in the present case the system (4.3-14) takes the form

$$C_{j*r}(z) = \frac{1}{r!}(-1)^{n_1+n_2-r}(n_1 + n_2)!(z_1 + z_2)^{r-2n_j}.$$

As a *first intermediate result* we note that for $\hat{f}_1 \in C^\omega(\Omega_1)$ and $\hat{f}_2 \in C^\omega(\Omega_2)$, the function w_{jn_j} defined by

$$w_{jn_j}(z) = (z_1 + z_2)^{-2n_j}\hat{f}_j(z_j)$$

(4.4-9)
$$+ \sum_{r=0}^{n_1+n_2} \frac{1}{r!}(-1)^{n_1+n_2-r}(n_1+n_2)!(z_1+z_2)^{r-2n_j} D_{j*}^r \hat{f}_{j*}(z_{j*})$$

is a solution of the equation $L_j^{(n_j)} w_{jn_j} = 0$, where

$$L_j^{(n_j)} = D_1 D_2 + \frac{2n_j}{z_1+z_2} D_j + \frac{n_j - n_{j*}}{z_1+z_2} D_{j*} + \frac{2n_j(n_j - n_{j*} - 1)}{(z_1+z_2)^2}. \tag{4.4-10}$$

Since $h_{j,n_j-1} = h_{j*n_j}$, we have

$$h_{j,n_j-1}(z) = (D_{j*}a_{j*n_j} + a_{1n_j}a_{2n_j} - a_{0n_j})(z) = \frac{n_1+n_2}{(z_1+z_2)^2}$$

and

$$D_{j*}h_{j,n_j-1}(z)/h_{j,n_j-1}(z) = -2/(z_1+z_2).$$

These formulas and (4.4-8) suggest that we set

$$h_{j,n_j-\nu}(z) = \frac{\nu(n_1+n_2-\nu+1)}{(z_1+z_2)^2} \tag{4.4-11}$$

where $\nu = 0, \cdots, n_j$. We thus have

$$h_{jn_j} = 0, \qquad h_{jr} \neq 0 \quad \text{if} \quad r = 0, \cdots, n_j - 1$$

$$h_{j0}(z) = n_j(n_{j*}+1)/(z_1+z_2)^2$$

$$D_{j*}h_{j,n_j-\nu}(z)/h_{j,n_j-\nu}(z) = -2/(z_1+z_2), \qquad \nu = 1, \cdots, n_j.$$

Using (4.2-9) and (4.4-8), and remembering our choice (4.4-6), we obtain an equation $L_j^{(0)}w = 0$ whose coefficients are $a_j = 0$, a_{j*} as in (4.4-6) and

$$a_0 = D_j a_j + a_1 a_2 - h_{j0} = -h_{j0}.$$

Hence this equation has the form

$$L_j^{(0)}w = \left(D_1 D_2 + \frac{n_j - n_{j*}}{z_1+z_2} D_{j*} - \frac{n_j(n_{j*}+1)}{(z_1+z_2)^2} \right) w = 0. \tag{4.4-12}$$

We substitute (4.4-9) into (4.4-4), use (4.4-7, 11), and set

$$D = \prod_{\lambda=0}^{n_j-1}\left(\frac{1}{h_{j\lambda}}D_j\right)$$

$$\hat{D} = \frac{n_{j*}!}{n_j!(n_1+n_2)!}q^{n_{j*}-n_j}\left(q^2D_j\right)^{n_j}, \qquad q = z_1 + z_2.$$

As a *second intermediate result* we now obtain the following representation of solutions of (4.4-12).

$$\begin{aligned}\tilde{w}(z) &= e_{j*}(z)D\left(\frac{w_{jn_j}}{e_{j*}}\right)(z)\\ &= \hat{D}\left[q^{-n_1-n_2}\hat{f}_j(z_j)\right]\\ &\quad+\hat{D}\left[\sum_{r=0}^{n_1+n_2}\frac{(n_1+n_2)!}{r!}(-q)^{r-n_1-n_2}D_{j*}^r\hat{f}_{j*}(z_{j*})\right].\end{aligned} \tag{4.4-13}$$

By induction with respect to n_j we see that the first term on the right can be written

(4.4-14)

$$\hat{D}\left[q^{-n_1-n_2}\hat{f}_j(z_j)\right] = \sum_{r=0}^{n_j}\frac{(n_1+n_2-r)!}{r!(n_j-r)!}(-q)^{r-n_j}D_j^r\left(\frac{1}{(n_1+n_2)!}\hat{f}_j\right)(z_j).$$

We consider the other term $\hat{D}[\Sigma\cdots]$ on the right-hand side of formula (4.4-13). Since

$$\left(q^2D_j\right)^{n_j}q^{r-n_1-n_2} = (-1)^{n_j}\left[\prod_{\nu=0}^{n_j-1}(n_1+n_2-r-\nu)\right]q^{r-n_{j*}}$$

this term becomes

$$\sum_{r=0}^{n_{j*}}\frac{(n_1+n_2-r)!}{r!(n_{j*}-r)!}(-1)^{n_{j*}-r}q^{r-n_j}D_{j*}^r\left(\frac{n_{j*}!}{n_j!}\hat{f}_{j*}\right)(z_{j*}). \tag{4.4-15}$$

Introducing

$$\tilde{f}_j = \hat{f}_j/(n_1+n_2)! \qquad \text{and} \qquad \tilde{f}_{j*} = n_{j*}!\hat{f}_{j*}/n_j!$$

from (4.4-13, 14, 15) we obtain an ***integral-free representation*** of solutions of

(4.4-12) in the form

(4.4-16) $$\tilde{w}(z) = q^{-n_j} \sum_{p=1}^{2} \sum_{m=0}^{n_p} \frac{(n_1 + n_2 - m)!}{m!(n_p - m)!} (-1)^{n_p - m} q^m D_p^m \tilde{f}_p(z_p)$$

where $q = z_1 + z_2$.

To both (4.4-12) and (4.4-16) we apply the transformation (b) in Lemma 4.2-7 given by

$$z_p = \psi_p(\check{z}_p), \qquad p = 1, 2.$$

Here we assume that $\psi_p \in C^\omega(\psi_p^{-1}(\Omega_p))$ and $\check{D}_p \psi_p(\check{z}_p) \neq 0$ for all $\check{z}_p \in \psi_p^{-1}(\Omega_p)$, where $\check{D}_p = \partial/\partial \check{z}_p$. We write

$$\check{z} = (\check{z}_1, \check{z}_2), \qquad \check{w}(\check{z}) = \tilde{w}(z), \qquad \check{f}_p(\check{z}_p) = \tilde{f}_p(\psi_p(\check{z}_p))$$

where $p = 1, 2$. This yields $\check{L}_j^{(0)} \check{w} = 0$, where

(4.4-17)

$$\check{L}_j^{(0)} = \check{D}_1 \check{D}_2 + \frac{(n_j - n_{j^*}) \check{D}_j \psi_j(\check{z}_j)}{\psi_1(\check{z}_1) + \psi_2(\check{z}_2)} \check{D}_{j^*} - \frac{n_j(n_{j^*} + 1) \check{D}_1 \psi_1(\check{z}_1) \check{D}_2 \psi_2(\check{z}_2)}{(\psi_1(\check{z}_1) + \psi_2(\check{z}_2))^2}.$$

Setting

$$\psi(\check{z}) = \psi_1(\check{z}_1) + \psi_2(\check{z}_2)$$

we may write this more simply as

(4.4-18)

$$\check{L}_j^{(0)} = \check{D}_1 \check{D}_2 + (n_j - n_{j^*}) \check{D}_j \log \psi(\check{z}) \check{D}_{j^*} + n_j(n_{j^*} + 1)(\check{D}_1 \check{D}_2 \log \psi(\check{z})).$$

By (4.4-16) a corresponding representation of solutions is

(4.4-19) $$\check{w} = \psi^{-n_j} \sum_{p=1}^{2} \sum_{m=0}^{n_p} \frac{(n_1 + n_2 - m)!}{m!(n_p - m)!} (-1)^{n_p - m} \psi^m \left[(\check{D}_p \psi_p)^{-1} \check{D}_p \right]^m \check{f}_p.$$

This is the representation of Bauer [1976a] which we wanted to derive. The corresponding partial differential equation is of considerable practical importance, as was mentioned at the beginning of Chap. 3.

DERIVATION OF BERGMAN POLYNOMIAL KERNELS

Bergman polynomial kernels for (4.4-12) and (4.4-18) can be obtained from Theorem 3.3-2 or Theorem 4.2-4. Indeed, let us consider (4.4-18) and (4.4-19), writing again z_1, z_2 for the independent variables and D_p instead of $\check{D}_p$, for simplicity. We set

$$T_p f_p = w_p = \check{w}_p, \qquad \check{w} = \check{w}_1 + \check{w}_2$$

where $\check{w}_1$ and $\check{w}_2$ are the expressions on the right-hand side of formula (4.4-19) corresponding to $p = 1$ and $p = 2$, respectively. We write

$$\tilde{D}_p = \left[D_p \psi_p(z_p)\right]^{-1} D_p. \tag{4.4-20}$$

If we order by powers of D_p, we obtain

(4.4-21)

$$\check{w}_p(z) = \psi(z)^{n_p - n_j} \sum_{m=0}^{n_p} \left[\sum_{\mu=0}^{m} \frac{(n_{3-p} + \mu)!}{\mu!(n_p - \mu)!} (-\psi(z))^{-\mu} \tilde{D}_{p,m\mu}(z_p) \right] D_p^{n_p - m} \check{f}_p(z_p)$$

where

$$\tilde{D}_{p,m\mu}(z_p) = \sum \tilde{D}_p^{r_1} \cdots \tilde{D}_p^{r_{n_p-\mu}} 1(z_p) \tag{4.4-22}$$

(summation from 0 to 1, over all $r_1, \cdots, r_{n_p - \mu}$ satisfying the condition $r_1 + \cdots + r_{n_p - \mu} = m - \mu$). Using Theorem 3.3-2, from formula (4.4-21) we see immediately that $\tilde{k}_{j1}$ and $\tilde{k}_{j2}$ in (4.4-23) are Bergman polynomial kernels for Type II representations $w = T_1 f_1 + T_2 f_2$ of solutions of $L_j^{(0)} w = 0$ with $L_j^{(0)}$ as in (4.4-17) or (4.4-18); here, $j = 1, 2$. Because of Remark 2.1-1, starting from a particular solution of a given equation of the form (4.4-1) and using Construction Principle 4.4-1, we thus obtain the following result.

4.4-2 *Theorem*

Suppose that the three equations

$$L^{(0)} w = D_1 D_2 w - n_1 (D_2 \log \psi) D_1 w - n_2 (D_1 \log \psi) D_2 w = 0$$

$$L_1^{(0)} u = D_1 D_2 u + (n_1 - n_2)(D_1 \log \psi) D_2 u + n_1 (n_2 + 1)(D_1 D_2 \log \psi) u = 0$$

$$L_2^{(0)} v = D_1 D_2 v + (n_2 - n_1)(D_2 \log \psi) D_1 v + n_2 (n_1 + 1)(D_1 D_2 \log \psi) v = 0$$

with $\psi(z) = \psi_1(z_1) + \psi_2(z_2)$ *have holomorphic coefficients in* Ω, *and let* $n_1, n_2 \in \mathbb{N}_0$. *Furthermore, let*

$$k_p(z, t) = \sum_{m=0}^{n_p} C_{pm}(z) \frac{2^{2m} m!}{(2m)!} (z_p - s_p)^m t^{2m}$$

where C_{pm} *is the expression in the brackets in* (4.4-21). *Then polynomial kernels for Type II representations of solutions involving Bergman operators with point of reference s for those three equations are*

$$\tilde{k}_1 = \psi^{n_1} k_1, \qquad \tilde{k}_2 = \psi^{n_2} k_2$$

(4.4-23)
$$\tilde{k}_{11} = k_1, \qquad \tilde{k}_{12} = \psi^{n_2 - n_1} k_2$$

$$\tilde{k}_{21} = \psi^{n_1 - n_2} k_1, \qquad \tilde{k}_{22} = k_2$$

respectively. Furthermore, let

(4.4-24)

$$\tilde{\tilde{T}}_p = \sum_{m=0}^{n_p} \left[\sum_{\mu=m}^{n_p} \frac{(n_1 + n_2 - \mu)!}{\mu!(n_p - \mu)!} (-\psi(z))^{\mu} (-1)^{n_p} \tilde{D}_{p, n_p - m, n_p - \mu}(z_p) \right] D_p^m$$

with $\tilde{D} \cdots$ *as defined in* (4.4-22). *Then corresponding differential operators for Type* II *representations for solutions of those three equations are*

$$\tilde{T}_1 = \tilde{\tilde{T}}_1, \qquad \tilde{T}_2 = \tilde{\tilde{T}}_2$$

(4.4-25)
$$\tilde{T}_{11} = \psi^{-n_1} \tilde{\tilde{T}}_1, \qquad \tilde{T}_{12} = \psi^{-n_1} \tilde{\tilde{T}}_2$$

$$\tilde{T}_{21} = \psi^{-n_2} \tilde{\tilde{T}}_1, \qquad \tilde{T}_{22} = \psi^{-n_2} \tilde{\tilde{T}}_2$$

respectively. The operator $\tilde{\tilde{T}}_p$ *may also be written* [*cf.* (4.4-19)]

(4.4-26)
$$\tilde{\tilde{T}}_p = \sum_{m=0}^{n_p} \frac{(n_1 + n_2 - m)!}{m!(n_p - m)!} (-\psi(z))^m (-1)^{n_p} \tilde{D}_p^m$$

with $\tilde{D}_p$ *as defined in* (4.4-20).

Formulas (4.4-25) and (4.4-26) include the aforementioned result by Bauer [1976a]. The representations in Examples 2.5-5 and 3.3-4 also follow from Theorem 4.4-2 by taking

$$\psi_1(z_1) = 1/\lambda z_1, \qquad \psi_2(z_2) = z_2, \qquad \lambda \neq 0 \text{ complex}; \quad n_1 = n_2 = n.$$

4.5 SECOND CONSTRUCTION PRINCIPLE

In this section we show that the ideas of the proof of Theorem 4.3-1 can be used for deriving another construction principle for equations (2.1-1) and representations of corresponding polynomial kernels. This principle has the advantage that it is simple, although its practical applicability is restricted to the (still rather large) class of equations for which the determinants appearing in the method can be evaluated by a reasonable amount of work. The principle also yields a general representation of the zero solution in a very simple fashion and thus a statement on the uniqueness of Type II representations of solutions by differential operators.

We start from the ordinary differential equation

$$N_j y_j = \sum_{r=0}^{n_1+n_2} \tilde{\alpha}_{jr}(z_{j*}) D_{j*}^r y_j + D_{j*}^{n_1+n_2+1} y_j = 0 \tag{4.5-1}$$

with z_{j*} as the independent variable; cf. (4.3-15). Let $\eta_{j1}, \cdots, \eta_{j, n_1+n_2+1} \in C^\omega(\Omega_{j*})$ be a basis of solutions of (4.5-1) and W_j the corresponding Wronskian. Let the functions γ_{jr} be defined by [cf. formula (4.3-14)]

$$\begin{aligned} \tilde{\alpha}_{j0} &= (D_{j*}\gamma_{j0})/\beta_{jn_j} \\ \tilde{\alpha}_{jr} &= (D_{j*}\gamma_{jr} + \gamma_{j,r-1})/\beta_{jn_j} \qquad r = 1, \cdots, n_1 + n_2 \\ \beta_{jn_j} &= \gamma_{j, n_1+n_2}. \end{aligned} \tag{4.5-2}$$

Then we have from (4.5-1)

$$D_{j*}\left(\sum_{r=0}^{n_1+n_2} \gamma_{jr} D_{j*}^r \eta_{j\mu} \right) = \beta_{jn_j} N_j \eta_{j\mu} = 0, \qquad \mu = 1, \cdots, n_1 + n_2 + 1.$$

The next idea is to integrate with respect to z_{j*}. In this process we can introduce arbitrary functions of z_j which we denote by $\xi_{j\mu}$. We may assume these functions to be holomorphic and linearly independent on Ω_j. Then we obtain a system of linear algebraic equations in the unknowns γ_{jr}, of the form

$$\sum_{r=0}^{n_1+n_2} \gamma_{jr} D_{j*}^r \eta_{j\mu} + \xi_{j\mu} = 0, \qquad \mu = 1, \cdots, n_1 + n_2 + 1.$$

Since $W_j \neq 0$, the system has the solution

$$\gamma_{jr} = (-1)^{r+1} W_j^{-1} \det(\delta_{jr\mu\nu}), \qquad r = 0, \cdots, n_1 + n_2 \tag{4.5-3a}$$

where μ is as before and

$$\delta_{jr\mu\nu} = \begin{cases} \xi_{j\mu} & \text{if } \nu = 1 \\ D_{j*}^{\nu-2} \eta_{j\mu} & \text{if } \nu = 2, \cdots, r+1 \\ D_{j*}^{\nu-1} \eta_{j\mu} & \text{if } \nu = r+2, \cdots, n_1 + n_2 + 1. \end{cases} \tag{4.5-3b}$$

Setting

$$C_{j0} = e_{jn_j}, \qquad C_{j*r} = e_{jn_j}\gamma_{jr}, \qquad r = 0, \cdots, n_1 + n_2$$

we thus obtain

$$w_{jn_j} = e_{jn_j}\left[F_j + \sum_{r=0}^{n_1+n_2} (-1)^{r+1} W_j^{-1} \det(\delta_{jr\mu\nu}) D_{j*}^r F_{j*}\right].$$

Let Γ_j denote the determinant obtained by augmenting $\det(\delta_{jr\mu\nu})$ by one row and one column, namely, by taking

$$\begin{pmatrix} F_j & F_{j*} & D_{j*}F_{j*} & \cdots & D_{j*}^{n_1+n_2}F_{j*} \end{pmatrix}$$

as an additional new first row and by inserting the column

$$\left(D_{j*}^r \eta_{j\mu}\right), \qquad \mu = 1, \cdots, n_1 + n_2 + 1$$

between the $(r+1)$th and $(r+2)$th columns of $\det(\delta_{jr\mu\nu})$. Then

$$w_{jn_j} = e_{jn_j}\Gamma_j / W_j. \tag{4.5-4}$$

If we choose a_{1n_j} and a_{2n_j} so that $\beta_{jn_j} = \gamma_{j, n_1+n_2}$ with γ_{j, n_1+n_2} given by (4.5-3), and

$$a_{0n_j} = D_j a_{jn_j} + a_{1n_j} a_{2n_j} \tag{4.5-5}$$

it follows from Theorem 4.2-4 that (4.5-4) is a solution of

$$L^{(n_j)} w_{jn_j} = 0$$

with $L^{(n_j)} \in P_{j0}^0$. [Note that because of the form of (4.5-4) we also have $L^{(n_j)} \in P_{j*, n_1+n_2}$.]

CONSTRUCTION OF A DIFFERENTIAL EQUATION AND SOLUTIONS

Under the additional assumption in Theorem 4.3-1b we are able to construct an equation

$$L^{(0)} w = 0$$

and a representation of solutions from the preceding functions $\xi_{j\mu}$, $\eta_{j\mu}$, F_j, and F_{j*}, as follows.

Lemma 4.2-3 and formula (4.5-4) suggest that we set

$$w = \sum_{m=0}^{n_j} \chi_{jm} D_j^m w_{jn_j} = \sum_{m=0}^{n_j} \chi_{jm} D_j^m \left(e_{jn_j} \Gamma_j / W_j \right).$$

This is of the form (4.3-7), the reasons being that (i) Γ_j is linear in the functions F_j and F_{j*} and their derivatives, and (ii) the order of differentiation of F_{j*} is at most n_{j*} [because otherwise the order of $w_{jn_j} = D_j^{(n_j)} w$ in (4.5-4) would exceed $n_1 + n_2$]. For $F_j = \xi_{j\mu}$ and $F_{j*} = \eta_{j\mu}$, $\mu \in \{1, \cdots, n_1 + n_2 + 1\}$, we have $\Gamma_j = 0$. Hence we obtain the system

$$\sum_{\rho=0}^{n_j} A_{j\rho} D_j^\rho \xi_{j\mu} + \sum_{r=0}^{n_{j*}} A_{j*r} D_{j*}^r \eta_{j\mu} = 0, \qquad \mu = 1, \cdots, n_1 + n_2 + 1$$

from which we can determine the $n_1 + n_2 + 2$ coefficient functions $A_{j\rho}$ and A_{j*r}. Since the $\xi_{j\mu}$'s and $\eta_{j\mu}$'s are linearly independent, that system has maximum rank. Hence there exist nontrivial solutions which are uniquely determined up to a constant, which we can choose equal to 1, without restriction. Then the solution w of the equation $L^{(0)} w = 0$ to be constructed is

$$w = \det\left(\varepsilon_{j\kappa\lambda} \right) \tag{4.5-6}$$

where, for $\lambda = 1, \cdots, n_j + 1$,

$$\varepsilon_{j1\lambda} = D_j^{\lambda-1} F_j, \qquad \varepsilon_{j\kappa\lambda} = D_j^{\lambda-1} \xi_{j,\kappa-1} \qquad \text{if } \kappa = 2, \cdots, n_1 + n_2 + 2$$

and for $\lambda = n_j + 2, \cdots, n_1 + n_2 + 2$,

$$\varepsilon_{j1\lambda} = D_{j*}^{\lambda - n_j - 2} F_{j*}, \qquad \varepsilon_{j\kappa\lambda} = D_{j*}^{\lambda - n_j - 2} \eta_{j,\kappa-1} \qquad \text{if } \kappa = 2, \cdots, n_1 + n_2 + 2.$$

From (4.5-6) the coefficients $A_{j\rho}$ and A_{j*r} are now obtained by developing the determinant by the first row. Substituting (4.5-6) into (2.1-1) and comparing coefficients, we have the conditions

$$\begin{aligned} L^{(0)} A_{j0} &= 0, \qquad & D_{j1} A_{jn_j} &= 0 \\ L^{(0)} A_{j*0} &= 0, \qquad & D_{j*1} A_{j*n_{j*}} &= 0. \end{aligned} \tag{4.5-7}$$

Here, A_{j0}, A_{jn_j}, A_{j*0}, and $A_{j*n_{j*}}$ cannot vanish identically, since this would contradict the linear independence of the $\xi_{j\mu}$ and $\eta_{j\mu}$. The additional condition from Theorem 4.3-1b guarantees the existence of

$$L^{(0)} \in P_{jn_j}^0 \cap P_{j*n_{j*}}.$$

In this way, from (4.5-7) we obtain the following representations of the coefficients:

$$
\begin{aligned}
a_{j0} &= -\left(D_{j^*}A_{jn_j}\right)/A_{jn_j} \\
a_{j^*0} &= -\left(D_jA_{j^*n_{j^*}}\right)/A_{j^*n_{j^*}} \\
a_{00} &= -\Big[D_1D_2A_{j0} - \left(D_jA_{j0}\right)\left(D_{j^*}A_{jn_j}\right)/A_{jn_j} \\
&\qquad -\left(D_{j^*}A_{j0}\right)\left(D_jA_{j^*n_{j^*}}\right)/A_{j^*n_{j^*}}\Big]/A_{j0}\,.
\end{aligned}
\tag{4.5-8}
$$

For a_{00} we also have an alternative formula which results by replacing A_{j0} with A_{j^*0}.

RESULT IN THIS SECTION

Summarizing these results, we have the following theorem and construction principle.

4.5-1 Theorem

(a) (i) *Let* $\xi_{j1},\cdots,\xi_{j,n_1+n_2+1} \in C^\omega(\Omega)$ *be linearly independent functions depending only on* z_j. *Let* $\eta_{j1},\cdots,\eta_{j,n_1+n_2+1} \in C^\omega(\Omega)$ *be a basis of solutions of an ordinary differential equation of the form* (4.5-1) *with* z_{j^*} *as the independent variable. Then there exists a partial differential equation*

$$
L^{(n_j)}w_{jn_j} = 0 \qquad \text{such that} \qquad L^{(n_j)} \in P_{j0}^0 \cap P_{j^*,\,n_1+n_2}\,,
$$

the coefficients being determined by $\beta_{jn_j} = \gamma_{j,\,n_1+n_2}$ [*cf.* (2.1-3), (4.2-6, 7), (4.4-2), (4.5-2)] *with* $\gamma_{j,\,n_1+n_2}$ *as in* (4.5-3) *and* a_{0n_j} *given by* (4.5-5).

(ii) *An integral-free representation of solutions* w_{jn_j} *in terms of the preceding functions* $\xi_{j\mu}$, $\eta_{j\mu}$ *and two functions* F_j, $F_{j^*} \in C^\omega(\Omega)$ *of the variables* z_j *and* z_{j^*}, *respectively, is given by* (4.5-4).

(b) *Under the additional assumption that* $w_{jn_j} = D_j^{(n_j)}w$ *with* $h_{jr} \neq 0$, *where* $r = 0,\cdots,n_{j-1}$ [*cf. Theorem* 4.3-1b], *there exists an equation*

$$
L^{(0)}w = 0 \qquad \text{such that} \qquad L^{(0)} \in P_{jn_j}^0 \cap P_{j^*n_{j^*}}\,.
$$

The coefficients of $L^{(0)}$ *are given by* (4.5-8), *and solutions of* $L^{(0)}w = 0$ *are represented by* (4.5-6). *Corresponding integral operators of class P for Type II representations have the kernels defined by*

$$\tilde{k}_j(z,t) = \sum_{m=0}^{n_j} \frac{2^{2m}m!}{(2m)!} A_{j,n_j-m}(z)(z_j - s_j)^m t^{2m}$$

and

$$\tilde{k}_{j^*}(z,t) = \sum_{m=0}^{n_{j^*}} \frac{2^{2m}m!}{(2m)!} A_{j^*,n_{j^*}-m}(z)(z_{j^*} - s_{j^*})^m t^{2m}.$$

Here

$$(4.5\text{-}9)\qquad \begin{aligned} A_{j\rho} &= (-1)^\rho \det(\varphi_{j\rho\mu\nu}), & \rho &= 0,\cdots,n_j,\\ A_{j^*r} &= (-1)^{n_j+1+r} \det(\psi_{j^*r\mu\nu}), & r &= 0,\cdots,n_{j^*}. \end{aligned}$$

The determinants have order $n_1 + n_2 + 1$. *For* $\mu = 1,\cdots,n_1+n_2+1$, *their elements are*

$$\varphi_{j\rho\mu\nu} = \begin{cases} D_j^{\nu-1}\xi_{j\mu} & \text{if } \nu = 1,\cdots,\rho \\ D_j^{\nu}\xi_{j\mu} & \text{if } \nu = \rho+1,\cdots,n_j \\ D_{j^*}^{\nu-n_j-1}\eta_{j\mu} & \text{if } \nu = n_j+1,\cdots,n_1+n_2+1 \end{cases}$$

and

$$\psi_{j^*r\mu\nu} = \begin{cases} D_j^{\nu-1}\xi_{j\mu} & \text{if } \nu = 1,\cdots,n_j+1 \\ D_{j^*}^{\nu-n_j-2}\eta_{j\mu} & \text{if } \nu = n_j+2,\cdots,n_j+1+r \\ D_{j^*}^{\nu-n_j-1}\eta_{j\mu} & \text{if } \nu = n_j+2+r,\cdots,n_1+n_2+1. \end{cases}$$

Investigating the conditions under which the determinant vanishes identically, we immediately obtain a uniqueness theorem for integral-free representations:

UNIQUENESS OF SOLUTIONS

4.5-2 Theorem

The integral-free form (4.5-6) *of solutions of* $L^{(0)}w = 0$ *with coefficients* (4.5-8) *given by* $A_{j\rho}$ *and* A_{j^*r} *as in* (4.5-9) *is unique up to an additive term*

$$w_0 = \sum_{\rho=0}^{n_j} A_{j\rho} \sum_{\mu=1}^{n_1+n_2+1} \kappa_\mu D_j^\rho \xi_{j\mu} + \sum_{r=0}^{n_{j^*}} A_{j^*r} \sum_{\mu=1}^{n_1+n_2+1} \kappa_\mu D_{j^*}^r \eta_{j\mu}$$

with arbitrary complex constants κ_μ.

Applications of Theorems 4.5-1 and 4.5-2 are included in the next section.

4.6 APPLICATIONS

In this section we demonstrate how the construction principle given by Theorem 4.5-1 can be utilized in practice. As a typical application, we derive from it the results of Theorem 4.4-2 in the self-adjoint case, corresponding to $n_1 = n_2 = n \in \mathbb{N}_0$, choosing $j = 1$ without loss of generality.

APPLICATION OF THEOREM 4.5-1

We start from an ordinary differential equation of the form (4.5-1), namely,

$$(4.6\text{-}1) \qquad N_1 y_1 = D_2^{2n+1} y_1 = 0 .$$

For this equation, $\tilde{\alpha}_{1r} = 0$ when $r = 0, \cdots, 2n = n_1 + n_2$ in (4.5-1). A basis of solutions is

$$(4.6\text{-}2) \qquad \eta_{1\mu}(z_2) = z_2^{\mu-1}, \qquad \mu = 1, \cdots, 2n+1 .$$

Furthermore, we can choose

$$(4.6\text{-}3) \qquad \xi_{1\mu}(z_1) = (-1)^{\mu-1} z_1^{\mu-1}, \qquad \mu = 1, \cdots, 2n+1 .$$

We now insert (4.6-2, 3) into (4.5-9) and observe that

$$(4.6\text{-}4) \qquad A_{2r}(z_1, z_2) = (-1)^{n+1+r} A_{1r}(-z_2, -z_1), \qquad r = 0, \cdots, n .$$

Since these determinants are zero when $z_1 = -z_2$, $n \geq 1$, we conclude that A_{1r} and A_{2r} are functions of $z_1 + z_2$. Hence for obtaining their form, it suffices to study the simplified case $z_2 = 0$.

Accordingly, we first consider $A_{1r}(z_1, 0)$. Without changing the order of the last $n+1$ columns, we move these columns so that they become the first columns of the rearranged determinant. The latter we develop by these first $n+1$ columns, whose elements are $(\nu - 1)!$ $(\nu = 1, \cdots, n+1)$ in the principal diagonal and zero otherwise. We then obtain

$$(4.6\text{-}5) \qquad A_{1r}(z_1, 0) = (-1)^r 0!1! \cdots n! \det S_r(\rho_{\kappa\lambda}), \qquad r = 0, \cdots, n .$$

Here S_r is the operator that maps $(\rho_{\kappa\lambda})$ onto the matrix obtained by deleting the $(r+1)$th column from $(\rho_{\kappa\lambda})$, where

$$\rho_{\kappa\lambda} = (-1)^{n+\kappa} z_1^{n+1+\kappa-\lambda} (n+\kappa)!/(n+1+\kappa-\lambda)! ,$$

$$\kappa = 1, \cdots, n; \quad \lambda = 1, \cdots, n+1 .$$

In (4.6-5) we take out $(-1)^{n+\kappa}z_1^{\kappa}(n+\kappa)!$ as common factors from the rows $(\kappa = 1,\cdots,n)$ and, similarly, $z_1^{n+1-\lambda}/(2n+1-\lambda)!$ from the columns where $\lambda = 1,\cdots,n+1$, $\lambda \neq r+1$. Then we have

$$\det S_r(\rho_{\kappa\lambda}) = (-1)^{n(3n+1)/2}\frac{(2n-r)!}{n!}z_1^{n^2+r}\det S_r(\tilde{\rho}_{\kappa\lambda})$$

where

$$\tilde{\rho}_{\kappa\lambda} = \frac{(2n+1-\lambda)!}{(n+1+\kappa-\lambda)!}, \quad \kappa = 1,\cdots,n; \quad \lambda = 1,\cdots,n+1.$$

The remaining determinant we can readily evaluate by first rewriting it as an nth order Vandermonde determinant. Indeed, by linearly combining the κth row and the subsequent rows, we can obtain the new κth row

$$(2n-1)^{n-\kappa},\cdots,(n-1)^{n-\kappa};$$

here we have to delete the element $(2n-r-1)^{n-\kappa}$, because of S_r. The usual form of a Vandermonde determinant is obtained by transposition and interchange of some rows and columns:

$$\det S_r(\tilde{\rho}_{\kappa\lambda}) = \det S_r^*(\rho_{\lambda\kappa}^*).$$

In this equation, S_r^* denotes the operation of deleting the $(r+1)$th *row*, and

$$\rho_{\lambda\kappa}^* = (n-2+\lambda)^{\kappa-1}, \quad \lambda = 1,\cdots,n+1; \quad \kappa = 1,\cdots,n.$$

This yields

$$\det S_r^*(\rho_{\lambda\kappa}^*) = \frac{1!2!\cdots n!}{r!(n-r)!}.$$

Inserting the preceding expressions into (4.6-5), we obtain

$$A_{1r}(z_1,0) = (-1)^{r+n(3n+1)/2}\left(\prod_{\nu=0}^{n-1}[\nu!(\nu+1)!]\right)\frac{(2n-r)!}{r!(n-r)!}z_1^{n^2+r}.$$

Hence, as previously described,

$$\text{(4.6-6a)} \quad A_{1r}(z) = \left[K(n)(z_1+z_2)^{n(n+1)}\right]\frac{(2n-r)!}{r!(n-r)!}\left(\frac{-1}{z_1+z_2}\right)^{n-r}$$

where

$$\text{(4.6-6b)} \qquad K(n) = (-1)^{n(n+1)/2}\prod_{\nu=0}^{n-1}[\nu!(\nu+1)!] = \mathit{const} \neq 0.$$

Also, by (4.6-4),

$$A_{2r}(z) = -A_{1r}(z), \qquad r = 0, \cdots, n. \tag{4.6-6c}$$

Now we are able to determine the coefficients of $L^{(0)}$ and to give an integral-free form of solutions according to (4.5-8) and (4.5-6), respectively, and polynomial kernels as in Theorem 4.5-1. From (4.6-6) and (4.5-8) we have

$$a_{10}(z) = a_{20}(z) = -n(n+1)(z_1 + z_2)^{-1}$$

$$a_{00}(z) = n^2(n+1)^2(z_1 + z_2)^{-2}.$$

Using Remark 2.1-1 we obtain two further equations and corresponding solutions (see below).

Our results can be summarized as follows.

4.6-1 Example

If we choose (4.6-1, 2, 3), then solutions of the equations

(4.6-7)

$$\text{(a)} \quad L^{(0)}w = D_1 D_2 w - \frac{n(n+1)}{z_1 + z_2}(D_1 w + D_2 w) + \frac{n^2(n+1)^2}{(z_1 + z_2)^2} w = 0$$

$$\text{(b)} \quad L_1^{(0)}u = L_2^{(0)}u = D_1 D_2 u - \frac{n(n+1)}{(z_1 + z_2)^2} u = 0$$

with $n \in \mathbb{N}_0$ are obtained from Theorem 4.5-1 in the form

(4.6-8)

$$\tilde{w}(z) = \left[K(n)(z_1 + z_2)^{n(n+1)}\right]\tilde{u}(z)$$

$$\tilde{u} = \tilde{T}_1 F_1 + \tilde{T}_2 F_2$$

$$\tilde{T}_j = (-1)^{j+1} \sum_{m=0}^{n} \frac{(2n-m)!}{m!(n-m)!} \left(\frac{-1}{z_1 + z_2}\right)^{n-m} D_j^m, \qquad j = 1, 2$$

with $K(n)$ given by (4.6-6b). Furthermore, (4.5-9) and (4.6-6) now yield corresponding polynomial kernels for Type II representations:

(4.6-9)

$$\tilde{k}_j(z, t) = K(n)(z_1 + z_2)^{n(n+1)} k_j(z, t) \qquad \text{in case (a)},$$

$$\tilde{k}_j(z, t) = k_j(z, t) \qquad \text{in case (b)},$$

$$k_j(z, t) = (-1)^{j+1} \sum_{m=0}^{n} \binom{n+m}{2m} \left(\frac{-4(z_j - s_j)}{z_1 + z_2}\right)^m t^{2m}, \qquad j = 1, 2.$$

These results agree with those obtained by Theorem 4.4-2, as can be seen by taking $n_1 = n_2 = n$, replacing $\tilde{T}_2$ by $-\tilde{T}_2$ and setting

$$\tilde{f}_1 = F_1, \qquad \tilde{f}_2 = -F_2$$

in (4.6-8). Lemma 4.2-7 permits the extension of our earlier results to more general equations, for instance, to the equations in Theorem 4.4-2 with $n_1 = n_2 = n$.

APPLICATION OF THEOREM 4.5-2

Theorem 4.5-2 can be applied in a similar fashion and yields the following

4.6-2 Example

Let a solution of (4.6-7b) be given by

$$\tilde{u} = \tilde{T}_1 \tilde{f}_1 - \tilde{T}_2 \tilde{f}_2$$

with $\tilde{T}_1, \tilde{T}_2$ as in (4.6-8). Then the most general pair $(\tilde{f}_1^*, \tilde{f}_2^*)$ that is mapped by $(\tilde{T}_1, -\tilde{T}_2)$ onto the same $\tilde{u}$ is

$$\tilde{f}_1^*(z_1) = \tilde{f}_1(z_1) + \sum_{\nu=0}^{2n} \kappa_\nu (-1)^\nu z_1^\nu, \qquad \tilde{f}_2^*(z_2) = \tilde{f}_2(z_2) - \sum_{\nu=0}^{2n} \kappa_\nu z_2^\nu$$

where the κ_ν's are arbitrary complex constants. In this way we have also obtained another proof of an earlier result by Bauer (cf. Bauer [1966b], Theorem 10).

ANOTHER APPLICATION OF THEOREM 4.5-1

4.6-3 Example

We choose $n_1 = 1$, $n_2 = 2$, $j = 1$, and

$$\zeta_{1\mu} = z_1^\mu, \qquad \eta_{1\mu} = z_2^{\mu-1}, \qquad \mu = 1, \cdots, 4.$$

From Theorem 4.5-1 we then obtain

$$L^{(0)}w = D_1 D_2 w + \frac{3}{z_1 - z_2} D_1 w + \frac{2(z_2 - 3z_1)}{z_1(z_1 - z_2)} D_2 w + \frac{6(z_2 - 3z_1)}{z_1(z_1 - z_2)^2} w = 0.$$

Let $s \in \Omega$, where Ω denotes a simply connected domain in $\mathbb{C}^2$ not containing z such that $z_1 = z_2$. Then, also by Theorem 4.5-1, corresponding differential operators for Type II representations are

$$\tilde{T}_1 = 2(z_1 - z_2)^2[z_2 - 4z_1 + z_1(z_1 - z_2)D_1]$$

$$\tilde{T}_2 = z_1^2(z_1 - z_2)^2[6 + 4(z_1 - z_2)D_2 + (z_1 - z_2)^2 D_2^2].$$

Polynomial kernels for these representations are given by

$$\tilde{k}_1(z, t) = 2z_1(z_1 - z_2)^3 + 4(z_1 - z_2)^2(z_2 - 4z_1)(z_1 - s_1)t^2$$

$$\tilde{k}_2(z, t) = z_1^2(z_1 - z_2)^4 + 8z_1^2(z_1 - z_2)^3(z_2 - s_2)t^2$$
$$+ 8z_1^2(z_1 - z_2)^2(z_2 - s_2)^2 t^4.$$

4.7 *SELF-ADJOINT EQUATIONS*

It is of practical importance to note that *self-adjoint* equations (2.1-1), which are of the form

$$Lw = D_1 D_2 w + a_0 w = 0 \tag{4.7-1}$$

admit a special approach with respect to the existence of polynomial kernels for Type II representations. This additional method to be developed in the present section is based on the idea of a direct representation of the coefficient a_0 by means of the functions h_{jr} occurring in (4.2-4).

In connection with adjoint equations we shall use asterisks as follows.

4.7-1 Notations

Let L^* denote the formal adjoint of L in (2.1-1), so that

$$L^* w^* = D_1 D_2 w^* + a_1^* D_1 w^* + a_2^* D_2 w^* + a_0^* w^* = 0 \tag{4.7-2a}$$

where

(4.7-2b)

$$a_1^* = -a_1, \qquad a_2^* = -a_2, \qquad a_0^* = a_0 - D_1 a_1 - D_2 a_2.$$

To the notations in 4.2-1 we attach an asterisk to mean that a_0, a_1, a_2 are to be replaced by a_0^*, a_1^*, a_2^*, respectively; in particular,

$$D_{jr}^* = D_{j*} + a_{j,r-1}^*, \qquad r \in \mathbb{N},$$

$$h_{jr}^* = D_j a_{jr}^* + a_{1r}^* a_{2r}^* - a_{0r}^*, \qquad r \in \mathbb{N}_0;$$

similarly for $L^{(0)*} = L^*$, w_{jr}^*, and so on.

Obviously, the results of Sec. 4.2 also hold for equation (4.7-2). Thus, if $h_{j0}^* \neq 0$, then for $r = 1$ we have

$$L^{(1)*} w_{j1}^* = 0\,, \qquad w_{j1}^* = D_{j1}^* w^*\,, \qquad D_{j1}^* = D_{j*} + a_{j0}^*$$

and, according to (4.2-10) and (4.7-2b),

$$\begin{aligned} a_{j1}^* &= a_{j0}^* - D_{j*}(\log h_{j0}^*) = -a_{j0} - D_{j*}(\log h_{j*0}) \\ a_{j*1}^* &= a_{j*0}^* = -a_{j*0} \\ a_{01}^* &= a_{00}^* - D_j a_{j0}^* + D_{j*} a_{j*0}^* - a_{j*0}^* D_{j*}(\log h_{j*0}^*) \\ &= a_{00} - 2D_{j*} a_{j*0} + a_{j*0} D_{j*}(\log h_{j0})\,. \end{aligned}$$

Here the equality in the last line follows from

$$h_{j0}^* = h_{j*0} = D_j a_{j0}^* + a_{10}^* a_{20}^* - a_{00}^*\,, \qquad j = 1, 2\,.$$

From the preceding representations of h_{jr}^* and of the coefficients a_{j1}^*, a_{j*1}^*, a_{01}^* we obtain

$$h_{j1}^* = h_{j*1}\,, \qquad j = 1, 2\,.$$

This shows that with respect to h_{j1} and h_{j1}^*, the application of D_{j1}^* to (4.7-2) has the same effect as the application of D_{j*1} to (2.1-1). From Theorem 4.2-4 and Remark 4.2-5 we thus obtain the following result.

4.7-2 Lemma

For $n_j \in \mathbb{N}_0$ and fixed $j = 1$ or 2 we have

$$L \in P_{jn_j}^0 \qquad \text{if and only if} \qquad L^* \in P_{j*n_j}^0\,.$$

EXISTENCE THEOREM FOR SELF-ADJOINT EQUATIONS

For self-adjoint equations, Lemma 4.7-2 implies:

4.7-3 Theorem

Let L be defined by (4.7-1) and $n \in \mathbb{N}_0$. Then:

(a) *Equivalent are $L \in P_{1n}^0$, $L \in P_{2n}^0$, and $L \in P_{1n}^0 \cap P_{2n}^0$.*

(b) *$L \in P_{1n}^0 \cap P_{2n}^0$ if and only if there exist functions h_{jr} with $h_{1n} = 0$ but $h_{1r} \neq 0$ for $r = 0, \cdots, n-1$, such that $L^{(n)} D_1^{(n)} w = 0$ follows from $Lw = 0$,* *and the coefficient a_0 of L satisfies the equation*

$$a_0 = -D_1D_2\left[\log\left((-a_0)^n\prod_{r=1}^{n-1}h_{1r}^{n-r}\right)\right]. \tag{4.7-3}$$

Proof. Part (a) is obvious by Lemma 4.7-2 and $a_1 = a_2 = 0$. Part (b) follows by combining (a) and Theorem 4.2-4. Finally, the formula for a_0 results from (4.2-8), $h_{1n} = 0$, and $h_{10} = h_{20} = -a_0$. ■

APPLICATIONS OF THEOREM 4.7-3

It is remarkable that we can use Theorem 4.7-3b to give the following motivation for the equation investigated by Bauer [1971].

4.7-4 Example

(a) We consider (4.4-12) with $n_1 = n_2 = n$, where $n \in N$, so that we have (4.7-1) with

$$a_0(z) = -n(n+1)(z_1+z_2)^{-2}. \tag{4.7-4}$$

In this case, from (4.4-11) we have

$$h_{1r}(z) = \frac{(n-r)(n+r+1)}{(z_1+z_2)^2} = -\frac{(n-r)(n+r+1)}{n(n+1)}a_0(z) \tag{4.7-5}$$

where $r = 0,\cdots,n-1$. Obviously, (4.7-4) and (4.7-5) satisfy (4.7-3).

(b) Part (a) suggests that instead of $(z_1+z_2)^2$ we substitute $F(z)$ into (4.7-4) and (4.7-5). This yields

$$a_0(z) = -n(n+1)/F(z)$$

$$h_{1r}(z) = (n-r)(n+r+1)/F(z), \qquad r = 0,\cdots,n-1.$$

Then equation (4.7-3) is equivalent to

$$FD_1D_2(\log F) + 2 = 0, \tag{4.7-6}$$

the condition appearing in Bauer [1971]. If we now set

$$F = e^{-\Psi}$$

then (4.7-6) takes the form

$$D_1D_2\Psi = 2e^{\Psi}. \tag{4.7-7}$$

Up to the constant factor 2, this is **Liouville's equation**, whose solutions are well known. It follows that the solution of (4.7-7) is given by

$$\Psi(z) = \log \frac{D_1\psi_1(z_1) D_2\psi_2(z_2)}{[\psi_1(z_1) + \psi_2(z_2)]^2}$$

the functions ψ_1 and ψ_2 being arbitrary but such that F becomes holomorphic in a prescribed domain Ω. Hence we have again obtained the self-adjoint case of the equations in Theorem 4.4-2.

(c) Clearly, the coefficient a_0 in part (b) can equally well be obtained by applying the transformation (b) in Lemma 4.2-7 to the present part (a). This again yields

$$a_0(z) = -n(n+1)\frac{D_1\psi_1(z_1) D_2\psi_2(z_2)}{[\psi_1(z_1) + \psi_2(z_2)]^2}$$

and, by (4.7-5) and the formulas in the proof of Lemma 4.2-7,

$$h_{1r}(z) = (n-r)(n+r+1)\frac{D_1\psi_1(z_1) D_2\psi_2(z_2)}{[\psi_1(z_1) + \psi_2(z_2)]^2}$$

where $r = 0, \cdots, n-1$. Hence we see that (4.7-3) remains valid.

ANOTHER CONSTRUCTION OF EQUATIONS AND SOLUTIONS

In addition to the approach given by Lemma 4.2-7, there is still another possibility, namely, the construction of equations and corresponding representations of solutions from a known equation and its solutions. In precise terms, we have the following.

4.7-5 Theorem

Let W_0 be a solution of a self-adjoint equation

$$L^{(0)}w = D_1D_2w + a_0w = 0$$

involving two functions F_1, F_2 such that F_j depends only on z_j $(j = 1, 2)$ and is arbitrary, except for regularity as previously characterized. Let $\omega_0 \neq const$ be obtained from W_0 by selecting a specific pair (F_1, F_2). Then

$$(4.7\text{-}8) \quad W_1 = \frac{1}{\omega_0}\int[(\omega_0 D_1W_0 - W_0D_1\omega_0)\,dz_1 - (\omega_0D_2W_0 - W_0D_2\omega_0)\,dz_2]$$

is a solution of

$$(4.7\text{-}9)\quad \tilde{L}^{(1)}W = D_1D_2W + \tilde{a}_{01}W = 0\,, \qquad \tilde{a}_{01} = -\omega_0 D_1 D_2(1/\omega_0)\,.$$

Proof. We substitute (4.7-8) into (4.7-9) and observe that the assumptions $L^{(0)}W_0 = 0$ and $L^{(0)}\omega_0 = 0$ imply the relation

$$\omega_0 D_1 D_2 W_0 - W_0 D_1 D_2 \omega_0 = \omega_0 (D_1 D_2 W_0 + a_0 W_0) = 0\,.$$

The assertion now follows by straightforward calculation. ■

If we apply this theorem n times to the equation $L^{(0)}w = 0$ given before, we obtain an equation of the form

$$(4.7\text{-}10)\quad \begin{aligned} &\tilde{L}^{(n)}W = D_1D_2W + \tilde{a}_{0n}W = 0 \\ &\tilde{a}_{0n} = 2D_1D_2\left(\log \prod_{r=0}^{n-1} \omega_r\right) + a_0\,. \end{aligned} \qquad n \in \mathbb{N}\,,$$

A corresponding solution is (4.7-8) with W_0 and ω_0 replaced by W_{n-1} and ω_{n-1}, respectively. Here W_r and ω_r, $r = 0,\cdots, n-1$, are assumed to be solutions of the equations

$$(4.7\text{-}11)\quad \begin{aligned} &\tilde{L}^{(r)}W = D_1D_2W + \tilde{a}_{0r}W = 0\,, \qquad r = 0,\cdots, n-1\,, \\ &\tilde{a}_{0r} = -\omega_{r-1}D_1D_2(1/\omega_{r-1})\,, \qquad r \geq 1\,; \quad \tilde{a}_{00} = a_0\,. \end{aligned}$$

To verify this assertion, we merely have to show that $\tilde{a}_{0n}$ has the form given in (4.7-10). This we can do as follows. From the equation $\tilde{L}^{(r)}\omega_r = 0$ we have

$$-\tilde{a}_{0r} = \frac{1}{\omega_r} D_1 D_2 \omega_r$$

and, by the definition of $\tilde{a}_{0,r+1}$ in (4.7-11), also

$$\tilde{a}_{0,r+1} = -\omega_r D_1 D_2\left(\frac{1}{\omega_r}\right) = \frac{1}{\omega_r} D_1 D_2 \omega_r - \frac{2}{\omega_r^2}(D_1\omega_r)(D_2\omega_r)$$

where $r = 0,\cdots, n-1$. Hence we obtain

$$2D_1D_2\left(\log \prod_{r=0}^{n-1}\omega_r\right) = \sum_{r=0}^{n-1}(\tilde{a}_{0,r+1} - \tilde{a}_{0r}) = \tilde{a}_{0n} - a_0\,.$$

This proves the second formula in (4.7-10).

CONSTRUCTION USING THE LAPLACE EQUATION

Let us illustrate this procedure and Theorem 4.7-5 by starting from the Laplace equation and constructing equations that admit polynomial kernels of arbitrary degree.

4.7-6 Example

Consider the Laplace equation

$$L^{(0)}w = D_1 D_2 w = 0$$

and choose

$$\omega_r(z) = (z_1 + z_2)^{r+1}, \qquad r = 0,\cdots, n-1 \tag{4.7-12}$$

where $n \in \mathbb{N}$. Also consider the equations

$$\tilde{L}^{(r)}W = D_1 D_2 W + \tilde{a}_{0r} W = 0, \qquad r = 0,\cdots, n-1 \tag{4.7-13a}$$

where

$$\tilde{a}_{00} = a_0 = 0$$

$$\tilde{a}_{0r}(z) = -(z_1 + z_2)^r D_1 D_2 (z_1 + z_2)^{-r} \tag{4.7-13b}$$

$$= -r(r+1)(z_1 + z_2)^{-2}.$$

We claim that a general solution W_r of this equation in the sense of Theorem 4.7-5 is given by

$$W_r(z) = \sum_{j=1}^{2} (-1)^{j+1} \sum_{m=0}^{r} \frac{(2r-m)!}{m!(r-m)!} \left(\frac{-1}{z_1 + z_2} \right)^{r-m} D_j^m F_{rj}(z_j) \tag{4.7-14}$$

where $r = 0,\cdots, n-1$. Let us prove this by induction, using Theorem 4.7-5.

The case $r = 0$ is obvious. Suppose now that (4.7-14) represents a general solution of (4.7-13) with $r - 1$ (≥ 0) instead of r. In order to be able to apply (4.7-8, 9), we have to show that

$$\omega_{r-1}(z) = (z_1 + z_2)^r$$

is a special case of W_{r-1}. In fact, ω_{r-1} is obtained by setting

$$F_{r-1,j}(z_j) = (-1)^{j+1} z_j^{2r-1}/(r-1)!, \qquad j = 1,2. \tag{4.7-15}$$

Inserting this into W_{r-1} as defined by (4.7-14), we have

$$W_{r-1}(z) = \sum_{m=0}^{r-1} \hat{A}_{r-1,m}(z_1 + z_2)^{m-r+1}\left(z_1^{2r-1-m} + z_2^{2r-1-m}\right)$$

$$= (z_1 + z_2)^{-r+1}\left[\sum_{\mu=0}^{r-1} \sum_{m=\mu}^{r-1} \hat{A}_{r-1,m}\binom{m}{\mu} z_1^{2r-1-\mu} z_2^{\mu}\right.$$

$$\left. + \sum_{\mu=r}^{2r-1} \sum_{m=2r-1-\mu}^{r-1} \hat{A}_{r-1,m}\binom{m}{2r-1-\mu} z_1^{2r-1-\mu} z_2^{\mu}\right]$$

where

$$\hat{A}_{r-1,m} = \frac{(-1)^{r-1-m}(2r-1)!}{m!(r-1-m)!(r-1)!(2r-1-m)}, \qquad m = 0,\cdots,r-1.$$

By using a well-known formula for sums of products of binomial coefficients [cf. Netto [1927], p. 251, (25)] we find that

$$\sum_{m=t}^{r-1} \hat{A}_{r-1,m}\binom{m}{t} = \binom{2r-1}{t}, \qquad t = 0,\cdots,r-1.$$

Thus, W_{r-1} with the special $F_{r-1,j}$ given by (4.7-15) takes the form

$$W_{r-1}(z) = (z_1 + z_2)^{-r+1} \sum_{\mu=0}^{2r-1} \binom{2r-1}{\mu} z_1^{2r-1-\mu} z_2^{\mu} \tag{4.7-16}$$

$$= (z_1 + z_2)^{r} = \omega_{r-1}(z).$$

This shows that ω_{r-1} is a special case of W_{r-1} in (4.7-14). The final step of our inductive proof consists in showing that

$$\tilde{W}_r = W_r \tag{4.7-17}$$

where $\tilde{W}_r$ is obtained from ω_{r-1} and W_{r-1}, as given by (4.7-8), and W_r is given by (4.7-14). Substituting W_{r-1} and ω_{r-1}, as resulting from (4.7-14) and (4.7-16), respectively, we have

$$\tilde{W}_r = \tilde{W}_{r1} + \tilde{W}_{r2}$$

where

$$\tilde{W}_{rj}(z) = (z_1 + z_2)^{-r} \int \Bigg\{ \sum_{m=0}^{r-1} \hat{B}_{r-1,m}(z_1 + z_2)^m \big[(m + 1 - 2r) D_j^m F_{r-1,j}(z_j) + (z_1 + z_2) D_j^{m+1} F_{r-1,j}(z_j)\big]\, dz_j - \sum_{m=0}^{r-1} \hat{B}_{r-1,m}(m + 1 - 2r)(z_1 + z_2)^m D_j^m F_{r-1,j}(z_j)\, dz_{j*} \Bigg\},$$

$$\hat{B}_{km} = \frac{(2k - m)!(-1)^{k-m}}{m!(k - m)!}.$$

If we set

$$F_{rj}(z_j) = (-1)^{j+1} \int F_{r-1,j}(z_j)\, dz_j$$

we see by straightforward calculation that $\tilde{W}_{rj}$ takes the form

$$\tilde{W}_{rj}(z) = (z_1 + z_2)^{-r}(-1)^{j+1} \times$$

$$\times \int \Bigg\{ \sum_{m=0}^{r} \hat{B}_{rm} \frac{r}{2r - m}(z_1 + z_2)^m D_j^{m+1} F_{rj}(z_j)\, dz_j + \sum_{m=0}^{r-1} \hat{B}_{rm} \frac{m - r}{2r - m}(z_1 + z_2)^m D_j^{m+1} F_{rj}(z_j)\, dz_{j*} \Bigg\}, \quad j = 1, 2. \tag{4.7-18}$$

On the other hand, setting $\lambda_r = \omega_{r-1} W_r$ with ω_{r-1} and W_r as in formulas (4.7-16) and (4.7-14), we can rewrite W_r in the form

$$W_r = W_{r1} + W_{r2}$$

where

$$W_{rj} = \frac{1}{\omega_{r-1}} \int \big[D_1 \lambda_{rj}\, dz_1 + D_2 \lambda_{rj}\, dz_2\big] \tag{4.7-19}$$

with

$$\lambda_{rj}(z) = (-1)^{j+1} \sum_{m=0}^{r} \hat{B}_{rm}(z_1 + z_2)^m D_j^m F_{rj}(z_j), \qquad j = 1, 2.$$

Obviously, the right-hand sides of (4.7-18) and (4.7-19) are equal. We thus have $\tilde{W}_{rj} = W_{rj}$, $j = 1, 2$. This is (4.7-17), and the proof of the assertion involving (4.7-14) is complete. (It is interesting that this also follows from Example 4.6-1, as the reader may show.)

Example 4.7-6 shows that from Theorem 4.7-5 we now have obtained a further construction principle for equations and corresponding kernels. In this principle we assume that an equation and a corresponding polynomial kernel are given.

To avoid misunderstandings, we should emphasize that in our present procedure the degree of minimal kernels of the equation to be constructed depends to a large extent on the special solution ω in (4.7-8). Let us illustrate this by a typical example.

4.7-7 Example

We start from (4.7-13) with $r = 1$; thus,

$$\tilde{L}^{(1)}w = D_1 D_2 w - 2(z_1 + z_2)^{-2} w = 0. \tag{4.7-20}$$

We then use (4.7-14), that is,

$$W_1(z) = \sum_{j=1}^{2} (-1)^{j+1} \left[D_j F_{1j}(z_j) - 2(z_1 + z_2)^{-1} F_{1j}(z_j) \right]. \tag{4.7-21}$$

In the previous example we have seen that the choice

$$F_{1j}(z_j) = (-1)^{j+1} z_j^3, \qquad j = 1, 2$$

leads to

$$\omega_1(z) = (z_1 + z_2)^2.$$

By Theorem 4.7-5, this entails $W = W_2$ and

$$\tilde{\tilde{L}}^{(1)}W = \tilde{L}^{(2)}W = D_1 D_2 W - 6(z_1 + z_2)^{-2} W = 0. \tag{4.7-22}$$

Hence $\tilde{\tilde{L}}^{(1)} \in P_{12}^0 \cap P_{22}^0$.

On the other hand, the choice

$$F_{1j}(z_j) = (-1)^{j+1} z_j^2/4, \qquad j = 1, 2$$

yields

$$\omega_1(z) = z_1 z_2 (z_1 + z_2)^{-1}.$$

From this and (4.7-21), by setting

$$F_{0j}(z_j) = (-1)^{j+1}\left[D_j F_{1j}(z_j) - 2z_j^{-1}F_{1j}(z_j)\right]$$

we now obtain $W = W_0$ and

$$\tilde{\tilde{L}}^{(1)}W = \tilde{L}^{(0)}W = D_1 D_2 W = 0\,. \tag{4.7-23}$$

Consequently, $\tilde{\tilde{L}}^{(1)} \in P_{10}^0 \cap P_{20}^0$.

Additional References

Bauer [1966a, 1966b, 1976a]
Bauer–Florian [1976]
Bauer–Jank [1971]
Bauer–Peschl [1966]
Bauer–Ruscheweyh [1980]
Berglez [1979, 1980]
Darboux [1972]
Florian–Jank [1971]
Heersink [1976b, 1983]
Jank [1973]
Kracht [1974a, 1974b]
Kracht–Kreyszig [1969, 1975, 1979]
Kracht–Kreyszig–Schröder [1982]
Kracht–Schröder [1973]
Kreyszig [1968, 1971, 1975]
Püngel [1978, 1984]
Tricomi [1968]
Watzlawek [1971b, 1972, 1973]

Chapter Five

Further Closed-Form Kernels

Operators whose kernels are obtained as power series solutions of the kernel equation (2.3-2a) may have the disadvantage that they yield only *local* solutions of a given partial differential equation. There are three exceptions:

1. A series terminates (as discussed in Chaps. 3 and 4).
2. A series does not terminate, but can still be expressed in terms of finitely many known functions. Such a kernel will be called a kernel of ***closed form*** or *finite form*.
3. The Riemann function of the equation is known globally. We discuss this, along with relations to the *complex* Riemann function, in the next chapter.

The situation just described suggests constructing closed-form kernels since the corresponding operators are suitable for obtaining *global* solutions.

In practice, for obtaining such a kernel $\tilde{k}_j$, one starts from an expression for k_j which involves finitely many undetermined coefficient functions (functions of some or all of the independent variables in the given equation). One then attempts to determine the coefficient functions by substituting the expression for k_j into the kernel equation. This approach requires some ingenuity and a deeper insight into the form of an equation in order to be able to discover what ansatz for the kernel might work and lead to kernels that are general enough for representing sufficiently large independent families of particular solutions. This makes the method mathematically attractive, its practical importance being obvious, in particular in connection with numerical work on boundary value problems.

In more detail, the situation presents itself as follows. By assuming a special form for the kernel (involving undetermined functions), just as in the case of class P operators, we generally impose conditions on the coefficients of the equation to be solved since these coefficients appear in the kernel equation. It might well be that we start from a form for k_j which is so special that we do not obtain a nontrivial solution of the kernel equation at all. But even if we do get a solution of (2.3-2a), it is not obvious that equations of importance will admit the corresponding operator for representing solutions. Hence in each case we are confronted with the problem of determining the class of equations (2.1-1) or (2.1-2) that will admit a certain kind of kernel and the task of finding out whether this class contains equations of practical interest.

After having studied kernels for class P operators in detail in the preceding chapters, in the present chapter we shall consider three further classes of closed-form kernels for Bergman operators. The first class consists of the so-called exponential operators, which have a number of remarkable and interesting properties (see Secs. 5.1 and 5.2). The second class (Sec. 5.3) is obtained by combining certain class P and exponential operators The combination of both types of operators in full generality is still an open problem which seems quite difficult. Finally, in Sec. 5.4, we shall consider operators whose kernels are rational functions.

We mention in advance that in this chapter we profit by our ***generalized definition of the Bergman operator*** T_j in (2.2-8) and Def. 2.2-1, and of the Representation Theorem 2.3-1. In the following sections of this chapter, for simplicity, we choose $\sigma_j = 0$ but assume $\rho_j \neq 0$ in general. For if we chose $\sigma_j = \rho_j$, where ρ_j is the center of the series expansion of $f_j \in C^{\omega}(\tilde{\Omega}_j)$, it would be possible that the solution obtained by a Bergman operator with a kernel of the aforementioned form is not holomorphic at $\sigma_j = \rho_j$. This is caused by the fact that, for instance, a kernel of an exponential operator may have a singularity at $z_j = \sigma_j$. In contrast to the kernel of the first kind and class P kernels, here singularities of the kernels may occur that are not caused by singularities of the coefficients of equation (2.1-1). According to the definition originally given by Bergman [1937b, 1971], such a kernel would not be a Bergman kernel, and thus the classes of equations (2.1-1) admitting the preceding closed-form kernels for Bergman operators would be restricted considerably. For instance, if we required s to be the point of reference of the operator (that is, $\rho_j = \sigma_j = s_j$), not even an equation (2.1-1) with constant coefficients would admit an exponential operator. For in this case the kernel would not be holomorphic at the point of reference (cf. Example 5.2-2) although the coefficients are entire functions.

Hence our generalized definitions provide for a very substantial extension of the power and applicability of the Bergman method.

5.1 EXPONENTIAL OPERATORS AND KERNELS

Ideally, closed-form kernels should not only correspond to large and interesting classes of partial differential equations but also enjoy specific properties that make them particularly useful and versatile. For class P operators (Chaps. 3 and 4) this is the case, since they have the property that their theory includes various differential operators. In the present section we define a class of operators with closed-form kernels which have the advantage that they yield solutions to which the Fuchs–Frobenius theory of *ordinary* differential equations can be applied in order to investigate general properties of those solutions.

The equations to be considered are of the form

$$Lu = \left[D_1 D_2 + b(z) D_2 + c(z)\right] u = 0 \tag{5.1-1}$$

with $C^{\omega}(\Omega)$-coefficients as before; that is, for simplicity we use the reduced form in which $D_1 u$ has been eliminated, without loss of generality. This corresponds to equation (2.1-2) with $j = 1$, but we have dropped j for convenience. As in Sec. 2.2 we use the operator $T = T_1$ with $\tilde{k}_1 = k_1 = k$ ($\eta_1 = 0$) and $C_1 = [-1, 1]$, defined by

$$Tf_1(z) = \int_{-1}^{1} k(z,t) f_1\left(\rho_1 + \tfrac{1}{2}(z_1 - \sigma_1)\tau\right)\tau^{-1/2}\, dt \tag{5.1-2}$$

where $\tau = 1 - t^2$. We consider Type I representations of solutions with $j = 1$ in a neighborhood of s, where $s \neq 0$ in general. Note that here we choose $\sigma_1 = 0$, for simplicity. The class of operators to which the present sections are devoted is defined in terms of the form of their kernel as follows.

5.1-1 Definition

A Bergman operator is called an ***exponential operator*** if its kernel k in (5.1-2) is of the form

$$k = \exp q\,, \qquad q(z,t) = \sum_{\mu=0}^{m} q_\mu(z) t^\mu\,, \qquad m \in \mathbb{N}\,. \tag{5.1-3}$$

The first major task in connection with a closed-form kernel is the determination of the corresponding class of equations, albeit by a sufficient condition only, which admit operators with those kernels of a specific form. In the present case we are even able to state relatively simple *necessary and sufficient* conditions for b and c in (5.1-1) and give the exact form of the q_μ in the exponent. Here we exclude the case $c = 0$ which is trivial in the present connection. For, in the case $c = 0$, equation (5.1-1) reduces to an ordinary differential equation for the function $D_2 u$, and, moreover, a Bergman operator with any kernel independent of z_2 would trivially generate a solution of (5.1-1) with $c = 0$.

EXISTENCE AND REPRESENTATION OF EXPONENTIAL OPERATORS

5.1-2 Theorem (Criterion for Exponential Operators)

Equation (5.1-1) *admits an exponential operator if and only if the coefficients* b *and* c *in* (5.1-1) *can be represented in the form* (A) *or in the form* (B):

Case (A).

$$b(z) = -q_0'(z_1) - z_1^{-1} q_2(z_1) \tag{5.1-4}$$

$$c(z) = -(2z_1)^{-1} q_1(z) D_2 q_1(z) \tag{5.1-5}$$

where $q_0 \in C^\omega(\Omega_1)$ *is arbitrary and*

$$q_1(z) = z_1^{1/2}\left(\alpha_0(z_2) + \sum_{\nu=1}^{M} \alpha_\nu z_1^\nu\right), \qquad M = \left[\frac{m-1}{2}\right],$$

$$q_2(z_1) = \sum_{\nu=1}^{N} \beta_\nu z_1^\nu, \qquad N = \left[\frac{m}{2}\right]$$

and α_ν, β_ν *are constant for* $\nu \geq 1$.

Case (B).

$$b(z) = -q_0'(z_1) - z_1^{-1} q_2(z) \tag{5.1-6}$$

$$c(z) = -(2z_1)^{-1} D_2 q_2(z) \tag{5.1-7}$$

with q_0 *as before and*

$$q_2(z) = \beta_1(z_2) + \sum_{\nu=2}^{N} \beta_\nu z_1^\nu$$

where β_ν *are constant for* $\nu \geq 2$.

The other coefficient functions of q *in* (5.1-3) *are then given by*

$$q_{2\mu+1}(z_1) = \frac{(-4)^\mu (2\mu+1)!}{(\mu!)^2} \sum_{\nu=\mu}^{M} \binom{\nu}{\mu} \alpha_\nu z_1^{\nu+1/2}, \quad \mu = 1, 2, \cdots, M$$

in case (A) *and* $q_{2\mu+1} = 0$ *in case* (B), *and in both cases*

$$q_{2\mu}(z_1) = (-1)^{\mu+1} \sum_{\nu=\mu}^{N} \frac{1}{\nu} \binom{\nu}{\mu} \beta_\nu z_1^\nu, \quad \mu = 2, 3, \cdots, N.$$

Proof. In the present setting, the kernel equation (Sec. 2.3) is

$$(5.1\text{-}8)\qquad \left(\tau D_2 D_t - t^{-1} D_2 + 2z_1 tL\right)k = 0.$$

Substitution of $k = e^q$ yields

$$\left(\tau D_2 D_t - t^{-1} D_2 + 2z_1 tL\right)q + \left[\left(\tau D_t + 2z_1 t D_1\right)q\right] D_2 q$$
$$+2z_1 tc(z)(1-q) = 0.$$

Substituting q given by (5.1-3) into this equation and equating the coefficient of each occurring power of t to zero, we obtain a system of $2m+3$ nonlinear partial differential equations $\langle -1\rangle, \langle 0\rangle, \cdots, \langle 2m+1\rangle$ for $b, c, q_0, \cdots, q_m$; here $\langle \lambda\rangle$ is the equation resulting from t^λ. From $\langle -1\rangle$ we have $D_2 q_0 = 0$, as asserted in the theorem. $\langle 0\rangle$ yields no condition. From $\langle -1\rangle$ and $\langle 1\rangle$ we have

$$(5.1\text{-}9)\qquad c(z) = -(2z_1)^{-1}(q_1 D_2 q_1 + D_2 q_2)(z).$$

The other $2m$ equations $\langle 2\rangle, \cdots, \langle 2m+1\rangle$ can be written

$$(5.1\text{-}10)\qquad P_\lambda + \sum_{\nu=0}^{m} Q_{\lambda+\nu} D_2 q_{m-\nu} = 0, \qquad \lambda = 2, \cdots, 2m+1$$

where

$$P_\lambda = \lambda D_2 q_{\lambda+1} + 2z_1 D_1 D_2 q_{\lambda-1} + (2z_1 b - \lambda + 1) D_2 q_{\lambda-1}$$
$$Q_\gamma = (2z_1 D_1 - \delta) q_\delta + (\delta + 2) q_{\delta+2}, \qquad \delta = \gamma - m - 1$$

with the understanding that $q_\mu = 0$ for $\mu \neq 0, \cdots, m$. Hence, also $P_\lambda = 0$ for $\lambda \geq m+2$. Accordingly, $\langle m+2\rangle, \cdots, \langle 2m+1\rangle$ have the slightly simpler form

$$(5.1\text{-}11)\qquad \sum_{\nu=0}^{m} Q_{\lambda+\nu} D_2 q_{m-\nu} = 0, \quad \lambda = m+2, \cdots, 2m+1.$$

These equations are taken in reverse order to show that $D_2 q_m = 0$. Indeed, assuming $D_2 q_m \neq 0$ yields from $\langle \lambda\rangle$

$$(5.1\text{-}12)\qquad Q_\lambda = 0, \qquad \lambda = m+2, \cdots, 2m+1$$

successively in decreasing order of λ. The next three equations, $\langle m+1\rangle$, $\langle m\rangle$, $\langle m-1\rangle$, are more complicated. From (5.1-10) and (5.1-12) it follows that $\langle m+1\rangle$ reduces to

$$P_{m+1} + Q_{m+1} D_2 q_m = 0.$$

Now $Q_{2m+1} = 0$ implies $D_2 Q_{2m+1} = 0$, so that the previous equation, when written out, becomes

$$H = q_2 + z_1 D_1 q_0 + z_1 b = 0 .$$

This implies a reduction of $\langle m \rangle$ to

$$q_1 D_2 q_m - (m-1) D_2 q_{m-1} + 2 z_1 D_1 D_2 q_{m-1} = 0 .$$

$D_2 Q_{2m} = 0$ reduces this further to $q_1 D_2 q_m = 0$; hence $q_1 = 0$. Using this as well as $H = 0$ and (5.1-11), we see that $\langle m-1 \rangle$ entails $D_2 q_m = 0$, a contradiction to $D_2 q_m \neq 0$. By the same idea, similar contradictions can be derived and lead to

$$D_2 q_\mu = 0 , \qquad \mu = 3, \cdots, m-1 . \tag{5.1-13}$$

Equation (5.1-9) now readily implies (5.1-7). But $q_1 = 0$ was derived from $D_2 q_m \neq 0$ which is false. Hence $q_1 \neq 0$ must also be considered. It is easily seen that (5.1-13) remains valid also in this case, so that $\langle 2 \rangle$ gives $D_2 q_2 = 0$ and (5.1-9) reduces to (5.1-5). $H = 0$ was proved assuming $D_2 q_m \neq 0$, but still follows from (5.1-13) and $\langle 2 \rangle$ if $q_1 \neq 0$, and from (5.1-13) and $\langle 3 \rangle$ if $q_1 = 0$. In both of these conclusions, an exception occurs, but one sees that these exceptions would lead to $c = 0$; this is the trivial case which is excluded once and for all. $H = 0$ implies (5.1-4) and (5.1-6). The formulas for the coefficients q_μ follow from (5.1-10) by integrating with respect to z_1, taking (5.1-13) into account, and proceeding recursively. This completes the proof. ■

This proof, a simplified version of an earlier one, is typical. It illustrates that the difficulties in introducing closed-form kernels can be formidable. Specifically, it will depend on the initially chosen form of the kernel whether one can attack the recursive system of partial differential equations for the coefficients in a stepwise fashion as in the present case. Except for trivial choices of kernels, the equations of the system obtained will always be nonlinear.

From Theorem 5.1-2 we readily have the following result.

5.1-3 Corollary

Every equation (5.1-1) *with constant coefficients admits an exponential operator.*

5.2 PROPERTIES OF EXPONENTIAL OPERATORS

Historically, a first suggestion of the use of special exponential functions as kernels for the Helmholtz equation and a few other special cases was given by Bergman [1937b] (see also his résumés [1937a]). Later attempts by

Nielsen–Ramsey [1943] and Nielsen [1944a, b] to define and investigate this type of operators in full generality were unsuccessful. Our present methods in this section not only solve the problem in full generality, but can even be extended to other integral operators (see below). Hence they may be regarded as typical of approaches to other classes of operators.

Exponential operators have the basic property that they can be used for obtaining solutions which, in addition to the *partial* differential equation, also satisfy an *ordinary* differential equation. In the construction of the latter, one is at first confronted with two problems:

(i) One has to choose surfaces in $\mathbb{R}^4$ on which one wants to consider solutions u as functions of a single complex variable. We choose $x_2 = const$. This has the advantage that u becomes an analytic function of x_1.

(ii) One has to choose suitable classes of associated functions. By differentiation and elimination, one can prove the following. If each term of a sum of sufficiently often differentiable functions satisfies a linear ordinary differential equation with algebraic coefficients, so does the sum. Accordingly, we can choose the "standard" associated functions (5.2-1) shown in the next theorem.

RELATION TO ORDINARY DIFFERENTIAL EQUATIONS

5.2-1 Theorem

If (5.1-1) admits an exponential operator T, then every solution $u = u_n = Tf_{1n}$, where

$$(5.2\text{-}1) \qquad f_1(\zeta_1) = f_{1n}(\zeta_1) = (\zeta_1 - s_1)^n, \qquad n \in \mathbb{N}_0, \quad \rho_1 = s_1, \quad \sigma_1 = 0$$

satisfies a homogeneous linear ordinary differential equation in the independent variable $x_1 = (z_1 + z_2)/2$ (and $x_2 = (z_1 - z_2)/2i = const$). The order of this equation is independent of n and does not exceed $m + 1$, where m is the degree of the exponent of the kernel as defined in (5.1-3). The minimum order is 2.

Proof. We consider the integrand in (5.1-2). Because of the form of the kernel and the choice of the associated function f_1 this integrand has the form

$$v(x_1, x_2, t) = \tilde{v}(z, t) = \left(\frac{z_1}{2}\right)^n \tau^{n-1/2} \exp q(z, t).$$

It suffices to show that v satisfies an ordinary differential equation of the form

$$\tilde{P}(v) = D_t[\tau^\nu p(z,t)v], \qquad \nu > 0; \tag{5.2-2}$$

here $\tilde{P}$ is a linear ordinary differential operator (differentiation with respect to x_1) of order κ with κ to be determined later, and the right-hand side of (5.2-2) is integrable with respect to $t \in C_1 = [-1, 1]$. Indeed, integrating (5.2-2) with respect to t from -1 to 1, since the right-hand side contains the factor τ^ν, we obtain a homogeneous linear ordinary differential equation as indicated in the theorem. To construct (5.2-2), one may choose $\nu = 1$ and a polynomial p, say,

$$p(z,t) = \sum_{\lambda=0}^{l} p_\lambda(z)t^\lambda.$$

We now observe that derivatives of v with respect to x_1 as well as the function $\tau D_t v$ can be represented in the form of products of v and certain polynomials in t. More precisely, $d^\mu v/dx_1^\mu$ is v times a polynomial of degree μm in t. Hence on the left-hand side of (5.2-2) the highest power in t is $t^{\kappa m}$. To have the same highest power on the right, we choose

$$l = (\kappa - 1)m - 1.$$

We then obtain a system of $\kappa m + 1$ linear equations from which we can determine the κ coefficient functions of $\tilde{P}$ (the coefficient of the κth derivative being 1, by assumption) as well as the p_λ's in p, provided that the determinant D of the system does not vanish identically. The number of equations equals the number of unknowns if we choose

$$\kappa = m + 1.$$

This proves the claim regarding the order of the differential equation. If D is identically zero, that order is smaller, but the procedure of construction remains the same. $\kappa \geq 2$ follows from $l \geq 0$. This proves the theorem ■

5.2-2 *Example*

If

$$k(z,t) = \exp\left(i\sqrt{z_1 z_2}\, t\right)$$

the ordinary differential equation in Theorem 5.2-1 is of second order and can be written

$$(x_1^2 + x_2^2)v'' + \left(x_1 - \frac{x_2^2}{x_1} + 2inx_2\right)v' - \left(n^2 + in\frac{x_2}{x_1} - x_1^2\right)v = 0$$

where primes denote derivatives with respect to x_1. More general equations are considered by Kreyszig [1956].

As an important application, the present theorem yields the possibility of continuing solutions u beyond the domain in which the validity of the representation (5.1-2) is guaranteed by the Representation Theorem 2.3-1.

Furthermore, one can establish relations between the coefficients of the given *partial* differential equation and those of an *ordinary* differential equation. In this way, theorems on solutions of ordinary differential equations with analytic coefficients become applicable to the investigation of solutions of partial differential equations. In this connection, a basic issue is that of relations between singularities. Such relations can be obtained in two steps:

(i) By first establishing properties of those coefficients in terms of properties of the q_μ's in (5.1-3), and then

(ii) by deriving from (i) direct relations in which the q_μ's no longer occur.

In particular, if the q_μ's in (5.1-3) are holomorphic at 0, so are the coefficients of (5.1-1), whereas a corresponding ordinary differential equation can be at most weakly singular (can have at most a regular singular point; cf. Ince [1956]) at 0. If, however, q_0, q_1, or q_2 has a pole at some point, then (5.1-1) has singular coefficients and so do those ordinary differential equations. More specifically, simple poles of b can cause only weak singularities of corresponding ordinary differential equations, whereas poles of order $r > 1$ cause poles of order $r(m + 1 - \mu)$ in the coefficient of the μth derivative of that equation. For proofs and details, see Kreyszig [1956].

5.3 FURTHER RESULTS ON EXPONENTIAL OPERATORS

A little reflection shows that the idea of the proof in the preceding section can be generalized to include (i) other kernels that lead to a more complicated ordinary differential equation, as well as (ii) associated functions which are solutions of homogeneous linear ordinary differential equations.

Of particular interest is an extension of Theorem 5.2-1 to solutions corresponding to rational associated functions, since by their use one can construct solutions with singularities of a given location and type, a task which is of interest in fluid flow and other applications. The simplest such f_1 would be of the form

$$f_1(\zeta_1) = (\zeta_1 - \alpha)^{-n}, \qquad \alpha \text{ constant}. \tag{5.3-1}$$

For this case, the method of the previous proof—with a suitably modified function on the right-hand side of the equation for the integrand—yields the existence of an ordinary linear differential equation of order

$$r \leq m + 3.$$

We see that the order r may be higher than before, but is still independent of n. Similarly, for a general rational function f_1 with poles at various points, one can still construct an ordinary differential equation for solutions, of order

$$r < p(m+3) + (n+1)(m+1)$$

where p is the number of partial fractions and n the degree of the polynomial part of the associated function.

It is interesting to note that in the case of associated functions (5.3-1), *simple* poles of b cause only weak singularities of corresponding ordinary differential equations, just as in the case of holomorphic associated functions discussed in the previous section.

At the end of Sec. 5.1 it was mentioned that equations (5.1-1) with constant coefficients admit exponential operators. In this case, for solutions corresponding to the "standard" associated functions (5.2-1) we can obtain a representation in terms of Bessel functions as follows.

5.3-1 Example

If the equation

$$D_1 D_2 u + b D_2 u + cu = 0 \tag{5.3-2}$$

has constant coefficients, the simplest possible exponential operator has the kernel given by

$$k(z,t) = \exp\left(-bz_1 + 2i\sqrt{cz_1 z_2}\, t\right).$$

This is obtained for $q_0(z) = -bz_1$ and $q_1(z) = 2i\sqrt{cz_1z_2}$ which agrees, according to (5.1-4, 5), with Case (A) of Theorem 5.1-2. The application of the Bergman operator $T = T_1$ defined by $\tilde{k}_1 = k$, $\rho_1 = s_1$, and $\sigma_1 = 0$, to the standard associated functions (5.2-1) yields solutions $u = u_n$, where

$$u_n(z) = e^{-bz_1}\left[\int_{-1}^{1} \cos\left(2\sqrt{cz_1z_2}\, t\right)\left(\frac{z_1}{2}\right)^n \tau^{n-1/2}\, dt \right.$$
$$\left. + i\int_{-1}^{1} \sin\left(2\sqrt{cz_1z_2}\, t\right)\left(\frac{z_1}{2}\right)^n \tau^{n-1/2}\, dt\right], \qquad n \in \mathbb{N}_0.$$

Since the integrand in the second integral is an odd function of t and, consequently, this integral vanishes, $\exp(bz_1)u_n(z)$ corresponds to a special case of $w_{1n}(z)$ in (2.2-4a). From this we have the result

$$u_n(z) = \frac{\pi(2n)!}{8^n n! c^{n/2}} \left(\frac{z_1}{z_2} \right)^{n/2} J_n\left(2\sqrt{c z_1 z_2}\right) e^{-b z_1}.$$

We observe that this could also have been obtained by transforming (5.3-2) into equation (2.2-2) with coefficient $2\sqrt{c}$, setting $w(z) = \exp(bz_1)u(z)$, and then using the approach of Sec. 2.2.

It is plain that the kernel equation (2.3-2a)—or (5.1-8) in our present notation and setting—can be simplified by transformations of the independent variables or the dependent variable or both. The practical advantage of this approach is not obvious. In fact, it is doubtful whether there is in fact an advantage at all, since that simplification may have an adverse effect on the system of nonlinear partial differential equations from which one must (recursively) determine coefficients in representations of kernels. However, for exponential operators, the proof of Case (B) of Theorem 5.1-2 and other parts of the theory in the case of *even m* can be simplified substantially if instead of t we introduce

$$\tilde{\tau} = 2\tau = z_1(1 - t^2)$$

as a new independent variable. Then (5.1-8) becomes simply

$$D_2 k + 2(\tilde{\tau} - z_1) L k = 0.$$

For further results and details related to the idea of transformations, see Azzam–Kreyszig [1981].

5.4 GENERAL REMARKS. RATIONAL KERNELS

The difficulties in introducing closed-form kernels are twofold.

(i) The conditions (nonlinear partial differential equations) for the unknown coefficient functions may be so complicated that one is unable to determine those functions explicitly. This is a disadvantage in applications.

(ii) New kernels may turn out to be applicable only to equations for which other closed-form kernels are already known. Then the new kernels may still be useful if they are simpler or more suitable for specific problems.

COMBINED EXTENSION OF CLASS P AND EXPONENTIAL OPERATORS

To avoid (i), one may have to settle for a more restrictive class than originally anticipated, hoping that in doing so, one may not drift toward (ii). A case at hand is the multiplicative combination of exponential and class P kernels, that is, kernels of the form

$$k(z, t) = p(z, t) \exp q(z, t) \tag{5.4-1}$$

where p and q are polynomials in t of degrees m and n, respectively, with coefficient functions p_μ and q_ν, respectively, depending on $z = (z_1, z_2)$. In this generality, the system of nonlinear partial differential equations for the coefficient functions is extremely complicated. This should not be unexpected if we remember the situations for class P and for exponential operators. On the contrary, it seems somewhat surprising that by restricting the p_μ to be functions of z_2 and the q_ν to depend on z_1 only, one can succeed in deriving explicit equations and corresponding kernels. For instance, let us consider equation (5.1-1) with coefficients

$$\begin{aligned} b(z) &= \frac{1}{2z_1}[1 - 2z_1 D_1 q_0(z_1) + q_1(z_1) p_1(z_2) - 2q_2(z_1)] \\ c(z) &= -\frac{1}{2z_1} q_1(z_1) D_2 p_1(z_2) \end{aligned} \tag{5.4-2}$$

where, apart from the usual conditions, q_0, q_1, q_2, and p_1 are arbitrary functions of their respective variables. Under certain assumptions on q_1 and q_2, for this equation a kernel k_{I} of the form (5.4-1) is determined involving coefficient functions given by a recursion system; see Azzam–Kreyszig [1980]. Further kernels of the form (5.4-1) were obtained by Wallner [1981b], namely

$$\begin{aligned} k_{\mathrm{II}}(z, t) = &\left\{ \frac{Q_1(z_1)}{q_1(z_1)} + p_1(z_2)t + \frac{Q_1(z_1)}{q_1(z_1)}\left[2z_1 \frac{D_1 Q_1(z_1)}{Q_1(z_1)} - 1\right]t^2 \right\} \times \\ &\times \exp\left[Q_0(z_1) + Q_1(z_1)t\right] \end{aligned} \tag{5.4-3}$$

or

$$k_{\mathrm{III}}(z, t) = \left\{ \frac{\sqrt{z_1}}{q_1(z_1)} + p_1(z_2)t \right\} \exp\left[Q_0(z_1) + \sqrt{z_1}\, t\right] \tag{5.4-4}$$

where Q_1 is arbitrary, e.g., $Q_1 = q_1$, and

$$Q_0(z_1) = q_0(z_1) + \int_0^{z_1} \tilde{z}_1^{-1} q_2(\tilde{z}_1)\, d\tilde{z}_1 .$$

It is interesting that the preceding example belongs to Case (ii) (above) because the operator L in (5.1-1) with coefficients (5.4-2) belongs to the class

P_{11}. This can be seen by introducing $\tilde{q}_{10}(z_1) \equiv 1$ and

$$\tilde{q}_{11}(z) = \frac{1}{z_1}[-1 - 2z_1 D_1 q_0(z_1) + q_1(z_1)p_1(z_2) \qquad (5.4\text{-}5)$$

$$+ 2z_1(D_1 q_1(z_1))/q_1(z_1) - 2q_2(z_1)].$$

Then (5.4-2) is obtained by (4.1-4), and a corresponding polynomial kernel is given by (4.1-3); that is, $\tilde{k}_1 = k_{IV}$ where

$$k_{IV}(z,t) = 1 + \tilde{q}_{11}(z)z_1t^2. \qquad (5.4\text{-}6)$$

For example, let us consider the four kernels $k_I, \cdots, k_{IV}$ for the special equation

$$D_1D_2u - z_2D_2u + u = 0 \qquad (5.4\text{-}7)$$

[cf. Example 3.1-8 with $j = 1$, $n = 1$, and $\alpha = 1$]. Comparison with (5.4-2) shows that $q_0(z_1) = \frac{1}{2}\ln z_1$, $q_1(z_1) = -2z_1$, $q_2(z_2) = 0$, and $p_1(z_2) = z_2$. According to Azzam–Kreyszig [1980] we thus have

$$k_I(z,t) = (1 + z_2t + t^2)\exp\left(\tfrac{1}{2}\ln z_1 - 2z_1t\right).$$

Choosing $Q_1 = q_1$, from (5.4-3) we get

$$k_{II}(z,t) = k_I(z,t)$$

and from (5.4-4)

$$k_{III}(z,t) = \left(-\tfrac{1}{2} + z_2\sqrt{z_1}\,t\right)\exp\left(\sqrt{z_1}\,t\right).$$

Finally, in the case of equation (5.4-7) from (5.4-5, 6) we obtain

$$k_{IV}(z,t) = 1 - 2z_1z_2t^2$$

in agreement with the corresponding result in Example 3.1-8.

These typical considerations show that the concept of (5.4-1) is a valuable tool for augmenting the variety of explicitly known kernels.

RATIONAL FUNCTIONS AS KERNELS

Another natural choice for closed-form kernels seems to be *rational functions* of t, not polynomials, with coefficient functions depending on $z = (z_1, z_2)$. Experimentation shows readily that only by starting from special partial fraction representations one can hope to obtain systems of nonlinear partial differential equations for the coefficient functions which yield explicit representations, or at least a characterization of general properties, of these functions. This idea was proposed in 1982. The first results in this direction will now be discussed.

We first mention that for

$$k = z_1 t k_0 \tag{5.4-8}$$

the kernel equation (2.3-2a) is satisfied if

$$\left[\tau D_2 D_t + 2z_1 t\left(L + \frac{1}{2z_1}\right)\right] k_0 = 0 \tag{5.4-9}$$

holds. We now set

$$\begin{aligned} k_0(z,t) &= \sum_{\nu=0}^{n-1} p_{n-\nu}(z) Q(z,t)^{-m-n+1+\nu} \\ Q(z,t) &= t - q(z) \end{aligned} \tag{5.4-10}$$

with $m, n \in \mathbb{N}$ and $p_{n-\nu}, q \in C^{\omega}(\Omega)$. Substituting, clearing fractions, and equating the sum of the coefficient functions of each power of t to zero yields a system of $n+3$ nonlinear partial differential equations for the coefficient functions. We notice that this system is not too large; however, each equation turns out to be extremely complicated. In the further analysis, one shows first that

$$q(z) = z_1^{-1/2} \Lambda(z)^{1/2}, \qquad \Lambda(z) = z_1 + \lambda(z_2) \tag{5.4-11}$$

with an arbitrary analytic function λ, and

$$p_{n-\nu}(z) = \kappa_{n-\nu}(z_1) q(z)^{m+n-\nu}, \qquad \nu = 0, \cdots, n-1. \tag{5.4-12}$$

This entails a substantial reduction of *all* the equations, but the resulting equations are still much too complicated to be reproduced here. One observes next that certain combinations of functions occur in many of the equations; in particular, this holds for

$$\begin{aligned} A &= (2\Lambda)^{-1} \\ B_{n-\nu} &= b + D_1 \log\left(\kappa_{n-\nu} z_1^{1/2} D_2 q\right) \\ C &= qc/D_2 q. \end{aligned}$$

One shows next that one can find

$$\kappa_\mu(z_1) = \gamma_\mu \kappa_1(z_1), \qquad \mu = 2, \cdots, n \tag{5.4-13}$$

with constant γ_μ such that the equations obtained in the previous steps are satisfied. This has the important effect that at the same time $B_{n-\nu}$ becomes a function which is independent of ν and which we denote by B. The determination of B is then accomplished by taking suitable linear combinations of the aforementioned equations. In doing so, one can manage to obtain a single equation in which all the C-terms as well as terms involving $\kappa_1, \cdots, \kappa_{n-1}$ are eliminated and κ_n occurs merely as a common factor, which can be suppressed. This equation yields the surprisingly simple expression

$$B = -(m+n)/\Lambda.$$

The rest of the derivation concerns the determination of the constants γ_μ. This process is somewhat cumbersome, but eventually yields

$$\gamma_\mu = \delta_\mu(n)(m+\mu-2)!/(m-1)!, \qquad \mu \geq 2 \tag{5.4-14}$$

with δ_μ expressible in terms of factorial functions (independent of m).

The result is the following condition sufficient for the kernels considered to be admissible (cf. Kreyszig [1982]).

Let the coefficients of (5.1-1) be represented in the form

$$\begin{aligned} b(z) &= \eta(z_1) - \tfrac{1}{2}(2m+2n-1)[z_1 + \lambda(z_2)]^{-1} \\ c(z) &= \tfrac{1}{4}(m+1)(m+2n)[z_1 + \lambda(z_2)]^{-2} D_2\lambda(z_2) \end{aligned} \tag{5.4-15}$$

where $\eta \in C^\omega(\Omega_1)$, $\lambda \in C^\omega(\Omega_2)$, and $m, n \in \mathbb{N}$; cf. (4.1-12). Then a Bergman operator is defined by (5.4-8, 10) if q satisfies (5.4-11) and if the coefficient functions $p_{n-\nu}$ are given by (5.4-12), where the coefficients $\kappa_{n-\nu}$ are defined by

$$\kappa_1(z_1) = \gamma_1 \exp \int_0^{z_1} \eta(\tilde{z}_1)\, d\tilde{z}_1, \qquad \gamma_1 = const, \neq 0$$

for $\nu = n-1$ and by (5.4-13, 14) elsewhere.

As Wallner [1985] observed later, the equation with the coefficients of (5.4-15) [with η eliminated by means of (2.1-2, 3, 4)] is also contained in the class of equations for which a Bergman kernel can be expressed as a function of a suitable auxiliary variable and satisfies a second-order ordinary differential equation with respect to that variable. (This class was introduced and characterized by Florian [1965]; cf. also Kreyszig [1973].) Realizing this, in the preceding example Wallner arrives at the Bergman kernel

$$k(z, t) = \kappa_1(z_1)\,{}_2F_1\left(\frac{m}{2} + \tfrac{1}{2}, \frac{m}{2} + n; \tfrac{1}{2}; Z\right) \tag{5.4-16}$$

where

$$Z = \frac{z_1}{z_1 + \lambda(z_2)} t^2$$

and κ_1 as before. Observe that for the present parameters, the hypergeometric function may be rewritten as a *rational* function of Z, and hence of t (cf. Magnus et al. [1966], pp. 39, 41).

Thus an operator has been obtained that is suitable for an equation to which none of the previously discussed closed-form operators is admissible (resp. explicitly known). The preceding example may serve as an introduction in order to encourage further efforts in the explicit determination of closed-form kernels and operators.

Additional References

Azzam–Kreyszig [1980, 1981]
Bergman [1937b, 1971]
Böhmer [1970]
Ecker–Florian [1967]
Florian [1965]
Kreyszig [1955, 1956, 1958a, 1972, 1973, 1982]
Nielsen [1944a, 1944b]

Chapter Six

Riemann–Vekua Representation and Further Methods Related to Bergman Kernels

In this chapter we discuss relations between the Bergman integral operator method and several other methods related to it. In the same year, 1937, in which Bergman published his first papers on his integral-operator approach to partial differential equations, I. N. Vekua [1937] independently proposed his method based on the ***complex Riemann function***. We show that the Riemann–Vekua integral representations thus obtained can also be generated by two suitably chosen ***Le Roux operators*** (particularly, in the complex domain). These operators were rediscovered and studied further in some articles by Diaz and Ludford in the 1950s. We give their definition in complex form in Sec. 6.1. In Sec. 6.2 we introduce the complex Riemann function. For the equations considered in this book, we prove the existence and uniqueness of this function. Following an idea by Vekua [1967], this will be done by solving the defining Volterra-type integral equation. Basic properties of the complex Riemann function are considered in Sec. 6.3.

The most important section in this chapter is Sec. 6.4. In Theorem 6.4-1 we prove a (not yet published) result, showing how the Riemann function can be obtained from a Bergman kernel of the first kind in a relatively simple fashion. Since for this purpose we need only *one* kernel ($\tilde{k}_1$ or $\tilde{k}_2$) and thus only the solution of a recursion with respect to *one* variable [cf. (2.4-2)], this is of special interest in the problem of the explicit determination of the Riemann function. Furthermore, this also establishes an integral relation between the two Bergman kernels $\tilde{k}_1$ and $\tilde{k}_2$ of the first kind for Type II representations. Generalizing results by Watzlawek [1971a, c] and Kracht [1974a], in Sec. 6.4 we also describe a way of deriving Bergman kernels from the Riemann function. The preceding relations enable us to use the Bergman theory (especially

Theorems 2.3-1 and 2.6-2) for obtaining the Riemann–Vekua representation of the solution of Goursat problems. By a combination with Sec. 6.2 they also yield another existence proof for Bergman kernels of the first kind.

A further method of utilizing the complex Riemann function for the representation of solutions was invented by Gilbert [1970]. Gilbert's theory will be considered in Sec. 6.5. It has the advantage that it can be extended to equations in any number of variables. It is remarkable that Gilbert's G-function, defining the so-called ***Bergman–Gilbert operator*** is independent of the number of variables. We shall see that this operator can be obtained from the Riemann function for the two-dimensional equation. In Sec. 6.6 we shall outline the idea of the interpretation of the Bergman–Gilbert operator in Carroll's theory of ***transmutations*** (cf. Carroll [1979, 1982, 1984a]).

6.1 FUNDAMENTAL DOMAINS AND LE ROUX OPERATORS

We consider the real elliptic equation (2.1-5), that is,

$$(6.1\text{-}1)\quad \hat{L}\hat{w} = \Delta\hat{w} + \hat{a}_1(x_1, x_2)\hat{w}_{x_1} + \hat{a}_2(x_1, x_2)\hat{w}_{x_2} + \hat{a}_0(x_1, x_2)\hat{w} = 0$$

involving two independent real variables x_1, x_2 and real-analytic coefficients $\hat{a}_0, \hat{a}_1, \hat{a}_2$ in some domain $\hat{\Omega}_1$ in the x_1x_2-plane (cf. Sec. 2.1). Applying the transformations (2.1-6, 7), we obtain the equation (2.1-1) with $z_2 = \bar{z}_1$, viz.,

$$(6.1\text{-}2)\qquad w_{z_1\bar{z}_1} + a_1(z_1, \bar{z}_1)w_{z_1} + a_2(z_1, \bar{z}_1)w_{\bar{z}_1} + a_0(z_1, \bar{z}_1)w = 0$$

whose coefficients (2.2-8) are analytic for $(z_1, \bar{z}_1) \in \Omega_1 \times \Omega_1^*$, where

$$(6.1\text{-}3)\qquad \begin{aligned} \Omega_1 &= \{z_1 = x_1 + ix_2 \in \mathbb{C} \mid (x_1, x_2) \in \hat{\Omega}_1\} \\ \Omega_1^* &= \{z_2 = x_1 - ix_2 \in \mathbb{C} \mid (x_1, x_2) \in \hat{\Omega}_1\}. \end{aligned}$$

It is clear that the coefficients of (6.1-1) can locally be analytically continued by introducing complex values of x_1, x_2. Hence there exists some neighborhood $\hat{U}$ of $\hat{\Omega}_1$ in $\mathbb{C}^2$ in which those coefficients are holomorphic functions of two complex variables x_1, x_2. From (6.1-1), using (6.1-2), we thus obtain

$$(6.1\text{-}4)\qquad Lw = D_1D_2w + a_1(z)D_1w + a_2(z)D_2w + a_0(z)w = 0$$

where $z = (z_1, z_2)$, our equation (2.1-1), which involves two independent complex variables z_1, z_2 and coefficients holomorphic in some neighborhood

U of $\{(z_1, \bar{z}_1) \mid z_1 \in \Omega_1\}$ in $\mathbb{C}^2$. This neighborhood may be taken as

$$U = \bigcup_{z_1 \in \Omega_1} \{U_1(z_1) \times U_1^*(z_1) \mid U_1(z_1) \subset \Omega_1\}$$

where $U_1(z_1)$ is a sufficiently small neighborhood of $z_1 \in \Omega_1$ and $U_1^*(z_1) = \{z_2 \in \mathbb{C} \mid z_1 \in U_1(z_1)\}$ with z_1, z_2 as before, is a neighborhood of $z_2 = \bar{z}_1 \in \Omega_1^*$. Hence we have $U \subset \Omega_1 \times \Omega_1^*$.

FUNDAMENTAL DOMAINS

Throughout this chapter we *assume* that $\hat{\Omega}_1$ is a fundamental domain for equation (6.1-1) in the sense of I. N. Vekua [1967] (cf. also Henrici [1957]). This concept is defined as follows.

6.1-1 Definition

$\hat{\Omega}_1$ is called a ***fundamental domain*** for (6.1-1) [and Ω_1 a fundamental domain for (6.1-4)] if $\hat{\Omega}_1$ is simply connected and the coefficients of (6.1-4) obtained by (2.1-8) are of class $C^\omega(\Omega)$, where $\Omega = \Omega_1 \times \Omega_2$ and $\Omega_2 = \Omega_1^*$.

Hence for $\hat{\Omega}_1$ or Ω_1 in this definition we require the existence of a *global* analytic continuation of the coefficients of the equation (6.1-2) with respect to the domain $\Omega_1 \times \Omega_1^*$.

6.1-2 Examples

(a) If those coefficients are entire, then $\Omega_1 = \mathbb{C}$ is a fundamental domain for (6.1-4).

(b) If those coefficients are analytic in an arbitrary domain, then every sufficiently small simply connected subdomain is a fundamental domain for (6.1-4).

(c) Every simply connected subdomain of a fundamental domain is again a fundamental domain.

(d) A fundamental domain for (6.1-4) is also a fundamental domain for the *adjoint equation*

$$\begin{aligned} &L^*w = D_1D_2w^* + a_1^*(z)D_1w^* + a_2^*(z)D_2w^* + a_0^*(z)w^* = 0 \\ (6.1\text{-}5) \quad &a_1^* = -a_1, \qquad a_2^* = -a_2, \qquad a_0^* = a_0 - D_1a_1 - D_2a_2. \end{aligned}$$

Because of (6.1-3), the corresponding results for the equation (6.1-1) are obvious. We should only mention that in order to obtain (6.1-5), the application of the transformation (2.1-6) to (6.1-1) and the transition to the adjoint equation may be interchanged.

LE ROUX OPERATORS

Let us now introduce *Le Roux operators* for generating solutions of equations of the form (6.1-4). These operators were originally applied to real hyperbolic equations (cf. Diaz–Ludford [1955] and the references given there), but their extension to *complex* equations provides no serious problem.

6.1-3 Definition

The operator

(6.1-6a) $$J_j\colon\ C^\omega(\Omega_j) \ \to\ C^\omega(\Omega), \qquad \Omega = \Omega_1 \times \Omega_2$$

defined by

(6.1-6b) $$J_j\Psi_j(z) = \int_{s_j}^{z_j} R_j(z, \xi_j)\Psi_j(\xi_j)\,d\xi_j$$

where $R_j \in C^\omega(\Omega \times \Omega_j)$, $R_j \neq 0$, $\Psi_j \in C^\omega(\Omega_j)$, $s_j \in \Omega_j$ arbitrary and fixed, is called a ***Le Roux operator*** (and R_j a ***Le Roux kernel***) in Ω if

(6.1-7) $$w_j(z) = J_j\Psi_j(z), \qquad j = 1,2$$

is a $C^\omega(\Omega)$-solution of (6.1-4).

A simple criterion for Le Roux kernels is at once obtained:

6.1-4 Theorem

R_j is a Le Roux kernel for (6.1-4) *in Ω, and w_j in* (6.1-7) *thus a solution of* (6.1-4) *in Ω, if R_j satisfies in $\Omega \times \Omega_j$ the conditions*

(6.1-8a) $$LR_j = 0$$

and

(6.1-8b) $$\left[(D_{j*} + a_j)R_j\right]\big|_{\xi_j = z_j} = 0, \qquad j = 1,2.$$

Proof. From (6.1-6) and (6.1-7) we obtain by straightforward calculation

$$Lw_j = (D_j\Psi_j)\left[(D_{j*} + a_j)R_j\right]\big|_{\xi_j = z_j} + \Psi_j\left[LR_j\right]\big|_{\xi_j = z_j} - \int_{s_j}^{z_j} \Psi_j \frac{\partial}{\partial \xi_j} LR_j\,d\xi_j. \quad \blacksquare$$

6.1-5 Remark

Suppose that R_j satisfies the conditions of Theorem 6.1-4 and Ω_j^0 is a simply connected domain for which the closure is contained in a fundamental domain Ω_j for (6.1-4). If $s_j \in \overline{\Omega_j^0}$ and $\Psi_j \in C^\omega(\Omega_j^0)$ and the integral in (6.1-6b) exists, then the statement of Theorem 6.1-4 remains valid; that is, then (6.1-7) yields a solution of (6.1-4) in $\Omega_1^0 \times \Omega_2$ (when $j = 1$) and $\Omega_1 \times \Omega_2^0$ (when $j = 2$), respectively.

An example of Le Roux kernels R_1 and R_2 for (6.1-4) is easily derived from the complex Riemann function for the adjoint equation (6.1-5); see below. Hence we can also obtain representation formulas for solutions of *Goursat problems*. As in the case of Bergman operators, for general Goursat problems one needs *pairs* of Le Roux operators. Moreover, since even a Type II representation of solutions by operators of the form (6.1-6b) would be such that $w(s) = 0$, when $\varphi_j(s_j) \neq 0$ we must add a suitable solution in order to achieve $w(s) = \varphi_1(s_1) = \varphi_2(s_2)$. The details will be given in (6.4-17).

6.2 COMPLEX RIEMANN FUNCTION: EXISTENCE AND UNIQUENESS

Following Vekua [1967], we now define the complex Riemann (–Vekua) function.

6.2-1 *Definition*

Let Ω_1 be a fundamental domain for (6.1-4). A function R of the variables z_1, z_2, s_1, s_2 defined on $G = \Omega_1 \times \Omega_1^* \times \Omega_1 \times \Omega_1^*$ is called a ***Riemann function*** for the equation $Lw = 0$ given by (6.1-4) if R is a solution of the Volterra-type integral equation

(6.2-1)

$$R - \int_{s_1}^{z_1} (a_2 R)\Big|_{z_1=\xi_1} d\xi_1 - \int_{s_2}^{z_2} (a_1 R)\Big|_{z_2=\xi_2} d\xi_2 + \int_{s_1}^{z_1}\int_{s_2}^{z_2} (a_0 R)\Big|_{z=\xi} d\xi_2\, d\xi_1 = 1\,;$$

here, $z = (z_1, z_2)$ and $\xi = (\xi_1, \xi_2)$.

We prove the existence and uniqueness of a solution of this integral equation in G. For this purpose we start from an integral equation of the form

$$v(z_1) - \int_{s_1}^{z_1} K(z_1, \xi_1) v(\xi_1)\, d\xi_1 = f(z_1) \qquad (6.2\text{-}2)$$

with kernel $K \in C^\omega(\Omega_1 \times \Omega_1)$, $f \in C^\omega(\Omega_1)$, and $s_1 \in \Omega_1$; here, Ω_1 is a simply connected domain.

6.2-2 *Lemma*

There exists a unique solution of (6.2-2) *in the domain* Ω_1.

Proof. We first prove the uniqueness. Suppose, to the contrary, that there exist two such solutions v_1 and v_2. Then (6.2-2) implies that their difference $v_0 = v_1 - v_2$ satisfies $v_0(s_1) = 0$. Since $(D_1^n v_0)(s_1)$ can be represented as a linear combination of $(D_1^k v_0)(s_1)$, $k = 0, \cdots, n-1$, the Taylor series of v_0 with center s_1 vanishes identically. Thus v_0 is identically zero in some neighborhood of s_1 and hence in Ω_1, by the identity theorem for holomorphic functions. This proves the uniqueness $v_1 = v_2$ in Ω_1.

To prove the existence of a solution in Ω_1, we construct a sequence $(v_n)_{n \in \mathbb{N}_0}$ of functions defined by

$$\text{(6.2-3)} \quad v_0(z_1) = f(z_1), \qquad v_n(z_1) = f(z_1) + \int_{s_1}^{z_1} K(z_1, \xi_1) v_{n-1}(\xi_1)\, d\xi_1 .$$

If $\lim v_n = v$ in Ω_1 uniformly on every compact subset, then v is a C^ω-solution of (6.2-2). Hence the remaining task is to show the uniformity of the convergence of (v_n). This can be done by the method of majorants, as follows. We first prove that the v_n's can also be represented in the form

$$\text{(6.2-4)} \qquad v_n(z_1) = f(z_1) + \int_{s_1}^{z_1} \Gamma_n(z_1, \xi_1) f(\xi_1)\, d\xi_1$$

where the Γ_n's are defined by

$$\text{(6.2-5a)} \qquad \Gamma_n(z_1, \xi_1) = \sum_{m=1}^{n} K_m(z_1, \xi_1), \qquad n \in \mathbb{N}$$

with

$$\text{(6.2-5b)} \quad K_1 = K, \qquad K_{m+1}(z_1, \xi_1) = \int_{\xi_1}^{z_1} K(z_1, t_1) K_m(t_1, \xi_1)\, dt_1, \quad m \in \mathbb{N}.$$

Indeed, this follows immediately from (6.2-3) by induction, in particular, by using Dirichlet's formula

$$\text{(6.2-6)} \quad \int_{s_1}^{z_1} \int_{s_1}^{\xi_1} F(z_1, \xi_1, t_1)\, dt_1\, d\xi_1 = \int_{s_1}^{z_1} \int_{t_1}^{z_1} F(z_1, \xi_1, t_1)\, d\xi_1\, dt_1 ;$$

here, $F(z_1, \xi_1, t_1) = K(z_1, \xi_1)\Gamma_n(\xi_1, t_1)$. Now every arbitrary (but fixed) compact subset of $\Omega_1 \times \Omega_1$ is contained in a domain $\Omega_1^0 \times \Omega_1^0$ with closure $\overline{\Omega_1^0} \subset \Omega_1$ compact in $\mathbb{C}$. Also, $\overline{\Omega_1^0}$ can be covered by finitely many bounded convex domains $\Omega_{1\kappa}$, $\kappa = 1, \cdots, \kappa_0$, such that

$$\overline{\Omega_1^0} \subset \bigcup_{\kappa=1}^{\kappa_0} \Omega_{1\kappa} \subset \bigcup_{\kappa=1}^{\kappa_0} \overline{\Omega_{1\kappa}} \subset \Omega_1 .$$

Thus the function

$$K_{m+1}(z_1, \xi_1) = \int_{\xi_1}^{z_1} K(z_1, t_1) \int_{\xi_1}^{t_1} K(t_1, t_2) \cdots \int_{\xi_1}^{t_{m-2}} K(t_{m-2}, t_{m-1}) \times$$

$$\times \int_{\xi_1}^{t_{m-1}} K(t_{m-1}, t_m) K(t_m, \xi_1)\, dt_m\, dt_{m-1} \cdots dt_2\, dt_1,$$

with $t_0 = z_1$, $(z_1, \xi_1) \in \overline{\Omega_1^0} \times \overline{\Omega_1^0}$, and the paths of integration in $\overline{\Omega_1^0}$, may be estimated by using Cauchy's theorem. This can be done by taking for every (z_1, ξ_1) as the path of integration a polygon of at most κ_0 straight-line segments of length not exceeding α, where

(6.2.7a) $$\alpha = \max_{\kappa} \max\left\{ |\tau_1 - \tau_2| \mid (\tau_1, \tau_2) \in \overline{\Omega_{1\kappa}} \times \overline{\Omega_{1\kappa}} \right\}.$$

Thus, setting

(6.2-7b) $$M = \max\left\{ |K(z_1, \xi_1)| \mid (z_1, \xi_1) \in \overline{\Omega_1^0} \times \overline{\Omega_1^0} \right\}$$

we obtain

(6.2-7c) $$|K_{m+1}(z_1, \xi_1)| \leq \kappa_0 M \frac{(\alpha M)^m}{m!} \qquad \text{for those } (z_1, \xi_1); \quad m \in \mathbb{N}.$$

It follows that the series $\Sigma K_m(z_1, \xi_1)$ has the convergent majorant $\kappa_0 M \exp(\alpha M)$. Hence the sequence (Γ_n) given by (6.2-5) converges uniformly, and the resolvent kernel $\Gamma = \lim \Gamma_n$ of the equation is holomorphic in Ω_1. The same holds for the v_n's in (6.2-4):

(6.2-8) $$v(z_1) = \lim_{n \to \infty} v_n(z_1) = f(z_1) + \int_{s_1}^{z_1} \Gamma(z_1, \xi_1) f(\xi_1)\, d\xi_1$$

where v is holomorphic in Ω_1. This completes the proof. ■

Further consequences of this proof are collected in the next lemma. In particular, statement (a) follows from the representation (6.2-8) and (c) from (6.2-5).

6.2-3 *Lemma*

(a) *The unique solution v of* (6.2-2) *is also holomorphic with respect to s_1, so that $v \in C^{\omega}(\Omega_1 \times \Omega_1)$ as a function of both variables.*

(b) *If the functions in* (6.2-2) *depend holomorphically on a parameter* ξ_0 *in a simply connected domain* H, *so that* K *becomes a holomorphic function of three variables and* f *of two variables, the same holds for* Γ *and* v, *respectively.*

(c) *For* $(z_1, \xi_1) \in \Omega_1 \times \Omega_1$ *the following equation is valid*:

$$\Gamma(z_1, \xi_1) - \int_{\xi_1}^{z_1} K(z_1, t_1)\Gamma(t_1, \xi_1)\, dt_1 = K(z_1, \xi_1).$$

EXISTENCE AND UNIQUENESS OF THE COMPLEX RIEMANN FUNCTION

The lemmas just obtained enable us to prove the analogous results for Volterra equations in two variables as needed in Def. 6.2-1. In fact, the following theorem holds.

6.2-4 Theorem

The integral equation

$$\begin{aligned} w(z) &- \int_{s_1}^{z_1} K_1(z,\zeta_1) w(z)\Big|_{z_1=\zeta_1} d\zeta_1 - \int_{s_2}^{z_2} K_2(z,\zeta_2) w(z)\Big|_{z_2=\zeta_2} d\zeta_2 \\ &- \int_{s_1}^{z_1}\int_{s_2}^{z_2} K(z,\zeta) w(\zeta)\, d\zeta_2\, d\zeta_1 = f(z), \quad \zeta = (\zeta_1, \zeta_2) \end{aligned} \tag{6.2-9}$$

with $(z, s) = (z_1, z_2, s_1, s_2) \in G = (\Omega_1 \times \Omega_1^*)^2$, *fixed* s, *and* $K_1, K_2, K, f \in C^\omega(G)$ *has a unique* $C^\omega(\Omega_1 \times \Omega_1^*)$*-solution* w. *This solution also depends holomorphically on* s_1 *and* s_2, *so that* $w \in C^\omega(G)$.

Proof. (a) We first consider the special case of the *reduced equation* with $K_1 = K_2 = 0$ and write $K = \tilde{K}$, so that we have

$$\tilde{w}(z) - \int_{s_1}^{z_1}\int_{s_2}^{z_2} \tilde{K}(z,\zeta)\tilde{w}(\zeta)\, d\zeta_2\, d\zeta_1 = f(z) \tag{6.2-10}$$

where $\tilde{K} \in C^\omega(G)$. For this equation, the proof of Theorem 6.2-4 is quite similar to that of Lemma 6.2-2. Indeed, the uniqueness of the solution follows by considering the Taylor series expansion of the difference of two solutions in a neighborhood of (s_1, s_2). The solution $\tilde{w}$ of (6.2-10) is given by

$$\tilde{w}(z) = f(z) + \int_{s_1}^{z_1}\int_{s_2}^{z_2} \tilde{\Gamma}(z,\zeta) f(\zeta)\, d\zeta_2\, d\zeta_1 \tag{6.2-11}$$

where

$$\tilde{\Gamma}(z,\zeta) = \lim_{n\to\infty} \tilde{\Gamma}_n(z,\zeta)$$

and

$$\tilde{\Gamma}_n(z,\zeta) = \sum_{m=1}^{n} \tilde{K}_m(z,\zeta)$$

$$\tilde{K}_1(z,\zeta) = \tilde{K}(z,\zeta)$$

$$\tilde{K}_{m+1}(z,\zeta) = \int_{\zeta_1}^{z_1}\int_{\zeta_2}^{z_2} \tilde{K}(z,t)\tilde{K}_m(t,\zeta)\,dt_2\,dt_1, \qquad t = (t_1, t_2).$$

$\tilde{w}$ in (6.2-11) is obtained from the sequence

$$\tilde{w}_0(z) = f(z)$$

$$\tilde{w}_n(z) = f(z) + \int_{s_1}^{z_1}\int_{s_2}^{z_2} \tilde{K}(z,\zeta)\tilde{w}_{n-1}(\zeta)\,d\zeta_2\,d\zeta_1, \qquad n \in \mathbb{N}.$$

Here $\tilde{w}_n$ can be converted to the form

$$\tilde{w}_n(z) = f(z) + \int_{s_1}^{z_1}\int_{s_2}^{z_2} \tilde{\Gamma}_n(z,\zeta) f(\zeta)\,d\zeta_2\,d\zeta_1$$

as can be proved by induction and two applications of the Dirichlet formula (6.2-6). The uniform convergence of the sequence $(\tilde{\Gamma}_n)$ is again shown by introducing a majorant on arbitrary (fixed) compact subsets of G. For instance, here one can take the series representing $\tilde{M}\exp(\alpha_1\alpha_2\tilde{M})$, where α_1 and α_2 are constants similar to α already given and $\tilde{M}$ is a sufficiently large constant. The representation (6.2-11) also guarantees the holomorphic dependence of $\tilde{w}$ on $(s_1, s_2) \in \Omega_1 \times \Omega_1^*$, in analogy to Lemmas 6.2-2 and 6.2-3. This proves the theorem for equation (6.2-10).

(b) We turn to the general case of (6.2-9). This case can be reduced to that of an equation of the form (6.2-10) as follows. We consider K_j as the kernel of an equation (6.2-2) in Ω_j, $j = 1, 2$, regarding z_{j*} as a parameter on which the kernel depends holomorphically in Ω_{j*}; here, $\Omega_2 = \Omega_1^*$. Then parts (b) and (c) of Lemma 6.2-3 imply that for the resolvent kernels Γ_1 and Γ_2 of those equations we have

$$\Gamma_j \in C^\omega(\Omega_1 \times \Omega_1^* \times \Omega_j) \subset C^\omega(G) \tag{6.2-12}$$

and

$$\Gamma_j(z,\xi_j) - \int_{\xi_j}^{z_j} K_j(z,t_j)\Gamma_j(z,\xi_j)\Big|_{z_j=t_j} dt_j = K_j(z,\xi_j), \qquad j = 1, 2. \tag{6.2-13}$$

Here Γ_1 and Γ_2 are given by

(6.2-14a) $$\Gamma_j = \lim_{n\to\infty} \Gamma_{jn} = \sum_{m=1}^{\infty} K_{jm}$$

and

$$K_{j1} = K_j$$

(6.2-14b) $$K_{j,m+1}(z,\xi_j) = \int_{\xi_j}^{z_j} K_j(z,t_j) K_{jm}(z,\xi_j)\Big|_{z_j=t_j} dt_j, \qquad j = 1,2.$$

We now consider

(6.2-15) $$w(z) = \tilde{w}(z) + \sum_{j=1}^{2} \int_{s_j}^{z_j} \Gamma_j(z,\xi_j)\tilde{w}(z)\Big|_{z_j=\xi_j} d\xi_j$$

and suppose that $\tilde{w}$ is to be determined. We substitute w into the original equation (6.2-9). Then 6 of the resulting 13 terms vanish because of (6.2-13) and the Dirichlet formula. To the two triple integrals we apply the Dirichlet formula. Then we obtain a reduced equation of the form (6.2-10). This equation is equivalent to (6.2-9) if we take the kernel $\tilde{K}$ defined by

(6.2-16)
$$\begin{aligned}\tilde{K}(z,\xi) = K(z,\zeta) &+ K_1(z,\xi_1)\Gamma_2(z,\xi_2) + K_2(z,\xi_2)\Gamma_1(z,\xi_1)\\ &+ \int_{\xi_1}^{z_1} K(z,t_1,\xi_2)\Gamma_1(t_1,\xi_2,\xi_1)\,dt_1\\ &+ \int_{\xi_2}^{z_2} K(z,\xi_1,t_2)\Gamma_2(\xi_1,t_2,\xi_2)\,dt_2.\end{aligned}$$

Now, according to part (a) of this proof, equation (6.2-10) with this kernel (6.2-16) has a unique solution $\tilde{w} \in C^{\omega}(G)$. From this and (6.2-15) the same follows for the solution w of (6.2-9), because Γ_1 and Γ_2 are holomorphic in the domain considered [cf. (6.2-12)]. This completes the proof of the theorem. ■

Setting

(6.2-17)
$$\begin{aligned} &K_1(z,\xi_1) = a_2(\xi_1,z_2), \qquad && K_2(z,\xi_2) = a_1(z_1,\xi_2)\\ &K(z,\xi) = -a_0(\xi), && f(z) = 1\end{aligned}$$

in (6.2-9), we obtain equation (6.2-1) as a special case. This yields the following result.

6.2-5 Corollary

Equation (6.2-1) *has a unique* $C^{\omega}(G)$-*solution*.

Consequently, the Riemann function given by Def. 6.2-1 is *well defined* as a $C^{\omega}(G)$-solution of (6.2-1).

6.3 PROPERTIES OF THE COMPLEX RIEMANN FUNCTION

Further characterizations of the Riemann function familiar from the theory of real hyperbolic partial differential equations carry over to the present setting in complex. Indeed, we have the following assertions.

6.3-1 Theorem

(a) $R \in C^{\omega}(G)$ *is the Riemann function of* (6.1-4) *if and only if the following two conditions hold in* G.

(6.3-1) *With respect to the first two variables,* R *satisfies the adjoint equation* $L^*R = 0$ *given by* (6.1-5);

and

$$R(z,s)|_{z_2=s_2} = \exp \int_{s_1}^{z_1} a_2(\zeta_1, s_2)\, d\zeta_1$$

(6.3-2)

$$R(z,s)|_{z_1=s_1} = \exp \int_{s_2}^{z_2} a_1(s_1, \zeta_2)\, d\zeta_2 .$$

(b) *The conditions in* (6.3-2) *are equivalent to*

(6.3-3)

$$\left[(D_{j^*} - a_j) R \right]\Big|_{z_j = s_j} = 0, \quad j = 1,2, \qquad \text{and} \qquad R(z,z) = 1 \text{ in } G .$$

(c) *The conditions in* (6.3-2) *are equivalent to*

(6.3-4)

$$\left[\left(\frac{\partial}{\partial s_{j^*}} + a_j \right) R \right] \Bigg|_{s_j = z_j} = 0, \quad j = 1,2, \qquad \text{and} \qquad R(z,z) = 1 \text{ in } G .$$

Proof. (6.3-1) and (6.3-3) are obtained by applying $D_1 D_2$ to (6.2-1). The equivalence of (6.3-2) to (6.3-4) is trivial. Integration of (6.1-5) from s_1 to z_1 and from s_2 to z_2 and substitution of (6.3-3) into the resulting equation yields the equation (6.2-1) for R, and the theorem is proved. ■

SYMMETRY OF THE COMPLEX RIEMANN FUNCTION

A basic property often used in applications is the following one.

6.3-2 Theorem

Let R denote the Riemann function for the equation (6.1-4) *and R^* the Riemann function for the adjoint equation* (6.1-5) *in the domain G. Then in G,*

$$R^*(z, s) = R(s, z). \tag{6.3-5}$$

In particular, in the self-adjoint case,

$$R(z, s) = R(s, z).$$

Proof. From the conditions (6.3-4) for R we can easily conclude that R^* given by (6.3-5) satisfies the conditions (6.3-3) with a_j replaced by $-a_j$, as required for R^*; here, $j = 1, 2$. Thus, according to Theorem 6.3-1, it remains to show that R^*, as a function of z, satisfies $LR^* = 0$, the adjoint of the adjoint equation. For this purpose we write $D_{\zeta_j} = \partial/\partial\zeta_j$, $j = 1, 2$, and $L_\zeta = L|_{z=\zeta}$. Then from the equation (6.3-1) for R it follows that (in the variables ζ_1, ζ_2)

$$D_{\zeta_1}D_{\zeta_2}(w^*R^*) - R^*L_\zeta w^* = D_{\zeta_1}\left[w^*(D_{\zeta_2} - a_1)R^*\right] + D_{\zeta_2}\left[w^*(D_{\zeta_1} - a_2)R^*\right]$$

for arbitrary functions $w^* \in C^\omega(\Omega_1 \times \Omega_1^*)$. We now integrate this equation with respect to ζ_1 from s_1 to z_1 and with respect to ζ_2 from s_2 to z_2 in $\Omega_1 \times \Omega_1^*$. At this stage we observe the conditions (6.3-4) for R^*, with $-a_j$ instead of a_j [see before regarding the validity of (6.3-3) for R^* and Theorem 6.3-1]. We then obtain

$$\begin{aligned} w^*(z) = w^*(s)R^*(z, s) &+ \int_{s_1}^{z_1} R^*(z, \zeta_1, s_2)\left[(D_{\zeta_1} + a_2)w^*\right](\zeta_1, s_2)\, d\zeta_1 \\ &+ \int_{s_2}^{z_2} R^*(z, s_1, \zeta_2)\left[(D_{\zeta_2} + a_1)w^*\right](s_1, \zeta_2)\, d\zeta_2 \\ &+ \int_{s_1}^{z_1}\int_{s_2}^{z_2} R^*(z, \zeta)L_\zeta w^*(\zeta)\, d\zeta_2\, d\zeta_1. \end{aligned} \tag{6.3-6}$$

In particular, this holds if we set $w^*(\zeta) = R^*(\zeta, s)$. Because of the condition (6.3-3) for R^*, the equation (6.3-6) then reduces to

$$R^*(z, s) = R^*(z, s) + \int_{s_1}^{z_1}\int_{s_2}^{z_2} R^*(z, \zeta)L_\zeta R^*(\zeta, s)\, d\zeta_2\, d\zeta_1.$$

Since R is holomorphic in G, so is R^*, and since $R^* \neq 0$, we conclude that

$L_t R^*(t, s)$ must vanish identically in $\Omega_1 \times \Omega_1^*$. This shows that R^* given by (6.3-5) satisfies the condition (6.3-1). In view of Theorem 6.3-1 this proves Theorem 6.3-2. ■

TRANSFORMATION OF DIFFERENTIAL EQUATIONS AND RIEMANN FUNCTIONS

Since only a few Riemann functions R are known explicitly (cf. Chap. 7), but since one can obtain further Riemann functions by suitable transformations, it is of practical importance to investigate the behavior of R under a transformation of the differential equation. In the following theorem we state some results which are of interest in applications. These can be proved by straightforward calculation and verification of the conditions (6.3-1) and (6.3-3).

6.3-3 Theorem

(a) *For a given equation* $Lw = 0$ *of the form* (6.1-4), *let* Ω_1 *be a fundamental domain and R the Riemann function. Let*

$$\tilde{w}(z) = w(z)/\lambda(z)$$

where $\lambda \in C^\omega(\Omega_1 \times \Omega_1^*)$ *and* $\lambda(z) \neq 0$ *in* $\Omega_1 \times \Omega_1^*$. *Then* Ω_1 *is also a fundamental domain for the equation*

$$\tilde{L}\tilde{w} = D_1 D_2 \tilde{w} + \left(a_1 + \frac{1}{\lambda} D_2 \lambda\right) D_1 \tilde{w} + \left(a_2 + \frac{1}{\lambda} D_1 \lambda\right) D_2 \tilde{w} + \left(\frac{1}{\lambda} L\lambda\right) \tilde{w} = 0$$

and the Riemann function for this equation is given by

$$\tilde{R}(z, s) = R(z, s)\lambda(z)/\lambda(s).$$

(b) *Let* Ω_1 *and R be as before,* B_1 *a simply connected domain,*

$$f\colon B_1 \to \Omega_1, \quad t_1 \mapsto z_1, \quad \sigma_1 \mapsto s_1$$

a conformal bijection, and

$$\bar{f}\colon B_1^* \to \Omega_1^*, \quad t_2 \mapsto z_2 = \bar{f}(t_2) = \overline{f(\bar{t}_2)}, \qquad \sigma_2 \mapsto s_2 = \overline{f(\bar{\sigma}_2)}.$$

Then B_1 *is a fundamental domain for the equation*

$$\tilde{L}\tilde{w} = D_{t_1} D_{t_2} \tilde{w} + \tilde{a}_1 D_{t_1} \tilde{w} + \tilde{a}_2 D_{t_2} \tilde{w} + \tilde{a}_0 \tilde{w} = 0$$

where $D_{t_j} = \partial/\partial t_j$, *with coefficients*

$$\tilde{a}_j(t) = a_j\big(f(t_1), \bar{f}(t_2)\big) D_{t_{j^*}} f(t_{j^*}), \qquad j = 1, 2,$$

$$\tilde{a}_0(t) = a_0\big(f(t_1), \bar{f}(t_2)\big) D_{t_1} f(t_1) D_{t_2} \bar{f}(t_2).$$

The Riemann function $\tilde{R}$ *for this equation is given by*

$$\tilde{R}(t, \sigma) = R\big(f(t_1), \bar{f}(t_2), f(\sigma_1), \bar{f}(\sigma_2)\big) = R(z, s)$$

where $\sigma = (\sigma_1, \sigma_2)$.

Results about the explicit determination of Riemann functions for equations of the form (6.1-4) are discussed in Chap. 7, and many examples are given there.

6.4 RIEMANN–VEKUA REPRESENTATION AND BERGMAN REPRESENTATION

In this section we develop a new approach to the Riemann–Vekua representation for the Goursat problem involving the equation (6.1-4). We derive this representation from the Bergman representation theorem with kernels of the first kind. This method has the advantage that the Riemann function can be obtained from a single Bergman kernel of the first kind. This fact will be of importance in the explicit determination of the Riemann function for a given equation.

DETERMINATION OF THE RIEMANN FUNCTION FROM BERGMAN KERNELS

6.4-1 Theorem

Let $\tilde{k}_j = k_j e_j \in C^\omega(\Omega \times \bar{B}_0)$, $j = 1, 2$, *be the kernels for Bergman operators of the first kind for Type II representations of solutions of* (6.1-4). *Let* Ω_1 *be a fundamental domain for* (6.1-4) *and let* $\Omega = \Omega_1 \times \Omega_1^*$. *Assume that the* $\tilde{k}_j$*'s depend holomorphically on* s_1 *and* s_2 *(that is, on the point of reference) in* $\Omega_1 \times \Omega_1^*$. *Furthermore, suppose that the functions* η_1 *and* η_2 *in* (2.1-3) *are chosen so that* $e_j = e_j^R$, *where*

$$(6.4\text{-}1)\quad e_j^R(z, s) = \exp\left[-\int_{s_j}^{z_j} a_{j^*}(\xi)\Big|_{\xi_{j^*} = s_{j^*}} d\xi_j - \int_{s_{j^*}}^{z_{j^*}} a_j(\xi)\Big|_{\xi_j = z_j} d\xi_{j^*}\right]$$

and $j = 1, 2$. *Then we have, for* $j = 1, 2$,

$$(6.4\text{-}2) \qquad R^*(z, s) = \frac{1}{\pi} e_j^R(z, s) \int_{-1}^{1} k_j(z, t) \tau^{-1/2}\, dt, \qquad \tau = 1 - t^2$$

where R^* *is the Riemann function for the adjoint equation* (6.1-5). *By the symmetry of the Riemann function (cf. Theorem* 6.3-2), *this yields a representation for the Riemann function* R *of* (6.1-4), *depending holomorphically on the variables* z_1, z_2, s_1, s_2 *in the domain* $G = (\Omega_1 \times \Omega_1^*)^2$. *Furthermore, those Bergman kernels satisfy the integral relation*

$$(6.4\text{-}3) \qquad \int_{-1}^{1} \tilde{k}_1(z, t) \tau^{-1/2}\, dt = \int_{-1}^{1} \tilde{k}_2(z, t) \tau^{-1/2}\, dt.$$

Proof. We show that R^* satisfies the conditions (6.3-1) and (6.3-2), rewritten for R^* instead of R; that is:

(6.4-4) With respect to the first two variables, R^* is a solution of the adjoint equation $LR^* = 0$ of (6.1-5);

and [cf. (6.4-1)]

$$(6.4\text{-}5) \qquad \begin{aligned} R^*(z, s)|_{z_{l^*} = s_{l^*}} &= \exp\left[-\int_{s_l}^{z_l} a_{l^*}(\tilde{z})\Big|_{\tilde{z}_{l^*} = s_{l^*}} d\tilde{z}_l\right] \\ &= e_j^R(z, s)|_{z_{l^*} = s_{l^*}}, \qquad l^* = 3 - l; \quad l = 1, 2. \end{aligned}$$

From the Bergman Representation Theorem 2.3-1 we immediately conclude that (6.4-2) represents a $C^\omega(G)$-solution of $LR^* = 0$ with respect to the variables z_1, z_2 and with $f_j = 1/\pi$. From the condition (2.4-1) in the definition of a kernel of the first kind it follows that, for $z_1 = s_1$ or $z_2 = s_2$, the right side of (6.4-2) equals the right side of (6.4-5). Consequently, (6.4-2) gives indeed the Riemann function for (6.1-5). The integral relation in (6.4-3) then follows from the uniqueness of the Riemann function; see Corollary 6.2.5. This completes the proof. ■

APPLICATIONS OF THEOREM 6.4-1

Obviously, the relation (6.4-3) may be utilized for deriving relations between solutions of partial differential equations of the form (6.1-4), which are not self-adjoint. Furthermore, (6.4-2) and Theorem 6.3-2 yield representations of the Riemann function; cf. Examples 6.4-2 and Corollary 6.4-3.

6.4-2 Examples

(a) We consider the Helmholtz equation

$$Lw = D_1 D_2 w + \tfrac{1}{4} c^2 w = 0$$

where $c = const \neq 0$. In Example 2.5-4 we have shown that for this equation the kernels of the first kind are given by

$$\tilde{k}_j(z,t) = \cos\left(c\sqrt{(z_1 - s_1)(z_2 - s_2)}\,t\right), \qquad j = 1,2.$$

Applying Theorems 6.4-1 and 6.3-2 and using formula (2.2-5), we obtain from this

$$\begin{aligned} R(z,s) &= R(s,z) = R^*(z,s) \\ &= \frac{1}{\pi}\int_{-1}^{1} \cos\left(c\sqrt{(z_1 - s_1)(z_2 - s_2)}\,t\right)\tau^{-1/2}\,dt \\ &= \frac{1}{\pi}\sum_{m=0}^{\infty}\frac{1}{(2m)!}(-1)^m\left(c\sqrt{(z_1 - s_1)(z_2 - s_2)}\right)^{2m} \times \\ &\quad \times \int_{-1}^{1} t^{2m}(1 - t^2)^{-1/2}\,dt \\ &= \sum_{m=0}^{\infty}\frac{1}{(m!)^2}\left(-\frac{1}{4}\right)^m\left(c\sqrt{(z_1 - s_1)(z_2 - s_2)}\right)^{2m} \\ &= J_0\left(c\sqrt{(z_1 - s_1)(z_2 - s_2)}\right). \end{aligned} \tag{6.4-6}$$

(b) As an equation which also illustrates the relation (6.4-3) we consider

$$Lw = D_1 D_2 w - z_2 D_2 w + w = 0.$$

By a simple calculation one can verify that the Bergman kernels of the first kind have the form

$$\begin{aligned} \tilde{k}_j &= e_j^R k_j \qquad \text{with} \\ e_1^R(z,s) &= \exp\left[s_2(z_1 - s_1)\right] \\ k_1(z,t) &= 1 - 2(z_1 - s_1)(z_2 - s_2)t^2 \\ e_2^R(z,s) &= \exp\left[z_2(z_1 - s_1)\right] \\ k_2(z,t) &= \sum_{m=0}^{\infty}\left[4^m(m+1)!/(2m)!\right]\left[-(z_1 - s_1)(z_2 - s_2)t^2\right]^m \\ &= {}_1F_1\left(2; \tfrac{1}{2}; -(z_1 - s_1)(z_2 - s_2)t^2\right) \end{aligned} \tag{6.4-7}$$

where ${}_1F_1$ is *Kummer's function* (*confluent hypergeometric function*; cf. also Sec. 2.7). Applying (6.4-2) and (2.2-5), we obtain from (6.4-7) for $j = 1$

$$R^*(z,s) = \left[1 - (z_1 - s_1)(z_2 - s_2)\right]\exp\left[s_2(z_1 - s_1)\right] \tag{6.4-8}$$

and for $j = 2$

$$(6.4\text{-}9)\quad R^*(z, s) = \sum_{m=0}^{\infty} \frac{m+1}{m!} [-(z_1 - s_1)(z_2 - s_2)]^m \exp[z_2(z_1 - s_1)].$$

Obviously, (6.4-8) and (6.4-9) represent the same function, in agreement with (6.4-3).

From Theorem 6.4-1 we may derive the following:

6.4-3 Corollary

Let e_j^R be given by (6.4-1), and let $j = 1$ or $j = 2$ be arbitrary but fixed.

(a) *Suppose that k_j in (6.4-2) has the series expansion*

$$k_j(z, t; s) = \sum_{m=0}^{\infty} q_{jm}(z, s) t^{2m}.$$

Then the Riemann function has the representation

$$(6.4\text{-}10)\quad \begin{aligned} R(z, s) &= e_j^R(s, z) \sum_{m=0}^{\infty} q_{jm}(s, z)(2m)!/\left[4^m (m!)^2\right] \\ &= e_j^R(s, z) \sum_{m=0}^{\infty} \left[4^m (m!)^2\right]^{-1} \left[D_t^{2m} k_j(s, t; z)\right]\Big|_{t=0}. \end{aligned}$$

(b) *Suppose that $\tilde{k}_j = e_j^R k_j$ is a polynomial kernel of the first kind of degree n_j (in t^2) and with point of reference s. Then the Riemann function is given by*

$$(6.4\text{-}11)\quad R(z, s) = e_j^R(s, z)\left(1 + \sum_{m=1}^{n_j} q_{jm}(s, z)(2m)!/\left[4^m (m!)^2\right]\right);$$

here, the q_{jm}'s are uniquely determined and satisfy

$$q_{jm}(s, z) = \frac{1}{(2m)!}\left[D_t^{2m} k_j(s, t; z)\right]\Big|_{t=0}$$

and

$$q_{jm}(s, z)|_{s_l = z_l} = 0, \qquad l = 1, 2; \quad m = 1, \cdots, n_j.$$

Proof. From (2.2-5) and (6.4-2) and the symmetry of the Riemann function, part (a) is obvious. The proof of part (b) immediately follows from (a) if one observes Defs. 3.1-1 and 2.4-1 and Corollary 3.1-5. ■

We mention that part (b) of this corollary agrees with an analogous result by Bauer [1984] which was obtained by starting from differential operators as introduced in Sec. 3.3.

BERGMAN KERNELS IN TERMS OF COMPLEX RIEMANN FUNCTIONS

We are now able to solve the problem of the determination of the function k_j in terms of a given R, as follows.

6.4-4 Theorem

Let $R^* \in C^\omega(G)$ be the Riemann function for the adjoint equation (6.1-5) *of* (6.1-4); *here $G = (\Omega_1 \times \Omega_1^*)^2$ and Ω_1 is some fundamental domain for* (6.1-4). *Then the uniquely determined even Bergman kernels of the first kind can be represented in a neighborhood of $(s_1, s_2) \in \Omega_1 \times \Omega_1^*$ by the series*

$$\tilde{k}_j(z, t) = e_j^R(z, s)\left[1 + \sum_{m=1}^{\infty} \tilde{q}_{jm}(z)(z_j - s_j)^m t^{2m}\right]; \tag{6.4-12a}$$

here, $j = 1, 2$, and for $m \in \mathbb{N}$,

$$e_j^R(z, s)\tilde{q}_{jm}(z) = \frac{(-4)^m m!}{(2m)!}\left[\frac{\partial^m}{\partial \xi_j^m}\left(e_{j*}^R(\xi, s)R(\xi, z)\right)\bigg|_{\xi_{j*} = s_{j*}}\right]_{\xi_j = z_j}. \tag{6.4-12b}$$

In this formula, e_j^R is given in (6.4-1) *and we have $R(\xi, z) = R^*(z, \xi)$, that is, R is the Riemann function for* (6.1-4). *The domain of convergence of the series in* (6.4-12) *is given in Theorem* 2.4-2. *The analytic continuation is provided by*

$$\tilde{k}_j(z, t) = e_j^R(z, s)k_j(z, t)$$

$$\begin{aligned} k_j(z, t) &= 1 + \left[e_j^R(z, s)\right]^{-1}(z_j - s_j)^{1/2} t \times \\ &\quad \times \int_{z_j}^{Z_j}\left[\frac{\partial}{\partial \xi_j}\left(e_{j*}^R(\xi, s)R(\xi, z)\right)\big|_{\xi_{j*} = s_{j*}}\right](\xi_j - Z_j)^{-1/2}\, d\xi_j \end{aligned} \tag{6.4-13}$$

where $Z_j = s_j + (z_j - s_j)(1 - t^2)$, $j = 1, 2$.

Proof. We take the conditions (6.4-4) and (6.4-5) for R^* into account and use the definition (6.4-1) of e_j^R. Then a formal calculation shows that the functions $\tilde{q}_{jm}$ in (6.4-12b) satisfy the system (2.5-9) and therefore also the kernel equation (2.3-2a). Furthermore, from (6.4-1) and (6.4-5) we obtain

$$\tilde{q}_{jm}(z)|_{z_{j^*}=s_{j^*}} = 0, \qquad m \in N; \quad j = 1,2.$$

This implies the condition (2.4-1) for kernels of the first kind. Since by Corollary 2.5-3 the even part of a kernel of the first kind is unique, it follows that the representation (6.4-12) coincides with that in Theorem 2.4-2. Furthermore, repeated integration by parts shows that (6.4-13) is the analytic continuation of (6.4-12). ■

APPLICATION OF THEOREM 6.4-4

As a typical example illustrating the usefulness of (6.4-12) in applications, we derive the kernels (2.5-14) from the Riemann function for the equation in Example 2-5.5.

6.4-5 Example

Let us consider the equation (2.5-13):

$$Lw = D_1 D_2 w + \frac{\lambda n(n+1)}{(1+\lambda z_1 z_2)^2} w = 0, \quad \lambda \in \mathbb{C}; \quad n \in \mathbb{N}.$$

Setting $\tilde{R}(\rho) = R(z, s)$, where

$$\rho(z,s) = \frac{\lambda(z_1 - s_1)(z_2 - s_2)}{(1+\lambda z_1 z_2)(1+\lambda s_1 s_2)} \tag{6.4-14}$$

we see that the conditions for the Riemann function R of the preceding self-adjoint partial differential equation reduce to the ***hypergeometric equation***

$$\rho(1-\rho)\tilde{R}'' + (1-2\rho)\tilde{R}' + n(n+1)\tilde{R} = 0, \qquad \rho(0) = 1.$$

The regular solution of this *ordinary* differential equation is given by

$$\tilde{R}(\rho) = {}_2F_1(n+1, -n; 1; \rho) = P_n(1-2\rho). \tag{6.4-15}$$

Here, ${}_2F_1$ is the *hypergeometric function* and P_n the *Legendre polynomial* of nth degree, as usual. This representation (6.4-15) is valid in every domain $G = (\Omega_1 \times \Omega_1^*)^2$, where Ω_1 is a fundamental domain. Hence it holds for every simply connected domain Ω_1 for which the coefficient of w in (2.5-13) is in $C^\omega(\Omega_1 \times \Omega_1^*)$.

If we choose $s = (0,0)$, then (6.4-12, 14, 15) yield

$$\tilde{k}_j(z,t) = 1 + \sum_{m=1}^{\infty} \frac{(-4)^m m!}{(2m)!}\left[\frac{\partial^m}{\partial \xi_j^m} P_n\left(1 + \frac{2\lambda z_{j*}}{1+\lambda z_1 z_2}(\xi_j - z_j)\right)\right]\Bigg|_{\xi_j = z_j} \times \tag{6.4-16}$$

$$\times (z_j - s_j)^m t^{2m}.$$

Now $C_\kappa^\nu(1) = (2\nu)_\kappa/\kappa!$ when $\kappa \in \mathbb{N}_0$ and $-2\nu \notin \mathbb{N}_0$, and

$$d^m P_n(\sigma)/d\sigma^m = 2^m\left(\tfrac{1}{2}\right)_m C_{n-m}^{m+1/2}(\sigma), \quad m \leq n;$$

cf. Magnus et al. [1966], pp. 218, 232. We thus obtain for $[\cdots]|_{\xi_j = z_j}$ in (6.4-16) the expression

$$\left(\frac{2\lambda z_{j*}}{1+\lambda z_1 z_2}\right)^m \left[\frac{d^m}{d\sigma^m} P_n(\sigma)\right]\Bigg|_{\sigma=1} = \frac{(n+m)!}{m!(n-m)!}\left(\frac{\lambda z_{j*}}{1+\lambda z_1 z_2}\right)^m \quad \text{when } m \leq n$$

and 0 for $m > n$; here $m \in \mathbb{N}$. Inserting this into (6.4-16), we arrive at

$$\tilde{k}_j(z,t) = \sum_{m=0}^{n} \binom{n+m}{2m}\left(\frac{-4\lambda z_1 z_2}{1+\lambda z_1 z_2}\right)^m t^{2m}, \qquad j = 1,2$$

in agreement with (2.5-14).

RIEMANN–VEKUA SOLUTION OF THE GOURSAT PROBLEM

We shall now proceed to the derivation of the desired representation of the solution to the *Goursat problem* (2.1-1), (2.6-1) in terms of the Riemann function. This will be done by the use of *Type II Bergman representations* with the kernels $\tilde{k}_j$ as given by (6.4-13). We call this the ***Riemann–Vekua representation***.

6.4-6 Theorem

Consider the Goursat problem consisting of the equation (6.1-4) *and the conditions* (2.6-1),

$$w(z_1, s_2) = \varphi_1(z_1), \qquad w(s_1, z_2) = \varphi_2(z_2).$$

Suppose that $\varphi_1(s_1) = \varphi_2(s_2)$ *and* $\varphi_j \in C^\omega(\Omega_j)$, *where* Ω_1 *is a fundamental domain and* $\Omega_2 = \Omega_1^*$. *Then this problem has the following solution* (***Riemann–Vekua representation***) *in* $\Omega_1 \times \Omega_1^*$:

$$w(z) = \phi_1(s_1)R(s,z) + \sum_{j=1}^{2} \int_{s_j}^{z_j} \phi_j'(\zeta_j) R(\zeta, z)\Big|_{\zeta_{j*} = s_{j*}} d\zeta_j. \tag{6.4-17}$$

Here,

$$\phi_j(\zeta_j) = \varphi_j(\zeta_j) + \int_{s_j}^{\zeta_j} a_{j*}(\xi)\big|_{\xi_{j*}=s_{j*}} \varphi_j(\xi_j)\, d\xi_j, \qquad j = 1,2 \tag{6.4-18}$$

so that, denoting the Riemann function of (6.1-4) *by* R,

$$\begin{aligned} w(z) &= \varphi_1(s_1)R(s,z) \\ &\quad + \sum_{j=1}^{2} \int_{s_j}^{z_j} \left(\left[\varphi_j'(\zeta_j) + a_{j*}(\zeta)\varphi_j(\zeta_j) \right] R(\zeta, z) \right) \Big|_{\zeta_{j*}=s_{j*}} d\zeta_j . \end{aligned} \tag{6.4-19}$$

This representation can also be generated by $w = T_1 f_1 + T_2 f_2$, where T_1 and T_2 are the Bergman operators of the first kind given by the kernels in (6.4-13) and the associated functions f_1, f_2 are chosen as follows:

$$\begin{aligned} f_j(\xi_j) &= \frac{1}{\pi}\beta_j \varphi_j(s_j) \\ &\quad + \frac{4}{\pi}(\xi_j - s_j) \int_0^{\pi/2} \left(\frac{\partial}{\partial \tilde{z}_j} \left[\frac{\varphi_j(\tilde{z}_j)}{e_j^R(\tilde{z}, s)|_{\tilde{z}_{j*}=s_{j*}}} \right] \right) \Bigg|_{\tilde{z}_j = \tilde{\eta}_j(\xi_j)} \sin\theta\, d\theta \end{aligned} \tag{6.4-20}$$

with

$$\tilde{\eta}_j(\xi_j) = 2(\xi_j - s_j)\sin^2\theta + s_j, \quad j = 1,2, \quad \beta_1 + \beta_2 = 1 .$$

Proof. We start from the Bergman Type II representation with kernels of the first kind as given by (6.4-13). If we choose the associated functions f_1 and f_2 according to (2.6-4, 5), then Theorem 2.6-2 guarantees that the solution generated is the solution to the given Goursat problem. Now, indeed, (6.4-20) is a special case of (2.6-4) with (2.6-5) because, in view of (6.4-1), we may choose $\tilde{\varphi}_j$ in (2.6-5) in the form

$$\tilde{\varphi}_j(\tilde{z}_j) = \varphi_j(\tilde{z}_j)/e_j^R(\tilde{z}, s)|_{\tilde{z}_{j*}=s_{j*}} - (1 - \beta_j)\varphi_j(s_j) . \tag{6.4-21}$$

Hence we must now only show that the solution w thus obtained can be converted to the form (6.4-17, 18), that is, (6.4-19).

Inserting (6.4-13, 20) into the representation $w = T_1 f_1 + T_2 f_2$ of the solution, we first have

$$w = \sum_{j=1}^{2} \sum_{k=1}^{4} w_{jk} \tag{6.4-22a}$$

where the w_{jk}'s are of the form

$$w_{j1}(z) = e_j^R(z, s)\int_{-1}^{1} \frac{1}{\pi}\beta_j\varphi_j(s_j)\tau^{-1/2}\,dt = \beta_j\varphi_j(s_j)e_j^R(z, s)$$

$$w_{j2}(z) = e_j^R(z, s)\int_{-1}^{1} \frac{4}{\pi}(\xi_j - s_j)\int_0^{\pi/2} \left.\frac{d\tilde{\varphi}_j(\tilde{z}_j)}{d\tilde{z}_j}\right|_{\tilde{z}_j = \tilde{\eta}_j(\xi_j)} \sin\theta\, d\theta\, \tau^{-1/2}\,dt$$

$$w_{j3}(z) = \int_{-1}^{1} \frac{1}{\pi}\beta_j\varphi_j(s_j)(z_j - s_j)^{1/2} t \int_{z_j}^{2\xi_j - s_j}\left[\frac{\partial}{\partial\zeta_j}\left(e_{j^*}^R(\zeta, s)\,\times\right.\right.$$

$$\left.\left.\times\, R(\zeta, z)\right)\big|_{\zeta_{j^*} = s_{j^*}}\right]\frac{d\zeta_j}{\left(\zeta_j - (2\xi_j - s_j)\right)^{1/2}}\tau^{-1/2}\,dt \tag{6.4-22b}$$

$$w_{j4}(z) = \int_{-1}^{1} (z_j - s_j)^{1/2} t \int_{z_j}^{2\xi_j - s_j}\left[\frac{\partial}{\partial\zeta_j}\left(e_{j^*}^R(\zeta, s)R(\zeta, z)\right)\big|_{\zeta_{j^*} = s_{j^*}}\right] \times$$

$$\times \frac{d\zeta_j}{\left(\zeta_j - (2\xi_j - s_j)\right)^{1/2}} \frac{4}{\pi}(\xi_j - s_j)\int_0^{\pi/2} \left.\frac{d\tilde{\varphi}_j(\tilde{z}_j)}{d\tilde{z}_j}\right|_{\tilde{z}_j = \tilde{\eta}_j(\xi_j)} \sin\theta\, d\theta\, \tau^{-1/2}\,dt;$$

here, $\tau = 1 - t^2$ and

$$\tilde{\eta}_j(\xi_j) = 2(\xi_j - s_j)\sin^2\theta + s_j$$

$$\xi_j = s_j + \tfrac{1}{2}(z_j - s_j)\tau.$$

As in the proof of Theorem 2.6-2 we now obtain

$$w_{j2}(z) = e_j^R(z, s)\left[\tilde{\varphi}_j(z_j) - \tilde{\varphi}_j(s_j)\right].$$

For the calculation of w_{j3} we introduce $\rho_j = (z_j - s_j)\tau + s_j$ as a new variable of integration in the first integral of w_{j3} and interchange the order of integration by applying Dirichlet's formula. Then

$$w_{j3}(z) = \beta_j\varphi_j(s_j)\int_{z_j}^{s_j}\left(\int_{\zeta_j}^{s_j}\frac{d\rho_j}{\pi(\rho_j - \zeta_j)^{1/2}(s_j - \rho_j)^{1/2}}\right) \times$$

$$\times \frac{\partial}{\partial\zeta_j}\left(e_{j^*}^R(\zeta, s)R(\zeta, z)\right)\big|_{\zeta_{j^*} = s_{j^*}}\,d\zeta_j.$$

The inner integral equals 1. Because of (6.4-1, 5) it follows that

$$\left(e^R_{j*}(\zeta, s)R(\zeta, z)\right)\Big|_{\zeta_j = z_j,\, \zeta_{j*} = s_{j*}} = e^R_j(z, s). \tag{6.4-23}$$

We thus obtain by integration

$$w_{j3}(z) = \beta_j \varphi_j(s_j)\left[R(s, z) - e^R_j(z, s)\right]. \tag{6.4-24}$$

We finally substitute $\eta_j = \tilde{\eta}_j(\xi_j)$ (as before) in the third integral and ρ_j (as before) in the first integral. Then we have

$$w_{j4}(z) = -\int_{z_j}^{s_j}\int_{z_j}^{\rho_j}\int_{s_j}^{\rho_j}\left[\frac{\partial}{\partial \zeta_j}\left(e^R_{j*}(\zeta, s)R(\zeta, z)\right)\Big|_{\zeta_{j*} = s_{j*}}\right]\frac{d\tilde{\varphi}_j(\eta_j)}{d\eta_j} \times$$

$$\times \frac{1}{\pi(\zeta_j - \rho_j)^{1/2}(\rho_j - \eta_j)^{1/2}}\, d\eta_j\, d\zeta_j\, d\rho_j.$$

We now apply Dirichlet's formula twice, interchanging the order of integration in the first two integrals, and then in the result, in the last two integrals. This yields

$$w_{j4}(z) = \int_{z_j}^{s_j}\left(\int_{s_j}^{\zeta_j}\left(\int_{\eta_j}^{\zeta_j}\frac{d\rho_j}{\pi(\rho_j - \eta_j)^{1/2}(\zeta_j - \rho_j)^{1/2}}\right)\frac{d\tilde{\varphi}_j(\eta_j)}{d\eta_j}\, d\eta_j\right) \times$$

$$\times \frac{\partial}{\partial \zeta_j}\left(e^R_{j*}(\zeta, s)R(\zeta, z)\right)\Big|_{\zeta_{j*} = s_{j*}}\, d\zeta_j$$

$$= \int_{z_j}^{s_j}\left[\tilde{\varphi}_j(\zeta_j) - \tilde{\varphi}_j(s_j)\right]\frac{\partial}{\partial \zeta_j}\left(e^R_{j*}(\zeta, s)R(\zeta, z)\right)\Big|_{\zeta_{j*} = s_{j*}}\, d\zeta_j.$$

Integrating by parts and noting (6.4-1, 21) we obtain

$$\begin{aligned} w_{j4}(z) &= \int_{s_j}^{z_j}\left(\left[\varphi_j'(\zeta_j) + a_{j*}(\zeta)\varphi_j(\zeta_j)\right]R(\zeta, z)\right)\Big|_{\zeta_{j*} = s_{j*}}\, d\zeta_j \\ &\quad - \left[\tilde{\varphi}_j(z_j) - \tilde{\varphi}_j(s_j)\right]\left(e^R_{j*}(\zeta, s)R(\zeta, z)\right)\Big|_{\zeta_j = z_j,\, \zeta_{j*} = s_{j*}}. \end{aligned} \tag{6.4-25}$$

Because of (6.4-23), the second term in (6.4-25) equals $-w_{j2}(z)$. By (6.4-18) we thus have

$$w_{j2}(z) + w_{j4}(z) = \int_{s_j}^{z_j}\phi_j'(\zeta_j)R(\zeta, z)\Big|_{\zeta_{j*} = s_{j*}}\, d\zeta_j. \tag{6.4-26}$$

We now insert (6.4-22b, 24, 26) into (6.4-22a), finding

$$w(z) = \sum_{j=1}^{2} \Bigg(\beta_j \varphi_j(s_j) e_j^R(z, s) + \beta_j \varphi_j(s_j) \big[R(s, z) - e_j^R(z, s) \big] + \int_{s_j}^{z_j} \phi_j'(\zeta_j) R(\zeta, z) \Big|_{\zeta_{j^*} = s_{j^*}} d\zeta_j \Bigg).$$

This, however, is the representation (6.4-17, 18) [hence also (6.4-19)], because from $\varphi_1(s_1) = \varphi_2(s_2)$ and $\beta_1 + \beta_2 = 1$ we see that

$$\sum_{j=1}^{2} \beta_j \varphi_j(s_j) = \varphi_1(s_1) = \phi_1(s_1).$$

Our result is that, by Theorems 6.4-4 and 2.3-1, the function w of the preceding form is a solution of the given equation (6.1-4) which assumes the prescribed initial data on $z_1 = s_1$ and $z_2 = s_2$, as follows from Theorem 2.6-2. This completes the proof. ■

6.5 BERGMAN–GILBERT OPERATOR

A further, very interesting representation of solutions in terms of the Riemann function was introduced by Gilbert [1970a]. This method also has the remarkable property that it can be utilized in obtaining representations of solutions for equations in any number of independent variables. This is accomplished by a *method of ascent*, as explained in Gilbert's well-known monographs [1969b, 1974], along with various other possibilities of extending integral operator methods to higher dimensional problems. In the present section we consider the Bergman–Gilbert operator which produces the preceding representations. Furthermore, R. W. Carroll has recently interpreted this operator in the framework of his theory of transmutations; this will be discussed in Sec. 6.6.

We start from the "radial case" for elliptic equations in two real variables x_1, x_2; that is,

$$\Delta_2 \hat{w} + \hat{a}_1(r^2) r \hat{w}_r + \hat{a}_0(r^2) \hat{w} = 0, \quad \Delta_2 = \partial^2/\partial x_1^2 + \partial^2/\partial x_2^2 \tag{6.5-1}$$

where the coefficients are real-analytic functions of $r^2 = x_1^2 + x_2^2$. To eliminate $\hat{w}_r$, we set

$$\hat{u}(x) = \hat{w}(x) \exp\left[-\frac{1}{2} \int_0^r \hat{a}_1(\rho^2) \rho \, d\rho \right], \qquad x = (x_1, x_2).$$

Then (6.5-1) reduces to

$$\Delta_2 \hat{u} + \hat{a}(r^2)\hat{u} = 0 \tag{6.5-2}$$

with coefficient

$$\hat{a} = -\frac{r}{2}\hat{a}_{1r} - \hat{a}_1 - \frac{r^2}{4}\hat{a}_1^2 + \hat{a}_0 .$$

We may thus study this reduced equation, assuming a to be analytic in a neighborhood of the origin.

A BERGMAN REPRESENTATION INVOLVING HARMONIC FUNCTIONS

To equation (6.5-2) we can apply the theory of Chap. 2. As point of reference we choose simply $s = (0, 0)$. In the argument of the associated function f_1 we replace $\frac{1}{2}z_1$ by z_1. Next we remember from Sec. 2.1 the idea of a *real solution*. For (6.5-2) these solutions can be obtained by a suitable Type II representation (intuitively, a generalized operator Re; cf. Lemma 2.1-3 and Theorem 2.3-1). The form of (6.5-2) allows us to require the kernel of the desired Bergman operator to be of the form

$$k_1(z, t) \equiv k(\tilde{r}, t) = 1 + \sum_{m=1}^{\infty} \kappa_m(\tilde{r}^2)t^{2m}. \tag{6.5-3}$$

Here

$$\tilde{r}^2 = z_1 z_2 \qquad \text{with} \qquad z_1 = x_1 + ix_2; \quad z_2 = x_1 - ix_2$$

is complex unless x_1 and x_2 are real. From Sec. 2.3 we now conclude that a real-*valued* solution of (6.5-2) analytic in a neighborhood of the origin is given by

$$\hat{u}(x) = \operatorname{Re}\left[\int_{-1}^{1} k(\tilde{r}, t) f_1(z_1\tau)\tau^{-1/2}\,dt\right], \quad \tau = 1 - t^2; \quad x = (x_1, x_2)$$

with f_1 holomorphic in a neighborhood of the origin. Furthermore, the kernel equation (Sec. 2.3) now takes the form

$$\tau k_{\tilde{r}t} + \left(t - \frac{1}{t}\right)k_{\tilde{r}} + \tilde{r}t\left[k_{\tilde{r}\tilde{r}} + a(\tilde{r}^2)k\right] = 0, \quad a(\tilde{r}^2) \equiv \hat{a}(r^2). \tag{6.5-4}$$

Since k is real for $\tilde{r} = r$ and real t, we may rewrite the representation of $\hat{u}$ in the form

$$\hat{u}(x) = \int_{-1}^{1} k(r, t) \operatorname{Re}\left[f_1(z_1\tau)\right]\tau^{-1/2}\,dt. \tag{6.5-5}$$

As suggested by Theorem 2.5-1, we next define

$$h(x) = \int_{-1}^{1} \operatorname{Re}[f_1(z_1\tau)]\tau^{-1/2}\,d\tau \tag{6.5-6}$$

with

$$\operatorname{Re}[f_1(z_1)] = \sum_{n=0}^{\infty} \alpha_n r^n Y_n(\theta) \tag{6.5-7}$$

where the Y_n are circular harmonics. This series converges uniformly on every compact disk on which the Maclaurin series of f_1 converges. Hence we may perform termwise integration in (6.5-6) as well as in (6.5-9) on the left. We then obtain

$$h(x) = \sum_{n=0}^{\infty} B\left(n+\tfrac{1}{2},\tfrac{1}{2}\right)\alpha_n r^n Y_n(\theta) \tag{6.5-8}$$

and

$$\begin{aligned}&\int_0^1 \sigma(1-\sigma^2)^{m-1} h(\sigma^2 x)\,d\sigma \\ &\quad = \sum_{n=0}^{\infty} B\left(n+\tfrac{1}{2},\tfrac{1}{2}\right)\alpha_n r^n Y_n(\theta)\tfrac{1}{2}B(n+1,m), \qquad m\in\mathbb{N}.\end{aligned} \tag{6.5-9}$$

Now for the beta function,

$$B\left(n+\tfrac{1}{2},\tfrac{1}{2}\right)B(n+1,m) = B\left(n+\tfrac{1}{2},m+\tfrac{1}{2}\right)B\left(m,\tfrac{1}{2}\right), \quad m\in\mathbb{N}; \quad n\in\mathbb{N}_0.$$

Hence if we substitute the developments (6.5-3, 7) into the integral representation (6.5-5) and use (2.2-5) and (6.5-8, 9), we finally obtain the following theorem.

6.5-1 Theorem

Let h be any harmonic function defined in a disk centered at the origin. Let the coefficients $\kappa_m(\tilde{r}^2)$ be given by the expansion of a solution k of (6.5-4), *of the form* (6.5-3). *Then a solution of* (6.5-2) *is given by*

$$\hat{u}(x) = h(x) + \sum_{m=1}^{\infty} \frac{2\kappa_m(r^2)}{B\left(m,\tfrac{1}{2}\right)} \int_0^1 \sigma(1-\sigma^2)^{m-1} h(\sigma^2 x)\,d\sigma. \tag{6.5-10}$$

A FURTHER REPRESENTATION

It is evident that we can use Theorem 6.4-4 for calculating the coefficients $\kappa_m(\tilde{r}^2)$ in (6.5-10) from the Riemann function. Here, $e_j^R = 1$ since (6.5-2) is self-adjoint, and we can choose $s = (0,0)$ as before. Thus, by comparing

(6.5-3) and (6.4-12a) we obtain from (6.4-12b) the representation

(6.5-11)

$$\kappa_m(z_1 z_2) = z_1^m \tilde{q}_{1m}(z) = \frac{m!}{(2m)!}(-4z_1)^m \left[\frac{\partial^m}{\partial \zeta_1^m} R(\zeta_1, 0, z_1, z_2) \right]_{\zeta_1 = z_1}.$$

Here R is the complex Riemann function for the equation

(6.5-12)
$$D_1 D_2 u + \tfrac{1}{4} a(\tilde{r}^2) u = 0$$

with $a(\tilde{r}^2) = \hat{a}(r^2)$ when $z_2 = \bar{z}_1$. Hence, in the case of the representations (6.5-5) and (6.5-10), for real solutions $\hat{u}$ of our equation (6.5-2) we obtain the following further representation.

6.5-2 Corollary

For equation (6.5-2), *a representation of solutions equivalent to* (6.5-10) *is given by*

(6.5-13)
$$\hat{u}(x) = h(x) - 2z_1 \int_0^1 \sigma \left[\frac{\partial}{\partial \zeta_1} R(\zeta_1, 0, z_1, \bar{z}_1) \right]_{\zeta_1 = \sigma^2 z_1} h(\sigma^2 x)\, d\sigma$$

involving a real integral and the Riemann function R for (6.5-12).

Proof. We insert (6.5-11) into (6.5-10) and regroup terms as the integral of a Taylor series, obtaining

$$\hat{u}(x) = h(x) - 2z_1 \int_0^1 \sigma \sum_{m=1}^{\infty} \frac{1}{(m-1)!} \left[\frac{\partial^{m-1}}{\partial \zeta_1^{m-1}} R_{\zeta_1}(\zeta_1, 0, z_1, \bar{z}_1) \right]_{\zeta_1 = z_1} \times$$

$$\times (\sigma^2 z_1 - z_1)^{m-1} h(\sigma^2 x)\, d\sigma. \quad \blacksquare$$

6.5-3 Example

We consider the equation in Example 6.4-4 with $\lambda = 1$; that is,

$$\Delta_2 \hat{u} + \frac{4\nu(\nu+1)}{(1+r^2)^2} \hat{u} = 0, \qquad \nu \in \mathbb{N}.$$

According to (6.4-2) and (2.5-14), in this case we get the following representations of solutions by an application of Theorem 6.5-1 and Corollary 6.5-2:

$$\hat{u}(x) = h(x) + 2(\nu+1) \sum_{m=1}^{\nu} \binom{\nu}{m} \binom{\nu+m}{m-1} \left(\frac{-r^2}{1+r^2} \right)^m \times$$

$$\times \int_0^1 \sigma (1-\sigma^2)^{m-1} h(\sigma^2 x)\, d\sigma$$

and, equivalently,

$$\hat{u}(x) = h(x) - 2(\nu+1)\nu\frac{r^2}{1+r^2} \times$$

$$\times \int_0^1 \sigma\, {}_2F_1\left(\nu+2, 1-\nu; 2; (1-\sigma^2)\frac{r^2}{1+r^2}\right) h(\sigma^2 x)\, d\sigma\,.$$

Numerous representations (6.5-10) and (6.5-13) for further equations can be obtained from the explicitly given Riemann functions in the next chapter.

EXTENSION TO EQUATIONS IN n VARIABLES

We shall now extend the present method to equations in any number n (≥ 2) of independent variables, of the form

$$\Delta_n \hat{u} + \hat{a}(r^2)\hat{u} = 0 \tag{6.5-14}$$

where Δ_n is the n-dimensional Laplacian and $\hat{a}$ is a real-analytic function of $r^2 = x_1^2 + \cdots + x_n^2$. In this case, an analog of Bergman's representation of solutions may be established by the following theorem; cf. Gilbert [1970a].

6.5-4 Theorem

Let $k(\cdot, \cdot; n)$ be a regular solution of the partial differential equation [*cf.* (6.5-4)]

$$\tau k_{rt} + \left[t + (n-3)\frac{1}{t}\right]k_r + rt\left[k_{rr} + a(r^2)k\right] = 0\,, \quad \tau = 1 - t^2 \tag{6.5-15}$$

satisfying the three conditions

$$\lim_{t \to 0+} \left[t^{n-3} k_r(r, t; n)\right] r^{-1} = 0$$

$$\lim_{t \to 1-} \left[\tau^{1/2} k_r(r, t; n)\right] r^{-1} = 0$$

$$\lim_{r \to 0+} k(r, t; n) = 1, \qquad n \geq 2\,.$$

Furthermore, let H be any harmonic function of n variables, defined in a starlike domain with respect to the origin. Then the function $\hat{u}$ given by

$$\hat{u}(x) = \int_0^1 k(r, t; n) H(x\tau) \tau^{-1/2} t^{n-2}\, dt \tag{6.5-16}$$

is a solution of equation (6.5-14).

Proof. Substitute (6.5-16) into (6.5-14) and integrate by parts. ■

In analogy to the two-dimensional case just considered, we seek a solution of (6.5-15) in the form

$$k(r, t; n) = 1 + \sum_{m=1}^{\infty} \kappa_m(r; n) t^{2m}. \tag{6.5-17}$$

By inserting this into (6.5-15) we obtain for the coefficients κ_m the recursion formulas

$$\begin{aligned} (n-1)\kappa_1' &= -r\hat{a}(r^2) \\ (2m+n-3)\kappa_m' &= (2m-3)\kappa_{m-1}' - r\kappa_{m-1}'' - r\hat{a}(r^2)\kappa_{m-1}, \quad m \geq 2 \end{aligned} \tag{6.5-18a}$$

with

$$\kappa_m(0; n) = 0 \qquad \text{for } m \in \mathbb{N}. \tag{6.5-18b}$$

If the coefficient $\hat{a}$ is an entire function in the r^2-plane, the method of majorants can be used to prove the existence of solutions of this system such that the series (6.5-17) converges absolutely and uniformly on each compact subset of $\mathbb{C}$ and for all t in the closed unit disk; cf. Gilbert [1970a].

GILBERT'S REPRESENTATION OF SOLUTIONS

We can now state a representation for solutions of (6.5-14), analogous to (6.5-10).

6.5-5 Theorem

Let $\hat{a}$ be an entire function of r^2 and $h(\cdot; n)$ any harmonic function of n variables defined in a spherical neighborhood of the origin. Let the coefficients $\kappa_m(r; n)$ be given by (6.5-18). *Then a solution of* (6.5-14) *is given by*

$$\begin{aligned} \hat{u}(x) &= h(x; n) \\ &+ \sum_{m=1}^{\infty} \frac{2\kappa_m(r; n)}{B\left(m, \frac{1}{2}(n-1)\right)} \int_0^1 \sigma^{n-1}(1-\sigma^2)^{m-1} h(\sigma^2 x; n)\, d\sigma. \end{aligned} \tag{6.5-19}$$

Proof. We write h in the form

$$h(x; n) = \int_0^1 H(x\tau)\tau^{-1/2} t^{n-2}\, dt, \qquad \tau = 1 - t^2 \tag{6.5-20}$$

with a suitable harmonic function H regular on an open ball centered at the origin (H can be explicitly determined from h). Let

$$H(l, m_1, \cdots, m_{n-3}, \pm m_{n-2}; x_1, \cdots, x_n) \equiv H(m_k; \pm; x)$$

be a homogeneous harmonic polynomial of degree l (cf. Erdélyi et al. [1953–1955], vol. 2, p. 240). In hyperspherical polar coordinates this may be rewritten as

$$H(m_k; \pm; x) = r^l Y(m_k; \theta; \varphi) \tag{6.5-21}$$

where θ denotes $\theta_1, \cdots, \theta_{n-2}$ and Y on the right is called a *surface harmonic of degree* l. Then H has an expansion of the form

$$H(x) = \sum_{l, m_k} \alpha_l(m_k; \pm) H(m_k; \pm; x). \tag{6.5-22}$$

This series is Abel-summable on the surface of the ball. On any set, relatively compact in this ball, the convergence is uniform. Hence, after inserting (6.5-21, 22) into (6.5-20), we may perform termwise integration. Using formula (2.2-5) for the beta function [with $m = (n-2)/2$ and $n = l$], we obtain the series representation

$$h(x; n) = \frac{1}{2} \sum_{l, m_k} B\left(\frac{n-1}{2}, l + \frac{1}{2}\right) \alpha_l(m_k; \pm) r^l Y(m_k; \theta; \varphi). \tag{6.5-23}$$

This series converges in a ball at least as large as that in which H converges. Within this ball, termwise integration of the series for h in the integral in (6.5-24) is also permissible. We thus obtain

$$\begin{aligned} &\frac{2}{B\left(m, \frac{1}{2}(n-1)\right)} \int_0^1 \sigma^{n-1}(1-\sigma^2)^{m-1} h(\sigma^2 x; n)\, d\sigma \\ &= \sum_{l, m_k} \alpha_l(m_k; \pm) r^l Y(m_k; \theta; \varphi) \frac{B\left(\frac{1}{2}(n-1), l + \frac{1}{2}\right) B\left(m, l + \frac{1}{2}n\right)}{2B\left(m, \frac{1}{2}(n-1)\right)}. \end{aligned} \tag{6.5-24}$$

The quotient on the right equals

$$\frac{1}{2} B\left(m + \frac{n-1}{2}, l + \frac{1}{2}\right) = \int_0^1 t^{2m} \tau^l \tau^{-1/2} t^{n-2}\, dt.$$

Hence, if we take (6.5-21, 22) into account, we see that (6.5-24) becomes

$$\int_0^1 t^{2m} H(x\tau) \tau^{-1/2} t^{n-2}\, dt.$$

Substitution into (6.5-19) now yields

$$\hat{u}(x) = h(x; n) + \sum_{m=1}^{\infty} \kappa_m(r; n) \int_0^1 t^{2m} H(x\tau) \tau^{-1/2} t^{n-2}\, dt. \tag{6.5-25}$$

Observing (6.5-20) and the abovementioned uniform convergence property of the series (6.5-17), we conclude that (6.5-19) coincides with the representation (6.5-16). ■

We remark that, incidentally, we have also proved that solutions of the form (6.5-16) with k as in (6.5-17) can be expanded into the series

$$\begin{aligned} \hat{u}(x) = \sum_{l,\, m_k} \alpha_l(m_k; \pm) Y(m_k; \theta; \varphi) \times \\ \times r^l \sum_{m=0}^{\infty} \kappa_m(r; n) \frac{1}{2} B\left(m + \frac{n-1}{2}, l + \frac{1}{2}\right) \end{aligned} \tag{6.5-26}$$

where $\kappa_0(r; n) \equiv 1$, if H possesses the expansion (6.5-22). (Cf. also Gilbert [1970a], Theorem 8, and Bergman [1971], pp. 66–67.)

The representation (6.5-19) suggests introducing ***Gilbert's G-function*** defined by the series

$$G(r, p) = \sum_{m=1}^{\infty} \frac{2\kappa_m(r; n)}{B(m, \frac{1}{2}(n-1))} p^{m-1}. \tag{6.5-27}$$

Then, (6.5-19) takes the form

$$\hat{u}(x) = h(x; n) + \int_0^1 \sigma^{n-1} G(r, 1 - \sigma^2) h(\sigma^2 x; n)\, d\sigma. \tag{6.5-28}$$

This defines the so-called ***Bergman–Gilbert operator***, which transforms harmonic functions into solutions of (6.5-14).

GILBERT'S G-FUNCTION INDEPENDENT OF DIMENSION

Representation (6.5-28) has the remarkable advantage that the G-function is independent of n:

6.5-6 Lemma

Gilbert's G-function (6.5-27) *satisfies the partial differential equation*

$$2(1 - p) G_{rp} - G_r + r\left[G_{rr} + \hat{a}(r^2) G\right] = 0. \tag{6.5-29}$$

Moreover, G does not depend on the parameter n.

Proof. The first part follows by substitution of the G-function into (6.5-29) and the use of the recursion formulas (6.5-18). From (6.5-18, 27) we conclude that G must satisfy the Goursat data

$$(6.5\text{-}30) \quad \text{(a)} \quad G(0, p) = 0 \qquad \text{(b)} \quad G(r,0) = -\int_0^r \hat{a}(\rho^2)\rho\, d\rho .$$

The solution of this problem, however, is independent of n, and so is the G-function. ■

When $n = 2$, it follows from Corollary 6.5-2 that there exists a relation between Gilbert's G-function and the *Riemann function*. Indeed, a comparison of (6.5-13) and (6.5-28) shows that for $n = 2$,

$$(6.5\text{-}31) \qquad G(r, 1 - \sigma^2) = -2z_1 \left[\frac{\partial}{\partial \zeta_1} R(\zeta_1, 0, z_1, \bar{z}_1) \right]_{\zeta_1 = \sigma^2 z_1} .$$

Since G is real-valued for real arguments, in the (multiple) Maclaurin expansion of the right-hand side of (6.5-31), each power of z_1 must combine with the same power of $\bar{z}_1$ in such a way that the series depends only on $r^2 = z_1 \bar{z}_1$. Therefore, (6.5-31) can be represented in the form

$$(6.5\text{-}32) \qquad G(r, 1 - \sigma^2) = -2r \left[\frac{\partial}{\partial \zeta_1} R(\zeta_1, 0, r, r) \right]_{\zeta_1 = \sigma^2 r} .$$

Since G is independent of n, we can represent solutions of an equation in n independent variables in terms of the Riemann function for an equation in two variables via Gilbert's G-function, and we may summarize:

6.5-7 Theorem

Under the assumptions of Theorem 6.5-5, *a solution of* (6.5-14) *is given by* (6.5-28). *Here G is Gilbert's G-function defined by* (6.5-27, 18), *and G can be represented in terms of the Riemann function R of the two-dimensional equation* (6.5-12) *as shown in* (6.5-32).

EXISTENCE OF THE KERNEL (6.5-17) AND RELATION TO THE G-FUNCTION

Observing that

$$B\left(m, \frac{n-1}{2}\right) = \int_0^1 (1 - \rho)^{(n-3)/2} \rho^{m-1}\, d\rho , \qquad n \geq 2$$

from (6.5-27) we obtain

$$\frac{1}{2}t^2 \int_0^1 (1-\rho)^{(n-3)/2} G(r, \rho t^2)\, d\rho = \sum_{m=1}^{\infty} \kappa_m(r; n) t^{2m}.$$

Thus we have

6.5-8 *Remark*

The function k in (6.5-17) can be expressed as an integral transform of the G-function in the form

$$(6.5\text{-}33) \qquad k(r, t; n) = 1 + \frac{1}{2}t^2 \int_0^1 (1-\rho)^{(n-3)/2} G(r, \rho t^2)\, d\rho, \qquad n \geq 2.$$

Accordingly, using the representation (6.5-32) of the G-function in terms of the Riemann function, we can prove the existence of the function k by means of Corollary 6.2-5 on the existence of the Riemann function. Furthermore, from the domain of holomorphy of the Riemann function we may determine the domain of convergence of the series of k in the case of a coefficient a in (6.5-14) which is not entire but merely analytic in r^2 in some domain containing the origin.

SPECIAL EQUATIONS IN n VARIABLES

For illustrating the present results, let us consider two examples, beginning with the n-dimensional analog of Example 6.5-3:

6.5-9 *Example*

Given

$$(6.5\text{-}34) \qquad \Delta_n \hat{u} + \frac{4\nu(\nu+1)}{(1+r^2)^2}\hat{u} = 0, \qquad \nu \in \mathbb{N}; \quad n \geq 2.$$

The formally hyperbolic, complex two-dimensional form of this equation is

$$D_1 D_2 u + \frac{\nu(\nu+1)}{(1+z_1 z_2)^2} u = 0.$$

This equation has the Riemann function (6.4-15), that is,

$$R(\zeta, z) = {}_2F_1\left(\nu+1, -\nu; 1; \frac{(z_1-\zeta_1)(z_2-\zeta_2)}{(1+z_1 z_2)(1+\zeta_1\zeta_2)}\right).$$

From this and (6.5-32) we obtain for the present equation Gilbert's G-function

$$G(r,1-\sigma^2) = -2r\left[\frac{\partial}{\partial\zeta_1}\,{}_2F_1\left(\nu+1,-\nu;1;\frac{(r-\zeta_1)r}{1+r^2}\right)\right]_{\zeta_1=\sigma^2 r}$$

$$= -2(\nu+1)\nu\frac{r^2}{1+r^2}\,{}_2F_1\left(\nu+2,1-\nu;2;(1-\sigma^2)\frac{r^2}{1+r^2}\right).$$

Furthermore, by (6.5-17, 27) or by (6.5-33), we know that

$$k(r,t;n) = {}_2F_1\left(\nu+1,-\nu;\frac{n-1}{2};\frac{r^2t^2}{1+r^2}\right).$$

After a suitable change of notation this agrees with our result in Example 6.4-4 in the two-dimensional case. Consequently, according to Theorems 6.5-4, 5, 7, as a final result we have the representations of solutions

$$\hat{u}(x) = \int_0^1 {}_2F_1\left(\nu+1,-\nu;\frac{n-1}{2};\frac{r^2t^2}{1+r^2}\right)H(x\tau)\tau^{-1/2}t^{n-2}\,dt$$

$$\hat{u}(x) = h(x;n) + 2(\nu+1)\sum_{m=1}^{\nu}\binom{\nu}{m}\binom{\nu+m}{m-1}\left(\frac{-r^2}{1+r^2}\right)^m \times$$

$$\times\int_0^1 \sigma^{n-1}(1-\sigma^2)^{m-1}h(\sigma^2x;n)\,d\sigma$$

$$\hat{u}(x) = h(x;n) - 2(\nu+1)\nu\frac{r^2}{1+r^2}\times$$

$$\times\int_0^1 \sigma^{n-1}{}_2F_1\left(\nu+2,1-\nu;2;(1-\sigma^2)\frac{r^2}{1+r^2}\right)h(\sigma^2x;n)\,d\sigma.$$

6.5-10 Example

We consider the ***n*-dimensional Helmholtz equation**

(6.5-35) $$\Delta_n\hat{u} + c^2\hat{u} = 0, \qquad c\in\mathbb{R};\quad n\geq 2.$$

When $n=2$, the corresponding transformed complex equation

$$D_1D_2u + \tfrac{1}{4}c^2u = 0$$

has the Riemann function (cf. Example 6.4-2a)

$$R(\zeta, z) = J_0\left(c\sqrt{(z_1 - \zeta_1)(z_2 - \zeta_2)}\right).$$

This yields as G-function for (6.5-35)

$$G(r, 1 - \sigma^2) = \frac{-cr}{\sqrt{1 - \sigma^2}} J_1\left(cr\sqrt{1 - \sigma^2}\right).$$

Hence, for k we obtain

$$k(r, t; n) = \Gamma\left(\frac{n-1}{2}\right)\left(\frac{2}{crt}\right)^{(n-3)/2} J_{(n-3)/2}(crt).$$

Here, J_ν is the Bessel function of the first kind of order ν. Clearly, for $n = 2$ this reduces to our previous result with $r = (z_1\bar{z}_1)^{1/2}$ and $s = (0, 0)$, as one can readily verify by comparing with Example 2.5-4. From the theorems used in the previous example we obtain for solutions of the Helmholtz equation (6.5-35) the three representations

$$\hat{u}(x) = \Gamma\left(\frac{n-1}{2}\right)\left(\frac{cr}{2}\right)^{-(n-3)/2} \int_0^1 t^{(n-1)/2} J_{(n-3)/2}(crt) H(x\tau)\tau^{-1/2}\, dt$$

$$\hat{u}(x) = h(x; n) + 2\sum_{m=1}^{\infty} \frac{1}{(m-1)!\, m!}\left(\frac{icr}{2}\right)^{2m} \times$$

$$\times \int_0^1 \sigma^{n-1}(1 - \sigma^2)^{m-1} h(\sigma^2 x; n)\, d\sigma$$

$$\hat{u}(x) = h(x; n) - cr\int_0^1 \frac{\sigma^{n-1}}{(1 - \sigma^2)^{1/2}} J_1\left(cr(1 - \sigma^2)^{1/2}\right) h(\sigma^2 x; n)\, d\sigma.$$

The Bergman–Gilbert operator is also quite efficient in solving boundary value problems. For details, we refer to Gilbert [1974], Chap. III.

6.6 RELATIONS TO CARROLL'S THEORY OF TRANSMUTATIONS

Chapter 6 is devoted to methods related to the theory of Bergman-type operators, and so far we have discussed the complex Riemann–Vekua function, Le Roux operators, and Gilbert's G-function from this point of view. Recently, Carroll [1984a] has discovered another interesting relation, namely, that between Bergman operators and transmutation operators, which we shall now explain.

TRANSMUTATIONS

The idea of transmutation operators goes back to Delsarte [1938], Delsarte–Lions [1957], and Lions [1956, 1959]. More recent contributions were made by R. W. Carroll, R. Hersh, M. Thyssen, and others; cf. Carroll [1979, 1982]. We first recall the definition of a transmutation. Let $D = \partial/\partial r$ and consider differential polynomials $P(D)$ and $Q(D)$ defined on suitable spaces of functions or distributions. P and Q may have variable coefficients and be of different orders. We say that an operator B is a ***transmutation*** which *transmutes* $P(D)$ into $Q(D)$ if and only if formally

$$Q(D)B = BP(D), \qquad \text{written} \qquad B\colon P(D) \to Q(D).$$

A transmutation B is usually an integral operator.

If B is such a transmutation, then, for example, $P(D)h = 0$ implies $Q(D)Bh = 0$, so that $u = Bh$ satisfies $Q(D)u = 0$. If B is an isomorphism from a space X onto a space Y, then from $Q(D)u = 0$ we get $P(D)B^{-1}u = B^{-1}Q(D)u = 0$ so that $h = B^{-1}u$ satisfies $P(D)h = 0$.

In our present case we consider the radial part of Δ_n multiplied by r^2 in the form

$$\tilde{P}_n = P(D) = r^2D^2 + (n-1)rD, \qquad D = \partial/\partial r.$$

We further let

$$\tilde{\tilde{P}}_n = \tilde{P}(\tilde{D}) = \rho^2\tilde{D}^2 + (n-1)\rho\tilde{D}, \qquad \tilde{D} = \partial/\partial\rho$$

and

$$\tilde{Q}_n = Q(D) = r^2D^2 + (n-1)rD + r^2\hat{a}(r^2).$$

Since

$$r^2\Delta_n = \tilde{P}_n + \text{terms independent of } r$$

we can see that a transmutation $B\colon \tilde{P}_n \to \tilde{Q}_n$ in the radial variables induces a transmutation

$$B\colon r^2\Delta_n \to r^2\Delta_n + r^2\hat{a}(r^2).$$

Transmutations can be characterized in various ways. In general they will be given as integral operators with distribution kernels in the form

$$Bf(r) = \langle \beta(r,\rho), f(\rho)\rangle.$$

where $\langle \cdot , \cdot \rangle$ denotes distributional pairing. The kernels β can be represented as spectral integrals by using eigenfunctions of $\tilde{P}_n$ and $\tilde{Q}_n$. However, in the case under consideration, Carroll [1984a] obtained the kernels directly as inverse Mellin-type transforms, as we shall now explain.

DETERMINATION OF A TRANSMUTATION BY THE BERGMAN–GILBERT REPRESENTATION

Setting $\sigma^2 = \rho/r$, we transform the Bergman–Gilbert representation (6.5-28) into a Volterra-type representation. Writing $\tilde{h}(r,(\theta,\varphi)) \equiv h(x;n)$, we obtain

$$\hat{u} = \tilde{h} + \int_0^r \tilde{K}(r,\rho)\tilde{h}(\rho,\cdot)\,d\rho \tag{6.6-1}$$

where

$$\tilde{K}(r,\rho) = \frac{1}{2r}\left(\frac{\rho}{r}\right)^{n/2-1} G\left(r, 1-\frac{\rho}{r}\right). \tag{6.6-2}$$

We now write the kernel $\tilde{K}$ in the form

$$\tilde{K}(r,\rho) = \rho^{n-3}K(r,\rho). \tag{6.6-3}$$

From (6.5-29, 30) we can readily conclude that K satisfies the differential equation

$$\tilde{Q}_n K = \tilde{\tilde{P}}_n K \qquad \text{for } \rho < r \tag{6.6-4}$$

with the condition (in general, a jump discontinuity)

$$K(r,r) = -\frac{1}{2}r^{2-n}\int_0^r \hat{a}(\rho^2)\rho\,d\rho \qquad \text{at } \rho = r. \tag{6.6-5}$$

From (6.6-2) and (6.6-3) we see that

$$\rho^{n-1}K_\rho \to 0 \quad \text{and} \quad \rho^{n-1}K \to 0 \text{ as } \rho \to 0. \tag{6.6-6}$$

For the integral in (6.6-1) with kernel (6.6-3) we set

$$J\tilde{h} = \int_0^r \rho^{n-3}K(r,\rho)\tilde{h}(\rho,\cdot)\,d\rho. \tag{6.6-7}$$

Using (6.6-4, 5) and integration by parts, we obtain

$$\begin{aligned}
&\int_0^r \rho^{n-3}\tilde{h}(\rho,\cdot)\tilde{Q}_n K(r,\rho)\,d\rho \\
&= \int_0^r \rho^{n-3}\tilde{h}(\rho,\cdot)\tilde{\tilde{P}}_n K(r,\rho)\,d\rho \\
&= \int_0^r \tilde{h}(\rho,\cdot)\left(\rho^{n-1}K_\rho\right)_\rho(r,\rho)\,d\rho \\
&= r^{n-1}\tilde{h}(r,\cdot)\left[K_\rho(r,\rho)\right]\Big|_{\rho=r} + \frac{1}{2}r\tilde{h}_r(r,\cdot)\int_0^r \hat{a}(\rho^2)\rho\,d\rho \\
&\quad + \int_0^r \left(\rho^{n-1}\tilde{h}_\rho\right)_\rho(\rho,\cdot)K(r,\rho)\,d\rho
\end{aligned}$$

and

$$(K_r + K_\rho)(r, \rho)\big|_{r=\rho} = -\frac{1}{2}r^{1-n}\left[r^2\hat{a}(r^2) + (2-n)\int_0^r \hat{a}(\rho^2)\rho\, d\rho\right].$$

This finally yields

$$\text{(6.6-8)} \quad \tilde{Q}_n J\tilde{h} = -r^2\hat{a}(r^2)\tilde{h}(r,\cdot) + \int_0^r \rho^{n-3}K(r,\rho)\tilde{\tilde{P}}_n\tilde{h}(\rho,\cdot)\, d\rho.$$

From this follows

$$\tilde{Q}_n\hat{u} = \left[\tilde{P}_n + r^2\hat{a}(r^2)\right]\tilde{h} + \tilde{Q}_n J\tilde{h} = \tilde{P}_n\tilde{h} + J\tilde{P}_n\tilde{h} = (I+J)\tilde{P}_n\tilde{h}$$

where I is the identity operator. Hence our present result is

$$\text{(6.6-9)} \qquad \tilde{Q}_n B\tilde{h} = B\tilde{P}_n\tilde{h} \qquad \text{where} \qquad B = I + J.$$

This shows that B is a transmutation.

Following Carroll [1984a], let us briefly discuss these ideas in the context of distributions. For this purpose we write (6.6-1) in the form

$$\text{(6.6-10)} \qquad \hat{u} = \tilde{h} + \langle \rho^{n-3}\check{K}(r,\rho), \tilde{h}(\rho,\cdot)\rangle$$

where

$$\text{(6.6-11)} \qquad \check{K}(r,\rho) = K(r,\rho)Y(r-\rho)$$

with $K(r,\rho)$ given by (6.6-2, 3) and Y denoting the Heaviside function. Repeating the calculations that led to (6.6-8), we then obtain

$$\tilde{Q}_n J\tilde{h} = \langle \tilde{Q}_n\{\rho^{n-3}\check{K}(r,\rho)\}, \tilde{h}\rangle.$$

Furthermore, in terms of the formal adjoint $\tilde{\tilde{P}}_n^*$ of $\tilde{\tilde{P}}_n$ we get

$$\left(\tilde{Q}_n - \tilde{\tilde{P}}_n^*\right)\{\rho^{n-3}\check{K}(r,\rho)\} = -r^2\hat{a}(r^2)\,\delta(r-\rho)$$

where δ is Dirac's delta. In this way we have as our final result:

6.6-1 Theorem

Formula (6.6-10) *represents a transmutation*

$$\check{B}\colon\ \tilde{P}_n \quad\to\quad \tilde{Q}_n, \qquad \hat{u} = \check{B}\tilde{h}.$$

INVERSE OF THE GENERALIZED MELLIN TRANSFORM AND EXTENDED BERGMAN–GILBERT KERNEL

We recall that the ***Mellin transform*** of a function ϕ with complex parameter λ is defined by

$$M[\phi(\rho), \lambda] = \Phi(\lambda) = \int_0^\infty \rho^{\lambda-1}\phi(\rho)\, d\rho.$$

See, e.g., Oberhettinger [1974]. Observe that we could also express Φ as a two-sided Laplace transform, using the substitution $\rho = e^{-\omega}$. To a suitable kind of distributions, the Mellin transform may be extended by simply defining

$$M[\phi(\rho), \lambda] = \langle \rho^{\lambda-1}, \phi(\rho) \rangle;$$

cf. Zemanian [1968], Chap. IV. Let us now write

$$\check{B}\tilde{h} = \langle \check{\beta}(r, \rho), \tilde{h}(\rho, \cdot) \rangle \tag{6.6-12}$$

and set

$$\hat{u}_\lambda = \check{B}(\rho^\lambda) = \langle \rho\check{\beta}(r, \rho), \rho^{\lambda-1} \rangle = M[\rho\check{\beta}(r, \rho), \lambda]. \tag{6.6-13}$$

For this $\hat{u}_\lambda$ we consider $\tilde{Q}_n \hat{u}_\lambda$, noting (6.6-10). We then have on the one hand

$$\tilde{Q}_n r^\lambda = \lambda(\lambda + n - 2) r^\lambda + \hat{a}(r^2) r^{\lambda+2}$$

and on the other hand

$$\begin{aligned} &\tilde{Q}_n \langle \rho^{n-3} \check{K}(r, \rho), \rho^\lambda \rangle \\ &\quad = -\hat{a}(r^2) r^{\lambda+2} + \langle \rho^{n-3} \tilde{\tilde{P}}_n \{ K(r, \rho) Y(r - \rho) \}, \rho^\lambda \rangle \\ &\quad = -\hat{a}(r^2) r^{\lambda+2} + \langle \check{K}(r, \rho), (\rho^{n-1} \{ \tilde{D}\rho^\lambda \}')' \rangle \\ &\quad = -\hat{a}(r^2) r^{\lambda+2} + \lambda(\lambda + n - 2) \langle \rho^{n-3} \check{K}(r, \rho), \rho^\lambda \rangle. \end{aligned}$$

Consequently, $\hat{u}_\lambda$ satisfies

$$\tilde{Q}_n \psi = \lambda(\lambda + n - 2)\psi. \tag{6.6-14}$$

Since we have assumed that $\hat{a}$ has no singularity at $r = 0$, we can expect that a solution of (6.6-14) behaves near $r = 0$ like r^γ, where

$$\gamma^2 + (n - 2)\gamma - \lambda(\lambda + n - 2) = 0.$$

Hence $\gamma = \lambda$ or $\gamma = -\lambda - n + 2$. Let $\check{\psi}(\cdot; \lambda, \tilde{Q}_n)$ denote solutions ψ of (6.6-14) satisfying the condition

$$\lim_{r \to 0} [\check{\psi}(r; \lambda, \tilde{Q}_n) r^{-\lambda}] = 1. \tag{6.6-15}$$

We now replace $\hat{u}_\lambda$ in (6.6-13) by such a "normalized" solution $\check{\psi}$ and apply the inverse of the generalized Mellin transform (cf. Zemanian [1968], p. 108).

This yields a representation of $\rho\check{\beta}(r, \rho)$ and thus of the kernel $\check{\beta}$. The latter we call "***extended Bergman–Gilbert kernel***" since it arises from the representation of solutions obtained by the Bergman–Gilbert operator. Our result may be formulated as follows.

6.6-2 Theorem

The extended Bergman–Gilbert kernel $\check{\beta}$ defining the transmutation $\check{B}$ is given by the spectral formula

$$\begin{aligned}\check{\beta}(r, \rho) &= \frac{1}{2\pi i}\int \rho^{-\lambda-1}\check{\psi}\left(r; \lambda, \tilde{Q}_n\right) d\lambda \\ &= \frac{1}{2\pi i}\int \rho^{n-3}\rho^{\omega}\check{\psi}\left(r; \omega, \tilde{Q}_n\right) d\omega, \qquad \omega = -\lambda - n + 2.\end{aligned} \tag{6.6-16}$$

where $\check{\psi}(r; \lambda, \tilde{Q}_n) \equiv \check{\psi}(r; \omega, \tilde{Q}_n)$ and the integrals are taken along suitable contours from $c - i\infty$ to $c + i\infty$.

One may confirm by direct calculation that $\check{\beta}$ in this theorem possesses standard transmutation character in the sense that

$$\tilde{Q}_n\check{\beta} = \tilde{\tilde{P}}_n^*\check{\beta}.$$

EXTENDED BERGMAN–GILBERT KERNEL FOR A SPECIAL n-VARIABLE EQUATION

We conclude this section by applying our results to equation (6.5-35).

6.6-3 Example

We consider the same equation as in Example 6.5-10, namely (6.5-35):

$$\Delta_n\hat{u} + c^2\hat{u} = 0, \qquad c \in \mathbb{R}\setminus\{0\}; \quad n \geq 2.$$

For an application of Theorem 6.6-2 we must determine a solution $\check{\psi}$ of (6.6-14) satisfying (6.6-15), where (6.6-14) now takes the special form

$$r^2D^2\psi + (n-1)rD\psi + c^2r^2\psi = \lambda(\lambda + n - 2)\psi, \quad D = \partial/\partial r.$$

Imposing the transformation

$$\psi = r^{(2-n)/2}\hat{\psi}$$

we obtain the ordinary differential equation

$$r^2D^2\hat{\psi} + rD\hat{\psi} + \left[(cr)^2 - \left(\tfrac{1}{2}n + \lambda - 1\right)^2\right]\hat{\psi} = 0.$$

A solution is

$$\hat{\psi}(r) = J_{n/2+\lambda-1}(cr)$$

where J_ν is the Bessel function of the first kind of order ν. Hence

$$\check{\psi}(r) = \left(\frac{2}{c}\right)^{\lambda} \Gamma\left(\frac{n}{2} + \lambda\right)\left(\frac{cr}{2}\right)^{1-n/2} J_{n/2+\lambda-1}(cr)$$

is a solution of (6.6-14). Moreover,

$$\begin{aligned}\check{\psi}(r)r^{-\lambda} &= \Gamma\left(\frac{n}{2} + \lambda\right)\left(\frac{cr}{2}\right)^{1-n/2-\lambda} J_{n/2+\lambda-1}(cr) \\ &= \sum_{m=0}^{\infty} \frac{(-1)^m \Gamma\left(\frac{n}{2} + \lambda\right)}{m!\,\Gamma\left(\frac{n}{2} + \lambda + m\right)}\left(\frac{cr}{2}\right)^{2m}\end{aligned}$$

is an entire function of r which obviously satisfies

$$\lim_{r \to 0}\left[\check{\psi}(r)r^{-\lambda}\right] = 1$$

that is, condition (6.6-15) holds. By (6.6-16) we now obtain the desired representation of the extended Bergman–Gilbert kernel for (6.5-35) in the form

$$\begin{aligned}\check{\beta}(r,\rho) &= \frac{1}{2\pi i}\int \rho^{-\lambda-1}\Gamma\left(\frac{n}{2} + \lambda\right)\left(\frac{2}{c}\right)^{n/2+\lambda-1} r^{1-n/2} \times \\ &\qquad \times J_{n/2+\lambda-1}(cr)\, d\lambda \\ &= \frac{1}{2\pi i}\int \rho^{n-3}\rho^{\omega}\Gamma\left(2 - \frac{n}{2} - \omega\right)\left(\frac{c}{2}\right)^{n/2+\omega-1} r^{1-n/2} \times \\ &\qquad \times J_{1-n/2-\omega}(cr)\, d\omega\end{aligned} \tag{6.6-17}$$

with contours of integration as in the theorem.

Our result includes that by Carroll for $n = 3$, because (6.6-17) may be rewritten as

$$\check{\beta}(r,\rho) = \frac{1}{2\pi i}\int \rho^{-\sigma-1}(cr)^{-1/2} J_{\sigma+1/2}(cr)\left(\frac{c}{2}\right)^{-\sigma} \sqrt{2}\,\Gamma\left(\sigma + \tfrac{3}{2}\right) d\sigma. \tag{6.6-18}$$

See Carroll [1984a], Theorem 4.4, formula (4.15).

Additional References

Bergman [1971]
Carroll [1979, 1982, 1984a]
Colton [1980]
Delsarte–Lions [1957]
Diaz–Ludford [1955, 1956, 1957]
Eisenstat [1974]
Florian–Püngel–Tutschke [1985]
Friedlander–Heins [1969]
Garabedian [1954]
Gilbert [1969b, 1970a, 1974]
Gilbert–Wei [1983]
Henrici [1952, 1953, 1957, 1960]
Kracht [1974a]
Kracht–Kreyszig [1988]
Lanckau [1980b, 1986]
Lewy [1959]
Lions [1956, 1959]
Mackie [1965]
Millar [1980, 1983]
Mülthei–Neunzert [1969]
Sreedharan [1965]
Vekua [1967]
Volkmer [1979]
Watzlawek [1971a, 1971c]
Yu [1974]

Chapter Seven

Determination of Riemann Functions

In applications, the practical usefulness of the Riemann representation formula depends crucially on whether an *explicit closed-form* representation of the Riemann function for a given equation can be found. Unfortunately, there is no general standard method for this purpose. This explains why the problem of determining Riemann functions for practically important equations has attracted the attention of mathematicians over many decades through the present. Nevertheless, the number of equations for which the Riemann function is explicitly known has remained rather small. Earlier special results up to about 1957 were surveyed in an article by Copson [1958].

The problem fits well into the general circle of ideas with which this book is concerned. Indeed, a first method for the determination of the Riemann function has already been given in Theorem 6.4-1. This method is based on the use of Bergman kernels of the first kind for Type I representations of solutions. *This approach seems very promising*, since such a kernel results from a recursion with respect to only *one* of the variables, whereas the standard proof of Theorem 6.2-4 rests on a Picard iteration involving *both* variables.

In the present chapter we shall discuss significant ideas and successful techniques in connection with the determination of Riemann functions, along with corresponding results and applications. Beginning with Riemann's original work, in Sec. 7.1 we consider some of the "classical" methods whose roots date back before 1960, along with some of their more recent generalizations. Section 7.2 is concerned with addition formulas for Riemann functions as proposed by Olevskij and extended in 1981 by Du. Major results obtained during the last 18 years will be discussed in Sec. 7.3, beginning with H. Cohn's work of 1970–1973 and leading to ongoing investigations.

7.1 CLASSICAL METHODS AND GENERALIZATIONS

The Riemann function was originally introduced by Riemann [1860] for representing solutions of characteristic [cf. (2.6-1)] or noncharacteristic Cauchy problems in the *real* case. This concerned hyperbolic equations in two real independent variables $\tilde{x}_1$, $\tilde{x}_2$:

$$(7.1\text{-}1)\quad \tilde{w}_{\tilde{x}_1\tilde{x}_1} - \tilde{w}_{\tilde{x}_2\tilde{x}_2} + \tilde{a}_1(\tilde{x}_1, \tilde{x}_2)\tilde{w}_{\tilde{x}_1} + \tilde{a}_2(\tilde{x}_1, \tilde{x}_2)\tilde{w}_{\tilde{x}_2} + \tilde{a}_0(\tilde{x}_1, \tilde{x}_2)\tilde{w} = 0 .$$

An alternative normal form of this equation is obtained by applying the transformation

$$(7.1\text{-}2)\quad x_1 = \tilde{x}_1 + \tilde{x}_2 , \qquad x_2 = \tilde{x}_1 - \tilde{x}_2 , \qquad \hat{w}(x_1, x_2) = \tilde{w}(\tilde{x}_1, \tilde{x}_2) .$$

The result is

$$(7.1\text{-}3)\quad \hat{w}_{x_1x_2} + \hat{a}_1(x_1, x_2)\hat{w}_{x_1} + \hat{a}_2(x_1, x_2)\hat{w}_{x_2} + \hat{a}_0(x_1, x_2)\hat{w} = 0$$

with coefficients

$$\hat{a}_j(x_1, x_2) = \frac{1}{4}\left[\tilde{a}_1\left(\frac{x_1 + x_2}{2}, \frac{x_1 - x_2}{2}\right) + (-1)^{j+1}\tilde{a}_2\left(\frac{x_1 + x_2}{2}, \frac{x_1 - x_2}{2}\right)\right]$$

$$\hat{a}_0(x_1, x_2) = \frac{1}{4}\tilde{a}_0\left(\frac{x_1 + x_2}{2}, \frac{x_1 - x_2}{2}\right), \qquad j = 1, 2 .$$

We see that equation (7.1-3) is of the form (6.1-4). If we assume that the coefficients of (7.1-1), and hence of (7.1-3), are real-analytic in some common domain, and that $\tilde{x}_1$, $\tilde{x}_2$ are continued to complex values, say, z_1, z_2, then we obtain, at least locally, a representation of the *complex* Riemann function for (6.1-4) by means of the Riemann function for (7.1-1) or (7.1-3).

Consequently, it makes no difference whether we consider the equation (6.1-4), which may have been obtained from a real elliptic equation of the form (6.1-1) via (2.1-6, 7, 8), or whether we start from a real hyperbolic equation (7.1-3) or (7.1-1). Although the results about the representation of the Riemann function also hold in the complex case, independently of the way in which they are obtained, as a matter of convenience we shall remain flexible, referring to equation (7.1-3) or to equation (6.1-4). In the case of (7.1-3) we denote the variables of the Riemann function by x_1, x_2, ξ_1, ξ_2, and in the case of (6.1-4) by z_1, z_2, s_1, s_2.

RIEMANN'S APPROACH

Riemann's original approach was based on the fact that the Riemann function does not depend on the curve on which the Cauchy data are given. If, by some other means, it is possible to solve the Cauchy problem for a special initial curve depending on a parameter, a comparison of the two solutions yields the Riemann function. In two cases, Riemann was able to solve the Cauchy problem by a Fourier cosine transform with Cauchy data on a straight line. By this method he obtained in the case of the equation

$$\tilde{w}_{\tilde{x}_1\tilde{x}_1} - \tilde{w}_{\tilde{x}_2\tilde{x}_2} + \frac{2\alpha}{\tilde{x}_1}\tilde{w}_{\tilde{x}_1} = 0, \qquad \alpha \text{ constant} \tag{7.1-4}$$

for the Riemann function $\tilde{R}$ the representation

$$\tilde{R}(\tilde{x}_1, \tilde{x}_2, \tilde{\xi}_1, \tilde{\xi}_2) = \frac{\tilde{x}_1^{\alpha+1/2}}{(\cos\alpha\pi)\tilde{\xi}_1^{\alpha-1/2}} \int_0^\infty \cos\left[\lambda(\tilde{\xi}_2 - \tilde{x}_2)\right]\left[J_{\alpha-1/2}(\lambda\tilde{x}_1) \times \right.$$
$$\left. \times J_{1/2-\alpha}(\lambda\tilde{\xi}_1) - J_{\alpha-1/2}(\lambda\tilde{\xi}_1) J_{1/2-\alpha}(\lambda\tilde{x}_1)\right] d\lambda$$

with $\tilde{\xi}_2 - \tilde{x}_2$ between $\pm(\tilde{\xi}_1 - \tilde{x}_1)$. He further noted that each Bessel function can be replaced by a definite integral. Consequently, $\tilde{R}$ can be expressed as a triple integral which reduces to a hypergeometric function. The final result is

$$\begin{aligned} \tilde{R}(\tilde{x}_1, \tilde{x}_2, \tilde{\xi}_1, \tilde{\xi}_2) &= \left(\frac{\tilde{x}_1}{\tilde{\xi}_1}\right)^\alpha P_{-\alpha}\left(1 + \frac{(\tilde{x}_1 - \tilde{\xi}_1)^2 - (\tilde{x}_2 - \tilde{\xi}_2)^2}{2\tilde{x}_1\tilde{\xi}_1}\right) \\ &= \left(\frac{\tilde{x}_1}{\tilde{\xi}_1}\right)^\alpha {}_2F_1\left(\alpha, 1 - \alpha; 1; -\frac{(\tilde{x}_1 - \tilde{\xi}_1)^2 - (\tilde{x}_2 - \tilde{\xi}_2)^2}{4\tilde{x}_1\tilde{\xi}_1}\right) \end{aligned} \tag{7.1-5}$$

where ${}_2F_1$ is the hypergeometric function. Using (7.1-2), we see that in the case of (7.1-4), equation (7.1-3) takes the form

$$\hat{w}_{x_1x_2} + \frac{\alpha}{x_1 + x_2}(\hat{w}_{x_1} + \hat{w}_{x_2}) = 0. \tag{7.1-6}$$

The corresponding Riemann function is (cf. also Ludford [1953])

$$\begin{aligned} \hat{R}(x, \xi) &= \left(\frac{x_1 + x_2}{\xi_1 + \xi_2}\right)^\alpha {}_2F_1(\alpha, 1 - \alpha; 1; \hat{\rho}_1(x, \xi)) \\ \hat{\rho}_1(x, \xi) &= -\frac{(x_1 - \xi_1)(x_2 - \xi_2)}{(x_1 + x_2)(\xi_1 + \xi_2)}, \quad x = (x_1, x_2); \quad \xi = (\xi_1, \xi_2). \end{aligned} \tag{7.1-7}$$

GENERALIZATION OF RIEMANN'S APPROACH

Riemann's approach generalizes to equations of the form

(7.1-8)

$$\tilde{w}_{\tilde{x}_1\tilde{x}_1} - \tilde{w}_{\tilde{x}_2\tilde{x}_2} + \tilde{a}_1(\tilde{x}_1)\tilde{w}_{\tilde{x}_1} + \tilde{a}_2(\tilde{x}_2)\tilde{w}_{\tilde{x}_2} + (\tilde{a}_{01}(\tilde{x}_1) + \tilde{a}_{02}(\tilde{x}_2))\tilde{w} = 0 .$$

Here the indicated dependence of the coefficients implies that the coefficients of (7.1-3) depend on $z_1 + z_2$ and $z_1 - z_2$, respectively. The Helmholtz equation in Example 6.4-2a with the Riemann function (6.4-6) is of this form, and so are the equations

$$\tilde{w}_{\tilde{x}_1\tilde{x}_1} - \tilde{w}_{\tilde{x}_2\tilde{x}_2} + \frac{2\alpha_1}{\tilde{x}_1}\tilde{w}_{\tilde{x}_1} - \frac{2\alpha_2}{\tilde{x}_2}\tilde{w}_{\tilde{x}_2} = 0$$

with constant α_1 and α_2, and

$$\tilde{w}_{\tilde{x}_1\tilde{x}_1} - \tilde{w}_{\tilde{x}_2\tilde{x}_2} - \frac{2\alpha_2}{\tilde{x}_2}\tilde{w}_{\tilde{x}_2} + \alpha_0^2\tilde{w} = 0$$

with constant α_0 and α_2. By (7.1-2), these equations transform into

$$\hat{w}_{x_1x_2} + \left(\frac{\alpha_1}{x_1 + x_2} - \frac{\alpha_2}{x_1 - x_2}\right)\hat{w}_{x_1} + \left(\frac{\alpha_1}{x_1 + x_2} + \frac{\alpha_2}{x_1 - x_2}\right)\hat{w}_{x_2} = 0$$

and

$$\hat{w}_{x_1x_2} - \frac{\alpha_2}{x_1 - x_2}(\hat{w}_{x_1} - \hat{w}_{x_2}) + \tfrac{1}{4}\alpha_0^2\hat{w} = 0$$

respectively. We then continue x_1 and x_2 to complex values z_1 and z_2, respectively, obtaining the equations

(7.1-9)

$$D_1D_2w + \left(\frac{\alpha_1}{z_1 + z_2} - \frac{\alpha_2}{z_1 - z_2}\right)D_1w + \left(\frac{\alpha_1}{z_1 + z_2} + \frac{\alpha_2}{z_1 - z_2}\right)D_2w = 0$$

(7.1-10) $$D_1D_2w - \frac{\alpha_2}{z_1 - z_2}(D_1w - D_2w) + \tfrac{1}{4}\alpha_0^2 w = 0 .$$

For (7.1-9), Henrici [1957] gives the Riemann function

(7.1-11)

$$R(z, s) = \left(\frac{z_1 + z_2}{s_1 + s_2}\right)^{\alpha_1}\left(\frac{z_1 - z_2}{s_1 - s_2}\right)^{\alpha_2} \times$$

$$\times F_3(\alpha_1, \alpha_2, 1 - \alpha_1, 1 - \alpha_2; 1; \rho_1(z, s), \rho_2(z, s)) .$$

Here,

$$\text{(7.1-12)} \qquad \begin{aligned} &\text{(a)} \quad \rho_1(z,s) = -\frac{(z_1 - s_1)(z_2 - s_2)}{(z_1 + z_2)(s_1 + s_2)} \\ &\text{(b)} \quad \rho_2(z,s) = \frac{(z_1 - s_1)(z_2 - s_2)}{(z_1 - z_2)(s_1 - s_2)} \end{aligned} \qquad \text{[cf. (7.1-7)]}$$

and F_3 is the hypergeometric function of two variables. F_3 has the expansion

$$F_3(a, a', b, b'; c; \rho_1, \rho_2) = \sum_{m,n=0}^{\infty} \frac{(a)_m (a')_n (b)_m (b')_n}{(c)_{m+n} m! n!} \rho_1^m \rho_2^n$$

which converges for $|\rho_1| < 1$, $|\rho_2| < 1$. For (7.1-10), the Riemann function is

$$\text{(7.1-13)} \quad R(z,s) = \left(\frac{z_1 - z_2}{s_1 - s_2}\right)^{\alpha_2} \Xi_2(\alpha_2, 1 - \alpha_2; 1; \rho_2(z,s), \rho_3(z,s)).$$

Here Ξ_2 is the confluent hypergeometric function of two variables, which has the expansion

$$\Xi_2(a, b; c; \rho_2, \rho_3) = \sum_{m,n=0}^{\infty} \frac{(a)_m (b)_m}{(c)_{m+n} m! n!} \rho_2^m \rho_3^n$$

with ρ_2 as before and

$$\rho_3(z,s) = -\tfrac{1}{4}\alpha_0^2 (z_1 - s_1)(z_2 - s_2).$$

This series converges for $|\rho_2| < 1$, $|\rho_3| < 1$. The representation (7.1-13) can be found in Henrici [1957]; for properties of the special functions involved, see Erdélyi et al. [1953–1955], Vol. I, pp. 224–229.

HADAMARD'S ELEMENTARY SOLUTION

A second method of obtaining the Riemann function for equations of the form (7.1-3) consists in determining *Hadamard's elementary solution* in the case of the hyperbolic equation (7.1-1) or (7.1-3) (cf. Hadamard [1952]) or the fundamental solution in the case of the elliptic equation (6.1-1). If one is able to accomplish this, one can then obtain the Riemann function from the *coefficient of the logarithmic term*. For instance, in the case of the equation

$$\text{(7.1-14)} \qquad \hat{w}_{x_1 x_2} + \frac{\alpha(1-\alpha)}{(x_1 + x_2)^2} \hat{w} = 0$$

this leads to the Riemann function

$$\hat{R}(x, \xi) = {}_2F_1(\alpha, 1 - \alpha; 1; \hat{\rho}_1(x, \xi)) = P_{-\alpha}(1 - 2\hat{\rho}_1(x, \xi)) \tag{7.1-15}$$

with $\hat{\rho}_1, x, \xi$ as in (7.1-7). See Copson ([1958], pp. 338–341). Observe that equations (7.1-14) and (7.1-4) as well as Example 6.4-5 are related by simple transformations. For relations between the Green's function and the Riemann function, see also Mackie [1965].

CHAUNDY'S GENERALIZED HYPERGEOMETRIC EQUATIONS

In a series of papers, T. W. Chaundy [1935, 1936, 1939] discussed in great detail what he called partial differential equations of "**generalized hypergeometric type.**" These are of the form

$$\prod_{r=1}^{n} f_r(\delta_r) Z = y_1 y_2 \cdots y_n \prod_{r=1}^{n} g_r(\delta_r) Z, \qquad \delta_r = y_r \, \partial/\partial y_r \tag{7.1-16}$$

where the f_r and g_r are polynomials. This is another class of equations for which the Riemann function is known; we will return to this. In 1938, Chaundy considered self-adjoint equations of ***Darboux type*** of the form

$$\hat{w}_{x_1 x_2} + \sum_{r=1}^{4} (-1)^r M_r h_r(x)^{-2} \hat{w} = 0 \tag{7.1-17}$$

where $M_r = m_r(m_r + 1)$, $m_r = const$, and

$$h_1(x) = x_1 + x_2, \qquad h_2(x) = x_1 - x_2, \qquad h_3(x) = 1 - x_1 x_2,$$
$$h_4(x) = 1 + x_1 x_2.$$

If in this equation we introduce the new variables

$$y_r = (-1)^r (x_1 - \xi_1)(x_2 - \xi_2)[h_r(x) h_r(\xi)]^{-1} \tag{7.1-18}$$

and use the definition of δ_r, it follows from (7.1-17) by straightforward calculation that the Riemann function $\hat{R}$ must satisfy the equation

$$\sum_{r=1}^{4} c_r(x) \left[\delta_r^2 + \delta_r - \frac{1}{y_r} \delta_r (\delta_1 + \delta_2 + \delta_3 + \delta_4) - M_r \right] \hat{R} = 0 \tag{7.1-19}$$

and the condition

$$\hat{R} = 1 \quad \text{when} \quad x_1 = \xi_1 \quad \text{or} \quad x_2 = \xi_2 \tag{7.1-20}$$

(which also means that $\hat{R} = 1$ when $y_r = 0$). Here $c_r = (-1)^{r+1} h_r^{-2}$. Obviously, (7.1-19) is fulfilled if $\hat{R}$ simultaneously satisfies the four equations

$$(7.1\text{-}21) \quad \delta_r(\delta_1 + \delta_2 + \delta_3 + \delta_4)\hat{R} = y_r(\delta_r - m_r)(\delta_r + m_r + 1)\hat{R}, \qquad r = 1, \cdots, 4.$$

These are of the form (7.1-16). A solution of (7.1-17) also satisfying (7.1-21) is given by the ***multiple hypergeometric series***

(7.1-22)

$$\hat{R}(x, \xi) = \sum_{n_1, \cdots, n_4 = 0}^{\infty} \left(\prod_{r=1}^{4} \left[(-m_r)_{n_r} (m_r + 1)_{n_r} \frac{y_r^{n_r}}{n_r!} \right] \right) \Big/ (n_1 + n_2 + n_3 + n_4)!.$$

This series converges for sufficiently small $|y_r|$, that is, in a neighborhood of $x = \xi$.

This theory is quite versatile, and special results of practical interest are obtained if we choose particular values of the parameters involved. Indeed, for $m_1 = -\alpha$ and $m_2 = m_3 = m_4 = 0$ we obtain from (7.1-17, 22) the equation (7.1-14) and its Riemann function (7.1-15). Furthermore, if we choose $m_1 = m$, $m_2 = n$ and the others zero, then (7.1-17) reduces to

$$(7.1\text{-}23) \qquad \hat{w}_{x_1 x_2} + \left[-\frac{m(m+1)}{(x_1 + x_2)^2} + \frac{n(n+1)}{(x_1 - x_2)^2} \right] \hat{w} = 0.$$

It is interesting that the corresponding Riemann function, as obtained from (7.1-22), can be converted to the relatively simple form

$$(7.1\text{-}24) \quad \hat{R}(x, \xi) = (\delta_1 + \delta_2 + 1) \int_0^1 P_m(1 - 2y_1 + 2y_1 t) P_n(1 - 2y_2 t)\, dt.$$

If all four parameters are equal and are then denoted by m, another remarkably simple representation of the Riemann function results directly from (7.1-22); in fact, with y_r as before, we obtain

$$(7.1\text{-}25) \qquad \hat{R}(x, \xi) = P_m\left(-1 + 2 \prod_{r=1}^{4} (1 - y_r) \right).$$

CONTOUR INTEGRATION

A fourth method for deriving Riemann functions is the use of contour integration techniques. This idea was probably first proposed by Chaundy [1936, 1939], and independently developed by Mackie in extension of his work

[1954, 1955]. We explain this idea in terms of the equation

$$\hat{w}_{x_1x_2} + \frac{1}{x_1 + x_2}\left(\alpha_1\hat{w}_{x_1} + \alpha_2\hat{w}_{x_2}\right) = 0 \tag{7.1-26}$$

with constant α_1 and α_2, using a simplified approach suggested by Copson [1958]. We first note that the Riemann function as a function of x satisfies the adjoint equation, and as a function of ξ the given equation. Now a particular solution of the latter, with x replaced by ξ and t denoting a complex parameter, is

$$\lambda(t, \xi) = (t - \xi_1)^{-\alpha_2}(t + \xi_2)^{-\alpha_1}.$$

Similarly, a solution of the adjoint equation is

$$\mu(x, t) = (x_1 + x_2)(t - x_1)^{\alpha_2 - 1}(t + x_2)^{\alpha_1 - 1}.$$

For this reason, we assume for $\hat{R}$ a representation of the form

$$\hat{R}(x, \xi) = \frac{1}{2\pi i}\int_C \lambda(t, \xi)\mu(x, t)f(t)\,dt \tag{7.1-27}$$

with a suitable function f and curve C. We now cut the t-plane from the branch point $-x_2$ along the real axis to the branch point $-\xi_2$, and similarly from x_1 to ξ_1. Then the integrand becomes single-valued. We choose that branch which is real for sufficiently large t. Now by Theorem 6.3-1, the function $\hat{R}$ must satisfy the conditions

$$\begin{aligned} \hat{R}(\xi_1, x_2, \xi) &= (\xi_1 + x_2)^{\alpha_1}(\xi_1 + \xi_2)^{-\alpha_1} \\ \hat{R}(x_1, \xi_2, \xi) &= (x_1 + \xi_2)^{\alpha_2}(\xi_1 + \xi_2)^{-\alpha_2}. \end{aligned} \tag{7-1.28}$$

We let $x_1, x_2, \xi_1, \xi_2 > 0$, without restriction. Then we can take for C a simple closed curve C_1 which contains the segment from x_1 to ξ_1 in its interior and the segment from $-x_2$ to $-\xi_2$ in its exterior. We determine f. For $x_1 = \xi_1$ the representation (7.1-27) becomes simply

$$\hat{R}(\xi_1, x_2, \xi) = \frac{\xi_1 + x_2}{2\pi i}\int_{C_1}\frac{(t + x_2)^{\alpha_1 - 1}}{(t + \xi_2)^{\alpha_1}}f(t)\frac{dt}{t - \xi_1}.$$

If f is regular inside and on C_1, then this yields

$$\hat{R}(\xi_1, x_2, \xi) = \frac{(\xi_1 + x_2)^{\alpha_1}}{(\xi_1 + \xi_2)^{\alpha_1}}f(\xi_1).$$

We now see that the first condition in (7.1-28) is satisfied if we choose $f(\xi_1) = 1$. But (7.1-27) with $C = C_1$ and $f = 1$, written out,

$$(7.1\text{-}29) \qquad \hat{R}(x, \xi) = \frac{x_1 + x_2}{2\pi i} \int_{C_1} \frac{(t - x_1)^{\alpha_2 - 1}(t + x_2)^{\alpha_1 - 1}}{(t - \xi_1)^{\alpha_2}(t + \xi_2)^{\alpha_1}}\, dt,$$

satisfies also the second condition in (7.1-28). This holds since in (7.1-29) we can replace C_1 by a simple closed curve C_2 which contains the segment with endpoints $-x_2$ and $-\xi_2$ in its interior and that other segment in the exterior. Hence we have the result that (7.1-29) is the desired Riemann function.

It is remarkable that this Riemann function can be expressed in terms of a *single hypergeometric function*, as follows. All we have to do is to substitute

$$t - x_1 = \frac{(\xi_1 - x_1)(x_1 + \xi_2)\tilde{t}}{(\xi_1 + \xi_2) - (\xi_1 - x_1)\tilde{t}}$$

into (7.1-29). Then

$$\hat{R}(x, \xi) = \frac{(x_1 + x_2)^{\alpha_1}(x_1 + \xi_2)^{\alpha_2 - \alpha_1}}{2\pi i(\xi_1 + \xi_2)^{\alpha_2}} \times$$

$$\times \int_{C_0} \tilde{t}^{\alpha_2 - 1}(\tilde{t} - 1)^{-\alpha_2}\left[1 - \hat{\rho}_1(x, \xi)\tilde{t}\right]^{\alpha_1 - 1} d\tilde{t}$$

with $\hat{\rho}_1$ as in (7.1-7) and C_0 a simple closed contour enclosing the points 0 and 1 and leaving $1/\hat{\rho}_1(x, \xi)$ outside. By means of a familiar integral representation of the hypergeometric function (cf. Erdélyi et al. [1953–1955], Vol. I, p. 60]) this gives the result

$$(7.1\text{-}30) \quad \hat{R}(x, \xi) = \left(\frac{x_1 + x_2}{x_1 + \xi_2}\right)^{\alpha_1} \left(\frac{x_1 + \xi_2}{\xi_1 + \xi_2}\right)^{\alpha_2} {}_2F_1(1 - \alpha_1, \alpha_2; 1; \hat{\rho}_1(x, \xi)).$$

Since equation (7.1-26) is symmetric in x_1 and x_2, in addition to (7.1-30) we also have

$$\hat{R}(x, \xi) = \left(\frac{x_1 + x_2}{x_2 + \xi_1}\right)^{\alpha_2} \left(\frac{x_2 + \xi_1}{\xi_1 + \xi_2}\right)^{\alpha_1} {}_2F_1(1 - \alpha_2, \alpha_1; 1; \hat{\rho}_1(x, \xi)).$$

Note that this follows also from (7.1-30) by applying the familiar formula

$${}_2F_1(\alpha, \beta; \gamma; \hat{\rho}_1) = (1 - \hat{\rho}_1)^{\gamma - \alpha - \beta}{}_2F_1(\gamma - \alpha, \gamma - \beta; \gamma; \hat{\rho}_1).$$

Practically more advantageous is a *symmetric* representation of $\hat{R}$ which can be obtained by introducing $1 - \rho_1$ into the hypergeometric function in (7.1-30);

cf. Magnus et al. [1966], p. 47 near the bottom. This gives, for noninteger $\alpha_1 - \alpha_2$,

$$\hat{R}(x, \xi) = \sum_{j=1}^{2} \frac{\Gamma(\alpha_j - \alpha_{j^*})}{\Gamma(\alpha_j)\Gamma(1 - \alpha_{j^*})} \left(\frac{x_1 + x_2}{x_j + \xi_{j^*}} \right)^{\alpha_j} \left(\frac{x_j + \xi_{j^*}}{\xi_1 + \xi_2} \right)^{\alpha_{j^*}} \times \tag{7.1-31}$$

$$\times {}_2F_1\left(1 - \alpha_j, \alpha_{j^*}; 1 - \alpha_j + \alpha_{j^*}; \frac{(x_1 + \xi_2)(\xi_1 + x_2)}{(x_1 + x_2)(\xi_1 + \xi_2)}\right)$$

where $j^* = 3 - j$.

COMPLEX INTEGRAL TRANSFORMS

A further method of deriving Riemann functions consists in finding direct ways of solving the Goursat problem in Theorem 6.3-1a by which the Riemann function is determined. As an example we consider the solution of the equation

$$\hat{w}_{x_1 x_2} - \hat{w} = 0 \tag{7.1-32}$$

by means of a complex Fourier integral. Substituting $\hat{x}_1 = x_1 - \xi_1$, $\hat{x}_2 = x_2 - \xi_2$ and writing $\hat{\hat{w}}(\hat{x}_1, \hat{x}_2) \equiv \hat{w}(x_1, x_2)$, we see from Theorem 6.3-1a that the Riemann function of our present equation is the solution of

$$\hat{\hat{R}}_{\hat{x}_1 \hat{x}_2} = \hat{\hat{R}} \tag{7.1-33}$$

which is 1 on the coordinate axes. Following Titchmarsh [1948], pp. 297–298, we set

$$V(\alpha, \hat{x}_2) = \frac{1}{\sqrt{2\pi}} \int_0^\infty \hat{\hat{R}}(\hat{x}_1, \hat{x}_2) e^{i\alpha \hat{x}_1} \, d\hat{x}_1 \tag{7.1-34}$$

where $\alpha = \alpha_1 + i\alpha_2$ and $\alpha_2 > c > 0$. We now apply integration by parts with respect to $\hat{x}_1$ and observe that $\partial \hat{\hat{R}} / \partial \hat{x}_2$ vanishes for $\hat{x}_1 = 0$. Then we obtain

$$V_{\hat{x}_2}(\alpha, \hat{x}_2) = -\frac{1}{i\alpha\sqrt{2\pi}} \int_0^\infty \hat{\hat{R}}_{\hat{x}_1 \hat{x}_2}(\hat{x}_1, \hat{x}_2) e^{i\alpha \hat{x}_1} \, d\hat{x}_1 .$$

From this and (7.1-33),

$$V_{\hat{x}_2} = -\frac{1}{i\alpha} V.$$

This yields

$$V(\alpha, \hat{x}_2) = W(\alpha) e^{i\hat{x}_2/\alpha} \tag{7.1-35}$$

where W can be found by comparing this with (7.1-34), letting $\hat{x}_2$ tend to zero and noting that $\hat{\hat{R}}(\hat{x}_1, 0) = 1$. The result is simply

$$W(\alpha) = \frac{1}{\sqrt{2\pi}} \int_0^\infty e^{i\alpha \hat{x}_1} \, d\hat{x}_1 = -\frac{1}{i\alpha\sqrt{2\pi}} . \tag{7.1-36}$$

From this and (7.1-34, 35) we finally obtain the one-sided Laplace transform of $\hat{\hat{R}}(\cdot, \hat{x}_2)$ given by

$$\frac{1}{\beta} e^{\hat{x}_2/\beta} = \int_0^\infty \hat{\hat{R}}(\hat{x}_1, \hat{x}_2) e^{-\hat{x}_1 \beta} \, d\hat{x}_1 , \qquad \beta = -i\alpha .$$

The inversion formula is

$$\hat{\hat{R}}(\hat{x}_1, \hat{x}_2) = \frac{1}{2\pi i} \int_{k - i\infty}^{k + i\infty} \frac{1}{\beta} e^{\hat{x}_2/\beta} e^{\hat{x}_1 \beta} \, d\beta . \tag{7.1-37}$$

This illustrates the Fourier integral method for obtaining Riemann functions. In the present case, comparing our result with the representation of the Riemann function of (7.1-32) in terms of the Bessel function J_0 in Example 6.4-2a, we see that (7.1-37) also gives an integral representation of $J_0(2i\sqrt{\hat{x}_1 \hat{x}_2})$.

7.2 *ADDITION OF RIEMANN FUNCTIONS*

At the beginning of this chapter we emphasized the importance of the explicit knowledge of the Riemann function and the general difficulty in solving this problem for a given equation. In certain cases, one may employ the following method, whose idea was proposed by Olevskij as early as 1952, but worked out in more detail only in 1977, by Papadakis and Wood.

ADDITION FORMULA

The idea underlying that method is as follows. Suppose that we know the Riemann functions for two equations of the form

$$\tilde{w}_{\tilde{x}_1 \tilde{x}_1} - \tilde{w}_{\tilde{x}_2 \tilde{x}_2} + \tilde{a}_{0j}(\tilde{x}_j) \tilde{w} = 0 , \qquad j = 1, 2 \tag{7.2-1}$$

with coefficients depending only on one of the two independent variables, as indicated. Then we can determine the Riemann function for the "composed equation"

$$\tilde{w}_{\tilde{x}_1\tilde{x}_1} - \tilde{w}_{\tilde{x}_2\tilde{x}_2} + \left[\tilde{a}_{01}(\tilde{x}_1) + \tilde{a}_{02}(\tilde{x}_2)\right]\tilde{w} = 0 \tag{7.2-2}$$

in terms of the Riemann functions for the two given equations. Such an "*addition formula*" is obtained from the following theorem.

7.2-1 Theorem

Let $\tilde{R}_1$ and $\tilde{R}_2$ be the Riemann functions of equations (7.2-1). *Then the Riemann function $\tilde{R}$ for* (7.2-2) *is given by*

$$\tilde{R}(\tilde{x}, \tilde{\xi}) = \tilde{R}_1(\tilde{x}, \tilde{\xi}) + \int_{\tilde{x}_2-\tilde{\xi}_2}^{\tilde{x}_1-\tilde{\xi}_1} \tilde{R}_1(\tilde{x}_1, t, \tilde{\xi}_1, 0)\frac{\partial}{\partial t}\tilde{R}_2(t, \tilde{x}_2, 0, \tilde{\xi}_2)\,dt \tag{7.2-3a}$$

or

$$\tilde{R}(\tilde{x}, \tilde{\xi}) = \tilde{R}_2(\tilde{x}, \tilde{\xi}) + \int_{\tilde{x}_1-\tilde{\xi}_1}^{\tilde{x}_2-\tilde{\xi}_2} \tilde{R}_2(t, \tilde{x}_2, 0, \tilde{\xi}_2)\frac{\partial}{\partial t}\tilde{R}_1(\tilde{x}_1, t, \tilde{\xi}_1, 0)\,dt. \tag{7.2-3b}$$

Proof. We apply Theorem 6.3-1a to our equations (7.2-1). By the uniqueness of the Riemann function (Corollary 6.2-5) we have identically

$$\begin{aligned} \tilde{R}_1(\tilde{x}, \tilde{\xi}) &= \tilde{R}_1(\tilde{x}_1, \tilde{x}_2 - \tilde{\xi}_2, \tilde{\xi}_1, 0) = \tilde{R}_1(\tilde{x}_1, \tilde{\xi}_2 - \tilde{x}_2, \tilde{\xi}_1, 0) \\ \tilde{R}_2(\tilde{x}, \tilde{\xi}) &= \tilde{R}_2(\tilde{x}_1 - \tilde{\xi}_1, \tilde{x}_2 0, \tilde{\xi}_2) = \tilde{R}_2(\tilde{\xi}_1 - \tilde{x}_1, \tilde{x}_2, 0, \tilde{\xi}_2). \end{aligned} \tag{7.2-4}$$

Then the representation (7.2-3b) follows from (7.2-3a) by integration by parts. Hence it only remains to prove (7.2-3a).

Now the conditions (7.2-4) imply that $\tilde{R}_1$ is an even function of $\tilde{x}_2 - \tilde{\xi}_2$, and $\tilde{R}_2$ is even in $\tilde{x}_1 - \tilde{\xi}_1$. From (7.2-3a), since the integrand is an odd function of t and the interval of integration is either symmetric or empty, we thus obtain

$$\tilde{R}(\tilde{x}, \tilde{\xi}) = \tilde{R}_1(\tilde{x}, \tilde{\xi}) \qquad \text{for} \qquad |\tilde{x}_1 - \tilde{\xi}_1| = |\tilde{x}_2 - \tilde{\xi}_2|. \tag{7.2-5}$$

As the next step, we calculate $\tilde{R}_{\tilde{x}_1\tilde{x}_1} + \tilde{a}_{01}(\tilde{x}_1)\tilde{R}$ from (7.2-3a) and $\tilde{R}_{\tilde{x}_2\tilde{x}_2} - \tilde{a}_{02}(\tilde{x}_2)\tilde{R}$ from (7.2-3b). Since the Riemann functions $\tilde{R}_1$ and $\tilde{R}_2$ must satisfy the self-adjoint equations (7.2-1), we can use them to introduce $\tilde{R}_{1,\tilde{x}_2\tilde{x}_2}$ into the first result and $\tilde{R}_{2,\tilde{x}_1\tilde{x}_1}$ into the second result. If we subtract the latter from the former and take (7.2-4) into account, the left side of (7.2-2) with $\tilde{R}$

instead of $\tilde{w}$ becomes

$$\tilde{R}_{1,\tilde{x}_1}(\tilde{x}_1, t, \tilde{\xi}_1, 0)\big|_{t=\tilde{x}_1-\tilde{\xi}_1} \tilde{R}_{2,\tilde{x}_1}(\tilde{x}_1 - \tilde{\xi}_1, \tilde{x}_2, 0, \tilde{\xi}_2)$$
$$- \tilde{R}_{2,\tilde{x}_2}(t, \tilde{x}_2, 0, \tilde{\xi}_2)\big|_{t=\tilde{x}_2-\tilde{\xi}_2} \tilde{R}_{1,\tilde{x}_2}(\tilde{x}_1, \tilde{x}_2 - \tilde{\xi}_2, \tilde{\xi}_1, 0)$$
$$+ \int_{\tilde{x}_2-\tilde{\xi}_2}^{\tilde{x}_1-\tilde{\xi}_1} [\tilde{R}_{1,t}(\tilde{x}_1, t, \tilde{\xi}_1, 0)\tilde{R}_{2,t}(t, \tilde{x}_2, 0, \tilde{\xi}_2)]_{,t}\, dt .$$

Evaluating the integral in this formula gives

$$\left(\tilde{R}_{\tilde{x}_1\tilde{x}_1} - \tilde{R}_{\tilde{x}_2\tilde{x}_2} + [\tilde{a}_{01}(\tilde{x}_1) + \tilde{a}_{02}(\tilde{x}_2)]\tilde{R}\right)(\tilde{x}_1, \tilde{x}_2, \tilde{\xi}_1, \tilde{\xi}_2)$$
$$= \tilde{R}_{2,\tilde{x}_1}(\tilde{x}_1 - \tilde{\xi}_1, \tilde{x}_2, 0, \tilde{\xi}_2)\left[\tilde{R}_{1,\tilde{x}_1}(\tilde{x}_1, t, \tilde{\xi}_1, 0) + \tilde{R}_{1,t}(\tilde{x}_1, t, \tilde{\xi}_1, 0)\right]\Big|_{t=\tilde{x}_1-\tilde{\xi}_1}$$
$$(7.2\text{-}6) \qquad - \tilde{R}_{1,\tilde{x}_2}(\tilde{x}_1, \tilde{x}_2 - \tilde{\xi}_2, \tilde{\xi}_1, 0)\left[\tilde{R}_{2,\tilde{x}_2}(t, \tilde{x}_2, 0, \tilde{\xi}_2) + \tilde{R}_{2,t}(t, \tilde{x}_2, 0, \tilde{\xi}_2)\right]\Big|_{t=\tilde{x}_2-\tilde{\xi}_2} .$$

Since $\tilde{R}_1$ and $\tilde{R}_2$ are Riemann functions for the self-adjoint equations (7.2-1), the expressions in the brackets on the right-hand side of (7.2-6) vanish. Furthermore, the right-hand side of (7.2-5) equals 1. Hence we have shown that $\tilde{R}$, as given by (7.2-3a), satisfies (6.3-1, 3), rewritten for equation (7.2-2). Recalling the relations between the equations (7.1-1, 3) and (6.1-4), we see that because of Theorem 6.3-1 we have shown that $\tilde{R}$ is indeed the Riemann function for (7.2-2), and the proof is complete. ■

It is not essential that we started from an equation involving $\tilde{x}_1, \tilde{x}_2$ since the transformation of our results for an equation in x_1, x_2 given by (7.1-2) is immediate; similarly for an equation (6.1-4) in complex variables z_1, z_2.

7.2-2 *Example*

In terms of $\tilde{x}_1, \tilde{x}_2$, equation (7.1-23) is

$$(7.2\text{-}7) \qquad \tilde{w}_{\tilde{x}_1\tilde{x}_1} - \tilde{w}_{\tilde{x}_2\tilde{x}_2} + \left[-\frac{m(m+1)}{\tilde{x}_1^2} + \frac{n(n+1)}{\tilde{x}_2^2}\right]\tilde{w} = 0 .$$

To find its Riemann function from Theorem 7.2-1, we may use that, by (7.1-24), equation (7.2-7) with $n = 0$, and again with $m = 0$, has the Riemann function

$$\tilde{R}_j(\tilde{x}, \tilde{\xi}) = P_{n_j}(1 - 2\tilde{y}_j)$$

where

$$\tilde{y}_j = (-1)^j \frac{(\tilde{x}_1 - \tilde{\xi}_1)^2 - (\tilde{x}_2 - \tilde{\xi}_2)^2}{4\tilde{x}_j^2 \tilde{\xi}_j^2}, \qquad \begin{array}{ll} j = 1; & n_1 = m \\ j = 2; & n_2 = n. \end{array}$$

Substituting this into (7.2-3a), we obtain the representation

$$\begin{aligned} \tilde{R}(\tilde{x}, \tilde{\xi}) &= P_m(1 - 2\tilde{y}_1) \\ &+ \int_{\tilde{x}_2 - \tilde{\xi}_2}^{\tilde{x}_1 - \tilde{\xi}_1} P_m\left(1 - 2\tilde{y}_1|_{(\tilde{x}_2, \tilde{\xi}_2) = (t, 0)}\right) \frac{\partial}{\partial t} P_n\left(1 - 2\tilde{y}_2|_{(\tilde{x}_1, \tilde{\xi}_1) = (t, 0)}\right) dt. \end{aligned} \tag{7.2-8}$$

The reader may confirm this result by a direct application of (7.1-24) to (7.2-7) with general m and n.

NON-SELF-ADJOINT EQUATIONS

From a practical point of view, it is remarkable that Theorem 7.2-1 can be generalized to non-self-adjoint equations. The idea of proof is similar, and the result is obtained by direct calculations and utilization of the defining conditions of the Riemann function. As an alternative, we can use the transformation behavior of the Riemann function in order to arrive at the same result, which can be stated as follows.

7.2-3 Corollary

Let $\tilde{R}_1$ and $\tilde{R}_2$ be the Riemann functions for

$$\tilde{w}_{\tilde{x}_1\tilde{x}_1} - \tilde{w}_{\tilde{x}_2\tilde{x}_2} + 2\tilde{a}_{11}(\tilde{x}_1)\tilde{w}_{\tilde{x}_1} + \tilde{a}_{01}(\tilde{x}_1)\tilde{w} = 0$$

and

$$\tilde{w}_{\tilde{x}_1\tilde{x}_1} - \tilde{w}_{\tilde{x}_2\tilde{x}_2} - 2\tilde{a}_{22}(\tilde{x}_2)\tilde{w}_{\tilde{x}_2} + \tilde{a}_{02}(\tilde{x}_2)\tilde{w} = 0$$

respectively. Then the Riemann function for

(7.2-9)

$$\tilde{w}_{\tilde{x}_1\tilde{x}_1} - \tilde{w}_{\tilde{x}_2\tilde{x}_2} + 2\tilde{a}_{11}(\tilde{x}_1)\tilde{w}_{\tilde{x}_1} - 2\tilde{a}_{22}(\tilde{x}_2)\tilde{w}_{\tilde{x}_2} + [\tilde{a}_{01}(\tilde{x}_1) + \tilde{a}_{02}(\tilde{x}_2)]\tilde{w} = 0$$

is given by

$$\begin{aligned} \tilde{R}(\tilde{x}, \tilde{\xi}) &= \tilde{R}_1(\tilde{x}, \tilde{\xi}) \exp\left[\int_{\tilde{\xi}_2}^{\tilde{x}_2} \tilde{a}_{22}(t)\, dt\right] \\ &+ \int_{\tilde{x}_2 - \tilde{\xi}_2}^{\tilde{x}_1 - \tilde{\xi}_1} \tilde{R}_1(\tilde{x}_1, t, \tilde{\xi}_1, 0) \frac{\partial}{\partial t} \tilde{R}_2(t, \tilde{x}_2, 0, \tilde{\xi}_2)\, dt. \end{aligned} \tag{7.2-10}$$

To further increase the utility of the addition method for Riemann functions, Du [1981] has suggested a study of transformations which leave the Riemann function of the composed equation conformally invariant. In this way, one is able to derive further equations to which the method is applicable.

MULTIPLICATION FORMULA

Du [1981] also proposed a "*multiplication formula*" for Riemann functions, expressing the Riemann function of

$$\tilde{w}_{\tilde{x}_1\tilde{x}_1} - \tilde{w}_{\tilde{x}_2\tilde{x}_2} + \tilde{a}_{01}(\tilde{x}_1)\tilde{a}_{02}(\tilde{x}_2)\tilde{w} = 0$$

in terms of the Riemann functions of the two equations (7.2-1). The underlying idea is to combine the addition method with the Picard iteration used in Sec. 6.2. The technical details are complicated, and it remains to be seen whether this approach will lead to the discovery of new Riemann functions. For details of both methods we refer to Du's paper [1981].

7.3 FURTHER RESULTS

More promising than the papers just mentioned seem to be results by Cohn, Daggit, and some others, based on ideas whose interrelation we shall now investigate.

Cohn [1947] considered equations (7.1-3) with $\hat{a}_1 = 0 = \hat{a}_2$ (implying self-adjointness) and additionally required that

$$\hat{a}_0(x_1, x_2) \equiv H(x_1 + x_2);$$

hence

$$\hat{w}_{x_1x_2} + H(x_1 + x_2)\hat{w} = 0. \tag{7.3-1}$$

The defining Volterra integral equation (6.2-1) for the Riemann function thus is of the form

$$\hat{R}(x, \xi) = 1 - \int_{\xi_1}^{x_1}\int_{\xi_2}^{x_2} H(t_1 + t_2)\hat{R}(t_1, t_2, \xi)\, dt_2\, dt_1 .$$

An iterative solution of this equation can be obtained by the Picard iteration (cf. Sec. 6.2):

$$\hat{R}(x, \xi) = \sum_{m=0}^{\infty} (-1)^m \hat{R}_m(x, \xi) \tag{7.3-2}$$

where

$$\hat{R}_0(x, \xi) \equiv 1$$

(7.3-3)
$$\hat{R}_{m+1}(x, \xi) = \int_{\xi_1}^{x_1} \int_{\xi_2}^{x_2} \hat{R}_m(t_1, t_2, \xi) H(t_1 + t_2)\, dt_2\, dt_1, \quad m \in \mathbb{N}_0.$$

Cohn now asked whether there were any functions H for which the Riemann function $\hat{R}$ of (7.3-1) depended, insofar as x_1 and x_2 were concerned, only on the function $\hat{R}_1$ defined by (7.3-3). He showed that the answer is positive if and only if

(7.3-4)
$$H(x_1 + x_2) = -\frac{\lambda(\lambda + 1)\mu^2}{\sinh^2[\mu(x_1 + x_2 + \nu)]}$$

where λ, μ, and ν are arbitrary real constants. For the self-adjoint equation (7.3-1) with (7.3-4) he obtained the Riemann function as

(7.3-5a)
$$\hat{R}(x, \xi) = {}_2F_1(1 + \lambda, -\lambda; 1; V)$$

where

(7.3-5b)
$$V = -\frac{\sinh[\mu(x_1 - \xi_1)] \sinh[\mu(x_2 - \xi_2)]}{\sinh[\mu(x_1 + x_2 + \nu)] \sinh[\mu(\xi_1 + \xi_2 + \nu)]}.$$

Here, as limiting cases of (7.3-4) and (7.3-5) include the following ones.

For $\mu \to 0$ we obtain (7.3-1) with

(7.3-6a)
$$H(x_1 + x_2) = -\frac{\lambda(\lambda + 1)}{(x_1 + x_2 + \nu)^2}$$

having the Riemann function

(7.3-6b)
$$\hat{R}(x, \xi) = {}_2F_1\left(1 + \lambda, -\lambda; 1; -\frac{(x_1 - \xi_1)(x_2 - \xi_2)}{(x_1 + x_2 + \nu)(\xi_1 + \xi_2 + \nu)}\right).$$

Changing λ to $\lambda e^{\mu\lambda}$ $(\lambda > 0)$ in (7.3-4) and letting $\nu \to \infty$ yields

(7.3-7a)
$$H(x_1 + x_2) = -\tilde{\lambda} \exp[\tilde{\mu}(x_1 + x_2)]$$

(with new constants $\tilde{\lambda}, \tilde{\mu}$ well-determined by λ and μ); by a confluence process, the corresponding Riemann function is

(7.3-7b)
$$\hat{R}(x, \xi) = J_0(2\tilde{V}^{1/2})$$

where

$$\tilde{V} = -\tilde{\lambda}\tilde{\mu}^{-2}\left(e^{\tilde{\mu}x_1} - e^{\tilde{\mu}\xi_1}\right)\left(e^{\tilde{\mu}x_2} - e^{\tilde{\mu}\xi_2}\right).$$

(7.3-7) contains the limiting case (for $\tilde{\mu} \to 0$)

$$H(x_1 + x_2) \equiv -\tilde{\lambda} \tag{7.3-8}$$

with the Riemann function (7.3-7b) and $\tilde{V}$ reducing to

$$\tilde{V} = -\tilde{\lambda}(x_1 - \xi_1)(x_2 - \xi_2)$$

in accordance with the result in Example 5.4-2a.

In his 1970 paper, Cohn showed that if one removes the restriction that $\hat{a}_0$ be a function of the variable $x_1 + x_2$, then essentially no new Riemann functions can be obtained by this particular method. The preceding condition, namely, that the unknown function $\hat{R}$ is functionally dependent on the first approximation $\hat{R}_1$ was later called a "*functionally stable iteration*" (see Cohn [1973]). In the latter paper, Cohn interprets the conditions for this functional stability in terms of a curvature-type condition. As a condition for functional stability he obtains

$$\frac{\partial^2}{\partial x_1\, \partial x_2} \log H(x_1, x_2) = \rho H(x_1, x_2) \tag{7.3-9}$$

with constant ρ, namely,

$$\rho = -2/[\lambda(\lambda + 1)] \qquad \text{in cases (7.3-4) and (7.3-6)}$$

and

$$\rho = 0 \qquad \text{in cases (7.3-7) and (7.3-8)}.$$

The reason for calling ρ a "curvature constant" results from analogies such as the Bergman invariant metric (cf. Cohn [1973], Bergman [1970b]).

Related to Cohn's work are papers by Daggit [1970], Wood [1976], Geddes–Mackie [1977], Florian–Püngel [1979], and Wallner [1981a].

Geddes and Mackie [1977] investigated the question of whether Cohn's method can be generalized by looking for functions H of an auxiliary variable V depending on x_1, x_2, ξ_1, ξ_2, so that $\hat{R}$ can be represented as a function of V (and possibly of ξ_1, ξ_2) and satisfies a second-order ordinary differential equation with respect to V. Their result is that this method does not produce self-adjoint equations other than those represented by (7.3-4), apart from those obtained by trivial changes of variables.

USE OF INFINITESIMAL TRANSFORMATIONS

In the paper by Daggit [1970] infinitesimal transformations are used for predicting the form of the Riemann function for equations that need not be self-adjoint. (For relations to (7.3-9), see Daggit [1970], pp. 98–100, especially Lemmas A and B.) Here, Riemann functions are obtained in the form of Bessel functions, hypergeometric, and Laguerre functions of suitable auxiliary variables. As an example involving the Laguerre function we mention the equation

$$\hat{w}_{x_1x_2} + \frac{1}{x_2}\left(\alpha - \frac{1}{\beta x_1 x_2}\right)\hat{w}_{x_1} + \frac{1}{(x_1x_2)^2}\hat{w} = 0 \tag{7.3-10}$$

whose Riemann function as obtained by Daggit is

$$\hat{R}(x,\xi) = \left(\frac{x_2}{\xi_2}\right)^{\alpha} \exp\left(\frac{\xi_2 - x_2}{\beta x_1 x_2 \xi_2}\right) L_{-\beta}\left(-\frac{(x_1-\xi_1)(x_2-\xi_2)}{\beta x_1 x_2 \xi_1 \xi_2}\right). \tag{7.3-11}$$

Here $L_{-\beta}$ denotes the Laguerre function satisfying the initial value problem $L_{-\beta}(0) = 1$ of the ordinary differential equation

$$\tilde{\tilde{V}} L''_{-\beta} + \left(1 - \tilde{\tilde{V}}\right) L'_{-\beta} - \beta L_{-\beta} = 0$$

where

$$\tilde{\tilde{V}} = -\frac{(x_1-\xi_1)(x_2-\xi_2)}{\beta x_1 x_2 \xi_1 \xi_2}.$$

SIMPLE RIEMANN FUNCTIONS IN THE SENSE OF D. H. WOOD

The papers by Cohn and Daggit gave the idea for the introduction of *simple* Riemann functions by Wood [1976]. By definition, a Riemann function for (7.1-3) is ***simple*** if $\hat{R}$ is given in the form $\hat{R} = gF$, where

$$\hat{R}(x,\xi) = g(x,\xi)F(x,\xi) \tag{7.3-12}$$

and g and F satisfy the following conditions (i) to (iii).

(i) Let

$$F_{x_1x_2} + \alpha(x,\xi)F_{x_1} + \beta(x,\xi)F_{x_2} + \gamma(x,\xi)F = 0 \tag{7.3-13}$$

denote the equation which is obtained when the right-hand side of (7.3-12) is introduced into the adjoint equation of (7.1-3). Assume that $\alpha_{x_1} = -\alpha\beta$ holds.

(ii) Let F be representable in the form

$$F(x, \xi) = f(V(x, \xi), \xi) \tag{7.3-14}$$

with a suitable auxiliary variable V; hence, (7.3-13) and (7.3-14) yield the equation

$$\tilde{\alpha}(x, \xi)\frac{\partial^2 f}{\partial V^2} + \tilde{\beta}(x, \xi)\frac{\partial f}{\partial V} + f = 0 .$$

Assume furthermore that $\tilde{\alpha}$ satisfies

$$\frac{\partial(\tilde{\alpha}, V)}{\partial(x_1, x_2)} = 0 .$$

(iii) Finally, assume

$$V(x_1, \xi_2, \xi) = V(\xi_1, x_2, \xi)$$

which is finite (otherwise V could be replaced by $1/V$).

Under these assumptions the following holds (Wood [1976]):
Any function g is a function of ξ_1 and ξ_2 times

$$\exp\left[\int_{\xi_2}^{x_2} \hat{a}_1(x_1, \hat{x}_2)\, d\hat{x}_2 + \int_{\xi_1}^{x_1} \hat{a}_2(\hat{x}_1, \xi_2)\, d\hat{x}_1\right]$$

[cf. (6.4-1), (2.1-3)] and determines

$$\beta(x, \xi) = \int_{\xi_2}^{x_2} \left[\hat{a}_{1, x_1}(x_1, \hat{x}_2) - \hat{a}_{2, \hat{x}_2}(x_1, \hat{x}_2)\right] d\hat{x}_2$$

$$\alpha = 0 , \qquad \gamma = \hat{a}_0 - \hat{a}_1\hat{a}_2 - \hat{a}_{2, x_2}$$

[cf. (6.1-5), (2.1-4)]; furthermore, any V must be functionally dependent on the special

$$V = \int_{\xi_2}^{x_2}\int_{\xi_1}^{x_1} \gamma(\hat{x}_1, \hat{x}_2, \xi) \exp\left[\int_{\xi_1}^{\hat{x}_1} \beta(\hat{\hat{x}}_1, \hat{x}_2, \xi)\, d\hat{\hat{x}}_1\right] d\hat{x}_1\, d\hat{x}_2 .$$

From his assumptions, Wood restates Daggit's results, and besides, in the self-adjoint case, he proves that simple Riemann functions exist if and only if $(\log \gamma)_{x_1 x_2}$ is proportional to γ, i.e., (7.3-9) is satisfied; cf. Geddes–Mackie [1977].

The form of the auxiliary variables of simple Riemann functions in the case of self-adjoint equations gives the idea for the derivation of simple Riemann functions for non–self-adjoint equations by Florian–Püngel [1979]. There, in particular, we find the Riemann functions for the equations [in the complex form (6.1-4)]

(7.3-15)

$$D_1 D_2 w + \frac{(\kappa - \lambda) D_1 \psi_1(z_1)}{\psi_1(z_1) + \psi_2(z_2)} D_2 w - \frac{\kappa(\lambda + 1) D_1 \psi_1(z_1) D_2 \psi_2(z_2)}{(\psi_1(z_1) + \psi_2(z_2))^2} w = 0$$

[cf. (4.4-17)] and

(7.3-16)

$$D_1 D_2 w + [D_1 \psi_1(z_1) \psi_2(z_2)] D_2 w + [(\lambda + 1) D_1 \psi_1(z_1) D_2 \psi_2(z_2)] w = 0$$

(cf. Example 6.4-2b) with complex κ and λ. The Riemann functions are

$$R(z, s) = \left(\frac{\psi_1(z_1) + \psi_2(s_2)}{\psi_1(s_1) + \psi_2(s_2)} \right)^{\kappa - \lambda} {}_2F_1(-\lambda, \kappa + 1; 1; \omega_0(z, s)) \qquad (7.3\text{-}17)$$

with [cf. (7.1-12)]

$$\omega_0(z, s) = -\frac{[\psi_1(z_1) - \psi_1(s_1)][\psi_2(z_2) - \psi_2(s_2)]}{[\psi_1(z_1) + \psi_2(z_2)][\psi_1(s_1) + \psi_2(s_2)]}$$

and

$$R(z, s) = \exp(\psi_2(s_2)[\psi_1(z_1) - \psi_1(s_1)]) {}_1F_1(-\lambda; 1; \omega_1(z, s)) \qquad (7.3\text{-}18)$$

with

$$\omega_1(z, s) = [\psi_1(z_1) - \psi_1(s_1)][\psi_2(z_2) - \psi_2(s_2)]$$

respectively. Observe that these results could also have been obtained from the simpler equations with $\psi_1(z_1) \equiv z_1$, $\psi_2(z_2) \equiv z_2$ by applying an analog of Theorem 6.3-3.

NONSIMPLE RIEMANN FUNCTIONS

A method related to the last method of Sec. 7.1 is to solve the Goursat problem defining the Riemann function (cf. Theorem 6.3-1a) by means of the Laplace transform technique. This was done by Scott [1973] and by Wahlberg [1977]. The latter considers the equation

$$\tilde{w}_{\tilde{x}_1 \tilde{x}_1} - \tilde{w}_{\tilde{x}_2 \tilde{x}_2} + (1 + \lambda \tilde{x}_2) \tilde{w} = 0, \qquad \lambda \in \mathbb{R} \qquad (7.3\text{-}19)$$

and gets the Riemann function

$$\tilde{R}(\tilde{x}, \tilde{\xi}) = \sum_{k=0}^{\infty} \frac{1}{k!} \tilde{\omega}_2^k(\tilde{x}, \tilde{\xi}) J_{3k}(\tilde{\omega}_3(\tilde{x}, \tilde{\xi})) \tag{7.3-20}$$

where J_{3k} is the Bessel function of the first kind of order $3k$ and

$$\tilde{\omega}_2(\tilde{x}, \tilde{\xi}) = \frac{\lambda^2}{96}\left\{\left[(\tilde{x}_1 - \tilde{\xi}_1)^2 - (\tilde{x}_2 - \tilde{\xi}_2)^2\right] \Big/ \left[1 + \tfrac{1}{2}\lambda(\tilde{x}_2 + \tilde{\xi}_2)\right]\right\}^{3/2}$$

$$\tilde{\omega}_3(\tilde{x}, \tilde{\xi}) = \left\{\left[1 + \tfrac{1}{2}\lambda(\tilde{x}_2 + \tilde{\xi}_2)\right]\left[(\tilde{x}_1 - \tilde{\xi}_1)^2 - (\tilde{x}_2 - \tilde{\xi}_2)^2\right]\right\}^{1/2}.$$

For $\lambda = 1$ and $x_1 = -\frac{1}{2}(\tilde{x}_1 + \tilde{x}_2)$, $x_2 = \frac{1}{2}(\tilde{x}_1 - \tilde{x}_2) - 1$, we obtain from (7.3-19, 20) the result that

$$\hat{w}_{x_1 x_2} + (x_1 + x_2)\hat{w} = 0 \tag{7.3-21}$$

has the Riemann function

$$\hat{R}(x, \xi) = \sum_{k=0}^{\infty} \sum_{m=0}^{\infty} \frac{1}{k!m!(m + 3k)!} \omega_2^k(x, \xi) \omega_3^m(x, \xi) \tag{7.3-22}$$

where

$$\begin{aligned} \omega_2(x, \xi) &= -\tfrac{1}{12}(x_1 - \xi_1)^3 (x_2 - \xi_2)^3 \\ \omega_3(x, \xi) &= -\tfrac{1}{2}(x_1 - \xi_1)(x_2 - \xi_2)(x_1 + x_2 + \xi_1 + \xi_2). \end{aligned} \tag{7.3-23}$$

In (7.1-11, 13, 22, 25) and (7.3-22) we have examples for Riemann functions in more than one auxiliary variable, as opposed to the simple Riemann functions depending on one auxiliary variable only.

Another early example of nonsimple Riemann functions was given by Vol'man–Pampu [1977], who generalized (7.1-10, 13). Indeed, for the equation

$$\Delta \hat{w} + \frac{1}{x_2}\hat{w}_{x_2} + \left(1 - \frac{k^2}{x_2^2}\right)\hat{w} = 0, \qquad k \in \mathbb{N}_0$$

or, in complex form,

$$D_1 D_2 w - \frac{1/2}{z_1 - z_2}(D_1 w - D_2 w) + \left[\frac{1}{4} + \frac{k^2}{(z_1 - z_2)^2}\right] w = 0 \tag{7.3-24}$$

they obtained the Riemann function

$$(7.3\text{-}25)\quad R(z,s) = \left(\frac{z_1 - z_2}{s_1 - s_2}\right)^{1/2} \Xi_2\left(\frac{1}{2} + k, \frac{1}{2} - k; 1; \rho_2(z,s), \check{\rho}_3(z,s)\right).$$

Note that for $k = 0$, this reduces to (7.1-13) with $\alpha_2 = \frac{1}{2}$ and $\alpha_0 = 1$. Recall that ρ_2 and $\check{\rho}_3$ are defined by

$$\rho_2(z,s) = \frac{(z_1 - s_1)(z_2 - s_2)}{(z_1 - z_2)(s_1 - s_2)} \qquad \text{[cf. (7.1-12b)]}$$

$$\check{\rho}_3(z,s) = -\frac{1}{4}(z_1 - s_1)(z_2 - s_2) = \left(\rho_3|_{\alpha_0 = 1}\right)(z,s).$$

Here, Ξ_2 is the confluent hypergeometric series of two variables given by

$$\begin{aligned}\Xi_2&\left(\tfrac{1}{2} + k, \tfrac{1}{2} - k; 1; \rho_2, \check{\rho}_3\right)\\ &= \sum_{m,n=0}^{\infty} \frac{\left(\frac{1}{2} + k\right)_m \left(\frac{1}{2} - k\right)_m}{(m+n)!\, m!\, n!} \rho_2^m \check{\rho}_3^n\\ &= \sum_{n=0}^{\infty} (n!)^{-2} \check{\rho}_3^n\, {}_2F_1\left(\tfrac{1}{2} + k, \tfrac{1}{2} - k; 1 + n; \rho_2\right).\end{aligned}$$

Consequently, (7.3-25) takes the form

(7.3-26)

$$R(z,s) = \sum_{n=0}^{\infty} \frac{1}{(n!)^2} \left(\frac{z_1 - z_2}{s_1 - s_2}\right)^{1/2} {}_2F_1\left(\tfrac{1}{2} + k, \tfrac{1}{2} - k; 1 + n; \rho_2(z,s)\right) \check{\rho}_3^n(z,s).$$

SIX PROMISING RECENT METHODS

Mainly since 1977, great efforts have been made in order to find further nonsimple Riemann functions. Here the following six methods have led to substantial results.

(i) Performing *several* stages of the Picard iteration for the defining Volterra equation for the Riemann function, one may be able to guess the auxiliary variables in the representation of the Riemann function for a special equation considered.

(ii) Another guess may be made from known simple or nonsimple Riemann functions insofar as their auxiliary variables may be combined, transformed, or generalized, and then one may investigate whether there exist equations whose Riemann functions depend on those new auxiliary variables.

(iii) One may look for Riemann functions possessing power series expansions involving coefficients which are polynomials or special functions of mathematical physics in suitable auxiliary variables [see, e.g., (7.3-20) and (7.3-26)].

(iv) From known, sufficiently general, nonsimple Riemann functions one may obtain results for more special equations by limiting or confluence processes.

(v) Further results may also be obtained by application of suitable transformations or by algebraic processes applied to known Riemann functions.

(vi) In view of Corollary 6.4-3b one may investigate the possibility of obtaining a representation of the Riemann function by a differential operator if the equation admits at least one Type I representation of solutions by means of class P Bergman operators.

WORKS ON APPLICATIONS

For examples of those methods (often combined or overlapping) we refer to the following investigations: (i) Wallner [1980, 1981a]; (ii) Püngel [1978], Bauer [1981], Florian–Püngel [1981]; (iii) Püngel [1981], Florian–Püngel [1981]; (iv) Lanckau [1971, 1979b], Florian–Püngel [1979]; (v) Püngel [1980, 1982], Bauer [1982b]; (vi) Bauer [1982b, 1984], Berglez [1983].

These works, notably those on transformations, are too extensive to be discussed systematically. For this reason we shall concentrate on some particularly interesting selected results.

RESULTS BY LANCKAU

In his booklet, Lanckau [1971] determined the Riemann function for the equation

$$D_1 D_2 w + \left[\frac{\alpha(1-\alpha)}{(z_1+z_2)^2} - \frac{\beta(1-\beta)}{(z_1-z_2)^2} - \gamma \right] w = 0 . \tag{7.3-27}$$

He showed that the Riemann function for this equation depends on three auxiliary variables (see below) and satisfies three coupled second-order partial differential equations in three variables. The Riemann function, as a solution of these equations, is then determined from a special Lauricella hypergeometric function in three variables by a confluence process with respect to the third variable. His final result is

$$R(z,s) = F_B^+(\alpha, \beta, 1-\alpha, 1-\beta; 1; \rho_1(z,s), \rho_2(z,s), \rho_3^*(z,s)) \tag{7.3-28a}$$

with

$$\begin{aligned}&\rho_3^*(z,s) = \gamma(z_1 - s_1)(z_2 - s_2) = \left(\rho_3|_{\alpha_0 = 2i\gamma^{1/2}}\right)(z,s),\\ (7.3\text{-}28\text{b}) \qquad & \\ &\rho_1, \rho_2 \text{ as given in (7.1-12)}\end{aligned}$$

and in a neighborhood of $z = s$, F_B^+ has the series expansion

$$\begin{aligned}&F_B^+(\alpha, \beta, 1-\alpha, 1-\beta; 1; \rho_1, \rho_2, \rho_3^*)\\ (7.3\text{-}29) \qquad &= \sum_{n_1, n_2, n_3 = 0}^{\infty} \frac{(\alpha)_{n_1}(1-\alpha)_{n_1}(\beta)_{n_2}(1-\beta)_{n_2}}{(n_1+n_2+n_3)!n_1!n_2!n_3!} \rho_1^{n_1}\rho_2^{n_2}\rho_3^{*n_3}.\end{aligned}$$

Moreover, in Lanckau [1971] relations of equation (7.3-27) to equations of mathematical physics and special cases are discussed when one or two of the parameters α, β, γ are zero so that (7.3-28) reduces to hypergeometric functions in one variable or two variables; cf. also Lanckau [1979b]. Furthermore, it is obvious that in the case $\gamma = 0$ (hence, $\rho_3^* = 0$) the result (7.3-28) for (7.3-27) coincides with Chaundy's result (7.1-22) for (7.1-23) if one chooses $m_1 = m = -\alpha$, $m_2 = n = -\beta$, $m_3 = m_4 = 0$ and extends (7.1-22, 23) to complex variables.

RESULTS BY PÜNGEL

While Lanckau's equation (7.3-27) generalizes the special case (7.1-23) of Chaundy's equation (7.1-17), Püngel [1978] considered a generalization of (7.1-17) which seems to be the most general self-adjoint equation for which the Riemann function was explicitly constructed, up to that time. That equation is of Darboux type, namely,

$$(7.3\text{-}30) \qquad D_1 D_2 w + \sum_{r=1}^{4} M_r [\chi_r(z)]^{-2} w = 0$$

where

$$\begin{aligned}&M_r = m_r(m_r + 1)\\ (7.3\text{-}31) \qquad &\\ &\chi_r(z) = \kappa_r z_1 z_2 + \lambda_r z_1 + \mu_r z_2 + \nu_r, \qquad r = 1, \cdots, 4.\end{aligned}$$

The constants κ_r, λ_r, μ_r, ν_r are assumed to satisfy

$$(\kappa_r \nu_k - \lambda_r \mu_k) + (\kappa_k \nu_r - \lambda_k \mu_r) = 2\delta_{rk}, \quad r, k = 1, \cdots, 4.$$

The Riemann function for (7.3-30) has the form of the Lauricella hypergeometric function

(7.3-32)
$$R(z, s) = F_B(-m_1, -m_2, -m_3, -m_4, m_1 + 1, m_2 + 1, m_3 + 1, m_4 + 1; 1; Z_1, Z_2, Z_3, Z_4)$$

with

$$Z_r = (z_1 - s_1)(z_2 - s_2)[\chi_r(z)\chi_r(s)]^{-1}, \qquad r = 1, \cdots, 4.$$

Here, for $|Z_r| < 1$, F_B in (7.3-32) has the series expansion

$$R(z, s) = \sum_{n_1, \cdots, n_4 = 0}^{\infty} \left(\prod_{r=1}^{4} \left[(-m_r)_{n_r} (m_r + 1)_{n_r} \frac{1}{n_r!} Z_r^{n_r} \right] \right) \times \tag{7.3-33}$$
$$\times \frac{1}{(n_1 + n_2 + n_3 + n_4)!}.$$

Obviously, (7.3-30, 33) specializes to (7.1-17, 22) for $z_1 = x_1$, $z_2 = x_2$:

$$\chi_{1+k} = ih_{1+k}, \qquad \chi_{2+k} = h_{2+k}, \qquad k = 0, 2.$$

This contains not only Chaundy's result but also Lanckau's result (7.3-27, 28). Indeed, the latter is obtained by a confluence process, by setting

$$\begin{aligned} m_1 &= -\alpha, & \chi_1(z) &= i(z_1 + z_2) \\ m_2 &= -\beta, & \chi_2(z) &= z_1 - z_2 \\ m_4 &= 0, & \chi_4(z) &= 1 + z_1 z_2 \end{aligned}$$

and letting $\varepsilon \to 0$ in (7.3–30, 33) when

$$m_3 = i/\varepsilon, \qquad \chi_3(z) = \gamma^{1/2}\varepsilon z_1 z_2 + (\gamma^{1/2}\varepsilon)^{-1}.$$

This follows immediately from

$$\lim_{\varepsilon \to 0} \left\{ \frac{i}{\varepsilon} \left(\frac{i}{\varepsilon} + 1 \right) \left[\gamma^{1/2}\varepsilon z_1 z_2 + (\gamma^{1/2}\varepsilon)^{-1} \right]^{-2} \right\} = -\gamma$$

and

$$\lim_{\varepsilon \to 0} \left\{ \left(-\frac{i}{\varepsilon} \right)_{n_3} \left(\frac{i}{\varepsilon} + 1 \right)_{n_3} \times \right.$$
$$\left. \times \left(\frac{(z_1 - s_1)(z_2 - s_2)}{\left[\gamma^{1/2}\varepsilon z_1 z_2 + (\gamma^{1/2}\varepsilon)^{-1} \right] \left[\gamma^{1/2}\varepsilon s_1 s_2 + (\gamma^{1/2}\varepsilon)^{-1} \right]} \right)^{n_3} \right\}$$
$$= \rho_3^{*n_3}(z, s) \qquad \text{[cf. (7.3-28b)]}.$$

Examples illustrating method (iii) can be found in Püngel [1981]. There, Riemann functions for self-adjoint equations are given in the form of a power series in the auxiliary variable ρ_0,

(7.3-34)

$$R(z,s) = \sum_{m=0}^{\infty} \frac{1}{(m!)^2} p_m(z,s)\rho_0^m(z,s), \qquad \rho_0(z,s) = (z_1 - s_1)(z_2 - s_2)$$

being valid in a neighborhood of $\rho_0 = 0$ and having coefficients that are polynomials depending on one, two, or three suitable auxiliary variables. Such representations hold in the case that in the equation

$$D_1 D_2 w + a_0(z) w = 0 \tag{7.3-35}$$

the coefficient a_0 takes the form

(7.3-36)

$$\text{(a)}\quad a_0(z) = \kappa z_1 z_2 + \lambda z_1 + \mu z_2 + \nu \qquad \text{[cf. (7.3-21)]}$$

$$\text{(b)}\quad a_0(z) = \kappa z_1 z_2 + \lambda(z_1 + z_2) + \nu + \mu(z_1 - z_2)^{-2}$$

$$\text{(c)}\quad a_0(z) = \kappa z_1 z_2 + \nu + \lambda(z_1 + z_2)^{-2} + \mu(z_1 - z_2)^{-2}$$

[cf. (7.3-27)].

In case (a), the polynomials p_m have the form

(7.3-37a)

$$p_m(z,s) = \sum_{k=0}^{m} \alpha_{km}(\kappa, \lambda, \mu, \nu)\rho_4^k(z,s), \qquad m \in \mathbb{N}_0,$$

$$\rho_4(z,s) = -\tfrac{1}{2}[a_0(z) + a_0(s)]$$

[compare $\rho_0\rho_4$ with ω_3 in (7.3-23)], with explicitly known coefficients α_{km} depending only on the parameters indicated. For Püngel's cases (b) and (c), recursive systems are given from which the polynomials can be determined. In case (b), the p_m's are assumed to be polynomials in auxiliary variables ρ_5 and ρ_6 given by

(7.3-37b)

$$\rho_5(z,s) = -\tfrac{1}{2}[\lambda(z_1 + z_2 + s_1 + s_2) + \kappa(z_1 z_2 + s_1 s_2)]$$

$$\rho_6(z,s) = (z_1 - z_2)^{-1}(s_1 - s_2)^{-1}.$$

In case (c), they are polynomials in ρ_6, ρ_7, and ρ_8, where

(7.3-37c)

$$\rho_7(z,s) = (z_1 + z_2)^{-1}(s_1 + s_2)^{-1}$$

$$\rho_8(z,s) = z_1 z_2 + s_1 s_2.$$

RESULTS BY FLORIAN AND PÜNGEL

Florian–Püngel [1981] continued, for self-adjoint equations (7.3-35), the preceding investigations of Riemann functions in two or three auxiliary variables. The form (7.3-34) is again assumed, but now the functions p_m, depending on two auxiliary variables, say, α and β, are determined by considering the generating function

$$S(\alpha, \beta; \rho) = \sum_{m=0}^{\infty} q_m(\alpha, \beta)\rho^m/m! \tag{7.3-38a}$$

for the coefficients p_m; here, the q_m's are introduced by

$$p_m(z, s) = q_m(\alpha(z, s), \beta(z, s)). \tag{7.3-38b}$$

In the case of (7.3-35, 36a) with $\kappa = 0$, for example, S has the simple form

$$S(\alpha, \beta; \rho) = \exp\left[(\alpha + \beta)\rho - \tfrac{1}{12}\lambda\mu\rho^3\right] \tag{7.3-39a}$$

where

$$\begin{aligned} &\alpha(z, s) = -\tfrac{1}{2}(\lambda z_1 + \mu z_2 + \nu) \\ &\beta(z, s) \equiv \alpha(s, z), \qquad \text{hence} \qquad \alpha + \beta = \rho_4|_{\kappa=0} = \tilde{\rho}_4 . \end{aligned} \tag{7.3-39b}$$

We thus have

$$q_m(\alpha, \beta) = \left\{\frac{\partial^m}{\partial\rho^m} \exp\left[(\alpha + \beta)\rho - \tfrac{1}{12}\lambda\mu\rho^3\right]\right\}\Bigg|_{\rho=0}$$

and hence, by (7.3-34, 38b), [cf. (7.3-22, 20)]

$$\begin{aligned} R &= \sum_{m=0}^{\infty} \frac{1}{(m!)^2} \sum_{k=0}^{[m/3]} \frac{m!}{k!(m-3k)!}\left(-\frac{1}{12}\lambda\mu\right)^k \tilde{\rho}_4^{m-3k}\rho_0^m \\ &= \sum_{k=0}^{\infty}\sum_{m=0}^{\infty} \frac{1}{k!m!(m+3k)!}\left(-\frac{1}{12}\lambda\mu\right)^k \tilde{\rho}_4^m \rho_0^{m+3k} \\ &= \sum_{k=0}^{\infty} \frac{1}{k!}\left[-\frac{1}{12}i\lambda\mu(\rho_0/\tilde{\rho}_4)^{3/2}\right]^k J_{3k}\left(2i(\rho_0\tilde{\rho}_4)^{1/2}\right). \end{aligned} \tag{7.3-40}$$

For further generating functions S, e.g., for the equations (7.3-35) with coefficients a_0 given by (7.3-36a) with $\kappa \neq 0$, (7.3-36c) with $\kappa = \mu = 0$, and

$\lambda = -m_1(m_1+1)$, or

(7.3-36d)
$$a_0(z) = \lambda(z_1+z_2)^2 + \nu,$$

we refer to Florian–Püngel [1981].

RESULTS BY WALLNER

Method (i) used by Wallner [1980, 1981a] yields power series expansions in three (resp. four) suitable auxiliary variables for the Riemann functions of the equations (7.3-35, 36a, b, c) investigated by Püngel [1981]. Moreover, Wallner [1980] also treats equation (7.3-35) with

(7.3-36e)
$$a_0(z) = z_1 + D_2\psi(z_2)$$

where ψ is a polynomial of degree $m \geq 2$, say,

$$\psi(z_2) = \sum_{\mu=0}^{m} \gamma_\mu z_2^\mu .$$

It is shown that the corresponding Riemann function has the expansion

(7.3-41)
$$R(z,s) = \sum_{n_1,\cdots,n_m=0}^{\infty} \left\{\left[\sum_{r=1}^{m}(2r-1)n_r\right]!\right\}^{-1} \prod_{k=1}^{m}\left[\frac{1}{n_k!}(-1)^{n_k}\lambda_k^{n_k}(z,s)\right]$$

where the auxiliary variables $\lambda_1,\cdots,\lambda_m$ are defined by

(7.3-42)
$$\begin{aligned}\lambda_1(z,s) &= \tfrac{1}{2}(z_1^2 - s_1^2)(z_2-s_2) + (z_1-s_1)\sum_{\mu=1}^{m}\gamma_\mu(z_2^\mu - s_2^\mu)\\ \lambda_k(z,s) &= 2^{1-k}(z_1-s_1)^{2k-1}(z_2-s_2)^{2k-1}\times\\ &\times\sum_{\mu=k}^{m}\gamma_\mu\sum_{\nu=0}^{\mu-k}\frac{(k+\nu-1)!(\mu-\nu-1)!\mu!}{(k-1)!\nu!(\mu-k-\nu)!(\mu+k-1)!}z_2^\nu s_2^{\mu-k-\nu}\\ &\qquad k=2,\cdots,m.\end{aligned}$$

Obviously, the representation (7.3-41, 42) is a generalization of (7.3-22, 23). Observe further that in the special case $\psi(z_2) = z_2^m/m$, formula (7.3-42)

simplifies to

$$\lambda_1(z,s) = \frac{1}{2}(z_1^2 - s_1^2)(z_2 - s_2) + \frac{1}{m}(z_1 - s_1)(z_2^m - s_2^m)$$

(7.3-43)
$$\lambda_k(z,s) = 2^{1-k}(z_1 - s_1)^{2k-1}(z_2 - s_2)^{2k-1} \times$$

$$\times \sum_{\nu=0}^{m-k} \frac{(k+\nu-1)!(m-\nu-1)!(m-1)!}{(k-1)!\,\nu!(m-k-\nu)!(m+k-1)!} z_2^{\nu} s_2^{m-k-\nu}.$$

USE OF CLASS P OPERATORS. RESULTS BY K. W. BAUER

The Riemann function R^* of the adjoint equation $L^*w^* = 0$ [cf. (6.1-5)] of an equation $Lw = 0$ satisfies the original equation, $LR^* = 0$, and can be determined as the unique solution of the corresponding Goursat problem given by (6.3-2), with a_1, a_2 replaced by $-a_1$, $-a_2$; see Theorem 6.3-1(a). The Riemann function R of the equation $Lw = 0$ is then obtained by using the symmetry of R (Theorem 6.3-2). Hence, if L admits class P Bergman operators (equivalently, Bauer differential operators) for Type II representations of solutions, Theorem 3.4-2 implies that the solution of a Goursat problem for $Lw = 0$ can be generated by differential operators. According to our previous considerations, this also holds for the Riemann function of such an equation. From Riemann functions thus obtained we may then proceed to further results by algebraic or transformation techniques.

To give some results obtained by methods (v) and (vi) (see above), we consider Bauer [1982b]. By simple algebraic methods, Bauer determines the Riemann function for two classes of equations. The first class is of the form

(7.3-44)

$$D_1D_2w + \left[D_2 \log \frac{(z_1+z_2)^{\lambda}}{D_2p(z)}\right]D_1w + \left[D_1 \log \frac{(z_1+z_2)^{\mu}}{D_1p(z)}\right]D_2w + a_0(z)w = 0$$

where

$$a_0(z) = \frac{1}{D_1p(z)D_2p(z)}\left[D_1^2p(z)D_2^2p(z) - \frac{\lambda}{z_1+z_2}D_2p(z)D_1^2p(z)\right.$$

$$\left. - \frac{\mu}{z_1+z_2}D_1p(z)D_2^2p(z)\right] - \frac{\lambda+\mu}{z_1+z_2}, \quad \lambda, \mu \in \mathbb{Z}; \quad \lambda\mu \geq 0$$

involving an arbitrary particular solution p of the Euler equation

(7.3-45)
$$(z_1+z_2)D_1D_2u + \lambda D_1u + \mu D_2u = 0$$

satisfying $(z_1 + z_2)D_1 p(z) D_2 p(z) \neq 0$ in the domain considered. The second class is of the form

$$(7.3\text{-}46) \quad D_1 D_2 w - [D_2 \log q(z)] D_1 w - [D_1 \log D_1 q(z)] D_2 w + a_0(z) w = 0$$

with

$$a_0(z) = \frac{D_1^2 q(z) D_2 q(z)}{q(z) D_1 q(z)} - \frac{n(n+1)}{(z_1 + z_2)^2}, \qquad n \in \mathbb{N}$$

and q denoting an arbitrary particular solution of the differential equation

$$(7.3\text{-}47) \qquad (z_1 + z_2)^2 D_1 D_2 v - n(n+1) v = 0$$

satisfying $(z_1 + z_2) q(z) D_1 q(z) \neq 0$ in the domain considered.

Observe that, by Theorem 4.4-2, the solutions of (7.3-45) and (7.3-47) can be obtained by Type II representations by means of class P operators or Bauer differential operators, and that these operators are explicitly known. Indeed, for (7.3-45) choose $L^{(0)} w = 0$ in Theorem 4.4-2, $\psi(z) = (z_1 + z_2)^{-1}$ if $\lambda \geq 0$, $\mu \geq 0$, and $\psi(z) = z_1 + z_2$ if $\lambda < 0$, $\mu < 0$. The case of (7.3-47) is obvious from $L_2^{(0)} v = 0$ with $\psi(z) = z_1 + z_2$, $n = n_1 = n_2$. The Riemann function for equations of the form (7.3-44) or (7.3-46) is built up as a quotient of sums of products of derivatives of p (resp. q) and special solutions of (7.3-45) [resp. (7.3-47)] represented by means of those differential operators.

For three further examples we refer to Bauer [1981]. Furthermore, the preceding assumption may be weakened insofar as the existence of only a Type I representation of solutions by means of differential operators may be required for the determination of the Riemann function; cf. Bauer [1984]. We also remark that, by our results in Sec. 6.4, from a Bauer differential operator for Type I representations which, by Theorem 3.3-2, yields a suitable Bergman kernel of the first kind, we obtain directly the Riemann function by the use of Corollary 6.4-3. This means that from a single Bauer differential operator $\tilde{T}_j$ given by

$$(7.3\text{-}48a) \qquad \tilde{T}_j = \sum_{m=0}^{n_j} A_{jm}(z, s) D_j^m$$

and satisfying

$$(7.3\text{-}48b) \qquad \begin{aligned} & A_{jm}(z, s)|_{z_{j^*} = s_{j^*}} \equiv 0, && m = 0, \cdots, n_j - 1, \\ & A_{jn_j} = e_j^R, && \text{with } e_j^R \text{ given in (6.4-1)} \end{aligned}$$

we obtain the Riemann function

$$R(z, s) = \sum_{m=0}^{n_j} \frac{1}{m!} A_{j, n_j - m}(s, z)(s_j - z_j)^m. \tag{7.3-49}$$

(For the existence and construction of operators $\tilde{T}_j$, see Chap. 4.)

We finally mention the possibility of also defining analogs of the Riemann function for higher-dimensional and higher-order equations. For attempts in this direction, see Vekua [1967] and Florian–Püngel–Wallner [1983a, b].

Additional References

Bauer [1981, 1982b, 1984]
Berglez [1983]
Chaundy [1935, 1936, 1938, 1939]
Cohn [1947, 1970, 1973]
Copson [1958, 1971]
Daggit [1970]
Du [1981]
Eichler [1950]
Florian–Püngel [1979, 1981]
Florian–Püngel–Wallner [1983a, 1983b]
Friedlander–Heins [1969]
Geddes–Mackie [1977]
Hadamard [1952]
Henrici [1953, 1957, 1960]
Kracht–Kreyszig [1988]
Kraft [1969]
Lanckau [1971, 1979b]
Ludford [1953]
Mackie [1955, 1965]
Olevskij [1952]
Papadakis–Wood [1977]
Püngel [1980, 1981, 1982]
Scott [1973]
Vekua [1967]
Volkmer [1979]
Vol'man–Pampu [1977]
Wahlberg [1977]
Wallner [1980, 1981a]
Wood [1976]

Chapter Eight

Coefficient Problem and Singularities of Solutions

If a function is given by a series development in terms of other functions (powers of the independent variables, etc.), it is clear that from the coefficients of the series (and the knowledge of the properties of those functions), one should be able to determine "all" the properties of the function. This problem of expressing properties of a function in terms of properties of the coefficients of some kind of series development is called the *coefficient problem*. Now asking for "all" the properties of that function would be much too ambitious. However, for certain types of series, namely, for power series in powers of a single (complex) variable, one has discovered a surprisingly large number of *coefficient theorems* that is, theorems that characterize properties of functions in terms of those of corresponding series (power series in the present case). An impression of the wealth of information in that respect can be obtained from Bieberbach [1955] and [1968], Vol. II, and Dienes [1957] and the references included in these books. For special questions, see also Hille [1959–1962], Vol. II, and Wittich [1968].

In this chapter we shall see how we can use coefficient theorems from complex analysis in order to obtain coefficient theorems for solutions of linear partial differential equations. Our main tool will be the Bergman integral operator of the first kind, in particular, the representation of solutions as given by Theorem 2.5-1. After the explanation of the general idea (Sec. 8.1), in Sec. 8.2 we obtain a criterion for the existence of "pole-like singularities" of the solution in terms of Hankel determinants of coefficients. A further Hankel determinant criterion is applied to establish a relation between a solution of the given partial differential equation and a corresponding ordinary differential equation satisfied by that solution. This and an application of the Leau–Faber theorem are given in Sec. 8.3. Section 8.4 is devoted to two topics. First, we investigate a relation between the numbers of singularities of certain subseries of the double series expansion of a given solution. Second, we derive an extension of the famous Hadamard theorem on poles. Finally in this chapter, in Sec. 8.5, we consider pole-like singularities caused by certain singularities of the associated functions of solutions.

8.1 GENERAL SETTING AND APPROACH

In connection with the coefficient problem for solutions $w = Tf$ (T a Bergman operator or another integral operator) of a linear partial differential equation in two independent variables, a possible general setting is as follows. Assume $0 \in \Omega$ and w to be given locally by a series of the form

$$w(z) = \sum_{m=0}^{\infty} \sum_{n=0}^{\infty} \omega_{mn} z_1^m z_2^n . \tag{8.1-1}$$

Then the problem is the determination of properties of w expressed in terms of (ω_{mn}). In the solution, the steps are as follows.

1. Determine $f = T^{-1}w$. Note that, in general, T^{-1} need not exist automatically since several, even infinitely many f may correspond to the same w. However, for every w, we can characterize a "minimal" f in order to accomplish the existence of T^{-1}. In most practical cases, this process of restricting the domain of T is quite simple.

2. Use a power series representation for f, say,

$$f(z_1) = \sum_{\nu=0}^{\infty} \gamma_\nu z_1^\nu \tag{8.1-2}$$

and express (γ_ν) in terms of (ω_{mn}).

3. Apply a coefficient theorem from complex analysis, which asserts that a certain property A of (ω_{mn}) implies some property B of f.

4. Apply T to get $w = Tf$ and see how T converts property B into some property of w.

This is the process, very roughly described, if we start from a representation of the form (8.1-1). Another, also very feasible, approach consists in starting from (8.1-2) and using the linearity of T to produce from (8.1-2) a representation of w of the form

$$w = \sum_{\nu=0}^{\infty} \gamma_\nu w_\nu \tag{8.1-3}$$

where

$$w_\nu = Tf_\nu , \qquad f_\nu(z_1) = z_1^\nu . \tag{8.1-4}$$

The general idea of obtaining coefficient theorems in the case of representations of the form (8.1-3) is similar to that just explained.

It is plain that in (8.1-3, 4) one may replace (w_ν) by a sequence of particular solutions whose associated functions are polynomials that have some special property, for instance, orthogonal polynomials, such as the Legendre polynomials. Such a choice will, however, complicate matters, and it seems that from the standpoint of complex analysis, (8.1-4) is a very natural choice of (w_ν).

8.2 USE OF OPERATORS OF THE FIRST KIND

It is clear that the relation between the coefficients ω_{mn} in a representation

$$w(z) = Tf(z) = \sum_{m=0}^{\infty} \sum_{n=0}^{\infty} \omega_{mn} z_1^m z_2^n \tag{8.2-1}$$

of a solution w of a given partial differential equation and the coefficients γ_m in a representation

$$f(\zeta_1) = \sum_{m=0}^{\infty} \gamma_m \zeta_1^m \tag{8.2-2}$$

(ζ_1 to be chosen suitably, e.g., $\zeta_1 = \xi_1 = \frac{1}{2} z_1 \tau$ in the case of an operator T of the first kind with $s = 0$) of an associated function f of w depends on the choice of the operator T. Whereas these relations will be complicated in general (cf. Bergman [1944b], pp. 537–538), for an *operator of the first kind* they become relatively simple. Indeed, let us consider the equation (2.1-1),

$$D_1 D_2 w + a_1(z) D_1 w + a_2(z) D_2 w + a_0(z) w = 0,$$

and complex solutions $w = Tf$ as well as real solutions W as given by (2.1-10). Here we assume that T is an operator of the first kind with $s = 0$ and $\eta_1 = 0$ in e_1; cf. (2.1-3). Considering real solutions, we furthermore assume that equation (2.1-1) is obtained from a real elliptic equation of the form (2.1-5) by means of (2.1-6, 7), so that Lemma 2.1-3 is applicable.

Suppose that w is represented in the form (8.2-1) and, similarly,

$$W(z) = \sum_{m=0}^{\infty} \sum_{n=0}^{\infty} \tilde{\omega}_{mn} z_1^m z_2^n. \tag{8.2-3}$$

Furthermore, let f be represented in the form (8.2-2). Then, under these assumptions, setting $z_2 = 0$, from (2.5-5, 6, 7) and (2.4-1), we readily obtain, once and for all, the coefficient relations [cf. (2.2-7)]

$$\gamma_m = \omega_{m0} \beta_m, \qquad \beta_m = (m!)^2 8^m / \pi (2m)!, \qquad m \in \mathbb{N}_0. \tag{8.2-4}$$

For the *real* solution W, by the same reasoning we first have

$$
\begin{aligned}
\sum_{m=0}^{\infty} \tilde{\omega}_{m0} z_1^m &= W(z_1, 0) = \tfrac{1}{2}\left[w(z_1, 0) + \bar{w}(0, z_1)\right] \\
&= \tfrac{1}{2} \sum_{m=0}^{\infty} \frac{\gamma_m}{\beta_m} z_1^m + \tfrac{1}{2}\pi\bar{\gamma}_0 \exp\left[-\int_0^{z_1} \bar{a}_1(0, \tilde{z}_1)\, d\tilde{z}_1\right].
\end{aligned} \tag{8.2-5}
$$

For the exponential term we have to use the Maclaurin series, which is of the form

$$
\exp\left[-\int_0^{z_1} \bar{a}_1(0, \tilde{z}_1)\, d\tilde{z}_1\right] = \sum_{m=0}^{\infty} \alpha_m z_1^m. \tag{8.2-6}
$$

We then obtain

$$
\gamma_m = 2\beta_m\left(\tilde{\omega}_{m0} - \tfrac{1}{2}\pi\bar{\gamma}_0\alpha_m\right), \qquad m \in \mathbb{N}_0. \tag{8.2-7}
$$

POLE-LIKE SINGULARITIES

As a typical and simple application of these formulas, let us prove a theorem on pole-like singularities. By definition, a ***pole-like singularity*** is a singularity plane $z_1 = const$ on which w becomes infinite as in the case of a pole and which is a logarithmic branch plane (thus of infinite order) for at least one finite value of z_2. Furthermore, a *Hankel determinant* $H_r^{(\rho)}(g)$ associated with a function g,

$$
g(z_1) = \sum_{m=0}^{\infty} g_m z_1^m
$$

or with a sequence $\{g_m\}_{m\in\mathbb{N}_0}$, is defined as usual by

$$
H_r^{(\rho)}(g) = \det\left(\left(\chi_{\lambda\mu}^{(r,\rho)}\right)_{\lambda,\mu=1,\cdots,\rho}\right) = H_r^{(\rho)}\left(\{g_m\}_{m\in\mathbb{N}_0}\right) \tag{8.2-8}
$$

where $\chi_{\lambda\mu}^{(r,\rho)} = g_{r+\lambda+\mu-2}$ and $r \in \mathbb{N}_0$; $\rho \in \mathbb{N}$.

8.2-1 Theorem

Let (2.1-1) *be obtained from* (2.1-5) *by means of* (2.1-6, 7) *and have entire coefficients. Then a real solution W of* (2.1-1) *given by* (8.2-3) *has at most finitely many pole planes of order not exceeding q if and only if for q at most finitely many of the $(q+1)$-rowed Hankel determinants $H_r^{(q+1)}(W|_{z_2=0})$, $r = 0, 1, \cdots,$ with elements*

$$
\chi_{\lambda\mu}^{(r,q+1)} = \tilde{\omega}_{r+\lambda+\mu-2,0} \qquad \lambda, \mu = 1, \cdots, q+1; \quad r \in \mathbb{N}_0 \tag{8.2-9}
$$

are different from zero. These pole planes are pole-like singularities apart from certain exceptional cases (e.g., possibly occurring if $L \in FP_{1n}$ *holds with* $n < q$*; details follow in Remark* 8.5-2 *and in the proof of Theorem* 8.5-1).

Proof. We have from (8.2-3)

$$W(z_1, 0) = \sum_{m=0}^{\infty} \tilde{\omega}_{m0} z_1^m .$$

By a criterion of Borel [1894] the condition of the theorem involving Hankel determinants is necessary and sufficient for $g = W|_{z_2=0}$ to be a rational function with the degree of the denominator not exceeding q. Since a_1 is entire, so is the exponential term in (8.2-5). Formulas (8.2-2, 4, 5) together now show that f has the same number and types of singularities as g. We may omit the odd part (in t) of $\tilde{k}_1$ without loss of generality, since application of (2.2-8) to it would only yield a zero term. Substituting f into the series representation of w in Theorem 2.5-1 and noting that the $\tilde{q}_{1m}$ are entire functions, we see that integration produces finitely many poles. The remainder of the proof can be seen from the proof of Theorem 8.5-1 (below), where the calculation for a (fixed) pole of f will be carried out explicitly. ■

8.3 FURTHER COEFFICIENT THEOREMS

The proof just presented illustrates the fact that the transition from a coefficient theorem in complex analysis to a similar result on solutions of partial differential equations is generally quite simple and direct. Of course, here we should keep in mind that not every such coefficient theorem will yield a nontrivial result on these solutions.

In the present section we give a few more results illustrating these facts; here we assume that T is of the first kind, as before.

APPLICATION OF A THEOREM BY BEKE

There are numerous coefficient theorems for analytic functions involving Hankel determinants (cf. Dienes [1957], Pólya–Szegö [1972–1976], Vol. II, or Bieberbach [1968], vol. II). From another one of them the following interesting result is readily obtained.

8.3-1 Theorem

If the coefficients ω_{mn} *in* (8.2-1) *are such that for some fixed* $z_2 = z_2^{(0)} = const$ *at most finitely many of the Hankel determinants*

$$H_0^{(\rho)}\left(\left\{ m! \sum_{n=0}^{\infty} \omega_{mn} \left(z_2^{(0)}\right)^n \right\}_{m \in \mathbb{N}_0}\right), \qquad \rho = 1, 2 \cdots$$

are different from zero, then w satisfies a homogeneous linear ordinary differential equation whose coefficients are independent of z_1 *(but may depend on* $z_2^{(0)}$*).*

Proof. This follows from a theorem by E. Beke (cf. Pólya–Szegö [1972–1976], Vol. II, pp. 99, 285), according to which a function g, given by

$$g(z_1) = \sum_{m=0}^{\infty} g_m z_1^m,$$

satisfies a homogeneous linear ordinary differential equation with constant coefficients if and only if all but finitely many of the Hankel determinants $H_0^{(\rho)}(\{m!g_m\}_{m \in \mathbb{N}_0})$, $\rho = 1, 2, \cdots$, vanish. Theorem 8.3-1 now follows by an application of Beke's theorem to the function $g = w|_{z_2 = z_2^{(0)}}$ which, by (8.2-1), entails

$$g_m = \sum_{n=0}^{\infty} \omega_{mn} \left(z_2^{(0)}\right)^n, \qquad m \in \mathbb{N}_0. \quad \blacksquare$$

APPLICATION OF THE LEAU–FABER THEOREM

Another result is obtained as follows. It is well known that for a given power series expansion

$$g(z_1) = \sum_{m=0}^{\infty} g_m z_1^m$$

there exist various functions $\tilde{g}$ such that $\tilde{g}(m) = g_m$. In particular, there exists an *entire* function with this property:

8.3-2 Lemma

For any complex sequence $\{g_m\}$ *there exists an entire function* $\tilde{g}$ *such that* $\tilde{g}(m) = g_m$.

This will be used in deriving the following result.

8.3-3 Theorem

Let the assumptions on L be as in Theorem 8.2-1. *Let W be a real solution of* (2.1-1) *which is represented in the form* (8.2-3),

$$W(z) = \sum_{m=0}^{\infty} \sum_{n=0}^{\infty} \tilde{\omega}_{mn} z_1^m z_2^n.$$

Let $\{\tilde{\omega}_{m0}\}_{m \in \mathbb{N}_0}$ *be such that* $\tilde{g}$ *with* $\tilde{g}(m) = \tilde{\omega}_{m0}$ *in Lemma* 8.3-2 *satisfies*

$$|\tilde{g}(re^{i\theta})| < e^{\varepsilon r} \tag{8.3-1}$$

for all r *greater than some* r_0 *and* $\varepsilon > 0$. *Then the only singularity set of* W *in* $\tilde{\Omega}_1 \times \{z_2 \mid |z_2| < \infty\}$, $\tilde{\Omega}_1$ *a domain containing* 0 *and* 1, *is the plane* $z_1 = 1$. *If* $\tilde{g}$ *is a polynomial of degree* $q - 1$, *then this plane is a pole plane of order* q.

Proof. From (8.3-1) and the Leau–Faber theorem (cf. Dienes [1957], p. 337) it follows that the function $g = W|_{z_2=0}$ has at $z_1 = 1$ its only singularity in $\tilde{\Omega}_1$. We observe (8.2-5) and see that the difference of g and h_1, where

$$h_1(z_1) = \sum_{m=0}^{\infty} \frac{\gamma_m}{2\beta_m} z_1^m, \tag{8.3-2}$$

is an entire function. Thus h_1 is singular at $z_1 = 1$. From this, in analogy to the proofs of Theorems 8.2-1 and 8.5-1, by means of Theorem 2.5-1, we get the first assertion. The second assertion also follows from the Leau–Faber theorem. ■

8.4 COEFFICIENT THEOREMS FOR SOLUTIONS

We have seen that operators of the first kind are particularly suitable in connection with the coefficient problem for solutions of linear partial differential equations. The theorems in the previous section illustrate the interesting fact that the assumptions always involve the subsequences (ω_{m0}) or $(\tilde{\omega}_{m0})$ of the coefficients in a representation of a complex solution of the form (8.2-1),

$$w(z) = \sum_{m=0}^{\infty} \sum_{n=0}^{\infty} \omega_{mn} z_1^m z_2^n$$

or a real solution of the form (8.2-3),

$$W(z) = \sum_{m=0}^{\infty} \sum_{n=0}^{\infty} \tilde{\omega}_{mn} z_1^m z_2^n .$$

Those results are typical; indeed, for other coefficient theorems obtained by the use of an operator of the first kind, the situation is exactly the same. This seems an undesirable limitation due to the approach, but not inherent in the problem itself. In this section we discuss methods for overcoming this obstacle.

For this purpose we introduce suitable auxiliary functions, namely,

$$w_n(z_1) = \sum_{m=0}^{\infty} \omega_{mn} z_1^m, \qquad n = 0, 1, \cdots \tag{8.4-1}$$

and

$$W_n(z_1) = \sum_{m=0}^{\infty} \tilde{\omega}_{mn} z_1^m, \qquad n = 0, 1, \cdots. \tag{8.4-2}$$

Then we can write the representations (8.2-1) and (8.2-3) in the form

$$w(z) = \sum_{n=0}^{\infty} w_n(z_1) z_2^n \tag{8.4-3}$$

and

$$W(z) = \sum_{n=0}^{\infty} W_n(z_1) z_2^n \tag{8.4-4}$$

respectively. We note that to each function w_n and W_n there corresponds a subsequence

$$(\omega_{mn}) \quad \text{and} \quad (\tilde{\omega}_{mn}), \qquad n \text{ fixed}$$

respectively. In particular, to w_0 and W_0 there correspond the above subsequences (ω_{m0}) and $(\tilde{\omega}_{m0})$. Hence we can solve our problem by establishing relations between w_n $(n > 0)$ and w_0, and between W_n $(n > 0)$ and W_0. We explain the method and general results for complex solutions w. For real solutions, the underlying idea is quite similar, so that it suffices to illustrate the situation in terms of a typical example, which will be done thereafter.

For a complex solution, the desired relations are obtained from a system of ordinary differential equations resulting from the partial differential equation (2.1-1) by substituting w given by (8.4-3) as well as the Maclaurin series of the coefficients written as power series in powers of z_2, say,

$$a_p(z) = \sum_{n=0}^{\infty} \alpha_{pn}(z_1) z_2^n, \qquad p = 0, 1, 2. \tag{8.4-5}$$

By equating the coefficient of each power of z_2 to zero we obtain

$$w_\nu' + a_{20} w_\nu = \frac{1}{\nu} \Phi_\nu(w_0, \cdots, w_{\nu-1}, w_0', \cdots, w_{\nu-1}'), \tag{8.4-6}$$

$$\nu = 1, 2, \cdots, n; \quad n > 0 \text{ and fixed}$$

where

$$\Phi_\nu = -\sum_{\lambda=0}^{\nu-2} \left[\alpha_{1,\nu-\lambda-1} w_\lambda' + (\lambda+1)\alpha_{2,\nu-\lambda-1} w_{\lambda+1} + \alpha_{0,\nu-\lambda-1} w_\lambda\right]$$

$$-\alpha_{10} w_{\nu-1}' + \alpha_{00} w_{\nu-1}.$$

A key to various results is:

8.4-1 Theorem

Suppose the coefficients of (2.1-1) *to be entire. Let w be a solution such that for some $n > 0$ the corresponding function w_n is singular at some point $z_1 = z_1^{(0)}$. Then w_0 is also singular at $z_1 = z_1^{(0)}$. Hence the number S_n of singular points of w_n in a fixed domain (each singular point counted just once) satisfies*

$$S_n \leq S_0, \qquad n \in \mathbb{N}.$$

Proof. Let w_n be singular at $z_1 = z_1^{(0)}$, where $n > 0$ is fixed. By (8.4-6) this singularity results from an expression that involves the functions $w_0, \cdots, w_{n-1}$ and their first derivatives. Since a_0, a_1, a_2 are entire, so are the α_{pn} in (8.4-5). Hence one or several of the w_μ with $\mu < n$ must be singular at $z_1 = z_1^{(0)}$. By repeating this argument for w_μ we obtain the assertion. ■

It is clear that by the use of this theorem we can now replace conditions on (ω_{m0}) by conditions on (ω_{mn}), $n > 0$ and fixed, in coefficient theorems for complex solutions. This can be done for the theorems in the previous section and other theorems in a straightforward fashion.

Note that S_n in Theorem 8.4-1 is *not* monotone in n. For instance, in the case of the equation

$$D_1 D_2 w + D_2 w + z_2^3 w = 0 \tag{8.4-7}$$

we see that a singularity of w_4 at some $z_1 = z_1^{(0)}$ corresponds to regular points of w_1, w_2, w_3.

Furthermore, it is basic to observe that the converse of Theorem 8.4-1 is not true; that is, a singular point of w_0 may very well correspond to a regular point of w_n. The reason is that poles may be compensated by zeros, and (8.4-6) shows that one needs additional conditions in order to prevent that a regular point of w_n for some n may correspond to a singular point of w_0.

GENERALIZED HADAMARD THEOREM

Let us finish this section with an application of the famous Hadamard theorem on poles (cf. Dienes [1957], p. 335). We formulate the extended theorem for *real* solutions, so that it simultaneously illustrates our procedure in the case of these solutions.

8.4-2 Theorem (Generalized Hadamard theorem)

Suppose that L satisfies the same assumptions as in Theorem 8.2-1. *Let W be a real solution of* (2.1-1). *Furthermore, let*

$$\Lambda_{np} = \lambda_{np}/\lambda_{n,\,p-1}$$

where

$$\lambda_{np} = \overline{\lim_{m\to\infty}} \left| \sqrt[m]{H_m^{(p+1)}(\{\tilde{\omega}_{\kappa n}\}_{\kappa\in\mathbb{N}_0})} \right|, \qquad n, p \in \mathbb{N}.$$

Then the following statements hold.

(a) *If there exist p_0 and n_0 such that $\Lambda_{n_0 p_0} = 0$, then W is singular on at least p_0 planes.*

(b) *If for some n_0,*

$$\Lambda_{n_0} = \lim_{p\to\infty} \Lambda_{n_0 p} = \frac{1}{\rho_0} > 0$$

then W is singular on infinitely many planes of the form $\{z_1^{(0)}\} \times \mathbb{C}$ with suitable $z_1^{(0)} \in N(\partial B_1(0, \rho_0), \delta)$, where

$$N(\partial B_1(0, \rho_0), \delta) = \{z_1 \in \mathbb{C} \mid \exists \tilde{z}_1 \in \partial B_1(0, \rho_0): |z_1 - \tilde{z}_1| < \delta\}$$

with arbitrarily small fixed $\delta > 0$.

Proof. The elements of $H_m^{(p+1)}(\{\tilde{\omega}_{\kappa n}\}_{\kappa\in\mathbb{N}_0})$ constitute the subset of coefficients that corresponds to W_n. Hadamard's theorem thus implies that W_{n_0} with fixed n_0 has p_0 poles. Now W_{n_0} is related to W_0 by a system of ordinary differential equations, which is obtained by substitution and is quite similar to (8.4-6). From this system and the assumptions on the coefficients of (2.1-1), we conclude that W_0 is singular at a point where W_{n_0} is singular. Hence w_0 is singular at such a point, so that the same holds for h_1 in (8.3-2). This means that h_1 is singular at those points at which we initially have singularities of W_{n_0}. It follows that w is singular on the corresponding planes. The same holds for W, and assertion (a) now follows. The idea of the proof of (b) is the same. ■

It is obvious that many other coefficient theorems for solutions of partial differential equations can be obtained from coefficient theorems in complex analysis by utilizing the method explained in this chapter. The sample of theorems included in this chapter will serve to illustrate the underlying principle. In many cases the whole analysis will be quite direct; consequently, this approach is rather powerful and general.

8.5 SINGULARITIES OF SOLUTIONS

As an appropriate general tool for investigating the type and location of singularities of solutions, we have the representation of solutions by means of Bergman operators of the first kind as given in Sec. 2.5. Thus we obtain theorems on singularities of solutions caused by certain singularities of the associated functions.

We consider Type I representations of solutions of equation (2.1-1), setting $j = 1$ and assuming $s = 0$, $0 \in \Omega$.

8.5-1 Theorem

Suppose that the assumptions of Theorem 2.4-2 *with* $j = 1$, $s = 0$ *hold. Let* $h_1 \in C^\omega(B_1(z_1^{(0)}, \tilde{r}_1) \setminus \{z_1^{(0)}\})$ $(\tilde{r}_1 > 0)$ *possess a pole of order* α $(\in \mathbb{N})$ *for* $z_1 = z_1^{(0)}$, $z_1^{(0)} \in B_1(0, \frac{1}{2}r_1) \setminus \{0\}$. *Then, if*

$$\tilde{q}_{1\alpha}|_{z_1 = z_1^{(0)}} \neq 0 \tag{8.5-1}$$

the solution $w = w_1$ *defined by* (2.5-1, 3) *has a pole plane of the same order for* $z_1 = z_1^{(0)}$ *which is a pole-like singularity in the sense of Sec.* 8.2.

Proof. Because of our choice of $z_1^{(0)} \in B_1(0, \frac{1}{2}r_1)$ we have from Secs. 2.4 and 2.5 that (2.5-1, 3) represents a solution having a domain of holomorphy which is a simply connected subdomain of $G_1 = B_1(0, \frac{1}{2}r_1) \times B_2(0, r_2)$ and has $z_1^{(0)}$ as a boundary point. We introduce the Laurent expansion of h_1 about $z_1^{(0)}$,

$$\sum_{\nu = -\alpha}^{\infty} \delta_\nu \left(z_1 - z_1^{(0)}\right)^\nu, \qquad \delta_{-\alpha} \neq 0 \tag{8.5-2}$$

into w_1 of the form (2.5-1, 3) and consider the special expressions $H_{1m\nu}$ given by (2.5-3),

$$H_{1m\nu}(z_1) = \int_0^{z_1} \int_0^{z_{1m}} \cdots \int_0^{z_{12}} \left(z_{11} - z_1^{(0)}\right)^\nu dz_{11} \cdots dz_{1,m-1}\, dz_{1m}$$

for negative ν, $\nu = -\beta$ $(\beta \in \mathbb{N})$, $m \in \mathbb{N}$. After some calculation we obtain

$$H_{1m,-\beta}(z_1) = \kappa_{m\beta}\left(z_1 - z_1^{(0)}\right)^{-(\beta - m)} + p_{m\beta}(z_1) \qquad \text{if } m < \beta \tag{8.5-3}$$

$$H_{1m,-\beta}(z_1) = \lambda_{m\beta}\left(z_1 - z_1^{(0)}\right)^{m-\beta} \log\left(1 - z_1/z_1^{(0)}\right) + \tilde{p}_{m\beta}(z_1) \qquad \text{if } m \geq \beta; \tag{8.5-4}$$

here, $\kappa_{m\beta}$ and $\lambda_{m\beta}$ are nonzero real constants and $p_{m\beta}$, $\tilde{p}_{m\beta}$ are certain polynomials in z_1 depending on $z_1^{(0)}$. Thus, from (2.5-1, 3) and (8.5-2, 3, 4) we see that

$$
\begin{aligned}
w_1(z) = e_1(z)\Bigg[& \sum_{\beta=1}^{\alpha} \delta_{-\beta}\left(z_1 - z_1^{(0)}\right)^{-\beta} \\
& + \sum_{\beta=2}^{\alpha} \delta_{-\beta} \sum_{m=1}^{\beta-1} \tilde{q}_{1m}(z) \frac{(2m)!}{4^m m!} \kappa_{m\beta}\left(z_1 - z_1^{(0)}\right)^{-(\beta-m)} \\
& + \sum_{\beta=1}^{\alpha} \delta_{-\beta} \sum_{m=\beta}^{\infty} \tilde{q}_{1m}(z) \frac{(2m)!}{4^m m!} \lambda_{m\beta}\left(z_1 - z_1^{(0)}\right)^{m-\beta} \log\left(1 - \frac{z_1}{z_1^{(0)}}\right) \\
& + H(z)\Bigg]
\end{aligned}
\tag{8.5-5}
$$

where H denotes a certain function holomorphic on $\{z_1^{(0)}\} \times B_2(0, r_2)$. Here, the first term in (8.5-5) assures that $w = w_1$ has a pole plane of order α for $z_1 = z_1^{(0)}$ because of (8.5-1) and $\delta_{-\alpha} \neq 0$, $e_1(z_1^{(0)}, z_2) \neq 0$. From the third term in (8.5-5) we conclude that on the plane $z_1 = z_1^{(0)}$ there is a pole-like singularity. For, from (8.5-1) and $\delta_{-\alpha} \neq 0$, $\lambda_{\alpha\alpha} \neq 0$, it follows that $\delta_{-\alpha}\lambda_{\alpha\alpha}\tilde{q}_{1\alpha}(z_1^{(0)}, z_2)$ $\neq 0$ for some $z_2 \neq 0$ so that w_1 in (8.5-5) has a logarithmic branch plane where $\tilde{q}_{1\alpha}(z_1^{(0)}, z_2) \neq 0$. ■

8.5-2 Remark

The branch plane property of w in Theorem 8.5-1 may fail without assumption (8.5-1), for instance, if some or all of the $\delta_{-\beta}$'s, $\beta = 1, \cdots, \alpha - 1$, are equal to zero and (or) if $L \in FP_{1n}$ for some $n < \alpha$. The latter condition implies that the $\tilde{q}_{1m}$'s vanish identically for $m > n$, so that $\tilde{k}_1$ reduces to a polynomial kernel of degree $< \alpha$. For example, the third term in (8.5-5) vanishes in the case of equations (2.1-1) with $a_0 = a_1 a_2 + D_1 a_1$ (hence, $h_{10} = 0$; cf. Remark 4.2-5) since then the Bergman kernel of the first kind is $\tilde{k}_1(z, t) = e_1(z) k_1(z, t)$ with $k_1(z, t) \equiv 1$.

Additional References

Avila–Gilbert [1967]

Bergman [1944b, 1952c, 1954b, 1957, 1963, 1971]

Brown [1970]

Colton–Gilbert [1968]

Dont [1973]

Gilbert [1960a, 1960b, 1969b]

Gilbert–Howard–Aks [1965]

Gronau [1976, 1981]

Ingersoll [1948]

Kreyszig [1956, 1957a, 1958b, 1962]

McCoy [1980]

Millar [1971, 1973, 1976]

Mitchell [1946]

Netanyahu [1954]

Rosenthal [1974a, 1974b]

Suschowk [1962]

White [1961, 1962]

Chapter Nine

Approximation of Solutions

Starting with Runge's classical theorem, approximation theory in complex has become an extended field of common interest. Various methods and results on complex approximation are included in the book by Walsh [1969], which first appeared in 1935; the 1969 edition includes some new material as well as references to recent work. Approximation by polynomials in complex is also treated in a book by Sewell [1942]. For more recent results, see the book by Gaier [1980] and the literature compiled there; also see the survey by Gaier [1984] and the articles by Gutknecht [1983] and Korevaar [1980].

In this chapter we shall use integral operators for deriving approximation theorems for solutions of partial differential equations from some famous approximation theorems of complex analysis. This will give us analogs of the theorems of Runge and Mergelyan (Sec. 9.1), Walsh (Sec. 9.2), and Sewell (Sec. 9.3). The transition from complex analysis to the theory of partial differential equations will be accomplished by the operators defined in Chap. 2. This will demonstrate the power and generality of the "translation principle" furnished by integral operators.

We shall devote the whole discussion to *complex* solutions, but emphasize that it is almost a routine matter to reformulate the results for real-valued solutions, in the case that equation (2.1-1) was obtained from a real elliptic equation as explained in Sec. 2.1.

The idea of using integral operators in this way was suggested by S. Bergman (cf. Bergman [1971], p. 23), who also gave some typical applications in several of his papers ([1936], p. 102; [1937b], pp. 1179–1181; [1940]; [1943b], p. 141; [1963]). See also Rosenthal [1970] and Reinartz [1974]. A systematic approach to the approximation problem by the integral operator method was made in a Ph.D. thesis by Hoefer [1974], which was supervised by J. Mitchell (see also Mitchell [1973]).

9.1 ANALOGS OF RUNGE'S AND MERGELYAN'S THEOREMS

In the next sections we shall investigate solutions of equation (2.1-1) with coefficients in $C^\omega(\Omega)$, where $\Omega = \Omega_1 \times \Omega_2$, and Ω_1 and Ω_2 are simply connected domains in $\mathbb{C}$, as usual. We shall consider Type I representations of solutions with $j = 1$. For $j = 2$ and for Type II representations the results would be quite similar. Also, from our theorems in this chapter, one can easily obtain corresponding results for *real* solutions of equation (2.1-1) satisfying the additional conditions (2.1-9). This can be seen directly from (2.1-10) and Lemma 2.1-3.

In all three sections of the chapter we shall use the Bergman operator T_1 of the first kind, with $s = (0,0) \in \Omega$, $j = 1$, $\eta_1 = 0$, and $C_1 = [-1, 1]$, assuming that its kernel is even in t, so that this kernel is unique. Our main tools in these sections will be Theorems 2.4-2 and 2.5-1. We shall denote the solution $T_1 f_1$ by w, instead of w_1, and shall then use subscripts of w to denote sequences of solutions generated by T_1. For special dicylinders we introduce the notation

$$(9.1\text{-}1) \qquad B_{12}(0, r) = B_1(0, r) \times B_2(0, r) \subset \mathbb{C}^2$$

where the radius r is some positive number.

GENERALIZED RUNGE'S THEOREM

In this section we shall first generalize Runge's theorem for functions of one complex variable. This classical theorem can be formulated as follows (see Behnke–Sommer [1972], p. 258, Theorem III.31, or Conway [1973], p. 200, Corollary VIII.1.19).

Runge's approximation theorem

In a finite simply connected domain of the plane, all holomorphic functions can be uniformly approximated by polynomials on every compact subset of the domain.

This theorem generalizes to solutions of (2.1-1) as follows.

9.1-1 Theorem

Let the coefficients of (2.1-1), $Lw = 0$, *be of class* $C^\omega(B_{12}(0, \tilde{r}))$ ($\tilde{r} > 0$ *fixed*) *and let* $f_1 \in C^\omega(B_1(0, \frac{1}{4}\tilde{r}))$. *Furthermore, let* $w = T_1 f_1$ *be generated by the Bergman operator* T_1 *of the first kind. Then there exists a sequence* $\{w_k\}_{k \in \mathbb{N}_0}$ *of solutions of* (2.1-1) *such that, on* $B_{12}(0, r) \cup \partial B_{12}(0, r)$, *the solution* w *can be uniformly approximated by linear combinations of finitely many* w_k ($k \in \mathbb{N}_0$); *that is, for every positive* ε *there exist* (*complex*) *constants* α_k ($k = 0, \cdots, n_0$;

$n_0 \in \mathbb{N}_0$ suitably large) such that for all $z \in B_{12}(0, r) \cup \partial B_{12}(0, r)$, $r \in (0, \frac{1}{2}\tilde{r})$,

$$\left| w(z) - \sum_{k=0}^{n_0} \alpha_k w_k(z) \right| < \varepsilon .$$

The sequence $\{w_k\}_{k \in \mathbb{N}_0}$ can be chosen as a sequence of "elementary solutions" of the form

(9.1-2a)
$$w_k = T_1 f_{1k}$$

where

(9.1-2b)
$$f_{1k}(\xi_1) = \beta_k \xi_1^k, \qquad \beta_k = \frac{8^k (k!)^2}{\pi (2k)!}, \qquad k \in \mathbb{N}_0 .$$

Proof. Theorem 2.4-2 guarantees that T_1 exists and that w is a $C^\omega(B_{12}(0, \frac{1}{2}r) \cup \partial B_{12}(0, \frac{1}{2}r))$-solution of (2.1-1) for arbitrary $r \in (0, \frac{1}{2}\tilde{r})$. Using Theorems 2.4-2 and 2.5-1, we obtain on $B_{12}(0, r) \cup \partial B_{12}(0, r)$

(9.1-3a)
$$w(z) = e_1(z) \sum_{m=0}^{\infty} \tilde{q}_{1m}(z) \frac{(2m)!}{4^m m!} H_{1m}(z_1)$$

where

(9.1-3b)
$$\begin{aligned} H_{10}(z_1) &= h_1(z_1) \\ H_{1m}(z_1) &= \int_0^{z_1} \int_0^{z_{1m}} \cdots \int_0^{z_{12}} h_1(z_{11})\, dz_{11} \cdots dz_{1,m-1}\, dz_{1m}, \quad m \in \mathbb{N}: \end{aligned}$$

and

(9.1-3c)
$$h_1(z_1) = \int_{-1}^{1} f_1\left(\tfrac{1}{2} z_1 \tau\right) \tau^{-1/2}\, dt, \qquad \tau = 1 - t^2 .$$

Since, by assumption, $f_1 \in C^\omega(B_1(0, \frac{1}{4}\tilde{r}))$, we have $h_1 \in C^\omega(B_1(0, \frac{1}{2}\tilde{r}))$. Hence, by Runge's theorem, h_1 can be uniformly approximated by polynomials:

$$\forall\, \varepsilon^* > 0\ \exists\, n_0 \in \mathbb{N}_0\ \exists\, \alpha_1, \cdots, \alpha_{n_0} \in \mathbb{C}\ \forall\, z_1 \in B_1(0, r) \cup \partial B_1(0, r):$$

(9.1-4)
$$\left| h_1(z_1) - \sum_{k=0}^{n_0} \alpha_k z_1^k \right| < \varepsilon^* .$$

Let us now consider the solutions w_k in (9.1-2). These solutions can, of course, be represented in the form (9.1-3) with h_1 replaced by h_{1k} where, by (2.2-5) and (9.1-2),

(9.1-5)
$$h_{1k}(z_1) = \int_{-1}^{1} f_{1k}\left(\tfrac{1}{2} z_1 \tau\right) \tau^{-1/2}\, dt = z_1^k .$$

Obviously, $\Sigma_{k=0}^{n_0}\alpha_k w_k$ is a solution of (2.1-1). Representing w and the w_k's in the form (9.1-3) with h_1 and h_{1k}, respectively, we thus obtain for all $z \in B_{12}(0, r) \cup \partial B_{12}(0, r)$,

$$\left| w(z) - \sum_{k=0}^{n_0} \alpha_k w_k(z) \right|$$

$$= \left| e_1(z) \left\{ h_1(z_1) - \sum_{k=0}^{n_0} \alpha_k z_1^k + \sum_{m=1}^{\infty} \tilde{q}_{1m}(z) \frac{(2m)!}{4^m m!} \times \right. \right.$$

$$\left. \left. \times \int_0^{z_1} \int_0^{z_{1m}} \cdots \int_0^{z_{12}} \left[h_1(z_{11}) - \sum_{k=0}^{n_0} \alpha_k z_{11}^k \right] dz_{11} \cdots dz_{1,m-1}\, dz_{1m} \right\} \right|$$

$$< \varepsilon^* M_0 \left[1 + \sum_{m=1}^{\infty} |\tilde{q}_{1m}(z)| \frac{(2m)!}{4^m m!} \frac{|z_1|^m}{m!} \right]$$

where $M_0 = \max \{ |e_1(z)| \mid z \in B_{12}(0, r) \cup \partial B_{12}(0, r) \} > 0$. Because of $(2m)! 2^{-m}/m! = 1 \cdot 3 \cdots (2m - 1) < 2 \cdot 4 \cdots (2m) = 2^m m!$, the expression in brackets $[\cdots]$ can be estimated by

$$1 + \sum_{m=1}^{\infty} |\tilde{q}_{1m}(z)| |z_1|^m .$$

According to Theorem 2.4-2a this series converges uniformly, and is therefore bounded by a constant $\tilde{M}_0 > 0$, on $B_{12}(0, r) \cup \partial B_{12}(0, r)$. Now, for an arbitrarily given $\varepsilon > 0$, we can choose $\varepsilon^* = \varepsilon/[M_0 \tilde{M}_0]$ and, according to (9.1-4), find an $n_0 \in \mathbb{N}_0$ and constants $\alpha_1, \cdots, \alpha_{n_0}$ (depending on ε^* and thus on ε) such that for all $z \in B_{12}(0, r) \cup \partial B_{12}(0, r)$,

$$\left| w(z) - \sum_{k=0}^{n_0} \alpha_k w_k(z) \right| < \varepsilon^* M_0 \tilde{M}_0 = \varepsilon.$$

This proves the theorem. ■

GENERALIZED MERGELYAN'S THEOREM

By the same method we can obtain an analog of the well-known approximation theorem of Mergelyan. This famous theorem originally asserts the following (see Rudin [1974], p. 423).

Mergelyan's theorem

If $g \in C^{\omega}(\mathring{K}) \cap C^0(K)$, where $\mathring{K}$ denotes the interior of K and K is a compact set of $\mathbb{C}$ whose complement $\mathbb{C} \setminus K$ is connected, then the following approxima-

tion property holds:

$$\forall\, \varepsilon > 0\ \exists \text{ polynomial } p_\varepsilon\ \forall\, z_1 \in K:$$

(9.1-6)

$$|g(z_1) - p_\varepsilon(z_1)| < \varepsilon.$$

Application of this theorem yields the following.

9.1-2 Theorem

Let the coefficients of (2.1-1), $Lw = 0$, *be of class* $C^\omega(B_{12}(0, \tilde{r}))$ [*cf.* (9.1-1)], $\tilde{r} > 0$. *Let* K *be a compact subset of* $B_{12}(0, \frac{1}{2}\tilde{r})$, *star-shaped with respect to the origin and such that the image* K_{z_1} *of the orthogonal projection of* K *into the* z_1*-plane is simply connected and has nonvoid interior* $(K_{z_1})^0$. *Let the solution* w *of* (2.1-1) *be a Type I representation by the Bergman operator* T_1 *of the first kind, as before, satisfying* $w|_{z_2=0} \in C^\omega((K_{z_1})^0) \cap C^0(K_{z_1})$. *Then there exists a sequence* $\{w_k\}_{k \in \mathbb{N}_0}$ *of solutions whose* T_1*-associated functions* f_{1k} *are polynomials, and* w *can be uniformly approximated on* K *by these solutions.*

Proof. By assumption, K_{z_1} is simply connected and $(K_{z_1})^0 \neq \varnothing$, and the solution w can be obtained by a Type I representation by means of T_1. Thus, according to Corollary 2.6-3, there exists a uniquely determined and explicitly known T_1-associated function f_1 of w such that $w = T_1 f_1$. From this function f_1 we obtain $h_1 \in C^\omega((K_{z_1})^0)$ by (9.1-3c). From $\eta_1 = 0$, the definition of a kernel of the first kind, and Theorem 2.5-1, we have

$$w(z_1, 0) = e_1(z_1, 0) h_1(z_1) = h_1(z_1) \qquad \text{for } z_1 \in \left(K_{z_1}\right)^0.$$

On the other hand, we have assumed that

$$w|_{z_2=0} \in C^\omega\left(\left(K_{z_1}\right)^0\right) \cap C^0\left(K_{z_1}\right).$$

Hence h_1 has an extension

$$\tilde{h}_1 \in C^\omega\left(\left(K_{z_1}\right)^0\right) \cap C^0\left(K_{z_1}\right).$$

From the assumptions on K we conclude that K_{z_1} is a compact subset of $B_1(0, \frac{1}{2}\tilde{r})$, star-shaped with respect to the origin, and possessing (simply) connected complement $\mathbb{C} \setminus K_{z_1}$. Thus the theorem of Mergelyan is applicable to $\tilde{h}_1$. The remainder of the proof now follows as in the proof of Theorem 9.1-1. ■

9.2 APPROXIMATION THEOREM OF WALSH TYPE

In this section we present a further typical illustration of the method of obtaining approximation theorems for solutions of partial differential equations (2.1-1) from approximation theorems in complex analysis. Also this transition is quite simple and direct. The assertions of our theorem (below) will be obtained from theorems by Walsh [1969] on maximal convergence; also see Hoefer [1974] and Mitchell [1973].

Following Walsh [1969], we first provide some auxiliary material from complex function theory.

Let K be a compact subset of $\mathbb{C}$ which has more than one point and a simply connected complement $C(K) = (\mathbb{C} \cup \{\infty\}) \setminus K$ in the extended plane. Then there exists a one-to-one conformal mapping

$$\varphi\colon\ C(K) \quad \to \quad (\mathbb{C} \cup \{\infty\}) \setminus (B_1(0,1) \cup \partial B_1(0,1))$$

with $\infty = \varphi(\infty)$. See Behnke–Sommer [1972], p. 352, or Hille [1959–1962], II, 322. Furthermore,

$$C_R = \left\{ z_1 \in \mathbb{C} \mid z_1 = \varphi^{-1}(\tilde{z}_1), |\tilde{z}_1| = R \right\}, \qquad R > 1$$

is the pre-image of a circle and C_{R_1} lies in the interior of C_{R_2} for $R_2 > R_1$. By $I(C_R)$ we denote the interior of the domain bounded by C_R.

Under the preceding assumptions the following result is known (Walsh [1969], pp. 75–76, Theorem IV.5):

Walsh's theorem

If $g \in C^\omega(I(C_R) \cup C_R)$, then there exist a sequence $\{p_n\}_{n \in \mathbb{N}_0}$ of polynomials p_n of respective degrees n and a constant $M(R) > 0$ such that

$$\text{(9.2-1)} \qquad \forall\, n \in \mathbb{N}_0\ \forall\, z_1 \in K\colon\ |g(z_1) - p_n(z_1)| \leq M(R)/R^n .$$

Moreover, the following holds.

9.2-1 Lemma

Let $g \in C^\omega(K)$. Then there exists a greatest $R_0 \in (1, \infty]$ such that $g \in C^\omega(I(C_{R_0}))$. Besides, if $R \in (1, R_0)$ is arbitrarily chosen, then there exists a sequence $\{p_n\}_{n \in \mathbb{N}_0}$ of polynomials, of respective degrees n, such that (9.2-1) *holds, but there exists no sequence of such polynomials such that* (9.2-1) *is valid when $R > R_0$.*

Proof. See Walsh [1969], p. 79, Theorem IV.7. There the proof is given under the more general assumption that $C(K)$ is connected and regular in the sense that $C(K)$ possesses a Green's function with pole at infinity. Our assumptions that K has more than one point and $C(K)$ is simply connected imply this regularity of $C(K)$; cf. Walsh [1969], p. 65. ∎

The polynomials in Lemma 9.2-1 do not depend on the choice of R $(< R_0)$. Such a sequence of polynomials is said to *converge maximally* to g on K. This concept was introduced by Walsh [1969].

GENERALIZED WALSH'S THEOREM

Applying Lemma 9.2-1, we now obtain

9.2-2 Theorem

Let the coefficients of (2.1-1) *be of class* $C^\omega(B_{12}(0, \tilde{r}))$, $\tilde{r} > 0$. *Let* K_1 *be a compact subset of* $B_1(0, \frac{1}{2}r)$, *where* $r \in (0, \tilde{r})$ *is arbitrary and fixed, containing more than one point and being star-shaped with respect to the origin. Let* w *be a* $C^\omega(B_1(0, \frac{1}{2}r) \times B_2(0, r))$*-solution of* (2.1-1) *represented in Type I form by the Bergman operator* T_1 *of the first kind, as before. Then there exists a sequence* $\{w_n\}_{n \in \mathbb{N}_0}$ *of solutions* $w_n = T_1 p_n$, *where* p_n *is a polynomial of degree* n, *and there exists a constant* $M > 0$ *such that for all* $n \in \mathbb{N}_0$, *all* $z \in K_1 \times K_2$, *and* $R \in (1, R_0)$ *the inequality*

$$|w(z) - w_n(z)| \leq MR^{-n} \tag{9.2-2}$$

holds. Here, $R_0 > 1$ *is the greatest number* (*or infinity*) *such that the function* h_1 *in the representation of the solution* w *by means of Theorem* 2.5-1 *is of class* $C^\omega(I(C_{R_0}))$; K_2 *denotes an arbitrarily chosen compact subset of* $B_2(0, r)$.

Proof. By Corollary 2.6-3 there exists a uniquely determined explicitly known T_1-associated function $f_1 \in C^\omega(B_1(0, \frac{1}{4}r))$ such that the solution w can be represented by Theorem 2.4-2c and hence by Theorem 2.5-1. That is, for w we have a representation of the form (9.1-3) with a uniquely determined explicitly known function $h_1 \in C^\omega(B_1(0, \frac{1}{2}r))$, and this representation is valid in $B_1(0, \frac{1}{2}r) \times B_2(0, r)$. By Walsh's Lemma 9.2-1 there exists a sequence of polynomials $\tilde{p}_n$ of respective degrees n, say,

$$\tilde{p}_n(z_1) = \sum_{k=0}^{n} \alpha_k z_1^k, \qquad \alpha_n \neq 0 \tag{9.2-3}$$

and a constant $M_1 > 0$ such that (9.2-1) holds for h_1, where M_1 is independent of $R \in (1, R_0)$. We thus have for all $n \in \mathbb{N}_0$ and for all $z_1 \in K_1$,

$$|h_1(z_1) - \tilde{p}_n(z_1)| \leq M_1 R^{-n}. \tag{9.2-4}$$

We let w_n denote the solutions obtained from (9.1-3) by inserting $\tilde{p}_n$, instead of h_1. Then from (9.1-3) and (9.2-4) it follows that for all $n \in \mathbb{N}_0$ and all

$z \in K_1 \times K_2$,

$$|w(z) - w_n(z)| \le |e_1(z)| \left[|h_1(z_1) - \tilde{p}_n(z_1)| + \sum_{m=1}^{\infty} |\tilde{q}_{1m}(z)| \frac{(2m)!}{4^m m!} \times \right.$$

$$\left. \times \left| \int_0^{z_1} \int_0^{z_{1m}} \cdots \int_0^{z_{12}} [h_1(z_{11}) - \tilde{p}_n(z_{11})] \, dz_{11} \cdots dz_{1,m-1} \, dz_{1m} \right| \right]$$

$$\le M_0 \left[M_1 R^{-n} + \sum_{m=1}^{\infty} |\tilde{q}_{1m}(z)| \frac{(2m)!}{4^m m!} M_1 R^{-n} \frac{|z_1|^m}{m!} \right]$$

$$\le M_0 M_1 R^{-n} \left[1 + \sum_{m=1}^{\infty} |\tilde{q}_{1m}(z)| |z_1|^m \right]$$

$$\le M_0 \tilde{M}_0 M_1 R^{-n}$$

where

$$M_0 = \max \{ |e_1(z)| \mid z \in K_1 \times K_2 \}$$

and

$$\tilde{M}_0 = \max \left\{ 1 + \sum_{m=1}^{\infty} |\tilde{q}_{1m}(z)| |z_1|^m \mid z \in K_1 \times K_2 \right\}.$$

$\tilde{M}_0$ exists, since the series converges *uniformly*; see the proof of Theorem 9.1-1. Hence if we take $M = M_0 \tilde{M}_0 M_1$ and set

$$p_n(\xi_1) = \sum_{k=0}^{n} \frac{8^k (k!)^2}{\pi (2k)!} \alpha_k \xi_1^k$$

then $w_n = T_1 p_n$, and the assertion of the theorem follows. ■

9.3 FURTHER APPROXIMATION THEOREMS FOR SOLUTIONS

In this section we shall extend the following theorem by Sewell [1935] involving a Lipschitz condition.

9.3-1 Theorem

Let $g \in C^{\omega}(B_1(0,1)) \cap C^0(B_1(0,1) \cup \partial B_1(0,1))$ satisfy a Lipschitz condition on $\partial B_1(0,1)$. Furthermore, let $\tilde{p}_n$ be the nth *partial sum of the Maclaurin series of g. Then there exists a constant $\tilde{M} > 0$ such that for all $n = 2, 3, \cdots$ and for* *all* $z_1 \in B_1(0,1) \cup \partial B_1(0,1)$,

$$(9.3\text{-}1) \qquad |g(z_1) - \tilde{p}_n(z_1)| \leq \tilde{M}\frac{\log n}{n}.$$

GENERALIZED SEWELL'S THEOREM

Sewell's theorem extends to solutions of partial differential equations as follows.

9.3-2 Theorem

Let the coefficients of (2.1-1) *be of class* $C^\omega(B_{12}(0,\tilde{r}))$ *with* $B_{12}(0,\tilde{r})$ *as defined in* (9.1-1). *Suppose that the solution* w *of equation* (2.1-1) *is of class*

$$C^\omega\left(B_1\left(0,\tfrac{1}{2}r\right) \times B_2(0,r)\right) \cap C^0\left(\left[B_1\left(0,\tfrac{1}{2}r\right) \cup \partial B_1\left(0,\tfrac{1}{2}r\right)\right] \times B_2(0,r)\right)$$

with $r \in (0,\tilde{r})$ *arbitrary and fixed. Let* w *be represented in Type I form by the Bergman operator* T_1 *of the first kind. Furthermore, let* $w|_{z_2=0}$ *satisfy a Lipschitz condition, say,*

$$(9.3\text{-}2) \qquad |w(z_1,0) - w(\tilde{z}_1,0)| < M_1|z_1 - \tilde{z}_1|$$

for all $z_1, \tilde{z}_1 \in \partial B_1(0,\frac{1}{2}r)$ *with suitable* M_1. *Then there exists a sequence of solutions* $w_n = T_1 p_n$ *of* (2.1-1) *with polynomials* p_n *of degrees not exceeding* n, *such that for all* $n \geq 2$ *and all* $z \in [B_1(0,\frac{1}{2}r) \cup \partial B_1(0,\frac{1}{2}r)] \times B_2(0,r)$ *we have*

$$(9.3\text{-}3) \qquad |w(z) - w_n(z)| < M\frac{\log n}{n}$$

where $M > 0$ *is a suitable constant.*

Proof. According to our assumptions, by Theorems 2.4-2 and 2.5-1, the solution w can be represented on $B_1(0,\frac{1}{2}r) \times B_2(0,r)$ in the form (9.1-3) with a uniquely determined and explicitly known function $h_1 \in C^\omega(B_1(0,\frac{1}{2}r))$. By our general assumptions on T_1 in this chapter, especially $\eta_1 = 0$ and $s = (0,0)$, we have

$$(9.3\text{-}4) \qquad w(z_1,0) = e_1(z_1,0)h_1(z_1) = h_1(z_1) \qquad \text{in } B_1\left(0,\tfrac{1}{2}r\right).$$

For $r < \tilde{r}$, the Bergman kernel of the operator T_1 is of class

$$C^\omega\left(\left[B_1\left(0,\tfrac{1}{2}r\right) \times B_2(0,r)\right] \cup \partial\left[B_1\left(0,\tfrac{1}{2}r\right) \times B_2(0,r)\right]\right)$$

with respect to the first two variables; cf. Theorem 2.4-2. By assumption, w is continuous on $[B_1(0,\frac{1}{2}r) \cup \partial B_1(0,\frac{1}{2}r)] \times B_2(0,r)$. Thus, by (9.3-4), h_1 can be continuously extended to the boundary of $B_1(0,\frac{1}{2}r)$ by $w|_{z_2=0}$. By assumption

(9.3-2), the extended function h_1 satisfies for $z_1, \tilde{z}_1 \in \partial B_1(0, \frac{1}{2}r)$ the inequality

$$|h_1(z_1) - h_1(\tilde{z}_1)| < M_1|z_1 - \tilde{z}_1|. \tag{9.3-5}$$

Moreover, representation (9.1-3) extends to $[B_1(0, \frac{1}{2}r) \cup \partial B_1(0, \frac{1}{2}r)] \times B_2(0, r)$. In order to apply Sewell's theorem 9.3-1 we introduce $z_1^* = 2z_1/r$ and write $h_1^*(z_1^*) = h_1(\frac{1}{2}rz_1^*) = h_1(z_1)$ and

$$\tilde{p}_n^*(z_1^*) = \sum_{k=0}^{n} \delta_k \left(\frac{r}{2}\right)^k z_1^{*k} = \sum_{k=0}^{n} \delta_k z_1^k = \tilde{p}_n(z_1)$$

where δ_k, $k \in \mathbb{N}_0$, denote the coefficients of the Maclaurin series of h_1. Now, by (9.3-5) we get

$$|h_1^*(z_1^*) - h_1^*(\tilde{z}_1^*)| < M_1 \frac{r}{2} |z_1^* - \tilde{z}_1^*|$$

for $z_1^*, \tilde{z}_1^* \in \partial B_1(0,1)$. From Theorem 9.3-1 it follows that (9.3-1) holds with $g = h_1^*$. Hence there is an $M_2 > 0$ such that

$$|h_1(z_1) - \tilde{p}_n(z_1)| = |h_1^*(z_1^*) - \tilde{p}_n^*(z_1^*)| \le M_2 \frac{\log n}{n} \tag{9.3-6}$$

for all $z_1^* \in B_1(0,1) \cup \partial B_1(0,1)$ and $z_1 \in B_1(0, \frac{1}{2}r) \cup \partial B_1(0, \frac{1}{2}r)$, respectively, and all $n = 2, 3, \cdots$. Setting $w_n = T_1 p_n$ with

$$p_n(\xi_1) = \sum_{k=0}^{n} \frac{8^k (k!)^2}{\pi (2k)!} \delta_k \xi_1^k \tag{9.3-7}$$

and proceeding as in the last part of the proof of Theorem 9.2-2, we obtain from (9.3-6) and Theorem 2.5-1 [cf. (9.1-3)] the inequality

$$|w(z) - w_n(z)| \le M_0 \tilde{M}_0 M_2 \frac{\log n}{n}.$$

Here the constants M_0 and $\tilde{M}_0$ (> 0) are the same as those in that proof, with $K_1 = B_1(0, \frac{1}{2}r) \cup \partial B_1(0, \frac{1}{2}r)$ and $K_2 = B_2(0, r) \cup \partial B_2(0, r)$. Taking $M = M_0 \tilde{M}_0 M_2$, we obtain the assertion of Theorem 9.3-2. ■

A FURTHER APPROXIMATION THEOREM

For a good understanding we want to add a few remarks on Sewell's theorem. From $g \in C^\omega(B_1(0,1))$ it does *not* follow that g satisfies a Lipschitz condition on $\partial B_1(0,1)$, not even that g is continuous on $\partial B_1(0,1)$. The point of Sewell's theorem is that, assuming the validity of a Lipschitz condition on $\partial B_1(0,1)$, we obtain the bound $(M \log n)/n$ on the whole closed unit disk. On a *smaller* disk $B_1(0, \tilde{R})$, where $\tilde{R} < 1$, we can do better by much simpler means, such as the *Cauchy inequalities*

$$\frac{1}{k!}\left|\left(D_1^k g\right)(0)\right| \leq \tilde{M}\tilde{R}^{-k}, \qquad k \in \mathbb{N}_0$$

where

$$\tilde{M} = \max\left\{|g(z_1)| \mid z_1 \in B_1(0, \tilde{R}) \cup \partial B_1(0, \tilde{R})\right\} \geq 0.$$

Indeed, for the nth partial sum $\tilde{p}_n$ of the Maclaurin series of g we then have

$$|g(z_1) - \tilde{p}_n(z_1)| \leq \sum_{k=n+1}^{\infty} \frac{1}{k!}\left|\left(D_1^k g\right)(0)\right| |z_1|^k$$

$$\leq \tilde{M} \sum_{k=n+1}^{\infty} \left(\frac{R}{\tilde{R}}\right)^k = \frac{\tilde{M}}{\tilde{R} - R} \frac{R^{n+1}}{\tilde{R}^n}$$

for all $n \in \mathbb{N}_0$ and all $z_1 \in B_1(0, R) \cup \partial B_1(0, R)$, $R \in (0, \tilde{R})$. For instance, if $R = \varepsilon\tilde{R}$, $\varepsilon \in (0,1)$, then for all those n and z_1,

$$|g(z_1) - \tilde{p}_n(z_1)| \leq \frac{\tilde{M}}{1 - \varepsilon}\varepsilon^{n+1}. \tag{9.3-8}$$

In particular, this holds on $\partial B_1(0, \varepsilon\tilde{R})$. For large enough n, this is certainly better than $(M \log n)/n$, but we should keep in mind that the present assumptions are stronger.

Further interesting facts related to this discussion are included in the chapter on lemniscates in the book by Hille [1959–1962], Vol. 2, Chap. 16.

As an application of the preceding discussion, let us prove:

9.3-3 Theorem

Let the assumptions of Theorem 9.3-2 *be fulfilled, except that* $w|_{z_2=0}$ *may not satisfy the Lipschitz condition* (9.3-2). *Let* $r_0 \in (0, r)$. *Then there exists a sequence of solutions* $w_n = T_1 p_n$ *of equation* (2.1-1) *with polynomials* p_n *of degree not exceeding* n, *such that for all* $n \in \mathbb{N}_0$ *and all* $z \in K_1 \times B_2(0, r)$, *we have*

$$|w(z) - w_n(z)| \leq M\varepsilon^n, \qquad \varepsilon = r_0/r, \qquad \varepsilon \in (0,1). \tag{9.3-9a}$$

Here the constant M *depends on* r_0, r, *and, via* T_1, *on the coefficients of* (2.1-1), *but is independent of* n, *and is given by*

$$M = M_0\tilde{M}_0 M_1 \frac{\varepsilon}{1 - \varepsilon}, \tag{9.3-9b}$$

with

$$M_0 = \max\{|e_1(z)| \mid z \in K_1 \times K_2\}$$

$$\tilde{M}_0 = \max\left\{1 + \sum_{m=1}^{\infty} |\tilde{q}_{1m}(z)||z_1|^m \mid z \in K_1 \times K_2\right\}$$

$$M_1 = \max\{|w(z_1, 0)| \mid z_1 \in \partial B_1(0, \tfrac{1}{2}r)\}$$

and

$$K_1 = B_1(0, \tfrac{1}{2}r_0) \cup \partial B_1(0, \tfrac{1}{2}r_0), \qquad K_2 = B_2(0, r) \cup \partial B_2(0, r).$$

[*For* e_1 *and* $\tilde{q}_{1m}$, *see* (2.1-3) *and Theorem* 2.4-2, *recalling that* $\eta_1 = 0$ *and* $s = (0, 0)$.]

Proof. The first part of the proof is similar to that of Theorem 9.3-2. It gives a representation for w of the form (9.1-3) with $h_1 = w|_{z_2=0}$ on $B_1(0, \frac{1}{2}r) \cup \partial B_1(0, \frac{1}{2}r)$ [cf. (9.3-4)], valid for $z \in [B_1(0, \frac{1}{2}r) \cup \partial B_1(0, \frac{1}{2}r)] \times B_2(0, r)$. Setting $z_1^* = z_1/\tilde{r}$, we see that the radii $\tilde{r}$, $\frac{1}{2}r$, and $\frac{1}{2}r_0$ correspond to 1, $\tilde{R} = \frac{1}{2}r/\tilde{r}$, and $R = \frac{1}{2}r_0/\tilde{r}$, respectively. Proceeding as in the derivation of (9.3-6) and using functions h_1^* and $\tilde{p}_n^*$ of z_1^*, from (9.3-8) we obtain

$$|h_1(z_1) - \tilde{p}_n(z_1)| \leq \frac{\tilde{M}_1}{1 - \varepsilon} \varepsilon^{n+1} \tag{9.3-10a}$$

for all $n \in \mathbb{N}_0$ and all $z_1 \in K_1$. Here,

$$\begin{aligned} \tilde{M}_1 &= \max\{|h_1(z_1)| \mid z_1 \in B_1(0, \tfrac{1}{2}r) \cup \partial B_1(0, \tfrac{1}{2}r)\} \\ &= \max\{|h_1(z_1)| \mid z_1 \in \partial B_1(0, \tfrac{1}{2}r)\} \\ &= \max\{|w(z_1, 0)| \mid z_1 \in \partial B_1(0, \tfrac{1}{2}r)\} \\ &= M_1. \end{aligned} \tag{9.3-10b}$$

From here on, we proceed as in the last part of the proof of Theorem 9.3-2: For $w_n = T_1 p_n$ with p_n obtained from $\tilde{p}_n$ by (9.3-7), we get on $z \in K_1 \times B_2(0, r)$,

$$\begin{aligned} |w(z) - w_n(z)| \leq {} & \max\{|e_1(z)| \mid z \in K_1 \times K_2\} \times \\ & \times \max\left\{1 + \sum_{m=1}^{\infty} |\tilde{q}_{1m}(z)||z_1|^m \mid z \in K_1 \times K_2\right\} \times \\ & \times \max\{|h_1(z_1) - \tilde{p}_n(z_1)| \mid z_1 \in K_1\}, \qquad n \in \mathbb{N}_0. \end{aligned}$$

Using (9.3-10) and the notations near the end of the theorem, we finally obtain

$$|w(z) - w_n(z)| \leq \left(M_0 \tilde{M}_0 M_1 \frac{\varepsilon}{1-\varepsilon} \right) \varepsilon^n .$$

This completes the proof. ■

ON THE CHOICE OF THE INTEGRAL OPERATOR IN APPROXIMATION THEORY

Our approximation theorems for solutions were derived by applying the Bergman operator of the first kind to polynomials that approximate the associated function. There are four main reasons for this choice of the operator.

1. This operator exists, at least locally, for all domains of $\mathbb{C}^2$ on which the coefficients of (2.1-1) are simultaneously holomorphic (see Sec. 2.4). In contrast, other operators may exist only for *certain subclasses* of equations (2.1-1) (see Chaps. 3, 4, and 5).

2. Possible singularities of the kernel of this operator can only be caused by corresponding singularities of one of the coefficients of (2.1-1). Hence, in this respect, this operator behaves more suitably on the common domain of holomorphy of the coefficients of (2.1-1) than other operators (see Chap. 5).

3. This operator has a relatively simple inverse. As a basic consequence, for the whole class of equations of the form (2.1-1), the associated function to a given solution is explicitly known (see Sec. 2.6).

4. Solutions obtained by an operator of the first kind admit several representations, as is shown in Secs. 2.4 and 2.5 and Chap. 6. Hence for different purposes, one has available a number of different representations.

However, for certain subclasses of equations (2.1-1), such as some of those discussed in Chaps. 3, 4, 5, and Sec. 10.4, other integral operators may be easier to apply or may yield better results. This situation may occur when the coefficients of an equation have singularities in the domain considered.

An important instance of this is the *Tricomi equation*, which we shall consider in Chap. 13. There we shall see (in Sec. 13.2) that the so-called *Bergman operator of the second kind* is preferable to that of the first kind. Bergman [1971] (pp. 106–122) pointed out that this also holds for certain generalizations of the Tricomi equation.

Some further remarks on the choice of operators in approximation theory are included in Rosenthal [1970] and Hoefer ([1974], pp. 36–44).

Additional References

Aupetit [1980]
Bergman [1940, 1943b, 1963, 1966b]
Colton [1975b, 1976a, 1976b, 1980]
Colton–Gilbert [1971]
Ďurikovič [1968]
Eisenstat [1974]
Fichera [1979]
Gaier [1980, 1984]
Gilbert–Atkinson [1970]
Gilbert–Lo [1971]
Gilbert–Wei [1983]
Hoefer [1974]
Kaucher–Miranker [1984]
McCoy [1979a, 1979b, 1982]
Mitchell [1973]
Reinartz [1974]
Saff [1971]
Walsh [1969]
Wendland [1979]

Chapter Ten

Value Distribution Theory of Solutions

Since the beginning of modern complex analysis, as initiated by Picard, Poincaré, Hadamard, Borel, and others, near the end of the last century, the theory of value distribution of analytic functions has grown into a large area of its own. It has attracted general interest over a long period of time during which deep and detailed results were obtained. This evolution was intensified by the appearance of the famous classic by R. Nevanlinna [1929]. For an introduction to this field, we also refer to the monographs by Dinghas [1961] and Hayman [1964] and to the survey article by Begehr [1983]. Meanwhile, this theory was applied to *ordinary* differential equations, resulting in an interesting theory for these equations; see, for instance, Wittich [1982] and Jank–Volkmann [1985], in particular, Chap. V and the literature given there.

Our present chapter will be devoted to some fundamental ideas in the theory of value distribution for solutions of *partial* differential equations (2.1-1). Our main tool will be the Type I representation of solutions by Bergman operators. These operators provide a translation principle for various results of the Nevanlinna theory and related theories, giving a detailed characterization of the value distribution of those solutions.

In Sec. 10.1 we investigate similarities between the value distribution of functions of one complex variable and that of those solutions; there we shall make use of Nevanlinna's First Fundamental Theorem. Section 10.2 deals with a generalization of Nevanlinna's Second Fundamental Theorem for solutions as obtained by H. Chernoff. An analog of the Little Picard Theorem for solutions is proved in Sec. 10.3. The remainder of the chapter is devoted to a generalization of the Great Picard Theorem. Generalizing the theory of polyanalytic functions by Balk [1966, 1983], Krajkiewicz [1973] has proved such a theorem for multianalytic functions. Now, in Sec. 10.4, we consider solutions generated by Bergman operators with polynomial kernels of the first kind (cf. Chap. 3), apply the Picard–Krajkiewicz theorem, and arrive at an extension of the Great Picard Theorem to solutions of partial differential equations.

10.1 GENERAL IDEA AND SETTING

In this chapter, we shall extend the theory of value distribution to partial differential equations, within the framework of integral operators. For this purpose, we utilize Type I representations of the form (2.2-8). With $j = 1$, point of reference $s = (0,0)$, and path of integration $C_1 = [-1,1]$, these are defined by the operator

$$T_1\colon V(\tilde{\Omega}_1) \quad \to \quad V(\Omega)$$

$$T_1 f_1(z) = \int_{-1}^{1} \tilde{k}_1(z_1, z_2, t) f_1\left(\tfrac{1}{2} z_1 \tau\right) \tau^{-1/2}\, dt, \quad \tau = 1 - t^2, \tag{10.1-1}$$

$$w = w_1 = T_1 f_1 .$$

Since we intend to relate our theory of solutions to Nevanlinna's theory and other results of complex analysis of *one* complex variable, it is natural to assume that equation (2.1-1) has been obtained from the real equation (2.1-5), as explained in Secs. 2.1 and 6.1, and that we investigate the behavior of w for the case $z_2 = \bar{z}_1$. For given f_1, formula (10.1-1) represents a solution of (2.1-1), as we know. For these solutions, with $z_2 = \bar{z}_1$, we explore the *value distribution*, based on theorems on the value distribution in complex analysis. We denote this class of solutions w by $K(\tilde{k}_1)$. Thus,

$$w \in K(\tilde{k}_1) = T_1(V(\tilde{\Omega}_1)).$$

We first recall a special equation and its Bergman operator that will be ideally suited for pointing out similarities of, and basic differences between, the classical theory of value distribution in complex analysis and the value distribution for solutions w (functions of $z_1, \bar{z}_1$) of equations of the form (2.1-1).

10.1-1 Example

For the Helmholtz equation (2.2-2) the Bergman kernel of the first kind is given by

$$\tilde{k}_1(z_1, \bar{z}_1, t) = \cos(crt), \qquad r = (z_1 \bar{z}_1)^{1/2}.$$

Setting

$$f_{1m}(\xi_1) = \frac{m!}{(2m)!\pi}(4c\xi_1)^m$$

we see that the corresponding Bergman operator T_1 yields the solution

$$(10.1\text{-}2)\quad w_{1m}(z_1, \bar{z}_1) = \left(\frac{z_1}{\bar{z}_1}\right)^{m/2} J_m(cr) = e^{im\phi} J_m(cr) = \frac{1}{r^m} J_m(cr) z_1^m$$

where $z_1 = re^{i\phi}$, $m \in \mathbb{N}_0$, and $c \neq 0$; cf. (2.2-4a, 6, 7).

The representation (10.1-2) illustrates that for solutions of partial differential equations the situation differs basically from that for complex analytic functions. Indeed, a solution $w \in K(\tilde{k}_1)$ will, in general, not have isolated singularities. For, if $\tilde{r}_0 = cr_0 \neq 0$ and $\tilde{r}_0$ is a zero of the Bessel function J_m, then w_{1m} in (10.1-2) vanishes on the circle $|z_1| = r_0$.

We shall demonstrate that, despite those and other basic differences, one can generalize certain concepts from the theory of analytic functions in such a way that extensions of classical theorems on value distribution become possible. For instance, just as in the classical case, the problem of the distribution of a-points mainly consists of relations between the behavior of

$$m\left[r, w(z_1, \bar{z}_1)\right] \quad \text{and} \quad n\left[r, \left(w(z_1, \bar{z}_1) - a\right)^{-1}\right]$$

as $r \to \infty$ (definitions below). In addition, as another matter of importance, we shall discover relations between these quantities and the coefficients of the series expansion of w at the origin.

GENERAL ASSUMPTIONS

In the following we make two ***assumptions***:

(i) The coefficients of L in (2.1-1) are entire functions of both variables.

(ii) The operator L in (2.1-1) admits a Bergman kernel $\tilde{k}_1$ such that there exists a function g of the variables[1] z_1, s, t which is an entire function of the complex variables z_1 and s and has the property that

$$(10.1\text{-}3)\quad \tilde{k}_1(z_1, \bar{z}_1, t) = g(z_1, s, t) \qquad \text{for } |z_1| = s \ (s \text{ real}, \geq 0); \quad t \in \bar{B}_0.$$

If $\tilde{k}_1$ satisfies (10.1-3), we indicate this by a cross and write $\tilde{k}_1^{\times}$ and $T_1^{\times}$.

Obviously, these conditions are essential, as a little reflection on the classical theory and on the general nature of Bergman operators shows. These assumptions are satisfied in Examples 10.1-1 and 3.1-8 as well as in various other cases of practical importance.

[1] Since the point of reference is $(0, 0)$ throughout this chapter, we can use s here as shown.

If we impose the additional condition that f_1 be entire, we have a class of entire solutions

$$w(z_1, \bar{z}_1) = \int_{-1}^{1} g(z_1, s, t) f_1(\tfrac{1}{2} z_1 \tau) \tau^{-1/2} \, dt, \qquad \tau = 1 - t^2 \tag{10.1-4}$$

of our given equation which have a most essential property that is crucial in making the "translation principle" work. This property is as follows. For $|z_1| = s$ the solution w is an entire function $v_{(s)}$ of the variable z_1. Hence for every nonnegative real s–geometrically: on the circle $|z_1| = s$–the function w takes on the values of an entire function of z_1.

From our assumptions on L, g, and f_1, together with (10.1-4), it follows that $v_{(s)}$ can be extended with respect to the variable s to the whole complex plane and has the expansion

$$v_{(s)}(z_1) = \sum_{m, n=0}^{\infty} c_{mn} s^m z_1^n . \tag{10.1-5}$$

ORDER OF GROWTH OF A FUNCTION

Let us first investigate the order of growth of w and establish relations between the coefficients of $v_{(s)}$ and w.

10-1.2 Lemma

Let V be an entire function of two complex variables s and z_1 having the expansion

$$V(s, z_1) = \sum_{m, n=0}^{\infty} C_{mn} s^m z_1^n \tag{10.1-6}$$

with ***nonnegative*** *real coefficients C_{mn}, and set*

$$\mu(S, Z) = \max_{|s|=S,\ |z_1|=Z} |V(s, z_1)|, \qquad S, Z > 0. \tag{10.1-7}$$

Let

$$\rho = \rho(k) = \limsup_{t \to \infty} \frac{\log \log \mu(t, kt)}{\log t}, \qquad 0 < \rho < \infty \tag{10.1-8}$$

be the order of growth of μ on

$$\{(s, z_1) \in \mathbb{C}^2 \mid |s| = t, \ |z_1| = kt, 0 < t < \infty\}, \qquad k > 0.$$

Then we have

$$(10.1\text{-}9) \qquad \rho = \rho_0, \qquad \rho_0 = \limsup_{m+n\to\infty} \frac{(m+n)\log(m+n)}{\log(1/C_{mn}) + n\log(1/k)}.$$

Proof. We show first that $\rho \geq \rho_0$ and then $\rho \leq \rho_0$. By Cauchy's inequality,

$$C_{mn} \leq \frac{\mu(|s|, |z_1|)}{|s|^m |z_1|^n}.$$

It follows that

$$C_{mn} k^n |s|^{m+n} \leq \mu(|s|, k|s|).$$

We now assume that k_0 is chosen so that for $|s| \geq S_0$ (S_0 sufficiently large) we have

$$(10.1\text{-}10) \qquad \mu(|s|, k|s|) \leq \exp(|s|^{k_0}).$$

This implies

$$C_{mn} k^n \leq \exp(|s|^{k_0})/|s|^{m+n} \qquad \text{for} \qquad |s| \geq S_0.$$

From this, for $|s| = [(m+n)/k_0]^{1/k_0}$ with sufficiently large $m+n$ we obtain

$$(10.1\text{-}11) \qquad C_{mn}^{1/(m+n)} k^{n/(m+n)} \leq \left(\frac{ek_0}{m+n}\right)^{1/k_0}.$$

By taking logarithms,

$$\frac{1}{m+n}[\log(1/C_{mn}) + n\log(1/k)] \geq \frac{1}{k_0}[\log(m+n) - \log k_0 - 1].$$

For sufficiently large $m+n$ we can neglect $\log k_0 + 1$, so that we obtain

$$(10.1\text{-}12) \qquad k_0 \geq \frac{(m+n)\log(m+n)}{\log(1/C_{mn}) + n\log(1/k)}.$$

Because of (10.1-8, 10) this yields $\rho \geq \rho_0$.

Let us now prove that $\rho \leq \rho_0$ by deriving a contradiction from the assumption that $\rho_0 < \kappa < \rho$. We write

$$(10.1\text{-}13) \qquad n_0(s) = 2^\kappa |s|^\kappa e\kappa.$$

Hence

$$\left(\frac{e\kappa}{m+n}\right)^{1/\kappa}|s| \le \frac{1}{2} \qquad \text{for } m+n \ge n_0(s). \tag{10.1-14}$$

Using $\rho_0 < \kappa$, we can work backwards from (10.1-12) to (10.1-11), with k_0 being replaced by κ and $m+n$ being sufficiently large. Therefore, using (10.1-6, 7, 11, 14), we have

$$\begin{aligned}\mu(|s|, k|s|) &\le \sum_{m,n=0}^{\infty} C_{mn} k^n |s|^{m+n} \\ &\le \sum_{m+n \le n_0(s)} C_{mn} k^n |s|^{m+n} + \frac{1}{2^{n_0(s)}}\left[n_0(s) + 3\right].\end{aligned}$$

For sufficiently large s and hence sufficiently large $n_0(s)$ we obtain

$$\mu(|s|, k|s|) \le \left[n_0(s) + 2\right]^2 c_0(s) \tag{10.1-15}$$

where

$$c_0(s) = \max_{m+n \le n_0(s)} \left(C_{mn} k^n |s|^{m+n}\right).$$

Again using (10.1-11), with κ instead of k_0, we conclude that

$$c_0(s) \le \max_{m+n \le n_0(s)} \left[\left(\frac{e\kappa}{m+n}\right)^{(m+n)/\kappa} |s|^{m+n}\right] \le \exp\left(|s|^\kappa\right);$$

here the second inequality follows from the fact that the maximum of the expression in the brackets $[\cdots]$ (as $m+n$ varies) is attained for $m+n = \kappa|s|^\kappa$. Together with (10.1-13, 15) this implies

$$\mu(|s|, k|s|) \le 2^{2\kappa}|s|^{2\kappa}e^2\kappa^2 \exp\left(|s|^\kappa\right)$$

for sufficiently large $|s|$. This, however, yields $\rho \le \kappa$, contrary to our previous assumption. Hence $\rho \le \rho_0$ must hold, and the proof is complete. ■

10.1-3 Remark

Let $\rho = \rho(k)$ be the order of growth of $\mu(|\cdot|, k|\cdot|)$, where $0 < k_1 \le k \le k_2 < \infty$, $\rho < \infty$. Then ρ is independent of k.

Proof. This follows immediately from (10.1-9) if we divide both the numerator and denominator by n; since $(1 + (m/n))\log(m+n)$ approaches infinity as

$m + n$ does, and similarly for $n^{-1}\log(1/C_{mn})$, we can neglect the term $\log(1/k)$. ■

COEFFICIENTS OF $v_{(s)}$ AND ORDER OF GROWTH OF SOLUTIONS w

Such relations between the entire functions $v_{(s)}$ obtained from (10.1-4) and solutions of $Lw = 0$ can be established as follows. Suppose that w is given by the series

$$w(z_1, z_2) = \sum_{\nu, n=0}^{\infty} w_{\nu n} z_1^{\nu} z_2^{n}. \tag{10.1-16}$$

Then we can express the coefficients C_{mn}, hence also ρ, in terms of the $w_{\nu 0}$'s with $\nu \in \mathbb{N}_0$. Indeed, we have:

10.1-4 Theorem

Suppose that the operator L satisfies the conditions (i) *and* (ii). *Let* $w = w_1 = T_1^{\times} f_1$ *with entire* f_1 *be the Type I representation by the Bergman operator* $T_1^{\times}$ *with kernel* $\tilde{k}_1^{\times}$ *and reference point* (0,0). *Furthermore, let* $\tilde{k}_1^{\times}$ *be of the first kind and let*

$$v_{(s)}(z_1) = \int_{-1}^{1} g(z_1, s, t) f_1\left(\tfrac{1}{2} z_1 \tau\right) \tau^{-1/2}\, dt \tag{10.1-17}$$

resulting from (10.1-4). *Here g is assumed to have the development*

$$g(z_1, s, t) = \sum_{m, \mu=0}^{\infty} g_{m\mu}(t) s^{m} z_1^{\mu}. \tag{10.1-18}$$

Then the series (10.1-5) *of* $v_{(s)}$ *has the coefficients*

$$c_{mn} = \sum_{\nu=0}^{n} \frac{4^{\nu}(\nu!)^2}{(2\nu)!\pi} w_{\nu 0} \int_{-1}^{1} g_{m, n-\nu}(t) \tau^{\nu - 1/2}\, dt \tag{10.1-19}$$

where in (10.1-16),

$$w_{\nu 0} = \frac{\pi(2\nu)!}{8^{\nu}(\nu!)^2} \gamma_{1\nu}, \qquad \gamma_{1\nu} = \frac{f_1^{(\nu)}(0)}{\nu!}, \qquad \nu \in \mathbb{N}_0. \tag{10.1-20}$$

Furthermore, for sufficiently large $|s|$,

$$\max_{|z_1|=|s|} |w(z_1, \bar{z}_1)| \le \exp\left(|s|^{\rho+\varepsilon}\right), \qquad \varepsilon > 0 \tag{10.1-21}$$

where

$$\rho = \limsup_{m+n\to\infty} \frac{(m+n)\log(m+n)}{\log|1/c_{mn}|}. \tag{10.1-22}$$

Proof. In Sec. 2.6 we obtained the coefficients of f_1 in terms of those of w (see Corollary 2.6-3). We substitute those expressions as well as (10.1-18) into (10.1-17), obtaining (10.1-19, 20). We now consider V in Lemma 10.1-2 with $C_{mn} = |c_{mn}|$. By this lemma and its proof, for sufficiently large $S > 0$ (depending on $\varepsilon > 0$), we obtain

$$\max_{|s|=|z_1|=S} \left| \sum_{m,n=0}^{\infty} |c_{mn}| s^m z_1^n \right| \le \exp\left(S^{\rho+\varepsilon}\right)$$

with ρ as in (10.1-22). Hence we obtain (10.1-21) if we remember that

$$\max_{|z_1|=S} |w(z_1, \bar{z}_1)| = \max_{|z_1|=S} |v_{(S)}(z_1)| = \max_{|z_1|=S} \left| \sum_{m,n=0}^{\infty} c_{mn} S^m z_1^n \right|$$

$$\le \max_{|s|=|z_1|=S} \left| \sum_{m,n=0}^{\infty} |c_{mn}| s^m z_1^n \right|. \quad \blacksquare$$

10.1-5 Remark

If the kernel $\tilde{k}_1^{\times}$ is not of the first kind, the result (10.1-21) still holds true, but then the coefficients of f_1 and w with $z_2 = 0$ are related in a more complicated way.

DISTRIBUTION OF VALUES

Let us now turn to the study of the *distribution of a-points of* w with $z_2 = \bar{z}_1$. More precisely, we shall first derive upper bounds in terms of the coefficients $w_{\nu 0}$ of w, for functions characteristic of the distribution of those points. In this connection we regard $v_{(s)}$ as a function of a single complex variable z_1 for some *fixed* (positive real) value of s.

For an analytic function f of a complex variable, the number of a-points (counting multiplicity) in a domain bounded by a finite set of Jordan curves C

on which there are no a-points is

$$d_f = \frac{1}{2\pi i}\int_C d\log\left[f(z_1) - a\right].$$

Analogously we may consider

$$\check{d} = \frac{1}{2\pi i}\int_C d\log\left[w(z_1, \bar{z}_1) - a\right] \tag{10.1-23}$$

for characterizing the domain with respect to the value a of the function

$$\check{w} = w|_{z_2 = \bar{z}_1} \tag{10.1-24}$$

if $\check{w}$ is continuously differentiable along the boundary and continuous in the domain. [By assumption, $\hat{w}(x_1, x_2) \equiv \check{w}(z_1, \bar{z}_1)$ holds for $z_1 = x_1 + ix_2$.]

The following properties can be proved if $\hat{w}$ is continuously differentiable in x_1 and x_2.

(A) $\check{d}$ is always an integer, either negative, positive, or zero. For functions analytic in the domain, always $\check{d} \geq 0$.

(B) If $\check{d}_1$ and $\check{d}_2$ correspond to two disjoint domains (for the same function and value a), their sum corresponds to the union of these domains.

(C) If $\check{d} \neq 0$, then $\check{w}$ assumes the value a at least once in the domain.

(D) $\check{d}$ is a characteristic of the a-points in the domain and does not depend on the domain in the sense that it remains unchanged under deformations of the domain during which no a-points slip in or out. Thus every set of a-points which can be isolated from other a-points by a domain with Jordan boundary has an *index*. In particular, an isolated a-point has an index, and for an analytic function of a complex variable, the index of an a-point is the multiplicity of the value a at that point.

We now define

$$n\left[s, \left(w(z_1, \bar{z}_1) - a\right)^{-1}\right] = \frac{1}{2\pi i}\int_{|z_1| = s} d\log\left[w(z_1, \bar{z}_1) - a\right]. \tag{10.1-25}$$

In classical function theory, if f is analytic in the disk $|z_1| \leq s$ and $f \neq a$, then $n[s, (f(z) - a)^{-1}]$ is defined to be the number of a-points (counting multiplicity) in that disk. In this case we shall always write $\tilde{n}[\cdots]$ instead of $n[\cdots]$. Definition (10.1-25) coincides with the present one if f has no a-points on $|z_1| = s$.

For the class of functions under consideration, if $n[s,(w(z_1, \bar{z}_1) - a)^{-1}]$ is defined, we have by Rouché's theorem

$$n\left[s,\left(w(z_1, \bar{z}_1) - a\right)^{-1}\right] = n\left[s,\left(v_{(s)}(z_1) - a\right)^{-1}\right] = \tilde{n}\left[s,\left(v_{(s)}(z_1) - a\right)^{-1}\right]. \tag{10.1-26}$$

10.1-6 Lemma

If $w \in K(\tilde{k}_1^{\times})$, then

$$n\left[s,\left(w(z_1, \bar{z}_1) - a\right)^{-1}\right] \geq 0$$

provided that it exists.

Proof. The proof is an immediate consequence of the definition (10.1-25), together with (10.1-26), which holds for $w \in K(\tilde{k}_1^{\times})$, and

$$\tilde{n}\left[s,\left(v_{(s)}(z_1) - a\right)^{-1}\right] \geq 0. \quad \blacksquare$$

APPLICATION

Let us illustrate these facts by Example 10.1-1 with $c = 1$. For w_{1m} we have

$$v_{(s)}(z_1) = \frac{1}{s^m} J_m(s) z_1^m.$$

If $a = 0$ and $J_m(s) \neq 0$, then $n[s, w_{1m}(z_1, \bar{z}_1)^{-1}]$ is defined and, by (10.1-26),

$$n\left[s, w_{1m}(z_1, \bar{z}_1)^{-1}\right] = \tilde{n}\left[s, \left(s^{-m} J_m(s) z_1^m\right)^{-1}\right] = m.$$

This means that the index of the point $a = 0$ at the origin is m. Now, $\check{w}_{1m}$ is also zero on the circles of radius s_ν, where $J_m(s_\nu) = 0$. We claim that the index of each of these circles is zero. For, if we take the annulus bounded by $|z_1| = s_\nu - \varepsilon$ and $|z_1| = s_\nu + \varepsilon$ (with sufficiently small $\varepsilon > 0$) as a domain isolating the circle $|z_1| = s_\nu$ from all the other zero points and take into account the orientation of the bounding circles of that annulus, we have indeed

$$n\left[s + \varepsilon, w_{1m}(z_1, \bar{z}_1)^{-1}\right] - n\left[s - \varepsilon, w_{1m}(z_1, \bar{z}_1)^{-1}\right] = 0. \tag{10.1-27}$$

In the case $a \neq 0$ we obtain

$$n\left[s,\left(w_{1m}(z_1, \bar{z}_1) - a\right)^{-1}\right] = \begin{cases} m & \text{if } |a| < |J_m(s)| \\ 0 & \text{if } |a| > |J_m(s)|. \end{cases} \tag{10.1-28}$$

On each circle with $|a| = |J_m(s)|$, there are m a-points, each of which has the index $+1$ or -1.

PROXIMITY FUNCTION

Now let us continue with our task of generalizing definitions and results from complex function theory. Consider a meromorphic function f of a single complex variable $z_1 = se^{i\phi}$. For such a function, the expression

$$m[s, f(z_1)] = \frac{1}{2\pi}\int_0^{2\pi} \log^+ |f(se^{i\phi})|\, d\phi$$

is defined. Here, as usual, for $x \geq 0$,

$$\log^+ x = \begin{cases} 0 & \text{if } 0 \leq x < 1 \\ \log x & \text{if } x \geq 1. \end{cases}$$

The function m is called *proximity function* (*Schmiegungsfunktion*). It is some kind of "average magnitude" of $\log|f|$ on arcs of $|z_1| = s$, where $|f|$ is large. Analogously, we can define for a function g depending on z_1 and $\bar{z}_1$,

$$m[s, g(z_1, \bar{z}_1)] = \frac{1}{2\pi}\int_0^{2\pi} \log^+ |g(se^{i\phi}, se^{-i\phi})|\, d\phi \tag{10.1-29}$$

provided that the integral exists. If $g(z_1, \bar{z}_1) = f(z_1)$ on $|z_1| = s$, where f is a meromorphic function, then $m[s, g(z_1, \bar{z}_1)]$ also exists. Consequently, $m[s, w(z_1, \bar{z}_1)]$ exists. Furthermore, if $w(z_1, \bar{z}_1) \not\equiv a$ on $|z_1| = s$, then $m[s, (w(z_1, \bar{z}_1) - a)^{-1}]$ also exists, and for functions $w \in K(\tilde{k}_1^{\times})$ we obtain

$$\begin{aligned} m[s, w(z_1, \bar{z}_1)] &= m[s, v_{(s)}(z_1)] \\ m[s, (w(z_1, \bar{z}_1) - a)^{-1}] &= m[s\, (v_{(s)}(z_1) - a)^{-1}]. \end{aligned} \tag{10.1-30}$$

Here we may equally well regard these two quantities as characterizations of the growth of $\check{w}$ and of the behavior of $\check{w}$ near the circle $|z_1| = s$, respectively.

GENERALIZATION OF NEVANLINNA'S FIRST THEOREM

We shall now derive a basic theorem on "entire solutions." In this result, s will be an element of a set of the form

$$M_1(a, c) = \left\{ s \geq 0 \;\middle|\; \left| \int_{-1}^{1} g(0, s, t)\tau^{-1/2}\, dt - a \right| \geq c \right\} \tag{10.1-31}$$

with g given by (10.1-3) and c any positive constant.

10.1-7 Theorem (Bounds for n and m)

Suppose that L satisfies the conditions (i) *and* (ii). *Let* $w \in K(\tilde{k}_1^{\times})$ *be a solution of* $Lw = 0$ *such that* $\check{w} = w|_{z_2 = \bar{z}_1}$ *given by*

$$\check{w}(z_1, \bar{z}_1) = \sum_{m, n=0}^{\infty} w_{mn} z_1^m \bar{z}_1^n$$
$$= \int_{-1}^{1} \tilde{k}_1^{\times}(z_1, \bar{z}_1, t) f_1\left(\tfrac{1}{2} z_1 \tau\right) \tau^{-1/2} \, dt$$

yields a function $\hat{w}$ *which is an entire function of* x_1 *and* x_2 *and has only isolated a-points in the real* $x_1 x_2$*-plane. Here,* $z_1 = x_1 + i x_2$; *cf.* (10.1-24). *Moreover, let* $f_1(0) \neq 0$. *Then for every* $\varepsilon > 0$ *and sufficiently large* $s \in M_1(a/f_1(0), c)$ *with* $c > 0$ *we have*

$$n\left[s, \left(w(z_1, \bar{z}_1) - a\right)^{-1}\right] \leq c_0 s^{\rho + \varepsilon} \tag{10.1-32}$$

and

$$m\left[s, \left(w(z_1, \bar{z}_1) - a\right)^{-1}\right] \leq c_0 s^{\rho + \varepsilon}, \qquad c_0 = const > 0. \tag{10.1-33}$$

Here, ρ *is defined in* (10.1-22). [*Representation* (10.1-20) *for the coefficients* w_{mn}, *however, only holds if, in addition,* $\tilde{k}_1^{\times}$ *is assumed to be of the first kind.*]

Proof. By (ii) there exists a function g satisfying (10.1-3) and such that the function

$$v_{(s)}(z_1) = \int_{-1}^{1} g(z_1, s, t) f_1\left(\tfrac{1}{2} z_1 \tau\right) \tau^{-1/2} \, dt$$

equals $\check{w}(z_1, \bar{z}_1)$ on $|z_1| = s$. To this function we apply ***Nevanlinna's First Fundamental Theorem*** (cf. Hayman [1964], p. 5), which states that

$$\begin{aligned} & m\left[r, \left(v_{(s)}(z_1) - a\right)^{-1}\right] + N\left[r, \left(v_{(s)}(z_1) - a\right)^{-1}\right] \\ & \quad = T\left[r, v_{(s)}(z_1)\right] - \log\left|v_{(s)}(0) - a\right| + \varepsilon(a, r) \end{aligned} \tag{10.1-34}$$

where $|\varepsilon(a, r)| \leq \log^+ |a| + \log 2$. By the definition of the *Nevanlinna's characteristic function* we have

$$T\left[r, v_{(s)}(z_1)\right] = m\left[r, v_{(s)}(z_1)\right] + N\left[r, v_{(s)}(z_1)\right]. \tag{10.1-35}$$

Since $v_{(s)}$, as an entire function, has no poles in $|z_1| = r$,

$$N\left[r, v_{(s)}(z_1)\right] = 0.$$

Now for $s \in M_1(a/f_1(0), c)$ we obtain from (10.1-31)

$$\log^+ \left| \frac{1}{v_s(0) - a} \right| \leq \log^+ \frac{|f_1(0)|}{c}.$$

From this and (10.1-34) we conclude that

$$\begin{aligned} m\left[r, \left(v_{(s)}(z_1) - a\right)^{-1}\right] + N\left[r, \left(v_{(s)}(z_1) - a\right)^{-1}\right] + \log^+ |v_{(s)}(0) - a| \\ \leq m\left[r, v_{(s)}(z_1)\right] + \log^+ \left| \frac{1}{v_{(s)}(0) - a} \right| + |\varepsilon(a, r)| \\ \leq m\left[r, v_{(s)}(z_1)\right] + \log^+ \frac{|f_1(0)|}{c} + \log^+ |a| + \log 2. \end{aligned} \tag{10.1-36}$$

Next we substitute $r = ks$, $k > 0$. Now for sufficiently large $s \in M_1(a/f_1(0), c)$ we conclude from Lemma 10.1-2 that

$$\begin{aligned} m\left[ks, v_{(s)}(z_1)\right] &\leq \log^+ \max_{|z_1| = ks} |v_{(s)}(z_1)| \\ &\leq \log^+ \exp\left(s^{\rho_0 + \varepsilon}\right) = s^{\rho_0 + \varepsilon} = s^{\rho + \varepsilon}, \qquad \varepsilon > 0. \end{aligned} \tag{10.1-37}$$

Here, ρ_0 is given in (10.1-9) with $C_{mn} = |c_{mn}|$ and c_{mn} as in (10.1-19) [but, in general, without (10-1.20)]; furthermore, ρ is defined in (10.1-22). The last equality holds because of Remark 10.1-3. Hence by (10.1-36) it follows that

$$N\left[ks, \left(v_{(s)}(z_1) - a\right)^{-1}\right] \leq s^{\rho + \varepsilon} + \tilde{c}_0 \leq (\tilde{c}_0 + 1)s^{\rho + \varepsilon} \tag{10.1-38}$$

for $\varepsilon > 0$ and sufficiently large $s \in M_1(a/f_1(0), c)$; here the constant $\tilde{c}_0 > 0$ depends on $f_1(0)$, a, and c. Now for $k > 3$,

$$\tilde{n}\left[s, \left(v_{(s)}(z_1) - a\right)^{-1}\right] \leq \frac{N\left[ks, \left(v_{(s)}(z_1) - a\right)^{-1}\right]}{\log k} \leq N\left[ks, \left(v_{(s)}(z_1) - a\right)^{-1}\right].$$

From this and (10.1-38) the desired result (10.1-32) follows. The other inequality is now obtained immediately. Indeed, by the idea that gave (10.1-38), from (10.1-36, 37) we conclude that for sufficiently large $s \in M_1(a/f_1(0), c)$,

$$m\left[ks, \left(v_{(s)}(z_1) - a\right)^{-1}\right] \leq c_0 s^{\rho + \varepsilon}$$

with positive constant c_0. This and the second formula in (10.1-30) yield (10.1-33). ■

10.2 GENERALIZED NEVANLINNA'S SECOND THEOREM

In the preceding section we derived upper bounds for $n[s, (w(z_1, \bar{z}_1) - a)^{-1}]$ and $m[s, (w(z_1, \bar{z}_1) - a)^{-1}]$ from the series developments of $w|_{z_2=0}$ and the kernel $\tilde{k}_1^{\times}$. These bounds were obtained by an application of Nevanlinna's First Fundamental Theorem. Using Nevanlinna's Second Fundamental Theorem, in the present section we shall obtain an upper bound for $m[s, w(z_1, \bar{z}_1)]$. This bound will depend on $n[s, (w(z_1, \bar{z}_1) - a_\lambda)^{-1}]$ for q values a_λ. Actually, we shall utilize a modification of Nevanlinna's theorem by Chernoff ([1946], pp. 475–478), whose proof is quite similar to that of the original theorem. In our theorem, Case (a) will mean that $\gamma_0 \neq 0$, and Case (b) that $\gamma_0 = 0$.

10.2-1 Modified Nevanlinna's Second Theorem

Let f_a be an entire function of z_1 with series

$$f_a(z_1) = \gamma_0 + \sum_{\nu=k}^{\infty} \gamma_\nu z_1^\nu, \quad k \geq 1; \quad \gamma_0 \neq 0; \quad \gamma_k \neq 0,$$

and set

$$f_b = f_a - f_a(0) = f_a - \gamma_0 .$$

To simplify notation, we shall write f for f_a in Case (a) *and for f_b in Case* (b). *Furthermore, let $a_1, \cdots, a_q$ $(q \geq 2)$ be distinct finite complex numbers such that for some $r_1 > 0$,*

$$\tilde{n}\left[r_1, (f(z_1) - a_\lambda)^{-1}\right] = 0, \qquad \lambda = 1, \cdots, q.$$

Finally, let $r_2 > r_1$. Then for $r \in (r_1, r_2)$ we have

(10.2-1)

$$(q-1) m[r, f(z_1)] \leq \left(\log \frac{r}{r_1}\right) \sum_{\lambda=1}^{q} \tilde{n}\left[r, (f(z_1) - a_\lambda)^{-1}\right] - N_1(r) + S_1(r)$$

where

$$N_1(r) \geq 0 \qquad \text{for } r \geq 1$$

and in Case (a):

$$S_1(r) = S_{1a}(r) = c_a + 8\log^+ r_2 + 6\log^+ \frac{1}{r_2 - r} + 4\log^+ \frac{1}{r}$$

(10.2-2)
$$+8\log^+ m[r_2, f(z_1)] + \log^+ \frac{1}{k|\gamma_k|} + \sum_{\lambda=1}^{q} \log|\gamma_0 - a_\lambda|$$

$$+6q\log^+ \max\left\{\frac{1}{|\gamma_0|}, \frac{1}{|\gamma_0 - a_\lambda|}, \lambda = 1, \cdots, q\right\}$$

with

$$c_a = c_a(q, a_1, \cdots, a_q)$$

(10.2-3)
$$= 56 + (q+4)\log^+ \max\{|a_\lambda|, \lambda = 1, \cdots, q\} + q\log 2$$

$$+2q\log[2q/\min\{1, |a_\kappa - a_\lambda|; \kappa, \lambda = 1, \cdots, q, \kappa \neq \lambda\}];$$

and in Case (b):

$$S_1(r) = S_{1b}(r) = c_b + 8\log^+ r_2 + 6\log^+ \frac{1}{r_2 - r} + 9\log^+ \frac{1}{r}$$

(10.2-4)
$$+8\log^+ m[r_2, f(z_1)] + \log^+ \frac{1}{k|\gamma_k|} + 5\log k$$

$$+6q\log^+ \max\left\{\frac{1}{|\gamma_k|}, \frac{1}{|a_\lambda|}, \lambda = 1, \cdots, q\right\}$$

with

(10.2-5)
$$c_b = c_b(q, a_1, \cdots, a_q) = c_a + 10 + q\log^+ \max\{|a_\lambda|, \lambda = 1, \cdots, q\}.$$

As in the preceding section, let $v_{(s)}$ be the entire function given by [cf. (10.1-17)]

(10.2-6)
$$v_{(s)}(z_1) = \int_{-1}^{1} g(z_1, s, t) f_1\left(\tfrac{1}{2}z_1\tau\right)\tau^{-1/2}\,dt.$$

We assume that L and $\tilde{k}_1^{\times}$ satisfy the conditions (i) and (ii) of Sec. 10.1 and that $f_1 \in C^\omega(\mathbb{C})$. Then

(10.2-7)
$$v_{(s)}(z_1) = \tilde{\gamma}_0(s) + \sum_{\gamma=k}^{\infty} \tilde{\gamma}_\nu(s) z_1^\nu, \qquad k \geq 1$$

where the $\tilde{\gamma}_\nu$'s ($\nu = 0, k, k+1, \cdots$) are entire functions of s and $\tilde{\gamma}_k$ is not identically zero. In analogy to the assumptions in Theorem 10.2-1, we shall refer to **Case (a)** if $\tilde{\gamma}_0(s) \not\equiv 0$ and to **Case (b)** otherwise. These assumptions imply that $v_{(s)}$ is not a constant for each fixed s which is not a zero of $g(z_1, \cdot, t)$. Since an entire function which is not identically zero can vanish at most at countably many points, we may apply Theorem 10.2-1 to $f = v_{(s)}$ whenever s is not such a point. We then obtain

(10.2-8)
$$(q-1)m\left[r, v_{(s)}(z_1)\right] \leq \left(\log \frac{r}{r_1}\right) \sum_{\lambda=1}^{q} \tilde{n}\left[r, \left(v_{(s)}(z_1) - a_\lambda\right)^{-1}\right] + S(r, s)$$

where $r_1 = r_1(s)$, $r > r_1(s)$, and $S(r, s)$ denotes $S_1(r) - N_1(r)$ corresponding to $f = v_{(s)}$.

SETS ASSOCIATED WITH $v_{(s)}$

For inequality (10.2-8) to be useful, it is important to keep the right-hand side bounded. Accordingly, we find it convenient to introduce five types of sets:

10.2-2 Definition

For $v_{(s)}$ given by (10.2-6, 7) and $c > 0$, let

$$M_1(a, c) = \left\{s \geq 0 \mid |v_{(s)}(0) - a| \geq c\right\}$$

$$= \left\{s \geq 0 \mid |\tilde{\gamma}_0(s) - a| \geq c\right\} \qquad \text{[cf. (10.1-31)]}$$

$$M_2(a, c) = \left\{s \geq 0 \mid |v_{(s)}(0) - a| \leq c\right\}$$

$$= \left\{s \geq 0 \mid |\tilde{\gamma}_0(s) - a| \leq c\right\}$$

$$M_3 = \begin{cases} \left\{s \geq 0 \mid \tilde{\gamma}_0(s) \neq 0, \tilde{\gamma}_k(s) \neq 0\right\} = M_{3a} & \text{in Case (a)} \\ \left\{s \geq 0 \mid \tilde{\gamma}_k(s) \neq 0\right\} = M_{3b} & \text{in Case (b)} \end{cases}$$

$$M_4(a, r_1) = \left\{s \geq 0 \mid v_{(s)}(z_1) \neq a \text{ for } |z_1| \leq r_1\right\}$$

$$M_5(0, c) = \left\{s \geq 0 \mid |\tilde{\gamma}_k(s)| \geq c\right\}.$$

Using these definitions, we obtain from Theorem 10.2-1 the following:

10.2-3 Lemma

If $f = v_{(s)}$ *and* $a_1, \cdots, a_q$ $(q \geq 2)$ *as in Theorem* 10.2-1, *and* $0 < r_1 < 1$, $r_2 > 1$, *then for* $r \in [1, r_2)$ *we have*:

(a) *In case* (a), *for*

$$s \in M_1(0, c) \cap \bigcap_{\lambda=1}^{q} M_1(a_\lambda, c) \cap M_5(0, c)$$

there exists a positive constant $\tilde{c}_a = \tilde{c}_a(k, c, q, a_1, \cdots, a_q)$ *such that*

$$\begin{aligned} S(r, s) &\leq \tilde{c}_a + 8 \log^+ r_2 + 6 \log^+ \frac{1}{r_2 - r} \\ &\quad + 8 \log^+ m\left[r_2, v_{(s)}(z_1)\right] + \sum_{\lambda=1}^{q} \log |\tilde{\gamma}_0(s) - a_\lambda|. \end{aligned} \tag{10.2-9}$$

(b) *In Case* (b), *for*

$$s \in \bigcap_{\lambda=1}^{q} M_1(a_\lambda, c) \cap M_5(0, c)$$

there exists a positive constant $\tilde{c}_b = \tilde{c}_b(k, c, q, a_1, \cdots, a_q)$ *such that*

$$S(r, s) \leq \tilde{c}_b + 8 \log^+ r_2 + 6 \log^+ \frac{1}{r_2 - r} + 8 \log^+ m\left[r_2, v_{(s)}(z_1)\right]. \tag{10.2-10}$$

Proof. We prove (a). By definition, $s \in M_1(0, c)$ implies $|\tilde{\gamma}_0(s)| \geq c$, hence $1/|\tilde{\gamma}_0(s)| \leq 1/c$. Similarly, from $s \in \bigcap_{\lambda=1}^{q} M_1(a_\lambda, c)$ it follows that $1/|\tilde{\gamma}_0(s) - a_\lambda| \leq 1/c$ for $\lambda = 1, \cdots, q$. Consequently,

$$\max\left\{1/|\tilde{\gamma}_0(s)|, 1/|\tilde{\gamma}_0(s) - a_\lambda|, \lambda = 1, \cdots, q\right\} \leq \frac{1}{c}.$$

Moreover, $s \in M_5(0, c)$ gives $1/|\tilde{\gamma}_k(s)| \leq 1/c$. Also, from $s \in M_1(0, c) \cap M_5(0, c)$ we obtain $s \in M_{3a}$, and Theorem 10.2-1 is applicable. Observing these inequalities and the definition of $S(r, s)$, and using $N_1(r) \geq 0$ for $r \geq 1$ [hence $S(r, s) \leq S_1(r)$], from (10.2-2, 3) we obtain (10.2-9), where $\tilde{c}_a$ takes the form

$$(10.2\text{-}11)\qquad \tilde{c}_a = c_a(q, a_1, \cdots, a_q) + \log^+ \frac{1}{kc} + 6q \log^+ \frac{1}{c}.$$

This shows that $\tilde{c}_a$ is a positive constant, depending on k, c, and $a_1, \cdots, a_q$. The proof of (b) is similar; Theorem 10.2-1 is again applicable, and (10.2-4, 5) implies (10.2-10) with

$$(10.2\text{-}12)\quad \tilde{c}_b = c_b(q, a_1, \cdots, a_q) + \log^+ \frac{1}{kc} + 5 \log k + 6q \log^+ \frac{1}{c}. \quad \blacksquare$$

APPLICATION OF THEOREM 10.2-1

We recall (10.1-26, 30). If we impose the additional condition that $s \in \bigcap_{\lambda=1}^{q} M_4(a_\lambda, \tilde{r})$, where $\tilde{r} > 0$, then $v_{(s)}(z_1) \neq a_\lambda$ $(\lambda = 1, \cdots, q)$ for $|z_1| \leq \tilde{r}$. We thus have

$$\tilde{n}\left[\tilde{r}, \left(v_{(s)}(z_1) - a_\lambda\right)^{-1}\right] = 0, \qquad \lambda = 1, \cdots, q$$

and $r_1(s) \geq \tilde{r}$. We now apply (10.2-8) and Lemma 10.2-3 to the case $r = s$. This yields the following

10.2-4 Theorem

Let $\tilde{r} > 0$, $s \geq 1$, and $r_2 > s > \tilde{r}$. Suppose that L satisfies the conditions (i) *and* (ii) *of Sec.* 10.1. *Furthermore, let w be a solution of $Lw = 0$, $\check{w} = w|_{z_2 = \bar{z}_1}$, such that $\check{w}$ coincides with $v_{(s)}$ [cf. (10.2-6, 7)] for $|z_1| = s$. Assume that no a_λ-points of $\check{w}$ are on $|z_1| = s$, $\lambda = 1, \cdots, q$.*

(a) *In Case* (a), *if*

$$s \in M_1(0, c) \cap M_5(0, c) \cap \bigcap_{\lambda=1}^{q} \left[M_1(a_\lambda, c) \cap M_4(a_\lambda, \tilde{r})\right],$$

then we have

$$\begin{aligned}(q - 1)m[s, w(z_1, \bar{z}_1)] \leq \left(\log \frac{s}{\tilde{r}}\right) \sum_{\lambda=1}^{q} n\left[s, (w(z_1, \bar{z}_1) - a)^{-1}\right] \\ (10.2\text{-}13) \qquad + \tilde{c}_a(k, c, q, a_1, \cdots, a_q) + 8 \log^+ r_2 + 6 \log^+ \frac{1}{r_2 - s} \\ + 8 \log^+ m\left[r_2, v_{(s)}(z_1)\right] + \sum_{\lambda=1}^{q} \log |\tilde{\gamma}_0(s) - a_\lambda|\end{aligned}$$

with $\tilde{\gamma}_0(s) = v_{(s)}(0)$.

(b) *In Case* (b), *if*

$$s \in M_5(0, c) \cap \bigcap_{\lambda=1}^{q} [M_1(a_\lambda, c) \cap M_4(a_\lambda, \tilde{r})]$$

then we have

$$(q-1)m[s, w(z_1, \bar{z}_1)] \le \left(\log \frac{s}{\tilde{r}}\right) \sum_{\lambda=1}^{q} n\left[s, (w(z_1, \bar{z}_1) - a_\lambda)^{-1}\right]$$

$$(10.2\text{-}14) \qquad + \tilde{c}_b(k, c, q, a_1, \cdots, a_q) + 8 \log^+ r_2 + 6 \log^+ \frac{1}{r_2 - s}$$

$$+ 8 \log^+ m\left[r_2, v_{(s)}(z_1)\right].$$

Here, $\tilde{c}_a$ *and* $\tilde{c}_b$ *are positive constants as in Lemma* 10.2-3 [*cf*. (10.2-11, 12)].

Note that in these results, the only restriction on r_2 is $r_2 > s$. These inequalities are basic. A special case of great importance arises when $r_2 = \sigma s$ with constant $\sigma > 1$. Then the term $1/(r_2 - s)$ is bounded; also $\log^+ r_2 \le \log^+ \sigma + \log^+ s$ and there is a constant $c_0 = c_0(\sigma) > 0$ such that

$$(10.2\text{-}15) \quad 8 \log^+ r_2 + 6 \log^+ \frac{1}{r_2 - s} \le c_0 + 8 \log^+ s \qquad \text{if } r_2 = \sigma s, \qquad \sigma > 1.$$

Here c_0 may be absorbed into $\tilde{c}_a$ or $\tilde{c}_b$, respectively.

It is important to emphasize that the inequalities (10.2-13, 14) do *not* rely on the property that w is a solution of $Lw = 0$, but only on the assumption that $w(z_1, \bar{z}_1) = v_{(s)}(z_1)$ on $|z_1| = s$, where $v_{(s)}$ is entire in z_1 and in s.

AN EXAMPLE RELATING TO BERGMAN OPERATORS

Our next task is the translation of the hypotheses of Theorem 10.2-4 into conditions for functions which are central to the Bergman operator theory, namely, $\tilde{k}_1^\times$ and g, both being related to the operator L of the partial differential equation. To collect ideas and gain experience, let us first observe what happens for a relatively simple special equation.

10.2-5 Example

We consider the equation

$$Lw = D_1 D_2 w - \tfrac{1}{4} w = 0.$$

From Sec. 2.1 we know that it can be obtained from

$$\Delta \hat{w} - \hat{w} = 0.$$

Example 10.1-1 with $c = i$ yields the Bergman kernel of the first kind given by

$$\tilde{k}_1(z_1, \bar{z}_1, t) = \cos(irt) = \cosh(rt), \qquad r = (z_1\bar{z}_1)^{1/2}.$$

Let f_1 be an entire function with the series development

$$f_1(\xi_1) = \gamma_{10} + \sum_{\nu=k}^{\infty} \gamma_{1\nu}\xi_1^{\nu}, \qquad \gamma_{10} \neq 0; \quad \gamma_{1k} \neq 0.$$

Then there exists an $\tilde{r}$ such that for $|\xi_1| \leq \tilde{r}$ we have

$$f_1(\xi_1) = \gamma_{10}\left[1 + \tilde{\varepsilon}_1 e^{i\tilde{\delta}_1}\right] \quad \text{where} \quad |\tilde{\varepsilon}_1| < \tfrac{1}{2}.$$

We thus obtain

$$v_{(s)}(z_1) = \tilde{\gamma}_0(s) + \sum_{\nu=k}^{\infty} \tilde{\gamma}_{\nu}(s) z_1^{\nu}$$

where

$$\begin{aligned} \tilde{\gamma}_{\nu}(s) &= \gamma_{1\nu}\frac{1}{2^{\nu}}\int_{-1}^{1}(\cosh st)\,\tau^{\nu-1/2}\,dt \\ &= \gamma_{1\nu}\frac{\pi(2\nu)!}{8^{\nu}\nu!}\sum_{\mu=0}^{\infty}[\mu!(\mu+\nu)!]^{-1}\left(\frac{s}{2}\right)^{2\mu}. \end{aligned} \tag{10.2-16a}$$

It follows that for $\nu = 0, k, k+1, \cdots$,

$$|\tilde{\gamma}_{\nu}(s)| \geq |\gamma_{1\nu}|\frac{\pi(2\nu)!}{8^{\nu}\nu!}\left[\frac{1}{2}\left(\frac{s}{2}\right)^{-2\nu} e^{s/2} - \sum_{\mu=0}^{\nu-1}\frac{1}{(2\mu)!}\left(\frac{2}{s}\right)^{2(\nu-\mu)}\right]. \tag{10.2-16b}$$

From this we conclude that $\tilde{\gamma}_0(s) \not\equiv 0$, $\tilde{\gamma}_k(s) \not\equiv 0$, and

$$|\tilde{\gamma}_0(s)| \to \infty, \qquad |\tilde{\gamma}_k(s)| \to \infty \qquad \text{as} \quad s \to \infty. \tag{10.2-17}$$

Furthermore, for $|z_1| \leq \tilde{r}$, from (10.2-16a) it follows that

$$v_{(s)}(z_1) = \tilde{\gamma}_0(s)\left[1 + \tilde{\varepsilon}_2 e^{i\tilde{\delta}_2}\right] \quad \text{where} \quad |\tilde{\varepsilon}_2| < \tfrac{1}{2}. \tag{10.2-18}$$

Thus (10.2-16b, 17, 18) imply that for s large enough,

$$s \in M_1(0, c) \cap M_5(0, c) \cap \bigcap_{\lambda=1}^{q}\left[M_1(a_{\lambda}, c) \cap M_4(a_{\lambda}, \tilde{r})\right] \tag{10.2-19}$$

where $a_1, \cdots, a_q$ and $c > 0$ are given. Now suppose that

$$\max_{|z_1|=\sigma s} |v_{(s)}(z_1)| \leq \exp s^{\rho(\sigma)+\varepsilon} \qquad \text{for every} \qquad \varepsilon > 0$$

[cf. Lemma 10.1-2]. Hence for $\sigma > 1$,

$$\log^+ m\left[\sigma s, v_{(s)}(z_1)\right] \leq \log^+ \log^+ \exp s^{\rho(\sigma)+\varepsilon} \leq \left[\rho(\sigma) + \varepsilon\right] \log^+ s\,.$$

Furthermore, from (10.2-16b) with $\nu = 0$ and (10.2-19) we obtain

$$\log|\tilde{\gamma}_0(s) - a_\lambda| \leq \tilde{c}_0 + s\,, \qquad \tilde{c}_0 = const > 0\,, \;\; \lambda = 1, \cdots, q\,.$$

Consequently, if we apply Theorem 10.2-4a and observe the remarks subsequent to that theorem, we see that (10.2-13) yields

$$(q-1)m[s, w(z_1, \bar{z}_1)] \leq \left(\log \frac{s}{\tilde{r}}\right) \sum_{\lambda=1}^{q} n\left[s, \left(w(z_1, \bar{z}_1) - a_\lambda\right)^{-1}\right] + \tilde{c} + qs$$

$$+8[1 + \rho(\sigma) + \varepsilon] \log^+ s,$$

$$\tilde{c} = \tilde{c}_a\left(k, c, q, a_1, \cdots, a_q\right) + c_0(\sigma) + q\tilde{c}_0$$

valid for large enough s such that $w(z_1, \bar{z}_1) \neq a_\lambda$ on $|z_1| = s$, $\lambda = 1, \cdots, q$. Hence if for any $\varepsilon > 0$,

$$\sum_{\lambda=1}^{q} n\left[s, \left(w(z_1, \bar{z}_1) - a_\lambda\right)^{-1}\right] = O(s^{\vartheta'+\varepsilon})$$

then for all positive s such that $w(z_1, \bar{z}_1) \neq a_\lambda$ we obtain

$$\text{(10.2-20)} \qquad m[s, w(z_1, \bar{z}_1)] = O(s^{\vartheta+\varepsilon})\,, \qquad \vartheta = \max(\vartheta', 1)\,.$$

If the set of these s is everywhere dense, as in the case when all a_λ-points of $\check{w}$ are isolated, then, by continuity, (10.2-20) holds for all positive s.

Formula (10.2-20) shows that

$$\sum_{\lambda=1}^{q} n\left[s, \left(w(z_1, \bar{z}_1) - a\right)^{-1}\right]$$

determines an upper bound of $m[s, w(z_1, \bar{z}_1)]$, unless $m[s, w(z_1, \bar{z}_1)] = O(s^{1+\varepsilon})$.

NEVANLINNA'S SECOND THEOREM GENERALIZED TO SOLUTIONS

The result in Example 10.2-5 is typical. It suggests the following

10.2-6 Theorem

Let conditions 1–4 *be satisfied*:

1. *Conditions* (i) *and* (ii) *for L*; *cf. Sec.* 10.1.

2. *There exist positive constants* r_0, s_0, *and c such that for* $|z_1| \le r_0$ *and* $s > s_0$,

$$g(z_1, s, t) = \tilde{g}(s, t)\left[1 + \varepsilon_0 e^{i\delta_0}\right]$$

where $|\varepsilon_0| \le c < 1$ *and* $\tilde{g}(s, t) = g(0, s, t)$ *is real and bounded below, for* $s > 0$ *and* $t \in [-1, 1]$.

3. f_1 *is entire and*

$$g(z_1, s, t)f_1(\tfrac{1}{2}z_1\tau) = \alpha_0(s, t) + \sum_{\nu=k}^{\infty} \alpha_\nu(s, t)z_1^\nu, \qquad k \ge 1$$

where $\alpha_0(s, t) = \gamma_{10}\tilde{g}(s, t) \not\equiv 0$ *and* $\alpha_k(s, t) \not\equiv 0$.

4. *For* $l = 0$ *and k*,

$$(10.2\text{-}21) \qquad |\tilde{\gamma}_l(s)| = \left|\int_{-1}^{1} \alpha_l(s, t)\tau^{-1/2}\,dt\right| \to \infty \qquad \text{as } s \to \infty.$$

We use the notations

$$(10.2\text{-}22) \qquad \mu(s) = \max_{|z_1|=s} |f_1(z_1)|$$

and

$$(10.2\text{-}23) \qquad M(s) = \max_{|z_1|=2s,\ |t|\le 1} |g(z_1, s, t)|.$$

Under these conditions there exists an $\tilde{r} > 0$ *such that*, *for s so large that* $w(z_1, \bar{z}_1) \ne a_1, \cdots, a_q$ *on* $|z_1| = s$, *we have*

$$(q-1)m[s, w(z_1, \bar{z}_1)] \le \left(\log \frac{s}{\tilde{r}}\right) \sum_{\lambda=1}^{q} n\left[s, (w(z_1, \bar{z}_1) - a_\lambda)^{-1}\right]$$

$$(10\text{-}2.24) \qquad + 8\log^+\log^+[\mu(s)M(s)] + q\log|\tilde{\gamma}_0(s)| + 8\log s + \tilde{c}$$

where $\tilde{c}$ *is a suitable positive constant.*

Proof. Let

$$f_1(\xi_1) = \sum_{\nu=0}^{\infty} \gamma_{1\nu}\xi_1^{\nu}, \qquad \gamma_{10} \neq 0.$$

Then there is an $r_3 > 0$ such that for $|\xi_1| \le r_3$,

$$f_1(\xi_1) = \gamma_{10}\left[1 + \varepsilon_1 e^{i\delta_1}\right] \qquad \text{where} \qquad |\varepsilon_1| < \tfrac{1}{16}(1-c).$$

For $|z_1| \le \tilde{r} = \min(r_0, r_3)$ and $s > s_0$ we thus obtain

$$v_{(s)}(z_1) = \int_{-1}^{1} \tilde{g}(s,t)\gamma_{10}\left[1 + \varepsilon_3 e^{i\delta_3}\right]\tau^{-1/2}\,dt$$

where

$$|\varepsilon_3| < \tfrac{2}{16} + \tfrac{14}{16}c < 1.$$

Now, since $\tilde{g}(s,t)$ is bounded below for $s > 0$,

$$\tilde{g}(s,t) = -m_1 + h(s,t)$$

for some constant m_1 and function $h(s,t) > 0$ for $s > 0$ and $t \in [-1,1]$. From this and (10.2-21) we obtain

$$\begin{aligned} h_0(s) &= \int_{-1}^{1} h(s,t)\tau^{-1/2}\,dt = m_1\pi + \int_{-1}^{1} \tilde{g}(s,t)\tau^{-1/2}\,dt \\ &= m_1\pi + \tilde{\gamma}_0(s)/\gamma_{10} \to \infty \qquad \text{as } s \to \infty. \end{aligned} \tag{10.2-25}$$

Now

$$\begin{aligned} v_{(s)}(z_1) &= -\gamma_{10}m_1\int_{-1}^{1}\left[1 + \varepsilon_3 e^{i\delta_3}\right]\tau^{-1/2}\,dt \\ &\quad + \gamma_{10}\int_{-1}^{1} h(s,t)\left[1 + \varepsilon_3 e^{i\delta_3}\right]\tau^{-1/2}\,dt \\ &= \gamma_{10}m_2(z_1) + \gamma_{10}h_0(s)\left[1 + \varepsilon_4 e^{i\delta_4}\right] \end{aligned}$$

where $m_2(z_1) \le 2m_1\pi$ and $|\varepsilon_4| \le \frac{2}{16} + \frac{14}{16}c$, for sufficiently large s and $|z_1| \le \tilde{r}$. Since $h_0(s) \to \infty$ as $s \to \infty$, we thus obtain

$$v_{(s)}(z_1) = \gamma_{10}h_0(s)\left[1 + \varepsilon_5 e^{i\delta_5}\right] \qquad \text{where} \qquad |\varepsilon_5| \le \tfrac{4}{16} + \tfrac{12}{16}c$$

or

$$v_{(s)}(z_1) = \tilde{\gamma}_0(s)\left[1 + \varepsilon_2 e^{i\delta_2}\right] \qquad \text{where} \qquad |\varepsilon_2| \le \tfrac{6}{16} + \tfrac{10}{16}c,$$

again for sufficiently large s and $|z_1| \leq \tilde{r}$. From this and (10.2-21) it follows that for sufficiently large s,

$$(10.2\text{-}26) \quad s \in M_1(0, c) \cap M_5(0, c) \cap \bigcap_{\lambda=1}^{q} \left[M_1(a_\lambda, c) \cap M_4(a_\lambda, \tilde{r}) \right].$$

Furthermore,

$$|v_{(s)}(z_1)| = \left| \int_{-1}^{1} g(z_1, s, t) f_1\left(\tfrac{1}{2} z_1 \tau\right) \tau^{-1/2} \, dt \right| \leq \mu(s) M(s) \pi$$

for $|z_1| = 2s$; cf. (10.2-22, 23). Setting $r_2 = 2s$, from the definition in Sec. 10.1 [after (10.1-28)], we get

$$\log^+ m\left[2s, v_{(s)}(z_1)\right] \leq \log^+ \log^+ \left[\mu(s) M(s)\right] + \log\log \pi .$$

Also, because of (10.2-26) there is a constant $\tilde{c}_0 > 0$ such that

$$\log |\tilde{\gamma}_0(s) - a_\lambda| \leq \tilde{c}_0 + \log |\tilde{\gamma}_0(s)| .$$

Introducing our preceding results into Theorem 10.2-4, (10.2-13) with $r_2 = 2s$, we obtain the desired inequality (10.2-24) with $\tilde{c}$ of the form

$$\tilde{c} = \tilde{c}_a(k, c, q, a_1, \cdots, a_q) + 8 \log 2 + 8 \log\log \pi + q\tilde{c}_0 .$$

This proves the theorem. ■

10.3 ANALOG OF THE LITTLE PICARD THEOREM

In this section and the next we shall discuss two methods of obtaining generalizations of the Little and the Great Picard Theorems to classes of solutions of partial differential equations. First, along the lines of the proof of Theorem 10.2-6 we shall derive an analog of the Little Picard Theorem for the class $K(\tilde{k}_1^\times)$ of solutions defined in Sec. 10.1. For another class not coinciding with $K(\tilde{k}_1^\times)$ we shall use the representation of solutions by differential operators as obtained in Chap. 3 and then apply the results on multianalytic functions in the sense of Krajkiewicz [1973]. This will give a generalization of the Great Picard Theorem to solutions in Sec. 10.4.

The well-known classical *Little Picard Theorem* states that a nonconstant entire function of one complex variable cannot omit two distinct values. It generalizes to solutions of the class $K(\tilde{k}_1^\times)$ as follows.

10.3-1 Generalized Little Picard Theorem

Suppose that conditions (i)–(v) *hold*:

(i) *and* (ii) *as in Sec.* 10.1, *with* $\tilde{k}_1$ *denoted by* $\tilde{k}_1^\times$, *as indicated in Sec.* 10.1.

(iii) *There exist positive constants* r_0, s_0, c *such that for* $|z_1| \leq r_0$ *and* $s > s_0$ *the function* g *may be represented by*

$$g(z_1, s, t) = \tilde{g}(s, t)\left[1 + \varepsilon_0 e^{i\delta_0}\right]$$

where $|\varepsilon_0| \leq c < 1$ *and* $\tilde{g}(s, t) = g(0, s, t)$ *is real and bounded below for* $s > 0$, $t \in [-1, 1]$.

(iv) *Let* $f_1 \in C^\omega(\mathbf{C})$ *have the series expansion*

$$f_1(\xi_1) = \sum_{\nu=k}^{\infty} \gamma_{1\nu}\xi_1^\nu, \qquad \gamma_{1k} \neq 0; \quad k \geq 1.$$

(v) *Let*

$$|\tilde{\gamma}_k(s)| = \left|\gamma_{1k}2^{-k}\int_{-1}^{1}\tilde{g}(s, t)\tau^{k-1/2}\,dt\right| \to \infty \qquad \text{as } s \to \infty.$$

Let $T_1^\times$ *be the operator defined by* (10.1-1) *with kernel* $\tilde{k}_1^\times$. *Then the solution*

$$\check{w} = w|_{z_2 = \bar{z}_1} \quad \text{where} \quad w = T_1^\times f_1 \qquad \text{[cf. (10.1-24)]}$$

assumes every finite complex value.

Proof. From our assumption (iv) it follows that there exists an $r_3 > 0$ such that for $|\xi_1| \leq r_3$,

$$f_1(\xi_1) = z_1^k\gamma_{1k}\left[1 + \varepsilon_1 e^{i\delta_1}\right] \quad \text{where} \quad |\varepsilon_1| < \tfrac{1}{16}(1 - c).$$

By an argument similar to that leading to (10.2-26), we obtain

$$v_{(s)}(z_1) = z_1^k\tilde{\gamma}_k(s)\left[1 + \varepsilon_2 e^{i\delta_2}\right] \quad \text{where} \quad |\varepsilon_2| \leq \tfrac{3}{8} + \tfrac{5}{8}c$$

valid for $|z_1| \leq \tilde{r} = \min(r_0, r_3)$ and sufficiently large s. Hence we have

$$\text{(10.3-1)} \quad \begin{aligned} &\text{(a)} \quad v_{(s)}(0) = 0 \\ &\text{(b)} \quad |v_{(s)}(z_1)| \geq \tilde{r}^k|\tilde{\gamma}_k(s)|[1 - |\varepsilon_2|] \geq \tfrac{5}{8}\tilde{r}^k(1 - c)|\tilde{\gamma}_k(s)| \end{aligned}$$

for $|z_1| = \tilde{r}$ and sufficiently large s.

Now suppose that $\check{w}$ does not attain some given value $a \in \mathbb{C}$. Then $n[s,(w(z_1, \bar{z}_1) - a)^{-1}]$ is defined for this a, and we have

$$n\left[s,\left(w\left(z_1,\bar{z}_1\right)-a\right)^{-1}\right]=\tilde{n}\left[s,\left(v_{(s)}\left(z_1\right)-a\right)^{-1}\right]=0. \tag{10.3-2}$$

But because of (10.3-1b) and assumption (v), for sufficiently large $s > \tilde{r}$ and $|z_1| = \tilde{r}$ we have

$$|v_{(s)}(z_1)| > |a| = \left|\left[v_{(s)}(z_1) - a\right] - v_s(z_1)\right|.$$

Hence, by Rouché's Theorem, the functions $v_{(s)}$ and $v_{(s)} - a$ have the same number of zeros in $|z_1| \leq \tilde{r}$; but because of (10.3-1a), this means that $v_{(s)}$ assumes the value a at least once in the disk $|z_1| \leq \tilde{r}$. It follows that

$$\begin{aligned} n\left[s,\left(w\left(z_1,\bar{z}_1\right)-a\right)^{-1}\right] &= \tilde{n}\left[s,\left(v_{(s)}\left(z_1\right)-a\right)^{-1}\right] \\ &\geq \tilde{n}\left[\tilde{r},\left(v_{(s)}\left(z_1\right)-a\right)^{-1}\right] \geq 1. \end{aligned}$$

This contradicts (10.3-2) and proves the theorem. ■

APPLICATIONS

Let us apply the theorem just proved to two of our earlier examples.

10.3-2 Example

We consider the equation

$$Lw = D_1 D_2 w - \tfrac{1}{4} w = 0 \qquad \text{(cf. Example 10.2-5)}.$$

A corresponding kernel is given by

$$\tilde{k}_1^{\times}(z,t) = \cos\left(i\sqrt{z_1 z_2}\, t\right)$$

so that

$$\tilde{k}_1^{\times}(z_1, \bar{z}_1, t) = \cosh(st), \qquad s = (z_1 \bar{z}_1)^{1/2}.$$

Hence it is obvious that the conditions (i)–(iii) of the preceding theorem are fulfilled. If we impose condition (iv), then condition (v) also holds, as can be seen from (10.2-16). Thus the statement of Theorem 10.3-1 is valid for solutions $w \in K(\tilde{k}_1^{\times})$, that is, $w = T_1^{\times} f_1$ obtained by the Bergman operator T_1 with kernel $\tilde{k}_1^{\times}$ as before and with f_1 satisfying condition (iv).

10.3-3 Example

Let the equation of Example 3.1-8 be given with $j = 1$, $n = 1$, and negative real α, that is,

$$Lw = D_1 D_2 w - \alpha z_2 D_2 w + \alpha w = 0, \qquad \alpha < 0.$$

From Example 3.1-8 we know a suitable Bergman kernel $\tilde{k}_1 = \tilde{k}_1^{\times}$, namely,

$$\tilde{k}_1^{\times}(z, t) = 1 - 2\alpha z_1 z_2 t^2.$$

It follows that

$$\tilde{k}_1^{\times}(z_1, \bar{z}_1, t) = 1 - 2\alpha s^2 t^2, \qquad s = (z_1 \bar{z}_1)^{1/2}.$$

Conditions (i) and (ii) are trivially satisfied. Obviously, also condition (iii) holds, with $\varepsilon_0 = 0$ and $\tilde{g}(s, t) = 1 - 2\alpha s^2 t^2$ is real and bounded below for $-\alpha > 0$, $s > 0$, and $t \in [-1, 1]$. If f_1 is an entire function satisfying condition (iv), then also condition (v) is easily proved:

$$|\tilde{\gamma}_k(s)| = \left| \gamma_{1k} 2^{-k} \int_{-1}^{1} (1 - 2\alpha s^2 t^2) \tau^{k-1/2} \, dt \right|$$

$$= |\gamma_{1k}| \frac{\pi(2k)!}{8^k (k!)^2} \left(1 + \frac{-\alpha}{k+1} s^2\right) \to \infty \qquad \text{as } s \to \infty.$$

Hence the statement of Theorem 10.3-1 holds for the solution $w = T_1^{\times} f_1$.

10.4 GENERALIZATION OF THE GREAT PICARD THEOREM

In this section we consider the reduced equation (2.1-2) with $j = 1$ [cf. (5.1-1)],

$$(10.4\text{-}1) \qquad Lu = [D_1 D_2 + b(z) D_2 + c(z)] u = 0$$

or, what amounts to the same, an equation of the form (2.1-1) with $a_1 = 0$ and a_2, a_0 replaced by b, c. We assume that

$$(10.4\text{-}2) \qquad L \in FP_{1n}$$

for some $n \in \mathbb{N}_0$; cf. Sec. 3.1. We deal with Type I representations

$$(10.4\text{-}3) \qquad u = T_1 f_1$$

where T_1 is a Bergman operator given by the polynomial kernel $\tilde{k}_1$ of the first kind with $C_1 = [-1,1]$, $\eta_1 = 0$, and $s = (0,0)$. Because of assumption (10.4-2), this kernel $\tilde{k}_1 = k_1$ exists and is uniquely determined by Corollary 3.1-5.

As has been mentioned at the beginning of Sec. 10.3, we shall apply a result on multianalytic functions in the sense of Krajkiewicz [1973]. For this purpose, let us first consider the basic definitions and the Picard–Krajkiewicz theorem, a generalization of the Great Picard Theorem to multianalytic functions.

In a series of papers, Balk [1963, 1965, 1966] (cf. also the survey by Balk in Lanckau–Tutschke [1983], pp. 68–84) and Bosch–Krajkiewicz [1970] consider polyanalytic functions of the $(n+1)$th order, especially the (Great) Picard Theorem for these functions. Here, a ***polyanalytic function*** g of $(n+1)$th order $(n \in \mathbb{N}_0)$ in a finite domain $\Omega_1 \subset \mathbb{C}$ is of the form

$$g(z_1) = \sum_{\mu=0}^{n} g_\mu(z_1)\bar{z}_1^\mu, \qquad g_n \neq 0 \tag{10.4-4}$$

where $g_0, \cdots, g_n$ are analytic functions in Ω_1. By definition, g has an *isolated essential singularity* at $z_1^{(0)} \in \Omega_1$ if there exists a punctured disk $\dot{B}_1(z_1^{(0)}, \rho) = B_1(z_1^{(0)}, \rho) \setminus \{z_1^{(0)}\} \subset \Omega_1$ $(\rho > 0)$ such that at least one of the functions g_μ has an isolated essential singularity at $z_1^{(0)}$. This definition is unambiguous since one can show that the coefficient functions g_μ in (10.4-4) are uniquely determined.

MULTIANALYTIC FUNCTIONS

The next step in generalizing analytic functions beyond (10.4-4) was done by Krajkiewicz [1973]. We consider Krajkiewicz's idea in connection with an extension of the Picard Theorem.

Let $z_1^{(0)}$ be arbitrarily chosen in $\mathbb{C} \cup \{\infty\}$, and let $\dot{B}_1(z_1^{(0)}, \rho)$ be defined as before in the case $z_1^{(0)} \neq \infty$, and $\dot{B}_1(\infty, \rho) = \mathbb{C} \setminus \{B_1(0,\rho) \cup \partial B_1(0,\rho)\}$.

10.4-1 Definition

A function g: $\dot{B}_1(z_1^{(0)}, \rho) \to \mathbb{C}$ is called ***multianalytic*** in $\dot{B}_1(z_1^{(0)}, \rho)$ if there exists an $n_0 \in \mathbb{N}_0$ and a sequence of functions g_μ analytic in $\dot{B}_1(z_0^{(0)}, \rho)$, $\mu \geq -n_0$, such that for all $z_1 \in \dot{B}_1(z_1^{(0)}, \rho)$,

$$g(z_1) = \sum_{\mu=-n_0}^{\infty} g_\mu(z_1)\overline{(z_1 - z_1^{(0)})}^\mu \qquad \text{for } z_1^{(0)} \neq \infty \tag{10.4-5a}$$

and

$$g(z_1) = \sum_{\mu=-n_0}^{\infty} g_\mu(z_1)(1/\bar{z}_1)^\mu \qquad \text{for } z_1^{(0)} = \infty\,. \tag{10.4-5b}$$

Here it is assumed that the series converges uniformly on every nonvoid compact subset of $\dot{B}_1(z_1^{(0)}, \rho)$. [For instance, assume that all g_μ in (10.4-5) are of class $C^\omega(B_1(z_1^{(0)}, \rho))$; hence (10.4-5) is almost uniformly convergent.] It can be shown that for a given function g the coefficient functions g_μ in (10.4-5) are uniquely determined.

One possibility of defining essential singularities in the case of multianalytic functions would be the adoption of the previous definition as given in the case of polyanalytic functions. A more general definition has been given by Krajkiewicz, as follows. He distinguishes three different possible cases:

(a)
$$g_\mu|_{\dot{B}_1(z_1^{(0)}, \rho)} = 0 .$$

(b) g_μ has an isolated essential singularity at $z_1^{(0)}$.

(c) There is a unique integer ν such that $(z_1 - z_1^{(0)})^\nu g_\mu$ (in the case $z_1^{(0)} \neq \infty$) and $z_1^{-\nu} g_\mu$ (in the case $z_1^{(0)} = \infty$), respectively, are holomorphic and nonzero at $z_1^{(0)}$.

Define

$$d_\mu(g_\mu) = \begin{cases} -\infty & \text{in Case (a)} \\ +\infty & \text{in Case (b)} \\ \nu & \text{in Case (c)}, \end{cases} \qquad \mu \geq -n_0 \tag{10.4-6a}$$

and

$$d(g) = \sup\{d_\mu(g_\mu) - \mu \mid \mu \geq -n_0\} . \tag{10.4-6b}$$

Hence,

$$d(g) \in \{\pm n \mid n \in \mathbb{N}_0\} \cup \{-\infty, +\infty\}$$

holds. Furthermore, $d(g) = -\infty$ is equivalent to $g = 0$ on $\dot{B}_1(z_1^{(0)}, \rho)$.

10.4-2 Definition

A multianalytic function g has an ***isolated essential singularity*** at $z_1^{(0)}$ if and only if $d(g) = +\infty$.

Thus, a sufficient (but not necessary) condition for the occurrence of an isolated essential singularity of a multianalytic function g at $z_1^{(0)}$ is that at least one of the g_μ $(\mu \geq -n_0)$ has an isolated essential singularity at $z_1^{(0)}$.

The definition of an exceptional value is extended in the following way.

10.4-3 Definition

Let g be a multianalytic function which does not have an essential singularity at $z_1^{(0)}$. Then $\tilde{g}$ is called an ***exceptional value*** (in the sense of Krajkiewicz) for the multianalytic function g at $z_1^{(0)}$ (or, equivalently, g has $\tilde{g}$ as an exceptional value at $z_1^{(0)}$) if there exists a $\rho' > 0$ such that $g - \tilde{g}$ is defined and nonvanishing in $\dot{B}_1(z_1^{(0)}, \rho')$. Two (such) exceptional values $\tilde{g}_1$, $\tilde{g}_2$ of g are said to be *different* if there is a $\rho'' > 0$ such that $\tilde{g}_1 - \tilde{g}_2$ is defined and nonvanishing on $\dot{B}_1(z_1^{(0)}, \rho'')$.

PICARD–KRAJKIEWICZ THEOREM

With those definitions the following extension of the Great Picard Theorem holds.

10.4-4 Theorem

Let g be a multianalytic function in $\dot{B}_1(z_1^{(0)}, \rho)$ ($z_1^{(0)}$ finite or infinite, $\rho > 0$), and let g possess an isolated essential singularity at $z_1^{(0)}$ (cf. Def. 10.4-2). Then g has at most one exceptional value (in the sense of Krajkiewicz [Def. 10.4-3]) at $z_1^{(0)}$.

For the *proof* we refer to Krajkiewicz [1973], pp. 426–438.

It is obvious that the well-known Great Picard Theorem for analytic functions (cf. Behnke–Sommer [1972], p. 490, Theorem V.39a, or Conway [1973], 302, Theorem XII.4.2) is contained in the version of Theorem 10.4-4 if one observes that an analytic function g is trivially also a multianalytic function [cf. representation (10.4-5)], that an isolated essential singularity of an analytic function satisfies Def. 10.4-2 [cf. (10.4-6a), Case (b)], and that for constant functions $\tilde{g}$, $\tilde{g}(z_1) \equiv \alpha$ ($\alpha = const \in \mathbb{C}$), we have $d(\tilde{g}) \neq +\infty$, namely, $d(\tilde{g}) = 0$ for $\alpha \in \mathbb{C} \setminus \{0\}$ and $d(\tilde{g}) = -\infty$ for $\alpha = 0$.

Secondly, observe that Theorem 10.4-4 would in general *not* hold if in (10.4-5) we had permitted infinitely many negative powers with nonzero coefficients. This can be seen from the counterexample

$$g(z_1) = \exp(2 \operatorname{Re} z_1) = \sum_{\mu=-\infty}^{0} \left(\frac{1}{(-\mu)!} e^{z_1} \right) (1/\bar{z}_1)^{\mu}, \qquad z_1^{(0)} = \infty .$$

GREAT PICARD THEOREM FOR SOLUTIONS

We now apply Theorem 10.4-4 to solutions u of the form (10.4-3) of equations $Lu = 0$ satisfying (10.4-1, 2). We assume that $b, c \in C^{\omega}(\Omega)$, $\Omega = \Omega_1 \times \Omega_2 \ni 0$, and

$$z_1^{(0)} \in B_1(0, \tfrac{1}{2}r) \setminus \{0\} \tag{10.4-7}$$

where $r > 0$ is such that $B_m(0, r) \cup \partial B_m(0, r) \subset \Omega_m$, $m = 1, 2$. Hence, in a suitable neighborhood of the origin, Theorem 2.5-1 is applicable. We further assume that $h_1 \in C^\omega(\dot{B}_1(z_1^{(0)}, r))$ has an isolated essential singularity at $z_1^{(0)}$ with the Laurent expansion

$$h_1(z_1) = \sum_{\varkappa=-\infty}^{\infty} \delta_\varkappa \left(z_1 - z_1^{(0)}\right)^\varkappa \tag{10.4-8a}$$

where

$$\delta_{-n} = \cdots = \delta_{-1} = 0. \tag{10.4-8b}$$

According to our assumptions, the uniquely determined kernel $\tilde{k}_1$ can be written in the form

$$\begin{aligned} \tilde{k}_1(z, t) &= \sum_{m=0}^{n} \tilde{q}_{1m}(z) z_1^m t^{2m} \\ &= \sum_{m=0}^{n} \left(\sum_{\mu=0}^{\infty} Q_{1m\mu}(z_1) z_2^\mu \right) z_1^m t^{2m} \end{aligned} \tag{10.4-9}$$

where $\tilde{q}_{10}(z) \equiv 1$ and $\tilde{q}_{1m}(z_1, 0) \equiv 0$ $(m \in \mathbb{N})$. The latter conditions, obtained from the assumption that $\tilde{k}_1$ be a kernel of the first kind, imply

$$\begin{aligned} &Q_{100}(z_1) \equiv 1 \\ &Q_{10\mu}(z_1) \equiv 0, \qquad \mu \in \mathbb{N}, \\ &Q_{1m0}(z_1) \equiv 0, \qquad m = 1, \cdots, n. \end{aligned} \tag{10.4-10}$$

Writing, as usual,

$$H_{10} = h_1$$

and

$$H_{1m}(z_1) = \int_0^{z_1} \int_0^{z_{1m}} \cdots \int_0^{z_{12}} h_1(z_{11})\, dz_{11} \cdots dz_{1,m-1}\, dz_{1m}$$

where the paths of integration are to be taken suitably in $B_1(0, \frac{1}{2}r) \setminus \{z_1^{(0)}\}$, $m = 1, \cdots, n$, and using Theorem 2.5-1, we thus obtain for the solution u generated by T_1 from h_1 in (10.4-8) the representation

$$u(z) = \sum_{\mu=0}^{\infty} \left(\sum_{m=0}^{n} Q_{1m\mu}(z_1) \frac{(2m)!}{4^m m!} H_{1m}(z_1) \right) z_2^\mu. \tag{10.4-11}$$

Restricting our considerations to the plane, that is, $z_2 = \bar{z}_1$, we see from (10.4-11) that $\check{u} = u|_{z_2=\bar{z}_1}$ is a multianalytic function which, according to the Theorems 2.4-2 and 2.5-1, is regular at least in every simply connected subdomain of $B_1(0, \frac{1}{2}r) \setminus \{z_1^{(0)}\}$. Moreover, because of (10.4-10) for $\mu = 0$ the coefficient function in (10.4-11) reduces to h_1. Hence the multianalytic function $\check{u}$ has an isolated essential singularity at $z_1^{(0)}$, by assumption (10.4-8a). Thus we can apply Theorem 10.4-4 and have proved the following result.

10.4-5 Theorem

Under the preceding assumptions on L, h_1, and $z_1^{(0)}$ [cf. (10.4-1, 2, 7, 8)], the solution u in (10.4-11), *obtained from* (2.5-1) *and* (10.4-8) *by means of the Bergman operator T_1 with polynomial kernel $\tilde{k}_1$ of the first kind, has the following property: $\check{u} = u|_{z_2=\bar{z}_1}$ has at $z_1^{(0)}$ at most one exceptional value (in the sense of Krajkiewicz).*

This entails the more familiar result that in every neighborhood of $z_1^{(0)}$, the function $\check{u} = u|_{z_2=\bar{z}_1}$ can omit at most one complex number, or, equivalently, it assumes all finite complex values with at most one possible exception.

APPLICATIONS

Let us reconsider some earlier examples from the viewpoint of the applicability of this basic theorem.

10.4-6 Example

From Examples 2.5-5, 3.1-7, 3.1-8, and from (4.1-12, 13) we conclude that, for suitably chosen domains Ω (e.g., a neighborhood of the origin), the following operators L satisfy the assumptions of Theorem 10.4-5:

$$L = D_1 D_2 + \lambda n(n+1)(1 + \lambda z_1 z_2)^{-2} \tag{10.4-12}$$

$$L = D_1 D_2 + 4z_1(1 + z_1^2 z_2)^{-2} \tag{10.4-13}$$

$$L = D_1 D_2 - \lambda z_1^{n-1} z_2^n D_2 + \lambda n (z_1 z_2)^{n-1} \tag{10.4-14}$$

$$\begin{aligned} L = {} & D_1 D_2 + \kappa z_2^m [\psi(z_2) + \lambda z_1 z_2^m]^{-1} D_2 \\ & + n[(n+1)\lambda - \kappa] D_1 \left(z_2^m [\psi(z_2) + \lambda z_1 z_2^m]^{-1} \right) \end{aligned} \tag{10.4-15}$$

where $m, n \in \mathbb{N}$, $\kappa, \lambda \in \mathbb{C}$, and ψ is a function holomorphic for $z_2 = 0$ with $\psi(0) \neq 0$. [Observe that (10.4-12) is a special case of (10.4-15), with $\kappa = 0$, $m = 1$, and $\psi(z_2) \equiv 1$.] Thus, Theorem 10.4-5 is applicable in the case of these operators.

Let us finally compare the results of Secs. 10.3 and 10.4, namely, Theorems 10.3-1 and 10.4-5. We assume that, in the case of entire coefficients of L, a theorem analogous to Theorem 10.4-5 is stated for $z_1^{(0)} = \infty$. Then a further analog of the Little Picard Theorem would be obtained. But this would not make Theorem 10.3-1 superfluous, for the following reasons.

(i) Theorem 10.3-1 states that $\check{u} = u|_{z_2=\bar{z}_1}$ omits *no* finite complex value, instead of none with one possible exception.

(ii) Theorem 10.3-1 also applies to certain operators which are not necessarily of class FP_{1n}, for instance to the Helmholtz operator in Example 10.3-2, which is not of this class for any $n \in \mathbb{N}_0$.

Additional References

Balk [1966, 1983]
Bauer–Ruscheweyh [1980]
Begehr [1983]
Bergman [1943b, 1944b, 1945b]
Bosch–Krajkiewicz [1970]
Chernoff [1946]
Frank–Hennekemper [1981]
Fryant [1977, 1979]
Habetha [1967]
Harke [1981]
Hayman [1964]
Jank–Volkmann [1985]
Krajkiewicz [1973]
Lanckau–Tutschke [1983]
McCoy [1979a]
McCoy–D'Archangelo [1976]
Nautiyal [1983]
Tonti–Trahan [1970]
Wittich [1982]

Chapter Eleven

Applications of Class P Operators. Function Theory of the Bauer–Peschl Equation

This chapter is concerned with the application of the results of Chaps. 2–4 and 6, in particular, on the representation of solutions by means of class P operators and differential operators, for obtaining function theoretic properties of solutions of the Bauer–Peschl equation (2.5-13) with $\lambda = -1$. The mathematical and physical importance of that equation, e.g., its relations to equations of mathematical physics such as the GASPT equation and to n-harmonic functions, is discussed in Sec. 11.1. A variety of representations of solutions by means of integral and differential operators is obtained in Sec. 11.2 for the special case of the Bauer–Peschl equation. A generalized Leibniz formula for Bauer differential operators for that equation will turn out to be a valuable tool in the proofs in Sec. 11.3. In passing, interesting representations of the Legendre functions are restated. Section 11.3 is devoted to properties of solutions which satisfy certain additional conditions. A maximum modulus theorem for solutions is proved. Then solutions are investigated which have the same real or imaginary parts, which are bounded on disks, or which are constant on some circles. The main results in Sec. 11.3 will be generalizations of the Schwarz lemma and of Privalov's uniqueness theorem. Further results of analogs of theorems in geometric function theory are proved for solutions in Sec. 11.4, e.g., analogs of the Great Picard theorem, Schottky's theorem, and Hadamard's three circles theorem. In Sec. 11.5, the foundation of a function theory as proposed by K. W. Bauer is developed for solutions which are representable by Type I representations and which therefore, in addition, also satisfy a generalized higher-order Cauchy–Riemann differential equation.

11.1 MATHEMATICAL AND PHYSICAL IMPORTANCE OF THE EQUATION

Throughout this chapter we shall consider function theoretic properties of solutions in the unit disk of the equation

$$(1 - z_1\bar{z}_1)^2 D_1\overline{D}_1 w - n(n+1)w = 0, \qquad n \in \mathbb{N} \tag{11.1-1}$$

where $\overline{D}_1 = \partial/\partial\bar{z}_1$. For $z_1 \in B_1(0,1)$ this equation is equivalent to equation (2.5-13) with $\lambda = -1$, $z_2 = \bar{z}_1$. The equation (11.1-1), which is called the Bauer–Peschl equation since the beginning of the 1970s (cf. Bauer–Ruscheweyh [1980]), is very important for both mathematical and physical reasons.

Since the time of G. Darboux and H. A. Schwarz, this equation, in particular with $n = 1$, has attracted interest from the differential geometric viewpoint in connection with minimal surfaces; cf. Schwarz [1890] and Bianchi [1910]. Then Dirichlet problems for equation (11.1-1) were studied by Agostinelli [1937] and Ganin [1957]. The investigation of differential invariants of certain families of functions and the application of resulting methods to elliptic partial differential equations by Peschl [1955, 1963] seem to mark the beginning of recent general interest in equation (11.1-1). The investigation of various explicit representations of solutions of equation (11.1-1) began with papers by Bauer [1965, 1966a] and Bauer–Peschl [1966]. Extensions of these representations to more general equations were initiated by the authors' paper [1969].

Equation (11.1-1) is closely related to several basic equations which are of mathematical and physical importance. It is related to the two-dimensional *wave equation*

$$\hat{w}_{x_1x_1} + \hat{w}_{x_2x_2} = \hat{w}_{x_3x_3} \tag{11.1-2}$$

from which it results by separation of variables and a suitable projection. It may also be obtained from the *ultrahyperbolic equation*

$$\hat{w}_{x_1x_1} + \hat{w}_{x_2x_2} = \hat{w}_{x_3x_3} + \hat{w}_{x_4x_4} \tag{11.1-3}$$

as well as from the *parabolic equation*

$$\hat{w}_{x_1x_1} + \hat{w}_{x_2x_2} = \hat{w}_{x_3x_3} + \hat{w}_{x_4}; \tag{11.1-4}$$

cf. Bauer [1965, 1966a].

Furthermore, $(1 - z_1\bar{z}_1)^2 D_1\overline{D}_1$ is the *Laplace operator of hyperbolic geometry*. Moreover, relations to the *special functions of mathematical physics* arise from the fact that the functions

$$P_n\left(\frac{1 + z_1\bar{z}_1}{1 - z_1\bar{z}_1}\right), \qquad Q_n\left(\frac{1 + z_1\bar{z}_1}{1 - z_1\bar{z}_1}\right)$$

(where P_n, Q_n denote the Legendre functions of the first and second kind, respectively) are particular solutions of (11.1-1).

By the use of

$$Z = i(1 + z_1)/(1 - z_1)$$

equation (11.1-1) can also be transformed into the form

$$(11.1\text{-}5) \qquad (Z - \bar{Z})^2 \check{w}_{Z\bar{Z}} + n(n + 1)\check{w} = 0;$$

cf. Part II by Ruscheweyh in Bauer–Ruscheweyh [1980]. This is the equation of certain *Eisenstein series*. Also, *nonanalytic automorphic forms* as solutions of this equation have stimulated investigations by several authors; see, e.g., Roelcke [1956].

RELATION TO GASPT

The transformation

$$\omega_n(Z, \bar{Z}) = (Z - \bar{Z})^{-n-1}\check{w}(Z, \bar{Z})$$

yields from (11.1-5) the equation

$$(11.1\text{-}6) \qquad (Z - \bar{Z})\omega_{n, Z\bar{Z}} + (n + 1)(\omega_{n, \bar{Z}} - \omega_{n, Z}) = 0$$

which is the complex form of the *equation of generalized axially symmetric potential theory* (GASPT). Indeed, consider the m-dimensional Laplace equation ($m \geq 2$) and set $x = x_1$, $y = +(x_2^2 + \cdots + x_m^2)^{1/2}$, then the GASPT equation is obtained:

$$(11.1\text{-}7) \qquad \hat{w}_{xx} + \hat{w}_{yy} + \frac{2\mu}{y}\hat{w}_y = 0 \quad \text{where} \quad 2\mu = m - 2;$$

cf. Gilbert [1969b]. According to (2.1-6, 7), choose

$$Z = x + iy, \qquad \bar{Z} = x - iy$$

and

$$m = 2n + 4.$$

Observing that $z_1\bar{z}_1 \neq 1$, $Z - \bar{Z} \neq 0$, and $y \neq 0$ correspond to each other, we have the equivalence of equations (11.1-6) and (11.1-7) for $z_1 \in B_1(0,1)$. Thus results on properties of solutions of (11.1-1) may be utilized to enrich GASPT. (For the introduction of analogous function theories for the heat, wave, Laplace, Euler–Poisson–Darboux, and the GASPT equations, see also the papers by Bragg–Dettman [1983, 1984].)

RELATION TO n-ANALYTIC AND n-HARMONIC FUNCTIONS

Finally, solutions of (11.1-1) are related to n-harmonic functions $\tilde{\omega}_n$ in the unit disk, as follows. The latter functions are solutions of

$$D_1\bar{D}_1\tilde{\omega}_n + \left[\frac{1}{2|z_1|}\frac{d}{d|z_1|}\log P_n\left(\frac{1+|z_1|^2}{1-|z_1|^2}\right)\right]\left(z_1D_1\tilde{\omega}_n + \bar{z}_1\bar{D}_1\tilde{\omega}_n\right) = 0 \tag{11.1-8}$$

and can be represented by *n-analytic functions* $\tilde{\omega}_{n1}, \tilde{\omega}_{n2}$ in the unit disk in the form

$$\tilde{\omega}_n = \tilde{\omega}_{n1} + \bar{\tilde{\omega}}_{n2}. \tag{11.1-9a}$$

Here, $\tilde{\omega}_{n1}$ and $\tilde{\omega}_{n2}$ are defined by

$$\tilde{\omega}_{nk}(z_1,\bar{z}_1) = \frac{D_1^n\bar{D}_1^n\left[g_k(z_1)(1-z_1\bar{z}_1)^{-1}\right]}{D_1^n\bar{D}_1^n\left[(1-z_1\bar{z}_1)^{-1}\right]} \tag{11.1-9b}$$

with suitable functions $g_k \in C^\omega(B_1(0,1))$. From representation (11.2-21) (below), one easily calculates that (11.1-9b) takes the form

$$\tilde{\omega}_{nk}(z_1,\bar{z}_1) = \tilde{T}_1\left(z_1^n g_k(z_1)\right)/\tilde{T}_1\left(z_1^n\right) \tag{11.1-10}$$

where $\tilde{T}_1$ is the differential operator obtained in (3.3-9), with $\lambda = -1$, $j = 1$, and $z_2 = \bar{z}_1$. Thus the n-analytic functions (11.1-9b) are representable by means of solutions of (11.1-1). Observing that

$$\begin{aligned}\tilde{T}_1\left(z_1^n\right) &= \sum_{m=0}^{n}\frac{(2n-m)!}{(n-m)!m!}\frac{n!}{(n-m)!}\left(\frac{z_1\bar{z}_1}{1-z_1\bar{z}_1}\right)^{n-m}\\ &= n!P_n\left(\frac{1+z_1\bar{z}_1}{1-z_1\bar{z}_1}\right),\end{aligned} \tag{11.1-11}$$

we have the result:

$\tilde{\omega}_{nk}$ is n-analytic in the unit disk if and only if

$$\tilde{\omega}_{nk}(z_1, \bar{z}_1) P_n\left(\frac{1+|z_1|^2}{1-|z_1|^2}\right)$$

is a solution of (11.1-1) of the form

$$\tilde{w}(z_1, \bar{z}_1) = \tilde{T}_1(z_1^n g_k(z_1)) \tag{11.1-12}$$

with a suitable function $g_k \in C^\omega(B_1(0,1))$ and $\tilde{T}_1$ as given by (3.3-9).

We shall not enter into a discussion of properties of solutions of (11.1-8), but refer to Ruscheweyh–Wirths [1976], where *Riemann's mapping theorem for n-analytic functions* is established, and to Ruscheweyh [1976], where *Hardy spaces of λ-harmonic functions* ($\lambda > 0$) are considered. (Cf. also Bauer–Ruscheweyh [1980], Part II.)

11.2 REPRESENTATIONS OF SOLUTIONS

We first recall the representations of $C^\omega(\Omega_0)$-solutions in a domain $\Omega_0 \subset B_1(0,1) \times B_2(0,1) \subset \mathbb{C}^2$ for the equation

$$D_1 D_2 w - n(n+1)(1 - z_1 z_2)^{-2} w = 0, \qquad n \in \mathbb{N} \tag{11.2-1}$$

resulting from our general theory in Chaps. 2, 3, 4, and 6. Obviously, the restriction of these results to the plane (when $z_2 = \bar{z}_1$) also holds; that is, they are valid for (11.1-1) with $z_1 \in B_1(0,1)$. Afterwards we shall derive some further useful representations of solutions of (11.1-1) which can be obtained from those others.

APPLICATION OF BERGMAN INTEGRAL OPERATORS

From Representation Theorem 2.3-1 and Example 2.5-5 it follows that

$$\tilde{k}_j(z,t) = \sum_{m=0}^{n} \binom{n+m}{2m} \left(\frac{4 z_1 z_2}{1 - z_1 z_2}\right)^m t^{2m}, \qquad j = 1,2 \tag{11.2-2}$$

are Bergman polynomial kernels of the first kind for Type II representations with $\eta_1 = \eta_2 = 0$ and $s = (0,0)$. They are uniquely determined according to Corollary 3.1-5. Moreover, they are minimal, as can be seen from Theorems 4.2-4 and 4.7-3a; cf. Lemma 4.2-7 an Example 4.7-4c.

Bergman integral type solutions are thus obtained by

$$(11.2\text{-}3a) \qquad w = w_1 + w_2, \qquad w_j = T_j f_j,$$

where

$$(11.2\text{-}3b) \quad T_j f_j(z) = \int_{-1}^{1} \tilde{k}_j(z,t) f_j\left(\tfrac{1}{2} z_j (1 - t^2)\right)(1 - t^2)^{-1/2}\, dt, \quad j = 1, 2.$$

Within $B_1(0,1) \times B_2(0,1)$ the domain of validity only depends on $\tilde{\Omega}_1, \tilde{\Omega}_2$ [cf. (2.2-8)]. The real solutions according to Lemma 2.1-3, i.e., with $f_2 = \overline{f_1}$, are real-valued if we assume $z_2 = \bar{z}_1$ as in Sec. 11.1.

APPLICATION OF BAUER DIFFERENTIAL OPERATORS

From (11.2-2, 3) the integral-free form of solutions by Bauer's differential operators according to Theorem 3.3-1 and Example 3.3-4 has the form

$$(11.2\text{-}4a) \qquad \tilde{w} = \tilde{w}_1 + \tilde{w}_2, \qquad \tilde{w}_j = \tilde{T}_j \tilde{f}_j$$

where

$$(11.2\text{-}4b) \quad \tilde{T}_j \tilde{f}_j(z) = \sum_{m=0}^{n} \frac{(2n-m)!}{m!(n-m)!} \left(\frac{z_{j*}}{1 - z_1 z_2} \right)^{n-m} D_j^m \tilde{f}_j(z_j), \quad j = 1, 2.$$

APPLICATION OF COMPLEX RIEMANN FUNCTION

The uniquely determined complex Riemann function of (11.2-1) is given in Example 6.4-5, namely,

$$(11.2\text{-}5a) \qquad R(z, \zeta) = \tilde{R}(\rho)$$

where

$$\tilde{R}(\rho) = {}_2F_1(n+1, -n; 1; \rho) = P_n(1 - 2\rho)$$

$$(11.2\text{-}5b) \qquad \rho(z, \zeta) = \frac{-(z_1 - \zeta_1)(z_2 - \zeta_2)}{(1 - z_1 z_2)(1 - \zeta_1 \zeta_2)}.$$

A given Goursat problem for (11.2-1) of the form

$$w(z_1, 0) = \varphi_1(z_1), \qquad w(0, z_2) = \varphi_2(z_2)$$

$$(11.2\text{-}6) \qquad \varphi_1(0) = \varphi_2(0), \qquad \varphi_j \in C^\omega(\Omega_j)$$

$$\Omega = \Omega_1 \times \Omega_2, \qquad \Omega_j = B_j(0,1), \qquad j = 1, 2$$

can be solved by the preceding methods and the use of Theorems 2.6-2, 3.4-2, 6.1-4, and 6.4-6.

SUMMARY OF RESULTS

We summarize these results as follows.

11.2-1 Theorem

(a) *The uniquely determined minimal polynomial kernels* $\tilde{k}_j$ *of the first kind for the Type II representation of solutions of* (11.2-1) *by means of Bergman integral operators* T_j [*with* $\eta_1 = \eta_2 = 0$, $s = (0,0)$] *are given by* (11.2-2). *For arbitrarily given functions* $f_j \in C^\omega(\tilde{\Omega}_j)$, *the function*

(11.2-7)

$$w(z) = \sum_{j=1}^{2} \sum_{m=0}^{n} \binom{n+m}{2m} \left(\frac{4z_1z_2}{1 - z_1z_2} \right)^m \int_{-1}^{1} t^{2m} f_j\left(\tfrac{1}{2}z_j(1-t^2)\right) \frac{dt}{(1-t^2)^{1/2}}$$

represents a $C^\omega(\Omega)$-*solution of* (11.2-1). *This solution is real-valued for* $z_2 = \bar{z}_1$ *if* $f_2(\bar{z}_1) = \overline{f_1(z_1)}$.

For the Goursat problem (11.2-1, 6) [*or for a given solution satisfying* (11.2-6) *with suitable functions* φ_j], *there exist* T_j-*associated functions* f_j, $j = 1,2$, *such that* (11.2-7) *solves the problem. The pair* (f_1, f_2) *satisfies* $\varphi_j(0) = \pi[f_1(0) + f_2(0)]$, *and the pair* $(D_1 f_1, D_2 f_2)$ *is uniquely determined. If we assume that*

$$f_j(0) = \frac{1}{\pi}\beta_j\varphi_j(0), \qquad j = 1,2; \qquad \beta_1 + \beta_2 = 1$$

then the solution of that Goursat problem is given by (11.2-7) *with*

$$(11.2\text{-}8) \qquad f_j(\xi_j) = \frac{1}{\pi}\beta_j\varphi_j(0) + \frac{4}{\pi}\xi_j \int_0^{\pi/2} \varphi_j'\left(2\xi_j \sin^2\theta\right) \sin\theta \, d\theta, \qquad j = 1,2.$$

(b) $\tilde{T}_j$ *in* (11.2-4b) *defines a Bauer differential operator for* (11.2-1), $j = 1,2$. *For arbitrarily given functions* $\tilde{f}_j \in C^\omega(\Omega_j)$, *the function*

$$(11.2\text{-}9) \qquad \tilde{w}(z) = \sum_{j=1}^{2} \sum_{m=0}^{n} \frac{(2n-m)!}{m!(n-m)!} \left(\frac{z_{j*}}{1 - z_1z_2} \right)^{n-m} D_j^m \tilde{f}_j(z_j)$$

represents a $C^\omega(\Omega)$-*solution of* (11.2-1).

For the Goursat problem (11.2-1, 6) [*or for a given solution satisfying* (11.2-6)], *there exist* $\tilde{T}_j$-*associated functions* $\tilde{f}_j$, $j = 1,2$, *such that* (11.2-9) *solves the problem. The pair* $(\tilde{f}_1, \tilde{f}_2)$ *is uniquely determined if one prescribes*

$$(11.2\text{-}10) \qquad \begin{aligned} \left(D_j^m \tilde{f}_j\right)(0) &= 0, \qquad && m = 0,\cdots, n-1, \\ \left(D_j^n \tilde{f}_j\right)(0) &= \beta_j\varphi_j(0), \qquad && j = 1,2; \quad \beta_1 + \beta_2 = 1. \end{aligned}$$

This yields as the solution of that Goursat problem representation (11.2-9) *with*

(11.2-11a) $$\tilde{f}_j(z_j) = \int_0^{z_j}\int_0^{z_{jn}}\cdots\int_0^{z_{j2}} h_j(z_{j1})\,dz_{j1}\cdots dz_{j,n-1}\,dz_{jn}$$

where

$$h_j(z_j) = \beta_j\varphi_j(0)$$

(11.2-11b) $$+\frac{2}{\pi}z_j\int_{-1}^{1}\int_0^{\pi/2}\varphi_j'\big(z_j(1-t^2)\sin^2\theta\big)\sin\theta\,d\theta\,(1-t^2)^{1/2}\,dt,$$

$$j = 1, 2.$$

(c) *R in* (11.2-5) *is the uniquely determined Riemann* (*–Vekua*) *function as well as a special Le Roux kernel* (*depending on a parameter*) *for* (11.2-1). *For arbitrarily given functions* $\hat{f}_j \in C^\omega(\Omega_j)$ *and constants* $\hat{\alpha}_j \in \mathbb{C}$, *the function*

(11.2-12) $$\tilde{\tilde{w}}(z) = \sum_{j=1}^{2}\left[\hat{\alpha}_j P_n\left(\frac{1+z_1z_2}{1-z_1z_2}\right) + \int_0^{z_j}\hat{f}_j(\zeta_j)P_n\left(1+2z_{j*}\frac{z_j-\zeta_j}{1-z_1z_2}\right)d\zeta_j\right]$$

represents a $C^\omega(\Omega)$*-solution of* (11.2-1).

For the Goursat problem (11.2-1, 6) [*or, for a given solution satisfying* (11.2-6)], *there exist pairs* $(\hat{f}_1, \hat{f}_2)$ *and* $(\hat{\alpha}_1, \hat{\alpha}_2)$ *such that* (11.2-12) *solves the problem. This holds for*

(11.2-13) $$\hat{f}_j(\zeta_j) = \varphi_j'(\zeta_j), \qquad \hat{\alpha}_j = \beta_j\varphi_j(0), \quad j = 1, 2; \quad \beta_1 + \beta_2 = 1$$

which yields the Riemann–Vekua representation of (11.2-1, 6).

Parts (a) and (b) of this preceding theorem deserve the following remark.

11.2-2 Remark

(a) As in the proofs of Theorems 2.6-2 and 6.4-6 we obtain for h_j in (11.2-11b) the expression

(11.2-14) $$h_j(z_j) = \varphi_j(z_j) - (1-\beta_j)\varphi_j(0)$$

so that (11.2-9) takes the form

$$\tilde{w}(z) = \sum_{j=1}^{2}\left[\varphi_j(z_j) - \beta_j\varphi_j(0)P_n\left(\frac{1+z_1z_2}{1-z_1z_2}\right)\right.$$

(11.2-15a)

$$\left. + \sum_{m=1}^{n}\frac{(n+m)!}{m!(n-m)!}\left(\frac{z_{j*}}{1-z_1z_2}\right)^m H_{jm}(z_j)\right]$$

where

(11.2-15b) $$H_{jm}(z_j) = \int_0^{z_j}\int_0^{z_{jm}}\cdots\int_0^{z_{j2}}\varphi_j(z_{j1})\,dz_{j1}\cdots dz_{j,m-1}\,dz_{jm}.$$

(b) The uniqueness of the $\tilde{T}_j$-associated functions $\tilde{f}_j$ as given by (11.2-11) was guaranteed by assumptions (11.2-10). If these conditions are dropped, from Example 4.6-2 (an application of Theorem 4.6-2) and Lemma 4.2-7 we obtain the following. The general representation of the identically vanishing solution by means of differential operators $\tilde{T}_1, \tilde{T}_2$, as shown in (11.2-9), is given by the pair $(\tilde{f}_{10}, \tilde{f}_{20})$, where

(11.2-16) $$\tilde{f}_{10}(z_1) = \sum_{\lambda=0}^{2n}\tilde{c}_\lambda z_1^\lambda, \qquad \tilde{f}_{20}(z_2) = -\sum_{\lambda=0}^{2n}\tilde{c}_{2n-\lambda}z_2^\lambda$$

with arbitrary constants $\tilde{c}_\lambda \in \mathbf{C}$ ($\lambda = 0,\cdots,2n$). Hence,

$$\tilde{w}(z) = \sum_{m=0}^{n}\frac{(n+m)!}{m![(n-m)!]^2}\left(\frac{z_1z_2}{1-z_1z_2}\right)^m \times$$

(11.2-17)

$$\times \sum_{\kappa=-m}^{n}(n+\kappa)!(\tilde{c}_{n+\kappa}z_1^\kappa - \tilde{c}_{n-\kappa}z_2^\kappa)$$

$$\equiv 0$$

in $B_1(0,1) \times B_2(0,1)$.

(c) Assume that a given solution $w_0 \in C^\omega(\Omega)$ can already be obtained by a Type I representation, say, by T_1 and $\tilde{T}_1$. Then the T_1- and $\tilde{T}_1$-associated functions $f_1, \tilde{f}_1$ in representations (11.2-7) and (11.2-9) are uniquely determined:

$$w_0 = w, \quad w \text{ as in (11.2-7) with } f_2 = 0 \quad \text{and}$$

(11.2-18)

$$f_1(\xi_1) = \frac{1}{\pi}w_0(0,0) + \frac{4}{\pi}\xi_1\int_0^{\pi/2}\left[\frac{d}{d\tilde{z}_1}w_0(\tilde{z}_1,0)\right]\Bigg|_{\tilde{z}_1=2\xi_1\sin^2\theta}\sin\theta\,d\theta$$

[cf. (11.2-8), Corollary 2.6-3]; and

$$w_0 = \tilde{w}, \quad \tilde{w} \text{ as in (11.2-9) with } \tilde{f}_2 = 0 \quad \text{and}$$

$$\tilde{f}_1(z_1) = \int_0^{z_1} \int_0^{z_{1n}} \cdots \int_0^{z_{12}} w_0(z_{11}, 0)\, dz_{11} \cdots dz_{1,n-1}\, dz_{1n} \tag{11.2-19}$$

[cf. (11.2-11a, 14)] or

$$\tilde{f}_1(z_1) = \frac{1}{(2n)!}\left[(1 - z_1 z_2)^2 D_2\right]^n w_0(z) \tag{11.2-20}$$

[derived from (11.2-9), $w_0 = \tilde{w}$].

(d) The solution $\tilde{w}_j = \tilde{T}_j \tilde{f}_j$ $(j = 1, 2)$ as given by (11.2-4b) may be converted to the form

$$\tilde{w}_j(z) = \frac{1}{n!}(1 - z_1 z_2)^{n+1} D_1^n D_2^n \left[\frac{\tilde{f}_j(z_j)}{z_j^n (1 - z_1 z_2)}\right]. \tag{11.2-21}$$

This follows by straightforward calculation; indeed,

$$\begin{aligned} D_1^n D_2^n \left[\frac{\tilde{f}_j(z_j)}{z_j^n (1 - z_1 z_2)}\right] &= D_{j^*}^n \left[n! \tilde{f}_j(z_j)(1 - z_1 z_2)^{-(n+1)}\right] \\ &= n! \sum_{m=0}^{n} \binom{n}{m} \left[D_{j^*}^{n-m}(1 - z_1 z_2)^{-(n+1)}\right] D_j^m \tilde{f}_j(z_j) \\ &= n! \sum_{m=0}^{n} \frac{n!}{m!(n-m)!} \frac{[n + (n-m)]!}{n!} z_j^{n-m} \times \\ &\quad \times (1 - z_1 z_2)^{-[n+1+(n-m)]} D_j^m \tilde{f}_j(z_j). \end{aligned}$$

Introduction of this expression into the right-hand side of (11.2-21) and comparison of the result with the right-hand side of (11.2-4b) proves formula (11.2-21). ■

FURTHER REPRESENTATIONS

Let us now, in the remainder of this section, concentrate our considerations on the case $z_2 = \bar{z}_1$, i.e., to equation (11.1-1) in the unit disk. We shall derive some further representations for $\tilde{w}_1 = \tilde{T}_1 \tilde{f}_1$.

We first observe that equation (11.1-1) is invariant under automorphisms of the unit disk. If such an automorphism is given by $\omega \mapsto \tilde{\omega} = \tilde{\omega}(\omega)$ and if we denote the operator $\tilde{T}_1$ in the respective variables $\omega, \tilde{\omega}$ by corresponding subscripts, namely, $\tilde{T}_{1(\omega)}, \tilde{T}_{1(\tilde{\omega})}$, we see by straightforward calculation that

$$\tilde{T}_{1(\tilde{\omega})}\big(\tilde{f}_1(\tilde{\omega})\big) = \tilde{T}_{1(\omega)}\left(\frac{\tilde{f}_1(\tilde{\omega}(\omega))}{[\tilde{\omega}'(\omega)]^n}\right). \tag{11.2-22}$$

Since $z_1\bar{z}_1 < 1$, a special automorphism of the unit disk is given by $\tilde{\omega}_0$, where

$$\tilde{\omega}_0(\omega) = \frac{\omega + z_1}{1 + \bar{z}_1\omega}. \tag{11.2-23}$$

Thus, (11.2-4b, 22, 23) yield

$$\begin{aligned}
\tilde{w}_1(z_1, \bar{z}_1) &= \tilde{T}_1\big(\tilde{f}_1(z_1)\big) = \left[\tilde{T}_{1(\tilde{\omega}_0)}\big(\tilde{f}_1(\tilde{\omega}_0)\big)\right]\Big|_{\omega=0} \\
&= \left[\tilde{T}_{1(\omega)}\big(\tilde{f}_1(\tilde{\omega}_0(\omega))(1 - z_1\bar{z}_1)^{-n}(1 + \bar{z}_1\omega)^{2n}\big)\right]\Big|_{\omega=0} \\
&= (1 - z_1\bar{z}_1)^{-n}\left[\frac{d^n}{d\omega^n}\left\{\tilde{f}_1\left(\frac{\omega + z_1}{1 + \bar{z}_1\omega}\right)(1 + \bar{z}_1\omega)^{2n}\right\}\right]\Bigg|_{\omega=0}.
\end{aligned} \tag{11.2-24}$$

If we set $\omega = z_1\beta/|z_1|$, for $\tilde{f}_1(z_1) = z_1^n g_1(z_1)$, the result (11.2-24) may also be rewritten as

$$\begin{aligned}
\tilde{w}_1(z_1, \bar{z}_1) &= \tilde{T}_1\big(z_1^n g_1(z_1)\big) \\
&= (1 - z_1\bar{z}_1)^{-n}\left[\frac{d^n}{d\beta^n}\left\{g_1\left(\frac{z_1 + (z_1/|z_1|)\beta}{1 + |z_1|\beta}\right) \times \right.\right. \\
&\qquad \left.\left. \times\big(|z_1| + \beta\big)^n(1 + |z_1|\beta)^n\right\}\right]\Bigg|_{\beta=0}.
\end{aligned} \tag{11.2-25}$$

We have thus proved the following corollary to Theorem 11.2-1b:

11.2-3 Corollary

(a) *For equation* (11.1-1), *solutions* $\tilde{w} = \tilde{w}_1$ *of the form* (11.2-9) (*with* $\tilde{f}_2 = 0$) *can be converted into the form* (11.2-24).

(b) *If* $\tilde{f}_1$ *satisfies* $\tilde{f}_1(z_1) = z_1^n g_1(z_1)$ ($g_1 \in C^\omega(\Omega_1)$, $z_1 \in \Omega_1 \subset B_1(0, 1)$), *then representation* (11.2-25) *also holds.*

By the use of Cauchy's formula in case (a) and of Schwarz's formula in case (b) (below), the differentiation in the representations (11.2-24) and (11.2-25) may be replaced by an integration. In particular, there also results a *generalization of Poisson's formula for disks* from harmonic functions to solutions of (11.1-1) [see (11.2-27) with $n = 0$ and (11.2-31), below].

11.2-4 Corollary

(a) *Let*

$$\tilde{f}_1 \in C^\omega(B_1(0,1)) \cap C^0(B_1(0,1) \cup \partial B_1(0,1)).$$

Then (11.2-24) *is equal to*

$$(11.2\text{-}26) \quad \tilde{T}_1(\tilde{f}_1(z_1)) = \frac{n!}{2\pi}\int_0^{2\pi} \tilde{f}_1\left(\frac{e^{i\varphi}+z_1}{1+\bar{z}_1 e^{i\varphi}}\right)\frac{\left(1+\bar{z}_1 e^{i\varphi}\right)^{2n}}{\left(1-z_1\bar{z}_1\right)^n}\frac{d\varphi}{e^{in\varphi}}.$$

(b) *For*

$$g_1 \in C^\omega(B_1(0,r_1)) \cap C^0(B_1(0,r_1) \cup \partial B_1(0,r_1)), \quad 0 < r_1 \le 1,$$

in $B_1(0, r_1)$ *representation* (11.2-25) *may be rewritten as*

$$(11.2\text{-}27) \quad \tilde{T}_1(z_1^n g_1(z_1)) = \frac{1}{2\pi}\int_0^{2\pi} g_1\left(r_1 e^{i\varphi}\right)\operatorname{Re}\left(\tilde{T}_1\left(z_1^n \frac{r_1 e^{i\varphi}+z_1}{r_1 e^{i\varphi}-z_1}\right)\right)d\varphi;$$

in particular for $r_1 = 1$,

$$(11.2\text{-}28) \quad \tilde{T}_1(z_1^n g_1(z_1)) = \frac{n!}{2\pi}\int_0^{2\pi} g_1(e^{i\varphi})\left[\frac{1-z_1\bar{z}_1}{|e^{i\varphi}-z_1|^2}\right]^{n+1} d\varphi.$$

Proof. Set

$$\tilde{F}(\omega) = \tilde{f}_1\left(\frac{\omega+z_1}{1+\bar{z}_1\omega}\right)(1+\bar{z}_1\omega)^{2n}$$

and use Cauchy's formula. Then

$$\left[\frac{d^n}{d\omega^n}\tilde{F}(\omega)\right]\Bigg|_{\omega=0} = \frac{n!}{2\pi i}\oint \tilde{F}(\breve{\omega})\frac{d\breve{\omega}}{(\breve{\omega}-\omega)^{n+1}}\Bigg|_{\omega=0}$$

$$= \frac{n!}{2\pi}\int_0^{2\pi}\tilde{F}(e^{i\varphi})e^{-in\varphi}\,d\varphi.$$

Introduction of this result into (11.2-24) yields statement (11.2-26) of case (a).

By an application of Schwarz's formula (cf. Lawrentjew–Schabat [1967], pp. 249–250), the function g_1 may be represented in the form

$$z_1^n g_1(z_1) = \frac{1}{2\pi}\int_0^{2\pi} \operatorname{Re}\left(g_1(r_1 e^{i\varphi})\right) z_1^n \frac{r_1 e^{i\varphi} + z_1}{r_1 e^{i\varphi} - z_1}\, d\varphi + i z_1^n \operatorname{Im}\left(g_1(0)\right).$$

We apply $\tilde{T}_1$ to both sides of this equation, use formula (11.2-21), and then apply the Re operator. Furthermore, we observe (11.1-11), which implies $\operatorname{Re}(i \operatorname{Im}(g_1(0))\tilde{T}_1(z_1^n)) = 0$. This proves

$$\begin{aligned} &\operatorname{Re} \tilde{T}_1(z_1^n g_1(z_1)) \\ &= \frac{1}{2\pi}\int_0^{2\pi} \operatorname{Re}\left(g_1(r_1 e^{i\varphi})\right) \operatorname{Re}\left(\tilde{T}_1\left(z_1^n \frac{r_1 e^{i\varphi} + z_1}{r_1 e^{i\varphi} - z_1}\right)\right) d\varphi \end{aligned} \tag{11.2-29}$$

that is, (11.2-27) insofar as the real parts are concerned. In order to show (11.2-27) also for the imaginary parts, we proceed as follows. We apply Schwarz's formula to $\operatorname{Re}(-ig_1) = \operatorname{Im}(g_1)$; then,

$$\begin{aligned} z_1^n(-ig_1(z_1)) &= \frac{1}{2\pi}\int_0^{2\pi} \operatorname{Re}\left(-ig_1(r_1 e^{i\varphi})\right) z_1^n \frac{r_1 e^{i\varphi} + z_1}{r_1 e^{i\varphi} - z_1}\, d\varphi \\ &\quad + i z_1^n \operatorname{Im}\left(-ig_1(0)\right). \end{aligned}$$

Similarly as in the case of (11.2-29) we thus get

$$\begin{aligned} \operatorname{Im} \tilde{T}_1(z_1^n g_1(z_1)) &= \operatorname{Re}\left(-i\tilde{T}_1(z_1^n g_1(z_1))\right) = \operatorname{Re} \tilde{T}_1(z_1^n(-ig_1(z_1))) \\ &= \frac{1}{2\pi}\int_0^{2\pi} \operatorname{Im}\left(g_1(r_1 e^{i\varphi})\right) \operatorname{Re}\left(\tilde{T}_1\left(z_1 \frac{r_1 e^{i\varphi} + z_1}{r_1 e^{i\varphi} - z_1}\right)\right) d\varphi. \end{aligned} \tag{11.2-30}$$

From (11.2-29, 30), representation (11.2-27) is obvious. Finally, representation (11.2-28) follows from (11.2-27) with $r_1 = 1$ if we observe that, according to (11.2-21),

$$\begin{aligned} &\operatorname{Re}\left(\tilde{T}_1\left(z_1^n \frac{e^{i\varphi} + z_1}{e^{i\varphi} - z_1}\right)\right) \\ &= \frac{1}{n!}(1 - z_1\bar{z}_1)^{n+1} \operatorname{Re}\left(D_1^n(e^{i\varphi} - z_1)^{-1} \bar{D}_1^n(e^{-i\varphi} - \bar{z}_1)^{-1}\right) \\ &= n!\left[\frac{1 - z_1\bar{z}_1}{|e^{i\varphi} - z_1|^2}\right]^{n+1}. \quad \blacksquare \end{aligned}$$

Let us take a look at the relation of (11.2-27) to the Poisson formula. It is obvious that the representations of solutions of (11.1-1) remain valid also in the case $n = 0$, that is, when (11.1-1) is equivalent to the Laplace equation in the unit disk. Hence from (11.2-4b) and (11.2-27) with $n = 0$, by introducing $z_1 = re^{i\psi}$, we obtain for $W(r, \psi) = \mathrm{Re}(g_1(re^{i\psi}))$

$$W(r, \psi) = \mathrm{Re}\left(\tilde{T}_1\left(g_1(re^{i\psi})\right)\right)$$

$$(11.2\text{-}31) \qquad = \frac{1}{2\pi}\int_0^{2\pi} \mathrm{Re}\left(g_1(r_1e^{i\varphi})\right) \mathrm{Re}\left(\frac{r_1e^{i\varphi} + re^{i\psi}}{r_1e^{i\varphi} - re^{i\psi}}\right) d\varphi$$

$$= \frac{1}{2\pi}\int_0^{2\pi} W(r_1, \varphi)\frac{r_1^2 - r^2}{r_1^2 - 2r_1r\cos(\psi - \varphi) + r^2}\, d\varphi.$$

That is, we obtain the Poisson formula for harmonic functions in disks. This agrees, for instance, with the result in Behnke–Sommer [1972], p. 169 (Theorem II.54); cf. also Conway [1973], 262 (formula 2.13).

Furthermore, (11.2-28) is related to the so-called *Laplace's second integral for P_n* (cf. Whittaker–Watson [1963], p. 314). Indeed, if we specialize $g_1(z_1) \equiv 1$ and use representation (11.1-11) for P_n, then (11.2-28) immediately yields

$$(11.2\text{-}32) \qquad P_n\left(\frac{1 + |z_1|^2}{1 - |z_1|^2}\right) = \frac{1}{2\pi}\int_0^{2\pi}\left[\frac{1 - |z_1|^2}{|e^{i\varphi} - z_1|^2}\right]^{n+1} d\varphi.$$

MULTIPLICATION THEOREM FOR BAUER DIFFERENTIAL OPERATORS

Establishing a *generalized Leibniz formula*, we can derive a multiplication theorem for differential operators $\tilde{T}_1$ which is a valuable tool in proving properties of solutions (see, e.g., Theorem 11.3-8; cf. also Ruscheweyh [1972b]).

11.2-5 Theorem

Let

$$(11.2\text{-}33) \qquad \tilde{T}_{1(\kappa)} = \sum_{m=0}^{\kappa} \frac{(2\kappa - m)!}{m!(\kappa - m)!}\left(\frac{\bar{z}_1}{1 - z_1\bar{z}_1}\right)^{\kappa - m} D_1^m;$$

that is, $\tilde{T}_{1(\kappa)}$ is a Bauer differential operator for the equation

$$(1 - z_1\bar{z}_1)^2 D_1\bar{D}_1 w - \kappa(\kappa + 1)w = 0, \qquad \kappa = 0, \cdots, n$$

especially $\tilde{T}_{1(n)} = \tilde{T}_1$ for (11.1-1). Define $\tilde{T}_{1(-1)}D_1 = I$, the identity. Let $\tilde{f}_{11}, \tilde{f}_{12} \in C^\omega(\Omega_1)$ and $\Omega_1 \subset B_1(0,1)$. Then

$$\tilde{T}_{1(n)}\left(z_1^n \tilde{f}_{11}(z_1)\tilde{f}_{12}(z_1)\right)$$

(11.2-34)

$$= \sum_{m=0}^{n} \binom{n}{m} \tilde{T}_{1(m)}\left(z_1^m \tilde{f}_{11}(z_1)\right) \tilde{T}_{1(n-m-1)}\left(z_1^{n-m} D_1 \tilde{f}_{12}(z_1)\right)$$

holds.

Proof. We use the well-known Lagrange–Bürmann formula (see, e.g., Henrici [1974–1986], vol. I, Sec. 1.9). As was shown by Ruscheweyh [1973a], this formula is equivalent to the generalized Leibniz formula

(11.2-35)

$$(D_1^n(uv))(z_1) = \sum_{m=0}^{n} \binom{n}{m} \left(D_1^{m-1}\left(\frac{1}{h^m} D_1 v\right)\right)(z_1)(D_1^{n-m}(uh^m))(z_1);$$

here, $D_1^{-1}D_1 = I$, and u, v, h are assumed to be sufficiently often differentiable, with $h(z_1) \neq 0$ in the domain considered. (Obviously, for $h(z_1) \equiv 1$ this formula becomes the familiar Leibniz formula.) We use representation (11.2-21) for $\tilde{T}_{1(\kappa)}$ and apply formula (11.2-35) in $\Omega_1 \setminus \{0\}$ to the functions given by

$$u(z_1) = \tilde{f}_{11}(z_1) z_1^n (1 - z_1\bar{z}_1)^{-(n+1)}$$

$$v(z_1) = \tilde{f}_{12}(z_1), \qquad h(z_1) = (1 - z_1\bar{z}_1) z_1^{-1}.$$

The result is (11.2-34), which also holds for $z_1 = 0$ if $\Omega_1 \ni 0$. ∎

APPLICATION TO LEGENDRE FUNCTIONS

Theorem 11.2-5 includes an interesting result from the theory of Legendre functions. To begin with, let us first consider a representation of the Legendre function Q_n of the second kind. We have

$$-\operatorname{Re}\tilde{T}_{1(n)}\left(z_1^n\frac{1}{n!}\log z_1\right)$$

$$= -\frac{1}{2}\left[\tilde{T}_{1(n)}\left(z_1^n\frac{1}{n!}\log z_1\right) + \overline{\tilde{T}_{1(n)}\left(z_1^n\frac{1}{n!}\log z_1\right)}\right]$$

$$(11.2\text{-}36)\qquad = -\frac{1}{2}\sum_{m=0}^{n}\frac{(2n-m)!}{m![(n-m)!]^2}\left(\frac{z_1\bar{z}_1}{1-z_1\bar{z}_1}\right)^{n-m}(\log z_1 + \log \bar{z}_1)$$

$$-2\cdot\frac{1}{2}\sum_{m=1}^{n}\frac{(2n-m)!}{m![(n-m)!]^2}\left(\frac{z_1\bar{z}_1}{1-z_1\bar{z}_1}\right)^{n-m}\times$$

$$\times\sum_{k=1}^{m}\frac{m!(n-m)!(-1)^{k-1}}{(m-k)!(n-m+k)!k}.$$

Setting

$$X = \frac{1+z_1\bar{z}_1}{1-z_1\bar{z}_1}$$

we obtain

$$(z_1\bar{z}_1)^{-1} = \frac{X+1}{X-1}, \qquad \frac{z_1\bar{z}_1}{1-z_1\bar{z}_1} = -\frac{1-X}{2}.$$

Defining

$$\sigma(0) = 0, \qquad \sigma(m) = \sum_{k=1}^{m}\frac{1}{k}, \qquad m \in \mathbb{N}$$

and using the identity

$$-\sum_{k=1}^{m}\frac{m!(n-m)!(-1)^{k-1}}{(m-k)!(n-m+k)!k} = \sigma(n-m) - \sigma(n)$$

we see that the right-hand side of (11.2-36) takes the form

(11.2-37)

$$\frac{1}{2}P_n(X)\log\frac{X+1}{X-1} + \sum_{m=0}^{n}\frac{(n+m)!(-1)^m}{(n-m)!(m!)^2}\left(\frac{1-X}{2}\right)^m[\sigma(m)-\sigma(n)].$$

This is a well-known representation of $Q_n(X)$ (cf. Erdélyi et al. [1953–1955], vol. I, pp. 149–150).

We now apply Theorem 11.2-5, setting

$$\tilde{f}_{11}(z_1) \equiv \frac{1}{n!}, \qquad \tilde{f}_{12}(z_1) = \log z_1 .$$

Using (11.1-11), we then obtain

(11.2-38)

$$\tilde{T}_{1(n)}\left(z_1^n \frac{1}{n!} \log z_1\right) = \sum_{m=0}^{n-1} \frac{1}{n-m} P_m(X) P_{n-m-1}(X) + P_n(X) \log z_1 .$$

Thus, from (11.2-36, 37, 38) it follows that

(11.2-39)

$$Q_n(X) = \frac{1}{2} P_n(X) \log \frac{X+1}{X-1} - \sum_{k=0}^{n-1} \frac{1}{k+1} P_k(X) P_{n-1-k}(X).$$

This is the *recursion formula for Legendre functions of Hermite and Schläfli* (cf. Whittaker–Watson [1963], p. 333, no. 28).

11.3 PROPERTIES OF SOLUTIONS SATISFYING ADDITIONAL CONDITIONS

In the preceding section we considered properties of representations of solutions, and we shall continue this discussion in the present section. We first emphasize that all results of Chaps. 8–10 remain also valid for solutions of equation (11.1-1), except for those which require entire coefficients of the equation in its canonical form (2.1-1), a condition imposed, for instance, in the first three sections of Chap. 10.

MAXIMUM MODULUS THEOREM FOR SOLUTIONS

We want to emphasize that, in addition to the reasons recalled in Sec. 11.1, there is a further important reason why we have restricted our demonstration in this chapter to a special class of equations, such as (11.1-1). The function theory of solutions may be expected to be the richer, the smaller the class of equations under consideration is. Let us show this by a simple example. For a moment let us consider the case that in equation (2.5-13) $\lambda = +1$ instead of $\lambda = -1$ is chosen; that is, we consider

(11.3-1)
$$(1 + z_1\bar{z}_1)^2 D_1\bar{D}_1 w + n(n+1)w = 0, \qquad n \in \mathbb{N} .$$

If we introduce

$$\tilde{f}_1(z_1) = (-1)^n (n-1)! z_1 / (2n-1)!$$

into the Type I representation of solutions, then by applying Theorem 3.3-1 and Example 2.5-5 we obtain the special solution of equation (11.3-1) given by

$$\tilde{w}_1(z_1, \bar{z}_1) = (z_1\bar{z}_1 - 1)\bar{z}_1^{n-1}/(1 + z_1\bar{z}_1)^n . \tag{11.3-2}$$

This solution is valid for all $z_1 \in \mathbf{C}$ and vanishes for $z_1\bar{z}_1 = 1$, but it does not vanish for $0 < z_1\bar{z}_1 < 1$. Thus the Maximum Modulus Theorem does not generally hold for solutions of (11.3-1). Such a theorem may, however, be proved in the case of equation (11.1-1).

11.3-1 Remark

A solution of (11.1-1) that is defined for z_1 in a subdomain Ω_1 of $B_1(0,1)$ and attains its maximum modulus at an interior point of Ω_1 is constant, and even vanishes identically.

For the *proof* associate the real-valued function ω to an arbitrarily given solution w of (11.1-1) by setting

$$\omega(x_1, x_2) = w(z_1, \bar{z}_1)\bar{w}(\bar{z}_1, z_1) = |w(z_1, \bar{z}_1)|^2, \quad z_1 = x_1 + ix_2 .$$

A necessary condition for ω to possess a maximum for some $(x_1^{(0)}, x_2^{(0)})$ with $z_1^{(0)} = x_1^{(0)} + ix_2^{(0)} \in B_1(0,1)$ is $\Delta_2\omega|_{(x_1^{(0)}, x_2^{(0)})} \leq 0$ or, what amounts to the same,

$$4D_1\bar{D}_1(w\bar{w})|_{z_1 = x_1^{(0)} + ix_2^{(0)}} \leq 0 . \tag{11.3-3}$$

On the other hand, from (11.1-1) it follows by a simple calculation that for $z_1 \in B_1(0,1)$,

$$4D_1\bar{D}_1(w\bar{w}) = 8|D_1 w|^2 + 8\frac{n(n+1)}{\left(1 - |z_1|^2\right)^2}|w|^2 \geq 0 . \tag{11.3-4}$$

From (11.3-3, 4) we get $w = 0$ as stated. ■

Thus we have proved a fundamental difference in the behavior of solutions of equations (11.3-1) and (11.1-1). From now on we consider exclusively equation (11.1-1). In the sequel we shall use the operators $\tilde{T}_1$ and $\tilde{T}_2$ given by (11.2-4b) in the plane $(z_2 = \bar{z}_1)$ and shall apply them to functions $\tilde{f}_1$ and $\tilde{f}_2$ [depending on z_1 and $\bar{z}_1$, respectively, where z_1 varies in a subset of the unit disk $B_1(0,1)$]. For the restrictions of the operators and functions to the plane we shall not introduce new notations, with one exception: Instead of $\tilde{T}_2$ we shall write $\tilde{\bar{T}}_1$, for obvious reasons.

REAL-VALUED SOLUTIONS

The explicit form of the special solution $\tilde{w}_1$ with $\tilde{w}_1(z_1, \bar{z}_1) = \tilde{T}_1(z_1^n)$ as given in (11.1-11) shows that domains of the plane may be mapped onto one-dimensional sets (e.g., an interval of the real axis) already in the case of Type I representations. In general we have the following.

11.3-2 Remark

A solution of (11.1-1) given in the form

$$(11.3\text{-}5)\quad \tilde{W}(z_1, \bar{z}_1) = \tfrac{1}{2}\left[\tilde{T}_1\left(\tilde{f}_1(z_1)\right) + \tilde{\bar{T}}_1\left(\tilde{\bar{f}}_1(\bar{z}_1)\right)\right] = \operatorname{Re}\tilde{T}_1\left(\tilde{f}_1(z_1)\right)$$

is real-valued. The same solution is also obtained by

$$(11.3\text{-}6)\quad \begin{aligned}\tilde{W}(z_1, \bar{z}_1) = \tfrac{1}{2}\Big[&\tilde{T}_1\left(\tilde{f}_1(z_1) + \Pi_{2n}(z_1)\right)\\ &+\tilde{\bar{T}}_1\left(\tilde{\bar{f}}_1(\bar{z}_1) - \bar{z}_1^{2n}\Pi_{2n}(1/\bar{z}_1)\right)\Big]\end{aligned}$$

where Π_{2n} is an arbitrary polynomial of degree $\leq 2n$. Here and in the sequel,

$$(11.3\text{-}7)\quad \tilde{\bar{T}}_1 = \sum_{m=0}^{n} \frac{(2n-m)!}{m!(n-m)!}\left(\frac{z_1}{1 - z_1\bar{z}_1}\right)^{n-m}\bar{D}_1$$

[cf. $\tilde{T}_2$ in (11.2-4b)]; $\tilde{\bar{f}}_1$ is given by $\tilde{\bar{f}}_1(\bar{z}_1) = \overline{\tilde{f}_1(z_1)}$.

The *proof* of this remark follows simply from Theorem 11.2-1a, b and Remark 11.2-2b.

SOLUTIONS WITH EQUAL REAL PARTS OR EQUAL IMAGINARY PARTS

From Remark 11.2-2b we can immediately conclude:

11.3-3 Remark

(a) Solutions

$$(11.3\text{-}8)\quad \tilde{w} = \tilde{T}_1\tilde{f}_{11} + \tilde{\bar{T}}_1\tilde{f}_{21}$$

and

$$\tilde{\tilde{w}} = \tilde{T}_1 \tilde{f}_{12} + \tilde{\bar{T}}_1 \tilde{f}_{22} \tag{11.3-9}$$

have the same real parts if the associated functions satisfy the equation

$$\tilde{f}_{12} + \bar{\tilde{f}}_{22} = \tilde{f}_{11} + \bar{\tilde{f}}_{21} + P$$

where P is a polynomial of degree $\leq 2n$ satisfying the condition

$$P(z_1) + z_1^{2n} \overline{P(1/\bar{z}_1)} = 0 .$$

(b) Solutions of the form (11.3-8) and (11.3-9) have the same imaginary parts if the associated functions satisfy the equation

$$\tilde{f}_{12} - \bar{\tilde{f}}_{22} = \tilde{f}_{11} - \bar{\tilde{f}}_{21} + Q$$

where Q is a polynomial of degree $\leq 2n$ satisfying the condition

$$Q(z_1) - z_1^{2n} \overline{Q(1/\bar{z}_1)} = 0 .$$

(c) The same results also hold in the case of Type I representations, if, for instance,

$$\tilde{f}_{21} = 0 \qquad \text{and} \qquad \tilde{f}_{22} = 0$$

are chosen.

SOLUTIONS BOUNDED ON DISKS OR CONSTANT ON CIRCLES

Let us now investigate solutions of (11.1-1) which are bounded on certain disks or which are constant on certain curves (circles). Here we first obtain

11.3-4 Theorem

Let $\tilde{w}$ be given by a Type II representation of solutions of equation (11.1-1) *by means of operators $\tilde{T}_1, \tilde{\bar{T}}_1$.*

(a) *If $\tilde{w}$ remains bounded for all $z_1 \in B_1(0,1)$ and if the associated functions $\tilde{f}_1, \tilde{f}_2$ are holomorphic on the closed unit disk, then*

$$\tilde{w} = 0 .$$

(b) *Let $\tilde{w}$ be defined for*

$$z_1 \in B_1(0, r) \cup \partial B_1(0, r), \qquad 0 < r < 1$$

and let $\tilde{w}(z_1, \bar{z}_1) = C$ *(C being an arbitrary complex constant) for* $z_1 \in \partial B_1(0, r)$. *Then* $\tilde{w}$ *is defined for all* $z_1 \in B_1(0, 1)$ *and can already be obtained by a Type I representation of the form*

$$\tilde{w}(z_1, \bar{z}_1) = \tilde{T}_1(\tilde{\gamma}(n, r) z_1^n) \tag{11.3-10a}$$

where

$$\tilde{\gamma}(n, r) = C\left[n! \sum_{\nu=0}^{n} \binom{2n-\nu}{n}\binom{n}{\nu}\left(\frac{r^2}{1-r^2}\right)^{n-\nu}\right]^{-1}. \tag{11.3-10b}$$

Proof. The proofs of both parts utilize the uniqueness of Fourier series expansions. From Theorem 11.2-1b and our assumptions concerning the boundedness of $\tilde{w}$ and $\tilde{f}_1, \tilde{f}_2$ in part (a), we conclude that in the representation

$$\tilde{w}(z_1, \bar{z}_1) = \sum_{m=0}^{n} \frac{(2n-m)!}{m!(n-m)!} \frac{1}{(1 - z_1\bar{z}_1)^{n-m}} \times$$
$$\times\left[\bar{z}_1^{n-m} D_1^m \tilde{f}_1(z_1) + z_1^{n-m} \bar{D}_1^m \tilde{f}_2(\bar{z}_1)\right]$$

the expression in brackets $[\cdots]$ must necessarily be zero of order n at least, for $m = 0$ and $z_1\bar{z}_1 = 1$; hence,

$$\bar{z}_1^n \tilde{f}_1(z_1) + z_1^n \tilde{f}_2(\bar{z}_1) = 0 \qquad \text{for} \qquad z_1\bar{z}_1 = 1. \tag{11.3-11}$$

From the Maclaurin expansions of $\tilde{f}_1$ and $\tilde{f}_2$, which is also valid on the boundary of the unit disk, we get

$$\tilde{f}_1(z_1) = \sum_{\nu=0}^{\infty} \tilde{\gamma}_{1\nu} e^{i\nu\varphi}, \qquad \tilde{f}_2(\bar{z}_1) = \sum_{\nu=0}^{\infty} \tilde{\gamma}_{2\nu} e^{-i\nu\varphi}$$

for $z_1 = e^{i\varphi} \in \partial B_1(0, 1)$, with uniquely determined constants $\tilde{\gamma}_{1\nu}, \tilde{\gamma}_{2\nu} \in \mathbb{C}$. Thus, for all $\varphi \in [0, 2\pi)$, equation (11.3-11) can be rewritten as

$$\begin{aligned} 0 &= \sum_{\nu=0}^{\infty} \tilde{\gamma}_{1\nu} e^{i(\nu-n)\varphi} + \sum_{\nu=0}^{\infty} \tilde{\gamma}_{2\nu} e^{i(n-\nu)\varphi} \\ &= \sum_{\mu=-\infty}^{\infty} \delta_\mu e^{i\mu\varphi} \end{aligned} \tag{11.3-12}$$

where

$$\delta_\mu = \begin{cases} \tilde{\gamma}_{2,\, n-\mu} & \text{if } \mu < -n \\ \tilde{\gamma}_{1,\, n+\mu} + \tilde{\gamma}_{2,\, n-\mu} & \text{if } -n \le \mu \le n \\ \tilde{\gamma}_{1,\, n+\mu} & \text{if } \mu > n. \end{cases} \tag{11.3-13}$$

Introducing $\delta_{1\mu} = \operatorname{Re} \delta_\mu$, $\delta_{2\mu} = \operatorname{Im} \delta_\mu$ into (11.3-12) and rearranging the terms in the result, we obtain

$$0 = \left\{ \delta_{10} + \sum_{\mu=1}^{\infty} \left[(\delta_{1\mu} + \delta_{1,-\mu}) \cos(\mu\varphi) - (\delta_{2\mu} - \delta_{2,-\mu}) \sin(\mu\varphi) \right] \right\}$$

$$+ i \left\{ \delta_{20} + \sum_{\mu=1}^{\infty} \left[(\delta_{2\mu} + \delta_{2,-\mu}) \cos(\mu\varphi) + (\delta_{1\mu} - \delta_{1,-\mu}) \sin(\mu\varphi) \right] \right\}.$$

The expressions in the braces $\{ \cdots \}$ are real. By the use of the uniqueness theorem for Fourier series expansions, a comparison of the coefficients thus yields the system

$$\begin{aligned} &\delta_{10} = 0, \qquad && \delta_{1\mu} + \delta_{1,-\mu} = 0, \qquad && \delta_{1\mu} - \delta_{1,-\mu} = 0, \\ &\delta_{20} = 0, && \delta_{2\mu} + \delta_{2,-\mu} = 0, && \delta_{2\mu} - \delta_{2,-\mu} = 0, \qquad \mu \in \mathbb{N}. \end{aligned}$$

Hence we conclude that $\delta_\mu = 0$, $-\infty < \mu < \infty$. From this, by the definition of the δ_μ's in (11.3-13), we get

$$\tilde{\gamma}_{2\nu} = -\tilde{\gamma}_{1,2n-\nu} \qquad \text{for } 0 \le \nu \le 2n$$

and

$$\tilde{\gamma}_{1\nu} = 0, \qquad \tilde{\gamma}_{2\nu} = 0 \qquad \text{for } \nu > 2n.$$

This means that the associated functions $\tilde{f}_1$, $\tilde{f}_2$ are of the form (11.2-16). Hence by Remark 11.2-2b, the solution $\tilde{w} = \tilde{T}_1 \tilde{f}_1 + \bar{\tilde{T}}_1 \tilde{f}_2$ vanishes identically. This proves part (a) of the theorem.

According to Theorem 11.2-1b and Remark 11.2-2b, for $\tilde{w}$ there exist two associated functions $\tilde{f}_1$, $\bar{\tilde{f}}_2$ defined and holomorphic at least on the closed disk $B_1(0, r) \cup \partial B_1(0, r)$ and being uniquely determined up to a polynomial of degree not exceeding $2n$ [cf. (11.2-16)]. Introducing the Maclaurin expansions of $\tilde{f}_1$ and $\tilde{f}_2$ into the representation of $\tilde{w}$, for $z_1 = re^{i\varphi} \in \partial B_1(0, r)$ we obtain

$$C = \tilde{w}(z_1, \bar{z}_1) = \sum_{m=0}^{n} \frac{(2n-m)!}{m!(n-m)!} \frac{1}{(1-r^2)^{n-m}} \times$$

$$\times \sum_{\nu=m}^{\infty} \frac{\nu!}{(\nu-m)!} r^{n+\nu-2m} \left[\tilde{\gamma}_{1\nu} e^{i(\nu-n)\varphi} + \tilde{\gamma}_{2\nu} e^{-i(\nu-n)\varphi} \right].$$

Here $\tilde{\gamma}_{1\nu}$ and $\tilde{\gamma}_{2\nu}$ denote the coefficients in the expansion of $\tilde{f}_1$ and $\tilde{f}_2$, respectively. By a simple calculation this equation is converted into the form

$$C = \sum_{\mu=-\infty}^{\infty} \delta_\mu e^{i\mu\varphi} \tag{11.3-14}$$

where

$$(11.3\text{-}15)\quad \tilde{\delta}_\mu = \begin{cases} \kappa_{-\mu}(n, r)\tilde{\gamma}_{2, n-\mu} & \text{if } \mu < -n \\ \kappa_{-\mu}(n, r)\tilde{\gamma}_{2, n-\mu} + \kappa_\mu(n, r)\tilde{\gamma}_{1, n+\mu} & \text{if } -n \leq \mu \leq n \\ \kappa_\mu(n, r)\tilde{\gamma}_{1, n+\mu} & \text{if } \mu > n \end{cases}$$

with real positive

$$\kappa_{\pm\mu}(n, r) = \sum_{m=0}^{N} \frac{(2n-m)!(n \pm \mu)!}{m!(n-m)!(n-m \pm \mu)!}\left(\frac{r^2}{1-r^2}\right)^{n-m} r^{\pm\mu}$$

where $N = \min(n, n \pm \mu)$. As in the case of equation (11.3-12), we split (11.3-14) into its real and imaginary parts. Thus, in the same way as in the proof of part (a), by the uniqueness theorem for Fourier series expansions and comparison of the coefficients, we get

$$\tilde{\delta}_\mu = 0 \qquad \text{for } \mu \neq 0$$

and

$$C = \tilde{\delta}_0 = \kappa_0(n, r)[\tilde{\gamma}_{1n} + \tilde{\gamma}_{2n}].$$

Because of (11.3-15) this yields

$$(11.3\text{-}16)\qquad \tilde{\gamma}_{1\nu} = 0, \qquad \tilde{\gamma}_{2\nu} = 0 \quad \text{for } \nu > 2n$$

and

$$(11.3\text{-}17)\quad C = (\tilde{\gamma}_{1n} + \tilde{\gamma}_{2n})n! \sum_{m=0}^{n} \binom{2n-m}{n}\binom{n}{m}\left(\frac{r^2}{1-r^2}\right)^{n-m}.$$

Moreover, for $-n \leq \mu < 0$, $0 < \mu \leq n$ we have

$$(11.3\text{-}18)\qquad \kappa_\mu(n, r)\tilde{\gamma}_{1, n+\mu} + \kappa_{-\mu}(n, r)\tilde{\gamma}_{2, n-\mu} = 0.$$

Multiplying (11.3-18) by $(1 - r^2)^n$ and expanding the result, by a formal computation we obtain

$$(11.3\text{-}19)\quad \begin{aligned} &\tilde{\gamma}_{1, n+\mu} n! \sum_{\rho=0}^{n+\mu} \binom{n+\mu}{\rho} \sigma_1(n, \mu, \rho) r^{2n-2\rho+\mu} \\ &+ \tilde{\gamma}_{2, n-\mu} n! \sum_{\rho=0}^{n} \binom{n-\mu}{\rho} \sigma_2(n, \mu, \rho) r^{2n-2\rho-\mu} = 0 \end{aligned}$$

for $-n \leq \mu < 0$. Here

$$\sigma_1(n, \mu, \rho) = \sum_{\omega=0}^{n+\mu-\rho} (-1)^{\omega} \binom{2n-\rho-\omega}{n} \binom{n+\mu-\rho}{\omega}$$

and

$$\sigma_2(n, \mu, \rho) = \sum_{\omega=0}^{n-\rho} (-1)^{\omega} \binom{2n-\rho-\omega}{n} \binom{n-\mu-\rho}{\omega}.$$

By the use of the formula

$$\sum_{\omega=0}^{\chi} (-1)^{\omega} \binom{p-\omega}{q} \binom{\chi}{\omega} = \begin{cases} \binom{p-\chi}{q-\chi} & \text{for } \chi \leq q \\ 0 & \text{for } \chi > q, \end{cases}$$

[cf. Netto [1927], p. 252, formula (27)], we can evaluate these expressions, so that we have from (11.3-19) for $-n \leq \mu < 0$,

$$\begin{aligned} 0 &= \tilde{\gamma}_{1,n+\mu} n! \sum_{\rho=0}^{n+\mu} \binom{n+\mu}{\rho} \binom{n-\mu}{\rho-\mu} r^{2n-2\rho+\mu} \\ &\quad + \tilde{\gamma}_{2,n-\mu} n! \sum_{\rho=-\mu}^{n} \binom{n-\mu}{\rho} \binom{n+\mu}{\rho+\mu} r^{2n-2\rho-\mu} \\ &= (\tilde{\gamma}_{1,n+\mu} + \tilde{\gamma}_{2,n-\mu}) n! \sum_{\rho=0}^{n+\mu} \binom{n+\mu}{\rho} \binom{n-\mu}{\rho-\mu} r^{2n-2\rho+\mu}. \end{aligned} \tag{11.3-20}$$

Since for $r > 0$ and $-n \leq \mu < 0$ the sum on the right-hand side is a positive real, (11.3-20) implies

$$\tilde{\gamma}_{2\nu} = -\tilde{\gamma}_{1,2n-\nu} \qquad \text{for } \nu = n - \mu = n+1, \cdots, 2n. \tag{11.3-21}$$

Similarly, from (11.3-18) one can derive the result (11.3-21) also in the case $0 < \mu \leq n$, i.e., for $\nu = n - \mu = 0, \cdots, n-1$. According to (11.3-16, 21) and Remark 11.2-2b, the generated solution $\tilde{w} = \tilde{T}_1 \tilde{f}_1 + \bar{\tilde{T}}_1 \tilde{f}_2$ thus reduces to

$$\begin{aligned} \tilde{w}(z_1, \bar{z}_1) &= \tilde{T}_1(\tilde{\gamma}_{1n} z_1^n) + \bar{\tilde{T}}_1(\tilde{\gamma}_{2n} \bar{z}_1^n) \\ &= \tilde{T}_1((\tilde{\gamma}_{1n} + \tilde{\gamma}_{2n}) z_1^n). \end{aligned} \tag{11.3-22}$$

Because of (11.3-17) the sum $\tilde{\gamma}_{1n} + \tilde{\gamma}_{2n}$ equals $\tilde{\gamma}(n, r)$ in (11.3-10b), and (11.3-22) is the desired representation (11.3-10a), which obviously implies that $\tilde{w}$ is defined for all $z_1 \in B_1(0, 1)$. This proves part (b) of the theorem. ■

11.3-5 Remark

According to Theorems 11.2-1b and 11.3-4b, all solutions of (11.1-1) that are continuous on $B_1(0, r) \cup \partial B_1(0, r)$, regular in $B_1(0, r)$ and constant on the circle $\partial B_1(0, r)$, $0 < r < 1$, can be represented in the form (11.3-10). By (11.1-11), this yields that (11.3-10) is representable as

$$\tilde{w}_1(z_1, \bar{z}_1) = \tilde{T}_1(\tilde{\gamma}(n, r) z_1^n) = \tilde{\gamma}(n, r) n! P_n\left(\frac{1 + z_1\bar{z}_1}{1 - z_1\bar{z}_1}\right) \tag{11.3-23}$$

where $\tilde{\gamma}(n, r)n! = const$ [cf. (11.3-10b)]. If we ask for all solutions of (11.1-1) that are constant on $\partial B_1(0, r)$, but not necessarily defined on all of $B_1(0, r)$, this means that we seek all solutions of the form

$$\tilde{w}_1(z_1, \bar{z}_1) = \omega(\chi) \tag{11.3-24}$$

where $\chi = (1 + z_1\bar{z}_1)/(1 - z_1\bar{z}_1)$. By the use of (11.3-24), for $z_1 \in B_1(0, 1)$, equation (11.1-1) takes the equivalent form

$$(\chi^2 - 1)\omega'' + 2\chi\omega' - n(n + 1)\omega = 0. \tag{11.3-25}$$

This is the Legendre differential equation (cf. also Example 5.4-5). Thus all solutions of the desired form turn out to be

$$\alpha_1 P_n(\chi) + \alpha_2 Q_n(\chi) \tag{11.3-26}$$

with constants $\alpha_1, \alpha_2 \in \mathbb{C}$, and P_n and Q_n denoting the Legendre functions of the first and second kind, respectively. Furthermore, the setting (11.3-24), which is motivated by the result (11.3-23), offers a second (shorter) proof of Theorem 11.3-4b.

Let us next consider solutions of (11.1-1) which are constant on circles

$$z_1\bar{z}_1 + (r - 1)(z_1 + \bar{z}_1) = 2r - 1, \qquad 0 < r < 1 \tag{11.3-27}$$

with center $z_1 = 1 - r$ and radius r, i.e, $\partial B_1(r - 1, r)$. Clearly, these are contained in $B_1(0, 1) \cup \{z_1 = 1\}$. Since for $z_1 \in B_1(0, 1)$, equation (11.3-27) is equivalent to

$$\frac{(z_1 - 1)(\bar{z}_1 - 1)}{1 - z_1\bar{z}_1} = \frac{r}{1 - r}$$

we introduce $\tilde{\omega}$ by setting

$$w(z_1, \bar{z}_1) = \tilde{\omega}(\tilde{\chi}), \qquad \tilde{\chi} = \frac{(z_1 - 1)(\bar{z}_1 - 1)}{1 - z_1\bar{z}_1}. \tag{11.3-28}$$

Thus, for $z_1 \in B_1(0,1)$, equation (11.1-1) may be rewritten as

$$\tilde{\chi}^2\tilde{\omega}'' + 2\tilde{\chi}\tilde{\omega}' - n(n+1)\tilde{\omega} = 0. \tag{11.3-29}$$

As is well known, this *Euler differential equation* has the general solution

$$\tilde{\omega}(\tilde{\chi}) = \alpha_1\tilde{\chi}^n + \alpha_2\tilde{\chi}^{-(n+1)}, \qquad \alpha_1, \alpha_2 \in \mathbb{C}.$$

Hence

(11.3-30)
$$w(z_1, \bar{z}_1) = \alpha_1\left(\frac{(z_1-1)(\bar{z}_1-1)}{1-z_1\bar{z}_1}\right)^n + \alpha_2\left(\frac{1-z_1\bar{z}_1}{(z_1-1)(\bar{z}_1-1)}\right)^{n+1}.$$

Here, the first term can be represented by means of a Type I representation [cf. (11.2-4b, 20)]:

$$\alpha_1\left(\frac{(z_1-1)(\bar{z}_1-1)}{1-z_1\bar{z}_1}\right)^n = \tilde{T}_1(\tilde{f}_{11}(z_1))$$

with

$$\tilde{f}_{11}(z_1) = \alpha_1\frac{(-1)^n n!}{(2n)!}(z_1-1)^{2n}. \tag{11.3-31}$$

The second term in (11.3-30) has a Type II representation,

$$\alpha_2\left(\frac{1-z_1\bar{z}_1}{(z_1-1)(\bar{z}_1-1)}\right)^{n+1} = \tilde{T}_1(\tilde{f}_{12}(z_1)) + \bar{\tilde{T}}_1(\tilde{f}_{22}(\bar{z}_1)) \tag{11.3-32}$$

with

(11.3-33)
$$\tilde{f}_{12}(z_1) = \frac{\alpha_2}{n!}\left[\frac{z_1^{n+1}}{1-z_1} + \beta z_1^n\right]$$

$$\tilde{f}_{22}(\bar{z}_1) = \frac{\alpha_2}{n!}\left[\frac{\bar{z}_1^{n+1}}{1-\bar{z}_1} + (1-\beta)\bar{z}_1^n\right], \qquad \beta \in \mathbb{C} \text{ arbitrary}$$

as can be seen immediately from (11.3-32) and Remark 11.2-2a. Thus we have proved the following:

11.3-6 Theorem

Let the solution $\tilde{w}$ *of* (11.1-1) *be continuous in the interior and on the boundary of the circle* (11.3-27), *with the possible exception* $z_1 = 1$; *that is, for*

$$z_1 \in [B_1(1-r, r) \cup \partial B_1(1-r, r)] \setminus \{z_1 = 1\}$$

and let $\tilde{w}$ *be regular in* $B_1(1-r, r)$. *If* $\tilde{w}$ *is constant on* $\partial B_1(1-r, r) \cap B_1(0,1)$, *then* $\tilde{w}$ *is defined for all* $z_1 \in B_1(0,1)$ *and has the form* (11.3-30). *Moreover,* $\tilde{w}$ *can be represented for* $z_1 \in B_1(0,1)$ *by*

$$\tilde{w}(z_1, \bar{z}_1) = \tilde{T}_1\big((\tilde{f}_{11} + \tilde{f}_{12})(z_1)\big) + \bar{\tilde{T}}_1\big(\tilde{f}_{22}(\bar{z}_1)\big)$$

where $\tilde{f}_{11}$, $\tilde{f}_{12}$, *and* $\tilde{f}_{22}$ *are given in* (11.3-31, 33).

APPLICATION AND EXTENSION OF SCHWARZ'S LEMMA

We shall now consider applications of two famous theorems of complex analysis, namely of the Schwarz lemma and of Privalov's uniqueness theorem.

We first introduce the following lemma.

11.3-7 **Lemma**

Let

$$\tilde{f}_1 \in C^\omega(B_1(0,1)) \cap C^0(B_1(0,1) \cup \partial B_1(0,1)).$$

Then for all $n \in \mathbb{N}$ *we have*

$$(11.3\text{-}34) \qquad \left|\tilde{T}_{1(n)}\big(\tilde{f}_1(z_1)\big)\right| \le \tilde{T}_{1(n)}(z_1^n) \qquad \text{for all } z_1 \in B_1(0,1)$$

if and only if

$$(11.3\text{-}35) \qquad |\tilde{f}_1(z_1)| \le 1 \qquad \text{for all } z_1 \in B_1(0,1).$$

Equality holds in (11.3-34) *if and only if*

$$(11.3\text{-}36) \qquad \tilde{f}_1(z_1) = z_1^n e^{i\delta}$$

for a suitable $\delta \in \mathbb{R}$. (*We remember that* $\tilde{T}_{1(n)}$ *is defined in* (11.2-33), *the subscript n indicates the order of the differential operator* $\tilde{T}_1$, *and the right-hand side of* (11.3-34) *equals*

$$\tilde{T}_{1(n)}(z_1^n) = n!P_n\left(\frac{1+|z_1|^2}{1-|z_1|^2}\right)$$

[*cf*. (11.1-11)].)

Proof. Let $n \in \mathbb{N}$ be arbitrary, but fixed. Let us first assume (11.3-35) and prove (11.3-34). For all $z_1 \in B_1(0,1)$ and all $\varphi \in [0, 2\pi]$ we have

$|(e^{i\varphi} + z_1)/(1 + \bar{z}_1 e^{i\varphi})| < 1$. Hence from (11.3-35) it follows that

(11.3-37)
$$\left| \tilde{f}_1\left(\frac{e^{i\varphi} + z_1}{1 + \bar{z}_1 e^{i\varphi}} \right) \right| \leq 1.$$

Because of the assumption

$$\tilde{f}_1 \in C^\omega(B_1(0,1)) \cap C^0(B_1(0,1) \cup \partial B_1(0,1))$$

we can apply Corollary 11.2-4a which, by the use of (11.3-37), now yields

(11.3-38)
$$\begin{aligned} \left|\tilde{T}_{1(n)}\left(\tilde{f}_1(z_1)\right)\right| &= \left| \frac{n!}{2\pi} \int_0^{2\pi} \tilde{f}_1\left(\frac{e^{i\varphi} + z_1}{1 + \bar{z}_1 e^{i\varphi}} \right) \frac{\left(1 + \bar{z}_1 e^{i\varphi}\right)^{2n}}{\left(1 - z_1\bar{z}_1\right)^n} \frac{d\varphi}{e^{in\varphi}} \right| \\ &\leq \frac{n!}{2\pi} \int_0^{2\pi} \frac{|1 + \bar{z}_1 e^{i\varphi}|^{2n}}{\left(1 - z_1\bar{z}_1\right)^n}\, d\varphi, \qquad z_1 \in B_1(0,1). \end{aligned}$$

Observing that

$$|1 + \bar{z}_1 e^{i\varphi}|^{2n} = \left(\frac{e^{i\varphi} + z_1}{1 + \bar{z}_1 e^{i\varphi}} \right)^n \left(1 + \bar{z}_1 e^{i\varphi}\right)^{2n} \frac{1}{e^{in\varphi}}$$

and applying Corollary 11.2-4a to $\tilde{T}_{1(n)}(z_1^n)$, we see that the expression on the right-hand side of (11.3-38) equals $\tilde{T}_{1(n)}(z_1^n)$, so that we have obtained inequality (11.3-34).

Let us now assume the validity of (11.3-34). For the proof of (11.3-35) we introduce the $C^\omega(B_1(0,1) \times \mathbb{C})$-function f defined by

(11.3-39)
$$f(z_1, R) = z_1^n(1 - R^2)^n \sum_{m=0}^{n} \frac{(n+m)!}{m!(n-m)!} \left(\frac{R^2}{1 - R^2} \right)^m \frac{1}{z_1^m} D_1^{n-m}\tilde{f}_1(z_1)$$

for all $z_1 \in B_1(0,1)$ and $R \in \mathbb{C}$. For arbitrarily fixed $R = R_0 \in (0,1)$, we can apply the Maximum Modulus Theorem to f on $B_1(0, R_0) \cup \partial B_1(0, R_0)$. If we furthermore observe that for $|z_1| = R_0$

$$f(z_1, R_0) = z_1^n\left(1 - R_0^2\right)^n \tilde{T}_{1(n)}\left(\tilde{f}_1(z_1)\right)$$

holds, and finally use assumption (11.3-34), we obtain

$$
\begin{aligned}
|f(z_1, R_0)| &\le R_0^n(1-R_0^2)^n \max_{|\tilde{z}_1|=R_0} \left|\tilde{T}_{1(n)}\left(\tilde{f}_1(\tilde{z}_1)\right)\right| \\
&\le R_0^n(1-R_0^2)^n \max_{|\tilde{z}_1|=R_0} \tilde{T}_{1(n)}(\tilde{z}_1^n) \\
&= R_0^n(1-R_0^2)^n \max_{\tilde{z}_1\bar{\tilde{z}}_1=R_0^2} \sum_{m=0}^{n} \frac{(n+m)!n!}{(n-m)!(m!)^2}\left(\frac{\tilde{z}_1\bar{\tilde{z}}_1}{1-\tilde{z}_1\bar{\tilde{z}}_1}\right)^m \qquad (11.3\text{-}40)\\
&= \sum_{m=0}^{n} \frac{(n+m)!n!}{(n-m)!(m!)^2} R_0^{n+2m}(1-R_0^2)^{n-m}
\end{aligned}
$$

for $z_1 \in B_1(0, R_0) \cup \partial B_1(0, R_0)$. Since R_0 was chosen arbitrarily in $(0,1)$, from (11.3-39, 40) for all $z_1 \in B_1(0,1)$ there follows that

$$
\begin{aligned}
|\tilde{f}_1(z_1)| &= \left|\frac{n!}{(2n)!}\lim_{R_0\to 1}\left[z_1^n(1-R_0^2)^n \sum_{m=0}^{n}\frac{(n+m)!}{m!(n-m)!}\left(\frac{R_0^2}{1-R_0^2}\right)^m \times \right.\right. \\
&\qquad \left.\left. \times \frac{1}{z_1^m} D_1^{n-m}\tilde{f}_1(z_1)\right]\right| \\
&= \frac{n!}{(2n)!}\left|\lim_{R_0\to 1} f(z_1, R_0)\right| \\
&\le \frac{n!}{(2n)!}\lim_{R_0\to 1}\left[\sum_{m=0}^{n}\frac{(n+m)!n!}{(n-m)!(m!)^2}R_0^{n+2m}(1-R_0^2)^{n-m}\right] \\
&= 1.
\end{aligned}
$$

This proves (11.3-35).

As in the first part of the proof, as a consequence of Corollary 11.2-4a we have

$$
\left|\tilde{T}_{1(n)}\left(\tilde{f}_1(z_1)\right)\right| = \left|\frac{n!}{2\pi}\int_0^{2\pi}\tilde{f}_1\left(\frac{e^{i\varphi}+z_1}{1+\bar{z}_1e^{i\varphi}}\right)\frac{(1+\bar{z}_1e^{i\varphi})^{2n}}{(1-z_1\bar{z}_1)^n}\frac{d\varphi}{e^{in\varphi}}\right|
$$

and

$$
\tilde{T}_{1(n)}(z_1^n) = \frac{n!}{2\pi}\int_0^{2\pi}\left(\frac{e^{i\varphi}+z_1}{1+\bar{z}_1e^{i\varphi}}\right)^n\frac{(1+\bar{z}_1e^{i\varphi})^{2n}}{(1-z_1\bar{z}_1)^n}\frac{d\varphi}{e^{in\varphi}}.
$$

Hence, the last statement of the lemma, the statement on equality in (11.3-34), is now obvious. ■

As mentioned above, by the use of the Schwarz lemma we can extend Lemma 11.3-7 and obtain a result on function theoretic properties of the $\tilde{T}_1$-associated function $\tilde{f}_1$.

We recall the *Schwarz lemma* (cf. Hille [1959–1962], vol. II, 235): If $\tilde{f}_1$: $B_1(0,1) \to B_1(0,1)$ is a holomorphic function satisfying $\tilde{f}_1(0) = 0$, then $|\tilde{f}_1(z_1)| \leq |z_1|$ holds for all $z_1 \in B_1(0,1)$.

From this and Theorem 11.2-5 and Lemma 11.3-7 we obtain:

11.3-8 Theorem

Let

$$\tilde{f}_1 \in C^\omega(B_1(0,1)) \cap C^0(B_1(0,1) \cup \partial B_1(0,1)).$$

Then for all $m, n \in \mathbb{N}$, the inequality

$$\text{(11.3-41)} \quad \left|\tilde{T}_{1(n)}\left(z_1^n \tilde{f}_1(z_1)\right)\right| \leq \left|\tilde{T}_{1(n)}\left(z_1^{n+m}\right)\right| \qquad \text{for all } z_1 \in B_1(0,1)$$

holds if and only if

$$\text{(11.3-42)} \qquad |\tilde{f}_1(z_1)| \leq 1 \qquad \text{for all } z_1 \in B_1(0,1)$$

and $\tilde{f}_1$ has at $z_1 = 0$ a zero of at least mth *order.*

Proof. Let $m, n \in \mathbb{N}$ be arbitrary but fixed. Let us first assume that

$$\tilde{f}_1 \in C^\omega(B_1(0,1)) \cap C^0(B_1(0,1) \cup \partial B_1(0,1))$$

satisfies (11.3-42) and possesses a zero at $z_1 = 0$ of order $\geq m$. Thus there exists a function

$$g \in C^\omega(B_1(0,1)) \cap C^0(B_1(0,1) \cup \partial B_1(0,1))$$

so that $\tilde{f}_1$ can be represented by

$$\text{(11.3-43)} \qquad \tilde{f}_1(z_1) = z_1^m g(z_1).$$

From (11.3-42, 43), by iterated application of Schwarz's lemma, we obtain that also

$$|z_1^k g(z_1)| \leq 1 \qquad \text{for all } z_1 \in B_1(0,1), \qquad k = 0, \cdots, m$$

holds, so that Lemma 11.3-7 is applicable to these functions. We now use Theorem 11.2-5 twice, in particular, in (11.2-34) setting first,

$$\tilde{f}_{11}(z_1) = g(z_1), \qquad \tilde{f}_{12}(z_1) = z_1^m$$

and second,

$$\tilde{f}_{11}(z_1) \equiv 1, \qquad \tilde{f}_{12}(z_1) = z_1^m.$$

If we finally observe that

$$(11.3\text{-}44) \quad z_1^{-m}\tilde{T}_{1(\kappa)}(z_1^{\kappa+m}) = \sum_{k=0}^{\kappa} \frac{(2\kappa-k)!(\kappa+m)!}{(\kappa-k)!k!(\kappa-k+m)!}\left(\frac{z_1\bar{z}_1}{1-z_1\bar{z}_1}\right)^{\kappa-k}$$

is real and nonnegative ($\kappa \in \mathbb{N}_0$), from the preceding results we conclude that

$$\begin{aligned} &\left|z_1^{-m}\tilde{T}_{1(n)}(z_1^{n+m}g(z_1))\right| \\ &\le \sum_{k=0}^{n-1}\binom{n}{k}\left|\tilde{T}_{1(k)}(z_1^k g(z_1))\right| \cdot \left|\frac{1}{z_1^m}\tilde{T}_{1(n-1-k)}(m z_1^{n-1-k+m})\right| \\ &\qquad + \left|\tilde{T}_{1(n)}(z_1^n g(z_1))\right| \\ &\le m\sum_{k=0}^{n-1}\binom{n}{k}\tilde{T}_{1(k)}(z_1^k)z_1^{-m}\tilde{T}_{1(n-1-k)}(z_1^{n-1-k+m}) + \tilde{T}_{1(n)}(z_1^n) \\ &= z_1^{-m}\tilde{T}_{1(n)}(z_1^{n+m}) \end{aligned}$$

for all $z_1 \in B_1(0,1)$. This proves (11.3-41).

Conversely, let us now assume the validity of (11.3-41). From Lemma 11.3-7 we then have

$$(11.3\text{-}45) \quad \left|\tilde{T}_{1(n)}(z_1^{n+m})\right| \le \tilde{T}_{1(n)}(z_1^n) \qquad \text{for all } z_1 \in B_1(0,1).$$

From (11.3-41, 45) it follows that

$$|z_1^n f_1(z_1)| \le 1 \qquad \text{for all } z_1 \in B_1(0,1)$$

again by utilizing Lemma 11.3-7. By applying the Schwarz Lemma n times we then obtain the boundedness statement (11.3-42) for $\tilde{f}_1$. Using the Maclaurin expansion of $\tilde{f}_1$,

$$\tilde{f}_1(z_1) = \sum_{\nu=0}^{\infty} \tilde{\gamma}_{1\nu} z_1^{\nu}, \qquad z_1 \in B_1(0,1)$$

the assumption (11.3-41) and representation (11.3-44) with $\kappa = n$, we get

$$\left| \sum_{k=0}^{n} \frac{(2n-k)!}{(n-k)!k!} \left(\frac{\bar{z}_1}{1 - z_1 \bar{z}_1} \right)^{n-k} \sum_{\nu=0}^{\infty} \frac{(n+\nu)!}{(n+\nu-k)!} \tilde{\gamma}_{1\nu} z_1^{\nu} \right|$$

$$= \left| \tilde{T}_{1(n)}\left(z_1^n \tilde{f}_1(z_1) \right) \right| \le \left| \tilde{T}_{1(n)}\left(z_1^{n+m} \right) \right|$$

$$= |z_1^m| \left| \sum_{k=0}^{n} \frac{(2n-k)!(n+m)!}{(n-k)!k!(n-k+m)!} \left(\frac{z_1 \bar{z}_1}{1 - z_1 \bar{z}_1} \right)^{n-k} \right|.$$

For $z_1 = 0$ this implies that

$$\tilde{\gamma}_{10} = \cdots = \tilde{\gamma}_{1,m-1} = 0.$$

Hence $\tilde{f}_1$ is of the form

$$\tilde{f}_1(z_1) = z_1^m \sum_{\nu=0}^{\infty} \tilde{\gamma}_{1,\nu+m} z_1^{\nu}, \qquad z_1 \in B_1(0,1).$$

This completes the proof of Theorem 11.3-8. ∎

APPLICATION OF PRIVALOV'S UNIQUENESS THEOREM

Finally in this section, we shall apply two theorems due to I. I. Privalov. For this purpose we recall these results of complex function theory; cf. Goluzin [1969], Chap. X, Secs. 1, 2.

First let us agree on some terminology. Suppose that a finite simply connected domain Ω_1 in the z_1-plane is bounded by a rectifiable closed Jordan curve of length $|\partial\Omega_1|$. An open set on $\partial\Omega_1$ (in the relative topology of the complex plane) may be written as the union of a countable set of arcs on $\partial\Omega_1$ and a measure of such an open set on $\partial\Omega_1$ may be defined as the sum of the lengths of the arcs of which it is composed. Furthermore, for any set E on $\partial\Omega_1$, we introduce an outer measure $m_e(E)$, namely, the greatest lower bound of the measures of all open sets on $\partial\Omega_1$ containing E. Furthermore, if $m_e(E) = |\partial\Omega_1| - m_e(\partial\Omega_1 \setminus E)$, we denote $m_e(E)$ by mes E, the measure of the set E on $\partial\Omega_1$. Now, let $z_1^{(0)}$ be a point on $\partial\Omega_1$ at which there is a unique tangent to $\partial\Omega_1$ and in a neighborhood of which $\partial\Omega_1$ lies on both sides of the normal. We shall refer to a continuous curve C_1 contained in Ω_1 and ending at

$z_1^{(0)} \in \partial\Omega_1$ as a *nontangential path* if that part of C_1 in a neighborhood of $z_1^{(0)}$ lies inside some angle of magnitude less than π with vertex at the point $z_1^{(0)}$ and with bisector coinciding with the inner normal to $\partial\Omega_1$. In the sequel when we speak of a set of points on $\partial\Omega_1$, we shall usually mean only points at which there exist tangents to $\partial\Omega_1$ possessing the property just described. Furthermore, if a function $\tilde{f}_1 \in C^\omega(\Omega_1)$ approaches a value a as z_1 approaches $z_1^{(0)} \in \partial\Omega_1$ along every nontangential path, we shall simply say that

$$\lim_{z_1 \to z_1^{(0)}} \tilde{f}_1(z_1) = a \qquad \text{along nontangential paths}.$$

Under the preceding assumptions and notations the following two theorems of complex function theory hold.

Theorem (Privalov's uniqueness theorem)

Let Ω_1, $\partial\Omega_1$, and $E \subset \partial\Omega_1$ satisfy the preceding assumptions and let mes $E > 0$. *For $\tilde{f}_1 \in C^\omega(\Omega_1)$ and all $z_1^{(0)} \in E$ suppose that*

$$\lim_{z_1 \to z_1^{(0)}} \tilde{f}_1(z_1) = 0 \qquad \text{along nontangential paths.}$$

Then $\tilde{f}_1|_{\Omega_1} = 0$ holds. (Cf. Goluzin [1969], pp. 428–429, Theorem X.2.1.)

Theorem (Privalov)

There does not exist a function in $C^\omega(\Omega_1)$ (Ω_1, as before, bounded by a rectifiable closed Jordan curve) that assumes the value ∞ along nontangential paths on a set E of positive measure on the boundary $\partial\Omega_1$. (Cf. Goluzin [1969], *pp.* 429–430, *Theorem* X.2.2.)

We are now in the position to formulate the following

11.3-9 Theorem

Let

$$\tilde{w}_1 = \tilde{T}_{1(n)}\tilde{f}_1 \qquad \textit{with} \qquad \tilde{f}_1 \in C^\omega(B_1(0,1))$$

satisfy

$$\forall\, z_1^{(0)} \in E: \lim_{z_1 \to z_1^{(0)}} \left[\tilde{w}_1(z_1, \bar{z}_1)(1 - z_1\bar{z}_1)^n\right] = 0 \tag{11.3-46}$$

along nontangential paths for some $E \subset \partial B_1(0,1)$ with mes $E > 0$. *Then*

$$\forall\, z_1 \in B_1(0,1): \ \tilde{w}_1(z_1, \bar{z}_1) = 0$$

holds.

Proof. For all $z_1 \in B_1(0,1)$ we have

$$\tilde{w}_1(z_1, \bar{z}_1)(1 - z_1\bar{z}_1)^n = \frac{(2n)!}{n!}\bar{z}_1^n\tilde{f}_1(z_1) + \sum_{m=1}^{n} \frac{(2n-m)!}{(n-m)!m!}\bar{z}_1^{n-m}(1 - z_1\bar{z}_1)^m D_1^m\tilde{f}_1(z_1). \tag{11.3-47}$$

Since from the preceding second theorem of Privalov it follows that

$$\forall\, m \in \mathbb{N}: \quad \lim_{|z_1| \to 1-0} \left[(1 - |z_1|^2)^m D_1^m \tilde{f}_1(z_1)\right] \neq 0$$

can hold at most on a set $E' \subset \partial B_1(0,1)$ with $\operatorname{mes} E' = 0$, from (11.3-46, 47) we obtain

$$\forall\, z_1^{(0)} \in E \setminus E': \quad \lim_{B_1(0,1) \ni z_1 \to z_1^{(0)}} \tilde{f}_1(z_1) = 0$$

where $\operatorname{mes}(E \setminus E') > 0$. Thus, by the Privalov's uniqueness theorem, $\tilde{f}_1|_{B_1(0,1)} = 0$ and $\tilde{w}_1(z_1, \bar{z}_1) = \tilde{T}_{1(n)}(\tilde{f}_1(z_1)) = 0$ for all $z_1 \in B_1(0,1)$ follow. ■

11.4 ANALOGS OF THEOREMS FROM GEOMETRIC FUNCTION THEORY

In this section we shall consider some analogs of theorems from geometric function theory. We have already given some results of this type at the end of Sec. 11.3, for instance, a generalized Schwarz lemma (Theorem 11.3-8). Presently we shall give an alternative proof of the Great Picard Theorem for solutions of (11.1-1) and, moreover, obtain a complete classification of the behavior of solutions in the neighborhood of an isolated singularity. Furthermore, a generalization of Schottky's theorem will be proved. Finally, for solutions of (11.1-1) an analog of Hadamard's three circles theorem will be established; the bound to be obtained is a linear combination of the Legendre functions P_n and Q_n, of the first and second kind. This is applied to the investigation of the behavior of solutions (possibly having singularities for some z_1 in the unit disk) as z_1 approaches the unit circle.

ANALOG OF THE GREAT PICARD THEOREM

Let us first consider an analog of the Great Picard Theorem for solutions of (11.1-1). For solutions satisfying condition (10.4-8) we have already proved such a theorem [cf. Theorem 10.4-5 and (10.4-12)] by rewriting the solution as a multianalytic function [cf. (10.4-11), (11.5-1)] and applying the Picard–Krajkiewicz theorem. Following Ruscheweyh [1974a], we shall now give a further proof of the Great Picard Theorem in the special case of solutions of (11.1-1), avoiding the assumption (10.4-8b).

11.4-1 Theorem

Let $\tilde{f}_1 \in C^{\omega}(\dot{B}_1(z_1^{(0)}, r))$ $[\dot{B}_1(z_1^{(0)}, r) = B_1(z_1^{(0)}, r) \setminus \{z_1^{(0)}\},\ z_1^{(0)} \in B_1(0,1),$ $0 < r < 1 - |z_1^{(0)}|]$ *have an essential singularity at* $z_1^{(0)}$. *Then in every neighborhood of* $z_1 = z_1^{(0)}$ *the solution* $\tilde{w}_1 = \tilde{T}_{1(n)}\tilde{f}_1$ *of* (11.1-1) *assumes all finite complex values with at most one exception.*

Proof. We assume that $z_1^{(0)} = 0$. [This means no loss of generality; for, otherwise, apply an automorphism of the unit disk transforming $z_1^{(0)}$ into 0 and observe the relation (11.2-22).] By assumption we thus have $\tilde{f}_1 \in C^{\omega}(\dot{B}_1(0, r_0))$, $0 < r_0 < 1$, and $\tilde{f}_1$ has an isolated essential singularity at $z_1^{(0)} = 0$. Similarly as in the proof of Lemma 11.3-7, we introduce a function $g(\cdot, R)$ by setting

$$g(z_1, R) = \sum_{m=0}^{n} \frac{(n+m)!}{m!(n-m)!}\left(\frac{R^2}{1-R^2}\right)^m \frac{1}{z_1^m} D_1^{n-m}\tilde{f}_1(z_1), \qquad 0 < R < r_0. \tag{11.4-1}$$

This function obviously satisfies $g(\cdot, R) \in C^{\omega}(\dot{B}_1(0, r_0))$ for fixed $R \in (0, r_0)$, and

$$g(z, R) = \tilde{T}_{1(n)}\big(\tilde{f}_1(z_1)\big) = \tilde{w}_1(z_1, \bar{z}_1) \qquad \text{for} \qquad |z_1| = R. \tag{11.4-2}$$

We now proceed to the actual proof of the theorem by deriving a contradiction from the assumption that $\tilde{w}_1$ omits *two* values α_1, α_2 $(\alpha_1 \neq \alpha_2)$ on some punctured disk, say, on $\dot{B}_1(0, r_1) \cup \partial B_1(0, r_1)$, $0 < r_1 < r_0$. We can assume that

$$\alpha_1 \neq D_1^n f_1(z_1) \neq \alpha_2 \qquad \text{for all} \qquad z_1 \in \partial B_1(0, r_1). \tag{11.4-3}$$

This is no loss of generality, as can be shown by use of the Bolzano–Weierstrass theorem and the identity theorem for holomorphic functions. Now, for $|z_1| = r_1$ we can represent $D_1^n\tilde{f}_1$ by means of (11.4-1), namely,

$$D_1^n\tilde{f}_1(z_1) = \lim_{R \to 0} g(z_1, R) \qquad \text{uniformly for } z_1 \in \partial B_1(0, r_1). \tag{11.4-4}$$

For subsequent application, we now recall the following *Hurwitz theorem* from complex function theory. Let $\{f_n\}_{n \in \mathbb{N}}$ be a sequence of holomorphic functions in some domain Ω_1 that converges uniformly in Ω_1 to a function f

which has a zero at $z_1 = z_1^{(1)}$. Then for every $\varepsilon \in (0, \varepsilon_0)$ ($\varepsilon_0 > 0$ suitable), there exists an $N(\varepsilon)$ such that for all $n > N(\varepsilon)$ the functions f_n and f have the same number of zeros in the disk $B_1(z_1^{(1)}, \varepsilon)$. (See Hille [1959–1962], vol. II, p. 205, Theorem 14.3.4.)

By this theorem, from (11.4-3, 4) we conclude that for sufficiently small R (> 0) we have

$$\alpha_1 \neq g(z_1, R) \neq \alpha_2 \qquad \text{for all } z_1 \in \partial B_1(0, r_1). \tag{11.4-5}$$

Because of (11.4-2), for such R's and $l = 1, 2$, we obtain

$$\begin{aligned} 0 &= \frac{1}{2\pi}\int_{|z_1|=r_1} d\arg(\tilde{w}_1 - \alpha_l) - \frac{1}{2\pi}\int_{|z_1|=R} d\arg(\tilde{w}_1 - \alpha_l) \\ &= A_l + N_{\alpha_l}[g(z_1, R);\ R < |z_1| < r_1] \qquad (11.4\text{-}6) \\ &\quad - \frac{1}{2\pi i}\int_{|z_1|=r_1} \frac{D_1 g(z_1, R)}{g(z_1, R) - \alpha_l}\, dz_1 \end{aligned}$$

where A_l denotes the first integral and $N_{\alpha_l}[\cdots]$ is the number of α_l-points (with respect to their order) of $g(\cdot, R)$ in the annulus $R < |z_1| < r_1$ (cf. Behnke–Sommer [1972], p. 207, Theorem III.16a, or Conway [1973], p. 119). Also, because of (11.4-4), we get

$$\frac{1}{2\pi i}\int_{|z_1|=r_1} \frac{D_1 g(z, R)}{g(z_1, R) - \alpha_l}\, dz_1 = \frac{1}{2\pi i}\int_{|z_1|=r_1} \frac{D_1^{n+1}\tilde{f}_1(z_1)}{D_1^n \tilde{f}_1(z_1) - \alpha_l}\, dz_1. \tag{11.4-7}$$

If we denote this expression by B_l and set $M_l = |A_l| + |B_l|$ we thus have proved the following: There exists an $r_2 \in (0, r_1)$ so that for all $R \in (0, r_2)$ we have

$$0 \leq N_{\alpha_l}[g(z_1, R);\ R < |z_1| < r_1] = B_l - A_l \leq M_l, \qquad l = 1, 2. \tag{11.4-8}$$

We now choose $r_3 \in (0, r_2)$ so that $D_1^n \tilde{f}_1$ has at least $M_l + 1$ α_l-points for at least one $l \in \{1, 2\}$ in the annulus $r_3 < |z_1| < r_1$. Since $D_1^n \tilde{f}_1$ has an essential singularity for $z_1^{(0)} = 0$, such an r_3 exists according to the classical Great Picard Theorem (cf. Behnke–Sommer [1972], p. 490, Theorem V.39a, or Conway [1973], p. 302). Finally, (11.4-4) and the Hurwitz theorem (above) yield that also $g(\cdot, R)$ has at least $M_l + 1$ α_l-points for that l and for sufficiently small R (> 0) in the annulus $r_3 < |z_1| < r_1$. This, however, contradicts (11.4-8), and thus the theorem is proved. ■

We remark that an exceptional value (admitted by Theorem 11.4-1) may actually occur. Indeed, choose

$$\tilde{f}_1(z_1) = e^{1/z_1} \qquad \text{and} \qquad n = 1;$$

then

$$\tilde{w}_1(z_1, \bar{z}_1) = \tilde{T}_{1(1)}\big(\tilde{f}_1(z_1)\big) = \frac{1}{z_1} e^{1/z_1}\left(\frac{2z_1\bar{z}_1}{1 - z_1\bar{z}_1} - \frac{1}{z_1}\right) \tag{11.4-9}$$

and $\tilde{w}_1$ omits the value 0 in a sufficiently small neighborhood of the origin.

SOLUTIONS NEAR SINGULARITIES OF $\tilde{T}_1$-ASSOCIATED FUNCTIONS

11.4-2 Remark

We mention that the behavior of a solution $\tilde{w}_1 = \tilde{T}_{1(n)}\tilde{f}_1$ which is defined and univalent for $z_1 \in \dot{B}_1(z_1^{(0)}, r) \subset B_1(0,1)$ can be completely characterized. There holds one and only one of the following three assertions:

(i) $|\tilde{w}_1(z_1, \bar{z}_1)| \le M < \infty$ for all $z_1 \in \dot{B}_1(z_1^{(0)}, r)$.

(ii) There exist an $\varepsilon > 0$ and an $m \in \mathbb{N}$ so that

$$0 < M_1 \le |z_1 - z_1^{(0)}|^{m+n} |\tilde{w}_1(z_1, \bar{z}_1)| \le M_2 < \infty$$

for all $z_1 \in \dot{B}_1(z_1^{(0)}, \varepsilon)$.

(iii) $\tilde{w}_1$ assumes all finite complex values with at most one possible exception in every neighborhood of $z_1 = z_1^{(0)}$.

For the $\tilde{T}_{1(n)}$-associated function $\tilde{f}_1$ in these cases we have, respectively:

(i) $\tilde{f}_1$ has a removable singularity at $z_1^{(0)}$.

(ii) $\tilde{f}_1$ has a pole of order m at $z_1^{(0)}$.

(iii) $\tilde{f}_1$ has an essential singularity at $z_1^{(0)}$.

The preceding three cases result from a theorem of Bauer–Peschl ([1966], Theorem 3) which asserts that a solution of (11.1-1) defined and univalent for $z_1 \in \dot{B}_1(z_1^{(0)}, r)$ $(r > 0)$ can be obtained from a $\tilde{T}_{1(n)}$-associated function $\tilde{f}_1$ which can be represented by a Laurent expansion

$$\tilde{f}_1(z_1) = \sum_{\nu=-\infty}^{\infty} \tilde{\gamma}_{1\nu}\left(z_1 - z_1^{(0)}\right)^{\nu}$$

in $\dot{B}_1(z_1^{(0)}, r)$. Hence, one only needs to consider the three possible types of singularities of $\check{f}_1$, in accordance with the three types given above. The statements on $\tilde{w}_1$ are then obvious in cases (i) and (ii). For case (iii), see Theorem 11.4-1.

We also mention that for the solutions in case (i) an *analog of Riemann's theorem on removable singularities* may be proved (see Ruscheweyh [1972b], Theorem 1).

ANALOG OF SCHOTTKY'S THEOREM

Next we give an analog of Schottky's theorem, a special version of which is as follows. Let $f \in C^{\omega}(B_1(0, r))$ fail to assume two (distinct) finite values α_1 and α_2 in the disk $B_1(0, r)$. Then there exists a bound $\tilde{S} = \tilde{S}(f(0); \alpha_1, \alpha_2; r^*)$ so that $|f(z_1)| \leq \tilde{S}$ for $|z_1| \leq r^* < r$ holds. (See Goluzin [1969], pp. 337–338.) This generalizes to solutions of (11.1-1) as follows.

11.4-3 Theorem

Let $\check{f}_1 \in C^{\omega}(B_1(0, r))$, $0 < r < 1$. *Let the solution* $\tilde{w}_1$ *of* (11.1-1) *be given by*

$$\tilde{w}_1(z_1, \bar{z}_1) = \tilde{T}_{1(n)}\big(z_1^n \check{f}_1(z_1)\big) \tag{11.4-10}$$

and suppose that $\tilde{w}_1$ *omits two finite values* α_1 *and* α_2 *for* $z_1 \in B_1(0, r)$. *Then there exists a bound* $S = S(\tilde{w}_1(0,0); \alpha_1, \alpha_2; r^*)$ *so that the inequality*

$$|\tilde{w}_1(z_1, \bar{z}_1)| \leq S \qquad \text{for all } z_1 \in B_1(0, r^*), \qquad 0 < r^* < r \tag{11.4-11}$$

holds.

Proof. We choose $n + 1$ numbers R_l $(l = 0, \cdots, n)$ such that

$$r > R_0 > \cdots > R_n > r^* .$$

If the functions $g(\cdot, R_l)$ $(l = 0, \cdots, n)$ are defined as in (11.4-1), they are holomorphic and different from α_1 and α_2 on $B_1(0, R_l) \cup \partial B_1(0, R_l)$, because of the assumptions on $\check{f}_1$ and $\tilde{w}_1$. Furthermore, from (11.4-1) we have [cf. (11.1-11)]

$$\begin{aligned} g(0, R_l) &= \left[\sum_{m=0}^{n} \frac{(n+m)!}{m!(n-m)!} \left(\frac{R_l^2}{1 - R_l^2} \right)^m \frac{1}{z_1^m} D_1^{n-m}\big(z_1^n \check{f}_1(z_1)\big) \right] \Bigg|_{z_1 = 0} \\ &= \check{f}_1(0) n! P_n\left(\frac{1 + R_l^2}{1 - R_l^2} \right) = \tilde{w}_1(0,0) P_n\left(\frac{1 + R_l^2}{1 - R_l^2} \right), \quad l = 0, \cdots, n . \end{aligned} \tag{11.4-12}$$

From Schottky's theorem thus follows the existence of constants $\tilde{S}_l = \tilde{S}_l(g(0, R_l);\, \alpha_1, \alpha_2;\, r^*)$ such that

$$|g(z_1, R_l)| \le \tilde{S}_l \le \tilde{S} \qquad \text{for } |z_1| \le r^*\,; \quad l = 0, \cdots, n\,, \tag{11.4-13}$$

where

$$\tilde{S} \ge \max\left\{\tilde{S}_0, \cdots, \tilde{S}_n\right\}$$

depends on $\tilde{w}_1(0,0)$ [cf. (11.4-12)], α_1, α_2, and r^*. We now introduce the functions h_m defined by

$$h_m(z_1) = \frac{(n+m)!}{m!(n-m)!}\frac{1}{z_1^m} D_1^{n-m}\left(z_1^n \check{f}_1(z_1)\right), \quad m = 0, \cdots, n\,. \tag{11.4-14}$$

Then, for $|z_1| \le r^*$, the equations given by (11.4-1) can be interpreted as a system of linear equations for the functions h_m, namely,

$$g(\cdot, R_l) = \sum_{m=0}^{n} \left(\frac{R_l^2}{1 - R_l^2}\right)^m h_m(\cdot)\,, \qquad l = 0, \cdots, n\,. \tag{11.4-15}$$

Since the determinant of the coefficients of this system is of Vandermonde's type and is nonzero because of those assumptions on the R_l's, there exists a unique solution $(h_0, \cdots, h_n)$, where

$$h_m = \sum_{l=0}^{n} \beta_{lm} g(\cdot, R_l) \tag{11.4-16}$$

with complex β_{lm}. From (11.2-9) and (11.4-10, 13, 14, 16) we finally conclude that

$$\begin{aligned}
|\tilde{w}_1(z_1, \bar{z}_1)| &= \left|\tilde{T}_{1(n)}\left(z_1^n \check{f}_1(z_1)\right)\right| \\
&= \left|\sum_{m=0}^{n} z_1^m h_m(z_1)\right| = \left|\sum_{m=0}^{n} z_1^m \sum_{l=0}^{n} \beta_{lm} g(z_1, R_l)\right| \\
&\le \sum_{m=0}^{n} |z_1|^m \sum_{l=0}^{n} |\beta_{lm}|\,|g(z_1, R_l)| \\
&\le \sum_{m=0}^{n} \sum_{l=0}^{n} |\beta_{lm}| \tilde{S} \qquad\qquad \text{for } z_1 \in B_1(0, r^*)\,.
\end{aligned}$$

Thus, if we denote the right-hand side by S, we have proved the statement (11.4-11) of the theorem since, by (11.4-12), the constant $\tilde{S}$ depends only on $\tilde{w}_1(0,0)$, α_1, α_2, and r^*, but not on other values of $\tilde{w}_1$. ■

ANALOG OF HADAMARD'S THREE CIRCLES THEOREM

In classical function theory, *Hadamard's three circles theorem* states the following.

Let f be a function holomorphic and bounded for $0 < r_1 < |z_1| < r_2$ whose limiting values $|f(z_1)|$ as z_1 approaches $\partial B_1(0, r_k)$ from inside the annulus, do not exceed m_k $(k = 1, 2)$. Then, for $r \in (r_1, r_2)$,

$$\max\{|f(z_1)| \mid z_1 \in \partial B_1(0, r)\} \leq m_1^{\lambda_1} m_2^{\lambda_2} \tag{11.4-17}$$

holds, where

$$\lambda_1 = \frac{\log(r/r_2)}{\log(r_1/r_2)}, \qquad \lambda_2 = \frac{\log(r/r_1)}{\log(r_2/r_1)};$$

cf. Goluzin [1969], pp. 343–344; Hille [1959–1962], vol. II, p. 410, Theorem 18.3.4.

A similar result holds for solutions of (11.1-1), as follows.

11.4-4 Theorem

Let $\tilde{w}$ be a solution of (11.1-1) *defined and bounded for all z_1 satisfying $0 < r_1 \leq |z_1| \leq r_2 < 1$. Furthermore, let*

$$|\tilde{w}(z_1, \bar{z}_1)| \leq m_k \qquad \text{for all } z_1 \in \partial B_1(0, r_k), \qquad k = 1, 2. \tag{11.4-18}$$

Then, for all $r \in [r_1, r_2]$ we have

$$\max\{|\tilde{w}(z_1, \bar{z}_1)| \mid z_1 \in \partial B_1(0, r)\} \leq W_0(r). \tag{11.4-19}$$

Here,

$$W_0(r) = \alpha_1 P_n(X) + \alpha_2 Q_n(X) \tag{11.4-20a}$$

where

$$\alpha_1 = \frac{m_1 Q_n(X_2) - m_2 Q_n(X_1)}{P_n(X_1) Q_n(X_2) - P_n(X_2) Q_n(X_1)}$$

$$\alpha_2 = \frac{m_2 P_n(X_1) - m_1 P_n(X_2)}{P_n(X_1) Q_n(X_2) - P_n(X_2) Q_n(X_1)} \tag{11.4-20b}$$

$$X = \frac{1 + r^2}{1 - r^2}, \qquad X_k = \frac{1 + r_k^2}{1 - r_k^2}, \qquad k = 1, 2$$

and P_n and Q_n denote the Legendre functions of the first and second kind, respectively.

Proof. We set $r^2 = z_1\bar{z}_1$. W_0 determines a function $\tilde{W}_0$ that depends on $z_1, \bar{z}_1$ and is defined on the whole annulus, and represents there a real-valued solution of (11.1-1) (cf. Remark 11.3-5) which, by definition (11.4-20), satisfies

$$(11.4\text{-}21)\quad \tilde{W}_0(z_1, \bar{z}_1) \equiv W_0(r_k) = m_k \qquad \text{for} \qquad z_1 \in \partial B_1(0, r_k), \ k = 1,2.$$

Since (11.1-1) is linear, for all $\varphi \in [0, 2\pi]$, the function $\operatorname{Re}(e^{i\varphi}\tilde{w}) - \tilde{W}_0$ is also a real-valued solution of (11.1-1). Because of (11.4-17, 21) it satisfies

$$\operatorname{Re}\left(e^{i\varphi}\tilde{w}\right) - \tilde{W}_0 \leq 0 \qquad \text{for} \qquad z_1 \in \partial B_1(0, r_k), \qquad k = 1,2.$$

This, however, implies that

$$(11.4\text{-}22)\qquad \operatorname{Re}\left(e^{i\varphi}\tilde{w}\right) - \tilde{W}_0 \leq 0 \qquad \text{for} \qquad r_1 \leq |z_1| \leq r_2 .$$

Indeed, assume that there exists a $z_1^{(1)} = x_1^{(1)} + ix_2^{(1)}$, $r_1 < |z_1^{(1)}| < r_2$, so that

$$\left(\operatorname{Re}\left(e^{i\varphi}\tilde{w}\right) - \tilde{W}_0\right)\left(z_1^{(1)}, \overline{z_1^{(1)}}\right) > 0 .$$

Then by continuity there must be a neighborhood $N(x_1^{(1)}, x_2^{(1)})$ of $(x_1^{(1)}, x_2^{(1)})$ which lies in the closed annulus and on whose boundary the expression $|\operatorname{Re}(e^{i\varphi}\tilde{w}) - \tilde{W}_0|$ vanishes. But this contradicts Remark 11.3-1, and thus (11.4-22) must be true.

Let us now consider an arbitrary but fixed $z_1^{(0)}$ with $|z_1^{(0)}| = r_0 \in [r_1, r_2]$. Choosing φ in (11.4-21) suitably, we then obtain

$$\begin{aligned} \left|\tilde{w}\left(z_1^{(0)}, \overline{z_1^{(0)}}\right)\right| &= \operatorname{Re}\left(\exp\left[-i \arg \tilde{w}\left(z_1^{(0)}, \overline{z_1^{(0)}}\right)\right] \tilde{w}\left(z_1^{(0)}, \overline{z_1^{(0)}}\right)\right) \\ &\leq \tilde{W}_0\left(z_1^{(0)}, \overline{z_1^{(0)}}\right) = W_0(r_0). \end{aligned}$$

This inequality holds for all $z_1 \in \partial B_1(0, r_0)$. Hence, (11.4-19) follows for $r = r_0$. Since $z_1^{(0)}$, and hence r_0, has been chosen arbitrarily within the given annulus, the proof of the theorem is complete. ■

Observe that in the case $n = 0$, inequality (11.4-19) for $\tilde{w}_1 = \tilde{T}_{1(0)}\tilde{f}_1 = \tilde{f}_1$ does not yield (11.4-17), but the alternative formula

$$(11.4\text{-}23)\qquad \max\left\{|\tilde{f}_1(z_1)| \mid z_1 \in \partial B_1(0, r)\right\} \leq m_1\lambda_1 + m_2\lambda_2$$

with λ_1, λ_2 as given in (11.4-17).

From Theorem 11.4-4 one may draw some conclusions concerning the behavior of class $\mathscr{B}_n$ solutions. Here, the class $\mathscr{B}_n$ denotes the solutions of (11.1-1) that are defined for $z_1 \in B_1(0,1)$, with the possible exception of finitely many isolated singularities, and remain bounded when $\partial B_1(0,1)$ is approached from inside the disk.

11.4-5 Corollary

(a) *Let* $\tilde{w} \in \mathscr{B}_n$ *have no singularity for* $0 \leq r_0 \leq |z_1| < 1$ *and define the function* $\mu\colon (r_0, 1) \to \mathbb{R}$ *by*

$$\mu(r) = \max\left\{\left|\tilde{w}(z_1, \bar{z}_1)/Q_n\left(\frac{1 + z_1\bar{z}_1}{1 - z_1\bar{z}_1}\right)\right| \;\middle|\; z_1 \in \partial B_1(0, r)\right\}. \tag{11.4-24}$$

Then μ *is monotone decreasing.*

(b) *If* $\tilde{w} \in \mathscr{B}_n$ *has no singularity, then* $\tilde{w} = 0$. (*Cf. Theorem* 11.3-4*a*.)

(c) *For all* $\tilde{w} \in \mathscr{B}_n$ *we have*

$$|\tilde{w}(z_1, \bar{z}_1)| = O\left[(1 - z_1\bar{z}_1)^{n+1}\right] \qquad \text{as } |z_1| \to 1. \tag{11.4-25}$$

Proof. Choose r_1 and r_2 in Theorem 11.4-4 so that $r_0 < r_1 < r_2 < 1$. For arbitrary fixed r_1, consider the limit for $r_2 \to 1$. Then we have $X_2 \to \infty$ and hence $P_n(X_2) \to \infty$, $Q_n(X_2) \to 0$ (cf. Magnus et al. [1966], p. 197), and m_2 remains bounded because of the definition of $\mathscr{B}_n$. Thus setting

$$m_k(r_k) = \max\{|\tilde{w}(z_1, \bar{z}_1)| \mid z_1 \in \partial B_1(0, r_k)\}, \qquad k = 1, 2$$

from (11.4-19, 20) for $r \in [r_1, 1)$ follows

$$\mu(r) \leq \frac{m_1(r_1)}{Q_n(X_1)} = \mu(r_1). \tag{11.4-26}$$

Since r_1 is arbitrarily chosen in the interval $(r_0, 1)$, part (a) of the corollary is proved.

The proof of part (b) follows from part (a) by setting $r_0 = 0$ and observing that

$$\lim_{r \to 0} \mu(r) = 0 \tag{11.4-27}$$

because of $Q_n(X) \to \infty$ for $X \to 1$ (i.e., for $r \to 0$) [cf. (11.2-37)]. For, since $\mu(r) \geq 0$, from (11.4-27) and part (a) we get $\mu = 0$ and thus, by (11.4-24), the statement of part (b).

Part (c) is an immediate consequence of the facts that $Q_n(X) = O(X^{-n-1})$ for $X \to \infty$ (i.e., for $r \to 1$) (cf. Magnus et al. [1966], p. 197) and that $\lim_{r \to 1} \mu(r)$ remains bounded [cf. (11.4-27)]. ■

11.5 A FUNCTION THEORY OF SOLUTIONS SATISFYING GENERALIZED CAUCHY–RIEMANN EQUATIONS

Since the beginning of this century there have been numerous attempts to generalize the powerful methods of function theory of one complex variable to other equations or systems and to higher dimensions. We mention here the *function theory in algebras* whose starting point is usually assumed to be a paper by Scheffers [1893–1894], where the complex differentiability was first generalized to the case of algebras of higher dimensions. Since that time, over one thousand papers on this and related subjects have appeared. Therefore it would be impossible to give here a survey of methods or of contributors. The function theories in quaternions and in the Clifford algebra have found particular interest. For a survey on the historical background as well as on the recent development, we refer to the article by Habetha [1983]. Moreover, we have included some papers on these topics in the bibliography, e.g., papers by F. Brackx, R. Delanghe, J. Edenhofer, W. Eichhorn, R. Fueter, K. Habetha, A. Kriszten, E. Kühn, E. Lammel, W. Nef, and H. H. Snyder.

Another branch of generalizations of analytic function theory is the *theory of pseudoanalytic or generalized analytic functions* in the sense of L. Bers and I. N. Vekua; cf. Bers [1956], Vekua [1963], Položiĭ [1960, 1965a, 1965b], and Tutschke [1977]. Ideas of the two aforementioned branches are combined in the theory of *generalized hyperanalytic functions* as investigated by R. P. Gilbert and his school. Here, the theory of elliptic systems in the plane whose investigations in the frame of a hypercomplex algebra was initiated by Douglis [1953] (so-called *theory of hyperanalytic functions*), is developed by following ideas which originated from the work of Bers and Vekua (see before). Thus, the relation between the hyperanalytic functions and the generalized hyperanalytic functions may be compared with the relation between the analytic and the generalized analytic functions. For details, we refer to the book by Gilbert–Buchanan [1983]; cf. also the report by Gilbert [1983].

Furthermore, in Sec. 10.4 we have already presented and utilized the idea of *polyanalytic functions* in the sense of M. B. Balk and their generalization, the *multianalytic functions* in the sense of Krajkiewicz [1973]. It is obvious that a Type I solution of equation (11.1-1) is also a multianalytic function. This follows from (11.2-9) by using the Maclaurin expansion with respect to z_2 and setting $\tilde{f}_2 = 0$, $z_2 = \bar{z}_1$, namely, by

$$\tilde{w}(z_1, \bar{z}_1) = \tilde{T}_1(\tilde{f}_1(z_1))$$

$$= \sum_{\mu=0}^{\infty} \left[\sum_{m=0}^{\min(\mu, n)} \frac{(n+m)!}{(n-m)!m!} \binom{\mu-1}{\mu-m} z_1^{\mu-m} D_1^{n-m} \tilde{f}_1(z_1) \right] \bar{z}_1^{\mu} \tag{11.5-1}$$

[cf. also (10.4-11)]. Thus one might try to generalize theorems from polyanalytic function theory (cf. Balk [1983]) to multianalytic functions (as was done by Krajkiewicz [1973] in the case of the Great Picard Theorem) and then to specialize those results to solutions (11.5-1) of (11.1-1). We shall not pursue this idea, but shall sketch a function theory proposed by Bauer [1967], which is better adapted to the present equation (11.1-1).

GENERALIZED CAUCHY–RIEMANN EQUATIONS

We remember that for a holomorphic function $\hat{w} = \hat{u} + i\hat{v}$ of one complex variable the real and imaginary parts satisfy the two-dimensional Laplace equation

$$\Delta_2 \hat{u} = 0 \qquad \text{and} \qquad \Delta_2 \hat{v} = 0 .$$

In our complex notation of Sec. 2.1, we have the equivalent form

$$D_1 \overline{D}_1 w = 0. \tag{11.5-2}$$

On the other hand, requiring that a solution of this equation satisfy the Cauchy–Riemann equations

$$\hat{u}_{x_1} = \hat{v}_{x_2} \qquad \text{and} \qquad \hat{u}_{x_2} = -\hat{v}_{x_1}$$

or in complex notation, equivalently,

$$\overline{D}_1 w = 0 \tag{11.5-3}$$

in some domain, we can sort out the holomorphic functions from among the harmonic ones.

We now proceed similarly. We replace equation (11.5-2) by our equation (11.1-1) under consideration. Instead of the additional condition (11.5-3), we then require that the solutions of (11.1-1) satisfy the condition

$$\left[(1 - z_1\bar{z}_1)^2 \overline{D}_1\right]^{n+1} w = 0 . \tag{11.5-4}$$

Since for $n = 0$, the equations (11.1-1) and (11.5-4) are in the unit disk equivalent to those in (11.5-2) and (11.5-3), respectively, it is obvious that we have thus defined a *kind of generalized holomorphic functions* which, in addition, are solutions of equation (11.1-1). For these solutions we shall develop a function theory in the remainder of this section.

Introducing the auxiliary variable Z defined by

$$Z = \bar{z}_1/(1 - z_1\bar{z}_1) \tag{11.5-5}$$

we see that the solutions of (11.5-4) which are $(n + 1)$ times continuously differentiable (with respect to z_1, $\bar{z}_1$) in the unit disk or some simply connected subdomain Ω_1 of it, must necessarily be a polynomial w_n in Z of degree not exceeding n and with coefficients g_m depending in Ω_1 holomorphically on z_1. On the other hand, every such polynomial, that is,

$$w_n(Z) = \sum_{m=0}^{n} g_m(z_1)Z^m \tag{11.5-6}$$

with arbitrary functions $g_m \in C^\omega(\Omega_1)$, $m = 0,\cdots, n$, is a solution of (11.5-4). Moreover, for a given solution of (11.5-4) the functions $g_0,\cdots, g_n$ are uniquely determined. Let us now consider solutions of (11.1-1) which are regular in Ω_1. According to Sec. 11.2 these solutions can be represented in the form

$$\begin{aligned}\tilde{w}(z_1, \bar{z}_1) &= \tilde{T}_1(\tilde{f}_1(z_1)) + \bar{\tilde{T}}_1(\tilde{f}_2(\bar{z}_1)) \\ &= \sum_{m=0}^{n} \frac{(n+m)!}{(n-m)!m!}\left[Z^m D_1^{n-m}\tilde{f}_1(z_1) + \bar{Z}^m \bar{D}_1^{n-m}\tilde{f}_2(\bar{z}_1)\right]\end{aligned} \tag{11.5-7}$$

where $(\tilde{f}_1, \bar{\tilde{f}}_2) \in C^\omega(\Omega_1) \times C^\omega(\Omega_1)$. A comparison of (11.5-6) and (11.5-7) shows that condition (11.5-4) selects those and only those solutions $\tilde{w}$ of (11.1-1) that are already representable by the Type I representation by means of the operator $\tilde{T}_1$.

CLASS $\mathscr{H}_n(\Omega_1)$

Let us denote the class of functions $\tilde{w}$ in (11.5-7) with $\tilde{f}_1 \in C^\omega(\Omega_1)$ and $\tilde{f}_2 = 0$ by $\mathscr{H}_n(\Omega_1)$ (Ω_1 as before). Furthermore, let $n \in \mathbb{N}$ be arbitrarily fixed. Obviously, then $\mathscr{H}_n(\Omega_1)$ is a linear space over $\mathbb{C}$. For functions $\tilde{w} \in \mathscr{H}_n(\Omega_1)$ we now define the *Bauer derivative* $\tilde{D}\tilde{w}$ of $\tilde{w}$ by

$$\tilde{D}\tilde{w} = (D_1 - \bar{z}_1^2\bar{D}_1)\tilde{w}\,. \tag{11.5-8}$$

This differentiation process produces again an element of $\mathscr{H}_n(\Omega_1)$ because for $\tilde{w} = \tilde{T}_1\tilde{f}_1$ we have

$$\tilde{D}\tilde{w} = \tilde{T}_1 D_1 \tilde{f}_1\,. \tag{11.5-9}$$

This also yields that, in order to obtain the Bauer derivative of $\tilde{w}$, in the representation of $\tilde{w}$ we merely have to replace the $\tilde{T}_1$-associated function $\tilde{f}_1$ by its classical derivative $\tilde{f}_1' = D_1\tilde{f}_1$. Our next step is to introduce *generalized kth powers* of

$$Z_n(z_1, z_1^{(0)}) = \tilde{T}_1((z_1 - z_1^{(0)}))\,, \qquad z_1^{(0)} \in \Omega_1$$

by

$$Z_n^k(z_1, z_1^{(0)}) = \tilde{T}_1((z_1 - z_1^{(0)})^k)\,, \qquad k \in \mathbb{Z}\,.$$

The expression

$$c_0 Z_n^0\left(z_1, z_1^{(0)}\right) = c_0 \frac{(2n)!}{n!} Z^n$$

(with an arbitrary complex constant c_0) will be called a *generalized constant.* The motivation for this terminology becomes obvious by the properties

$$\tilde{D} Z_n^k\left(z_1, z_1^{(0)}\right) = k Z_n^{k-1}\left(z_1, z_1^{(0)}\right)$$

and

$$\tilde{D}\left(c_0 Z_n^0\left(z_1, z_1^{(0)}\right)\right) \equiv 0 .$$

We emphasize that the functions $Z_n^k(z_1, z_1^{(0)})$ are contained in $\mathscr{H}_n(\Omega_1)$.

A FUNCTION THEORY OF SOLUTIONS OF CLASS $\mathscr{H}_n(\Omega_1)$

We shall now give some typical results on the function theory of $\mathscr{H}_n(\Omega_1)$; that is, on the *function theory of solutions of* (11.1-1) *satisfying the generalized Cauchy–Riemann equation* (11.5-4). All the subsequent theorems may easily be proved by utilizing the one-to-one correspondence between a given solution $\tilde{w}$ and its $\tilde{T}_1$-associated function $\tilde{f}_1$ [see Remark 11.2-2c, in particular (11.2-20)]. One applies the familiar theorems of classical function theory of one complex variable to the associated function and "translates" them into analogous theorems of solutions in $\mathscr{H}_n(\Omega_1)$.

Let $B_1(z_1^{(0)}, r)$ $(r > 0)$ be contained in Ω_1. Then, using the generalized powers, we have the following *expansion theorem*:

11.5-1 Theorem

Every function $\tilde{w} \in \mathscr{H}_n(\Omega_1)$ has a series expansion of the form

$$\tilde{w}(z_1, \bar{z}_1) = \sum_{k=0}^{\infty} \omega_k Z_n^k\left(z_1, z_1^{(0)}\right)$$

which converges in $B_1(z_1^{(0)}, r) \subset \Omega_1$ and involves uniquely determined complex constants ω_k $(k \in \mathbb{N}_0)$.

This theorem is the main tool for the proof of the analog of the *identity theorem* for holomorphic functions.

11.5-2 Theorem

If two functions $\tilde{w}_1, \tilde{w}_2 \in \mathscr{H}_n(\Omega_1)$ coincide on some open disk $B_1(z_1^{(0)}, r) \subset \Omega_1$, then $\tilde{w}_1 = \tilde{w}_2$ holds on Ω_1.

However, we point out that, in contrast to the case of holomorphic functions, in the preceding theorem the disk $B_1(z_1^{(0)}, r)$ can, in general, not be replaced by a path originating from $z_1^{(0)}$. To see this, consider the associated function

$$\tilde{f}_1(z_1) = z_1^{n+1} + z_1^{n-1}.$$

An application of $\tilde{T}_1$ yields the function

(11.5-10)

$$\tilde{w}(z_1, \bar{z}_1) = \tilde{T}_1\big(\tilde{f}_1(z_1)\big) = \frac{(z_1 + \bar{z}_1)n!}{(1 - z_1\bar{z}_1)^n} \sum_{m=1}^{n} \binom{n+1}{m}\binom{n-1}{n-m}(z_1\bar{z}_1)^{n-m}$$

which is of class $\mathscr{H}_n(B_1(0,1))$. The function $\tilde{w}$ in (11.5-10) vanishes in $B_1(0,1)$ for $z_1 = -\bar{z}_1$, that is, on the imaginary axis, and is real-valued for all $z_1 \in B_1(0,1)$. This example shows that, in general, for elements of $\mathscr{H}_n(\Omega_1)$ the zeros (and hence the *a*-points) are not isolated. Incidentally, this example is also a *counterexample against* the validity of the *open mapping theorem* for functions of $\mathscr{H}_n(\Omega_1)$.

As in the case of clasical function theory, from Theorem 11.5-2 we can derive a theorem concerning *analytic continuation* and a *monodromy theorem* for functions of $\mathscr{H}_n(\Omega_1)$.

11.5-3 Theorem

Let Ω_{11}, Ω_{12} be simply connected domains in $B_1(0,1)$ with $\Omega_{11} \cap \Omega_{12} \neq \varnothing$. Then, for every pair $\tilde{w}_1 \in \mathscr{H}_n(\Omega_{11})$, $\tilde{w}_2 \in \mathscr{H}_n(\Omega_{12})$ satisfying $\tilde{w}_1(z_1, \bar{z}_1) = \tilde{w}_2(z_1, \bar{z}_1)$ for all $z_1 \in \Omega_{11} \cap \Omega_{12}$, there exists a uniquely determined function $\tilde{w} \in \mathscr{H}_n(\Omega_{11} \cup \Omega_{12})$ such that $\tilde{w}$ coincides with $\tilde{w}_1$ on Ω_{11} and with $\tilde{w}_2$ on Ω_{12}.

The result of this theorem gives reason in the case $\Omega_{11} = B_1(z_1^{(1)}, r_1)$ and $\Omega_{12} = B_1(z_1^{(2)}, r_2)$ $(r_1, r_2 > 0)$ to regard $\tilde{w}_1$ and $\tilde{w}_2$ as regular *function elements* of the function $\tilde{w}$ which are obtained by analytic continuation from each other. The explicit realization of the continuation and the determination of further function elements of $\tilde{w}$ may, as before, be reduced to the analogous problem for the uniquely determined $\tilde{T}_1$-associated function. We have thus obtained:

11.5-4 Theorem

Let the regular function element $\tilde{w}_0$, that means, $\tilde{w}_0 \in \mathscr{H}_n(B_1(z_1^{(0)}, r_0))$ with suitable $z_1^{(0)} \in \Omega_1$ and $r_0 > 0$, admit unrestricted analytic continuation in Ω_1 according to Theorem 11.5-3. *Then there exists a uniquely determined function $\tilde{w} \in \mathscr{H}_n(\Omega_1)$ satisfying $\tilde{w}(z_1, \bar{z}_1) = \tilde{w}_0(z_1, \bar{z}_1)$ for all $z_1 \in B_1(z_1^{(0)}, r_0)$. In particular, if Γ_1 and Γ_2 are two paths in Ω_1 from $z_1^{(0)}$ to $\tilde{z}_1^{(0)}$ which are fixed-end-*

point homotopic, then the analytic continuations along Γ_1 *and* Γ_2 *produce the same function element* $\tilde{\tilde{w}}_0 \in \mathscr{H}_n(B_1(\tilde{z}_1^{(0)}, \tilde{r}_0))$ $(\tilde{r}_0 > 0)$.

Furthermore, we mention the possibility of defining contour integrals and primitives for functions of class $\mathscr{H}_n(\Omega_1)$. Let $\tilde{w} \in \mathscr{H}_n(\Omega_1)$ possess the $\tilde{T}_1$-associated function $\tilde{f}_1 \in C^\omega(\Omega_1)$ and let $z_1^{(0)}$, $z_1 \in \Omega_1$ be connected by a given rectifiable path Γ in Ω_1. Then we define as *generalized contour integral* for $\tilde{w}$ the expression

$$I(\Gamma; z_1^{(0)}, z_1)(\tilde{w}) = \tilde{T}_1\left(\int_{\Gamma(z_1^{(0)}, z_1)} \tilde{f}_1(z_{11})\, dz_{11}\right). \tag{11.5-11}$$

This is an element of $\mathscr{H}_n(\Omega_1)$, of course. For $z_1^{(0)}$ and z_1 fixed, say, $z_1 = \tilde{z}_1^{(0)}$, it represents a generalized constant, namely,

$$\begin{aligned} I(\Gamma; z_1^{(0)}, \tilde{z}_1^{(0)})(\tilde{w}) &= \left(\int_{\Gamma(z_1^{(0)}, \tilde{z}_1^{(0)})} \tilde{f}_1(z_{11})\, dz_{11}\right) Z_n^0(z_1, z_1^{(0)}) \\ &= c_n(\Gamma; z_1^{(0)}, \tilde{z}_1^{(0)}; \tilde{f}_1) Z^n \end{aligned} \tag{11.5-12}$$

where

$$c_n(\Gamma; z_1^{(0)}, \tilde{z}_1^{(0)}; \tilde{f}_1) = \frac{(2n)!}{n!} \int_{\Gamma(z_1^{(0)}, \tilde{z}_1^{(0)})} \tilde{f}_1(z_{11})\, dz_{11}$$

is a constant (in the usual sense) and Z is defined in (11.5-5).

A *generalized primitive* of $\tilde{w} \in \mathscr{H}_n(\Omega_1)$ is a function $\tilde{W} \in \mathscr{H}_n(\Omega_1)$ which is a primitive with respect to the Bauer differentiation process given by (11.5-8), that is, we require $\tilde{W}$ to satisfy the condition

$$\tilde{D}\tilde{W} = \tilde{w}. \tag{11.5-13}$$

Because of (11.5-9), for a given function $\tilde{w} \in \mathscr{H}_n(\Omega_1)$, generalized primitives of $\tilde{w}$ can only differ by a generalized constant.

A generalization of *Cauchy's theorem* as well as the relation between generalized contour integrals and generalized primitives are obtained in the subsequent theorems. Recall that we have assumed Ω_1 to be simply connected!

11.5-5 Theorem

For every function $\tilde{w} \in \mathscr{H}_n(\Omega_1)$ *there exists the generalized contour integral*

$$I(\Gamma; z_1^{(0)}, \tilde{z}_1^{(0)})(\tilde{w}) = \tilde{T}_1\left(\int_{\Gamma(z_1^{(0)}, \tilde{z}_1^{(0)})} \tilde{f}_1(z_{11})\, dz_{11}\right)$$

for every path Γ *in* Ω_1 *from* $z_1^{(0)}$ *to* $\tilde{z}_1^{(0)}$; *here,* $\tilde{f}_1 \in C^\omega(\Omega_1)$ *is the unique*

$\tilde{T}_1$-associated function of $\tilde{w}$. The resulting generalized constant is the same for all paths in Ω_1 with the same endpoints.

The last statement implies that this integral over closed paths is zero.

11.5-6 Theorem

If the generalized contour integral $I(\Gamma; z_1^{(0)}, z_1)(\tilde{w})$ of $\tilde{w} \in \mathscr{H}_n(\Omega_1)$ is independent of path for all $z_1 \in \Omega_1$, then it represents a generalized primitive $\tilde{W}$ for $\tilde{w}$. In particular, if $\tilde{F}_1$ is a primitive (in the usual sense) for the $\tilde{T}_1$-associated function $\tilde{f}_1$ of $\tilde{w}$, then for a path Γ from $z_1^{(0)}$ to $\tilde{z}_1^{(0)}$ we have

$$I\left(\Gamma; z_1^{(0)}, \tilde{z}_1^{(0)}\right)(\tilde{w}) = \tilde{T}_1\left(\tilde{F}_1\left(\tilde{z}_1^{(0)}\right)\right) - \tilde{T}_1\left(\tilde{F}_1\left(z_1^{(0)}\right)\right).$$

We have just outlined how a foundation of a function theory of class $\mathscr{H}_n(\Omega_1)$ may be obtained. For further results in this direction, for instance, concerning a generalization of Laurent expansions, residue theorems, and automorphic functions, we refer to Bauer [1965, 1967].

Let us conclude this section by taking a final look at some possibilities of how to obtain results concerning function theoretical properties of solutions of equation (2.1-1), possibilities revealed by the consideration of the special equation (11.1-1).

(i) The solutions can be investigated in the frame of classical function theory of one complex variable when applied to the T_1-associated function f_1 of w or to the $\tilde{T}_1$-associated function $\tilde{f}_1$ of $\tilde{w}$, or of two variables when directly applied to solutions w or $\tilde{w}$ along the lines shown in the preceding chapters.

(ii) One can try to establish a function theory of multianalytic functions in the sense of Krajkiewicz in generalizing the function theory of polyanalytic functions (cf. Balk [1983]). A first step has been done by Krajkiewicz [1973] in the case of the Great Picard Theorem. We used this in Sec. 10.4.

(iii) One also employs other theories of "generalized analytic functions" where the Cauchy–Riemann equations are replaced by more general equations. In this respect, there exist a great variety of possibilities; as mentioned at the beginning of this section. The next step is the investigation to what classes of equations (2.1-1) or classes of solutions such a theory applies. A special example using the "generalized Cauchy–Riemann equation" (11.5-4) has been treated throughout this section.

The first method has been intensively studied by many authors and is applicable to all equations of the form (2.1-1); cf., in particular, Chaps. 8–10. The second method seems to be promising and will be applicable to all equations (2.1-1) when restricted to $z_2 = \bar{z}_1$. The third method on the one hand offers a great variety of possibilities, but on the other hand will generally restrict the class of equations (2.1-1), for instance, to subclasses of equations admitting Type I representations by means of Bergman operators with polynomial kernels.

Additional References

Balk [1983]
Bauer [1965, 1966a, 1967, 1968, 1976b]
Bauer–Peschl [1966, 1967]
Bauer–Ruscheweyh [1980]
Begehr [1976]
Bers [1956]
Blohina [1970]
Brackx–Delanghe–Sommen [1982]
Bragg–Dettman [1983, 1984]
Case [1985]
Delanghe [1970]
Douglis [1953]
Edenhofer [1973, 1976]
Eichhorn [1961, 1969]
Fueter [1934]
Gilbert [1969b, 1983]
Gilbert–Buchanan [1983]
Gilbert–Wendland [1974–1975]
Goluzin [1969]
Goman [1984]
Gronau–Reich [1975]
Habetha [1983]
Henrici [1953, 1974–1986]
Horváth [1963]
Jank–Ruscheweyh [1975]
Kracht–Kreyszig [1969, 1970]
Kracht–Kreyszig–Schröder [1982]
Krajkiewicz [1973]
Kriszten [1952]
Kühn [1974]
Lammel [1950–1951]
Lieb [1970]
Malonek [1980]
Masuda [1974]
Nef [1942–1944]
Peschl [1955, 1963, 1968]
Položiĭ [1960, 1965a, 1965b]
Reich [1972]
Roelcke [1956]
Ruscheweyh [1972b, 1974a, 1976, 1982]
Ruscheweyh–Wirths [1976, 1985]
Snyder [1968]
Spindelböck–Reich [1975]
Tutschke [1977]
Tutschke–Withalm [1983]
Vekua [1963]
Withalm [1972, 1985]

Chapter Twelve

Application to Compressible Fluid Flow

Beginning with this chapter, we turn to physical applications of the integral operator method, namely, to *compressible* (nonviscous) fluid flow. Historically, this is one of the roots of the operator method, and Bergman [1945c, d, 1946a, b, 1947b, 1948, 1949b, 1952a, b, 1966a, 1973] and others have shown that this method is a powerful tool in treating and solving flow problems in the compressible case. See also Mises–Schiffer [1948], Mises [1958], Krzywoblocki [1960], and Nazarov [1964].

To keep the presentation self-contained, we devote the first two sections to some physical facts and principles and a derivation of the equations needed. In Sec. 12.3 we take a look at the application of pseudoanalytic functions to flow problems. Then in Sec. 12.4 we explain the basic idea of Chaplygin's approach. Section 12.5 is devoted to Bergman's operator method.

12.1 BASIC CONCEPTS AND EQUATIONS

A fluid and its flow are called *compressible* if its *density* ρ is variable, and *incompressible* if ρ is constant.

Fluids in the restricted sense, or *liquids*, have small compressibility, which can be neglected in many practical problems, so that they may then be regarded as incompressible. The concept of fluid state also includes gases and vapors, which have large compressibility. Accordingly, except when a flow of gas has small speed (a small fraction of the speed of sound), the flow must be regarded as compressible. The theory and application of phenomena for which compressibility is significant is called *gas dynamics*. As a special case, it includes the classical hydrodynamics of liquids (incompressible fluids), but the typical problems of gas dynamics involve considerable variability of density.

A fluid flow in a region R of space can be described mathematically by a vector field and two scalar fields. The vector field is the *velocity field* determined by the velocity vector function **v**, which gives the magnitude and direction of the velocity at time t and every point $\mathbf{x} = (x_1, x_2, x_3)$ in R.

One of the two scalar fields is the *density field* determined by the density function ρ, which gives mass per unit volume at time t and point **x**.

The other scalar field is the *pressure field* determined by the pressure function p, which gives the pressure at t and **x**.

A third scalar field, the *temperature field*, will not appear in our consideration since we shall assume the temperature to be constant, for simplicity.

We see that those fields are given by five functions, namely, ρ, p, and the three components of **v**. These functions are related by five equations. Two of them, the *equation of state* and the *equation of continuity*, will be discussed in this section, and three more equations, resulting from *Newton's second law*, in the next section.

EQUATION OF STATE

In this derivation, we assume that the fluid is *nonviscous*, that is, has no internal friction (no internal tangential stresses); such a fluid is also known as an *ideal* or *perfect fluid*. This assumption holds in the case of many situations and problems of practical importance, whereas it excludes the theory of boundary layers and of a few other phenomena.

The first equation, the *equation of state*, relates ρ and p (and temperature T, which is assumed to be constant in our consideration). Hence, it is an equation $F(\rho, p, T) = 0$, which, by our assumption, becomes $F(\rho, p) = 0$, and we can assume that it defines

$$p = h(\rho)$$

explicitly.

A physically important case is

$$p = \alpha\rho, \qquad \alpha > 0, \text{constant}.$$

In this case the motion of an ideal gas is called *isothermic*.

Another important type of equation of state is

$$p = \alpha\rho^{\kappa}$$

where $\kappa = 1.40$ for dry air; the corresponding motion of an ideal gas is called *isentropic*.

CONTINUITY EQUATION

The second of those five equations relates the velocity field and the density field:

$$\frac{\partial\rho}{\partial t} + \operatorname{div}(\rho\mathbf{v}) = 0. \tag{12.1-1}$$

This is called the *continuity equation* of a compressible fluid flow. div is the divergence. Equation (12.1-1) can be obtained as follows. The mass in a subregion R_1 of the region R of flow at time t is

$$M(t) = \iiint_{R_1} \rho(\mathbf{x}, t)\, d\mathbf{x}.$$

During a short time interval $(t, t + \delta t)$ the mass increases due to the flow through the boundary ∂R_1 of R_1, the amount of increase being

$$\delta M(t) = \delta t \iint_{\partial R_1} \rho \mathbf{v}_n \, dS$$

where n denotes the interior normal direction. From this and the principle of conservation of mass,

$$\frac{d}{dt} \iiint_{R_1} \rho \, d\mathbf{x} = \iint_{\partial R_1} \rho \mathbf{v}_n \, dS.$$

By Gauss's theorem, assuming that there are no *sources* or *sinks* in R (points at which fluid is produced or disappears), we obtain

$$\iiint_{R_1} \left[\frac{\partial \rho}{\partial t} + \operatorname{div}(\rho \mathbf{v}) \right] d\mathbf{x} = 0.$$

Since this must hold for every subregion, (12.1-1) follows. For another derivation, see Kreyszig [1988], pp. 493–495.

A flow is called *steady* if it is time-independent. Then $\partial \rho / \partial t = 0$ and (12.1-1) becomes

$$\operatorname{div}(\rho \mathbf{v}) = 0. \qquad (12.1\text{-}2)$$

If the flow is also incompressible, it follows that

$$\operatorname{div} \mathbf{v} = 0.$$

This is called the *incompressibility condition*.

VELOCITY POTENTIAL. COMPLEX POTENTIAL

A flow is called *irrotational* if

$$\operatorname{curl} \mathbf{v} = \mathbf{0}.$$

In this case, the velocity $\mathbf{v}$ of the flow is the gradient of a scalar function $-\phi$

(the minus sign being conventional), that is,

(12.1-3) $$\mathbf{v} = -\operatorname{grad} \phi .$$

ϕ is called the *velocity potential.*

Hence if the flow is *two-dimensional* (independent of x_3, say), writing $x_1 = x$, $x_2 = y$ for the Cartesian coordinates, we see that (12.1-3) becomes

$$\mathbf{v} = (v_1, v_2) = -\operatorname{grad} \phi = -(\phi_x, \phi_y) .$$

Also, if this flow is steady, then (12.1-1) takes the form

(12.1-4) $$(\rho\phi_x)_x + (\rho\phi_y)_y = 0 .$$

Note that in the incompressible case, this reduces to Laplace's equation

$$\Delta\phi = 0 .$$

This can be regarded as an integrability condition of the Cauchy–Riemann equations

$$\phi_x = \psi_y , \qquad \phi_y = -\psi_x .$$

ψ is called the *stream function* of the flow. The curves $\psi(x, y) = const$ are the *streamlines* of the flow (curves along which the particles move, curves to which the velocity vectors are tangential). The analytic function

$$f(z) = \phi(x, y) + i\psi(x, y)$$

($z = x + iy$) is called the ***complex potential*** of the flow.

More generally, equation (12.1-4) is the integrability condition of the system

(12.1-5) $$\rho\phi_x = \psi_y , \qquad \rho\phi_y = -\psi_x$$

which generalizes the Cauchy–Riemann equations.

12.2 BERNOULLI'S LAW, CHAPLYGIN'S EQUATION

In the previous section we discussed two of the five equations relating the five variables v_1, v_2, v_3, ρ, and p. Three more equations (a single vector equation) are obtained by considering the forces acting on the elements of fluid and applying Newton's second law. The result is

(12.2-1a) $$\rho \frac{d^2\mathbf{x}}{dt^2} = \rho\mathbf{F} - \operatorname{grad} p$$

or

$$\frac{d^2\mathbf{x}}{dt^2} + \frac{1}{\rho}\operatorname{grad} p = \mathbf{F} \tag{12.2-1b}$$

where $\mathbf{F}$ is the density per mass of external forces. Indeed, the resultant of all forces acting on a subregion $R_1 \subset R$ is

$$\begin{aligned} \mathbf{P} &= \iiint_{R_1} \rho \mathbf{F}\, d\mathbf{x} + \iint_{\partial R_1} p n\, dS \\ &= \iiint_{R_1} (\rho \mathbf{F} - \operatorname{grad} p)\, d\mathbf{x}. \end{aligned}$$

By Newton's second law, this equals the mass $M(t)$ (see above) times the acceleration. From this, since R_1 is any subregion of R, we obtain (12.2-1a).

The vector equation (12.2-1b) can be cast into a somewhat more convenient form as follows. Noting that $d\mathbf{x}/dt = \mathbf{v}$, where $\mathbf{v} = (v_1, v_2, v_3)$, as before, and $dx_j/dt = v_j$, we first have

$$\frac{d^2\mathbf{x}}{dt^2} = \frac{d\mathbf{v}}{dt} = \frac{\partial \mathbf{v}}{\partial t} + \sum_{j=1}^{3} \frac{\partial \mathbf{v}}{\partial x_j} v_j .$$

Straightforward calculation shows that the sum on the right can be written

$$(\operatorname{curl} \mathbf{v}) \times \mathbf{v} + \tfrac{1}{2} \operatorname{grad} (|\mathbf{v}|^2).$$

Using this expression, we see that (12.2-1b) takes the form

$$\frac{\partial \mathbf{v}}{\partial t} + (\operatorname{curl} \mathbf{v}) \times \mathbf{v} + \tfrac{1}{2} \operatorname{grad} (|\mathbf{v}|^2) + \frac{1}{\rho} \operatorname{grad} p = \mathbf{F}. \tag{12.2-2}$$

We now consider the important case of an irrotational steady flow, assuming the external forces to be conservative, so that they have a potential U, and

$$\mathbf{F} = -\operatorname{grad} U.$$

Then (12.2-2) takes the form

$$\tfrac{1}{2} \operatorname{grad} (|\mathbf{v}|^2) + \frac{1}{\rho} \operatorname{grad} p + \operatorname{grad} U = 0.$$

By integration,

$$\tfrac{1}{2}|\mathbf{v}|^2 + \int_{p_0}^{p} \frac{dp}{\tilde{\rho}} + U = const. \tag{12.2-3}$$

This is called *Bernoulli's law*. Using (12.1-3) we can write this

$$\frac{1}{2}|\text{grad}\,\phi|^2 + \int_{\rho_0}^{\rho} \frac{dp}{\tilde{\rho}} + U = const. \tag{12.2-4}$$

From (12.1-2) we have

$$\text{div}\,(\rho\,\text{grad}\,\phi) = 0. \tag{12.2-5}$$

Equations (12.2-4) and (12.2-5) are two equations for the two unknown functions ρ and ϕ.

HODOGRAPH PLANE. CHAPLYGIN'S EQUATION

Furthermore, if there are no external forces, we see from (12.2-3) and (12.2-4) that ρ becomes a function of $|\mathbf{v}|^2$, that is, of $|\text{grad}\,\phi|$. Concentrating again on two-dimensional flow, we see that then ρ is a function of $\phi_x^2 + \phi_y^2$. With such a function ρ, the system [cf. (12.1-5)]

$$\rho\phi_x = \psi_y, \qquad \rho\phi_y = -\psi_x$$

is *nonlinear*. However, it can be transformed into a *linear* system by a transition from the physical plane (the xy-plane) to the *hodograph plane* (the $v_1 v_2$-plane). Indeed, by using polar coordinates

$$q = |\mathbf{v}| = \sqrt{v_1^2 + v_2^2}, \qquad \theta = \arctan\frac{v_2}{v_1}$$

in the hodograph plane as new variables, equations (12.1-5) take the ***linear*** form

$$\phi_q = \frac{M^2 - 1}{\rho q}\psi_\theta, \qquad \phi_\theta = \frac{q}{\rho}\psi_q \tag{12.2-6}$$

where $M = q/a$ is the ***Mach number*** and $a = (dp/d\rho)^{1/2}$ is the local speed of sound.

Differentiating and eliminating ϕ, we can obtain a second-order differential equation for the stream function, namely

$$\left(\frac{q}{\rho}\psi_q\right)_q = \frac{M^2 - 1}{\rho q}\psi_{\theta\theta}$$

or

$$\psi_{qq} + \frac{\rho}{q}\left(\frac{q}{\rho}\right)_q \psi_q - \frac{M^2 - 1}{q^2}\psi_{\theta\theta} = 0.$$

In order to simplify the coefficient of ψ_q we differentiate Bernoulli's law, finding

$$q\frac{dq}{dp}+\frac{1}{\rho}=0 \qquad \text{or} \qquad \frac{dp}{dq}=-\rho q.$$

Using this and $d\rho/dp = 1/a^2$, we obtain the famous ***Chaplygin equation***

$$\psi_{qq}+\frac{1-M^2}{q^2}\psi_{\theta\theta}+\frac{1+M^2}{q}\psi_q=0. \tag{12.2-7}$$

The transition from the *physical plane* (the xy-plane in which the flow takes place) to the hodograph plane is called the ***hodograph method***. It is practically useful in cases in which it yields simple boundary conditions. These are primarily plane flows such that on the boundary the flow has constant direction or constant speed (hence constant pressure, by Bernoulli's law). Generally speaking, the advantages resulting from the linearization are frequently greater than the disadvantages encountered by complications in the boundary conditions.

12.3 METHOD OF PSEUDOANALYTIC FUNCTIONS

There are several approaches to the determination and investigation of solutions of (12.2-6) or (12.2-7). Before we turn to the approach by integral operators, we shall explain two other methods which are of interest in connection with integral operators. The first of them is the method of pseudoanalytic functions suggested by É. Picard, E. Beltrami, T. Carleman and others, and developed in detail by L. Bers (cf. Bers et al. [1954]; see also Courant–Hilbert [1962], pp. 374–406) and Vekua [1963]. In a sense, this method is the opposite of that of integral operators. Indeed, whereas in the latter we use complex analysis as it is and apply a mapping (an operator), in the former we replace the Cauchy–Riemann equations by more general equations and develop a corresponding generalized complex analysis along the lines of the usual one.

In the application of the Bers–Vekua method to (12.2-6) it is convenient to retain θ but replace q by a new independent variable λ such that (12.2-6) takes a form which is very similar to the Cauchy–Riemann equations, namely,

$$\phi_\theta = l(\lambda)\psi_\lambda, \qquad \phi_\lambda = -l(\lambda)\psi_\theta. \tag{12.3-1}$$

By comparing this with (12.2-6) we readily find that the desired transformation is

$$\lambda=\int_{q_0}^{q}\frac{1}{\tilde{q}}\left(1-M^2(\tilde{q})\right)^{1/2}d\tilde{q}, \qquad l=\frac{1}{\rho}(1-M^2)^{1/2}. \tag{12.3-2}$$

Thus, l is a known function of λ, depending on the physical nature of the fluid.

A flow is called ***subsonic*** if $M = q/a < 1$ and ***supersonic*** if $M > 1$.

Thus the transformation (12.3-2) is real if and only if the flow is subsonic everywhere in the region R.

In the method to be discussed, one combines a pair of solutions ϕ, ψ of (12.3-1) as in complex analysis; that is, one defines

$$f(z) = \phi(\theta, \lambda) + i\psi(\theta, \lambda), \qquad z = \theta + i\lambda .$$

f is called a ***pseudoanalytic function***.

As in complex analysis, from a given solution f we can obtain new solutions by (suitably defined) differentiation and integration.

Differentiation is defined by

$$f'(z) = \phi_\theta + i\psi_\theta = l\psi_\lambda - i\phi_\lambda/l .$$

If f is pseudoanalytic in a domain Ω_1 and $\phi, \psi \in C^2(\Omega_1)$, then f' is pseudoanalytic in Ω_1.

A proof follows easily by differentiation, since l depends only on one variable.

Note that if l is analytic, and is positive so that the system under consideration is elliptic, then $\phi, \psi \in C^2(\Omega_1)$ implies $\phi, \psi \in C^\omega(\Omega_1)$ by a well-known theory initiated by É. Picard (cf. Miranda [1970], pp. 212–216). Hence in this case, by differentiation one can obtain arbitrarily many new solutions from a given one.

Integration of f along a rectifiable arc $C \subset \mathbb{C}$ from a fixed point to a variable point z is defined by

$$F(z) = \int_C f(z)\,dz = \int_C (\phi\,d\theta - l\psi\,d\lambda) + i\int_C (\psi\,d\theta + (\phi/l)\,d\lambda) . \tag{12.3-3}$$

If f is pseudoanalytic in a simply connected domain Ω_1, so is F. Indeed, the integrals on the right-hand side of (12.3-3) are independent of path in Ω_1, and, setting

$$F(z) = \Phi(\theta, \lambda) + i\Psi(\theta, \lambda)$$

we find by differentiation that Φ and Ψ satisfy (12.3-1).

Hence integration is another process by which we can obtain new solutions of (12.3-1), that is, of (12.2-6).

In particular, one can start from a complex constant and by successive integration obtain solutions ("formal powers") which play a role similar to that of ordinary powers of z in complex analysis. The whole theory of solutions may then be developed along the lines of classical complex analysis, so that we can derive analogs of Cauchy's, Morera's, and other theorems. This theory may then be applied to boundary value problems. For details and extensions as well as references to the original literature, see Bers–Gelbart [1944], Bers [1952, 1956], Bers et al. [1954], pp. 69–94, and Courant–Hilbert [1962], pp. 374–406.

12.4 CHAPLYGIN'S APPROACH

The second of the two methods to be discussed before the integral operator method is by Chaplygin, and is of interest here, since Bergman's integral operator method is intended to overcome some of the difficulties and shortcomings of Chaplygin's method. The latter is essentially an approach by separating variables (which Bergman's method is *not*). It is suitable for obtaining approximate solutions of boundary value problems in two-dimensional compressible flow. The essential steps are as follows.

We start from Chaplygin's equation, in which $M^2 = q^2/a^2$ (Sec. 12.2), so that we can write

$$\psi_{qq} + \frac{1}{q^2}\left(1 - \frac{q^2}{a^2}\right)\psi_{\theta\theta} + \frac{1}{q}\left(1 + \frac{q^2}{a^2}\right)\psi_q = 0. \tag{12.4-1}$$

(Another possibility would be a start from (12.3-1) and elimination of ϕ by differentiation, the further steps and the final outcome being quite similar to our present approach.) We first look for particular solutions of the form

$$\psi_n^{(1)} = h_n(q)\cos(n\theta), \qquad \psi_n^{(2)} = h_n(q)\sin(n\theta).$$

Here we assume $n = 0, 1, \cdots$, so that the resulting solutions are single-valued. (Multivalued solutions would be obtained by choosing noninteger n.) By substitution into (12.4-1) we readily see that h_n must satisfy

$$h_n'' + \frac{1}{q}\left(1 + \frac{q^2}{a^2}\right)h_n' - \frac{n^2}{q^2}\left(1 - \frac{q^2}{a^2}\right)h_n = 0. \tag{12.4-2}$$

The form of the h_n depends on the equation of state of the fluid. For the isentropic case (Sec. 12.1)

$$p = \alpha\rho^\kappa, \qquad \kappa = c_p/c_v > 1$$

Chaplygin has shown that the h_n are *hypergeometric functions*. Indeed, in this case,

$$a^2 = \frac{dp}{d\rho} = \alpha\kappa\rho^{\kappa-1} = \kappa\frac{p}{\rho}.$$

Hence in Bernoulli's law (12.2-3),

$$\int_{\rho_0}^{\rho} \frac{dp}{\tilde{\rho}} = \alpha\kappa \int_{\rho_0}^{\rho} \tilde{\rho}^{\kappa-2}\, d\tilde{\rho} = \left(\frac{\kappa}{\kappa-1}\right)\frac{p}{\rho} = \frac{a^2}{\kappa-1}$$

so that (12.2-3) with $U = 0$ becomes

$$\frac{q^2}{2} + \frac{a^2}{\kappa-1} = K = const. \tag{12.4-3}$$

Let q^* denote the "*critical speed*", that is, the speed equal to the speed of sound, $q = a = q^*$. In terms of q^* in (12.4-3) we have

$$K = \frac{\kappa+1}{2(\kappa-1)} q^{*2}.$$

Hence (12.4-3) implies

$$a^2 = \tfrac{1}{2}(\kappa+1)q^{*2} - \tfrac{1}{2}(\kappa-1)q^2.$$

We now introduce in (12.4-2) the new variable

$$r = \left(\frac{\kappa-1}{\kappa+1}\right)\frac{q^2}{q^{*2}}.$$

Then

$$\frac{q^2}{a^2} = \frac{2r}{(\kappa-1)(1-r)}.$$

Denoting h_n as a function of r again by h_n, for simplicity, we obtain by straightforward calculation the hypergeometric equation

$$r(1-r)h_n'' + \left(1 - \frac{\kappa-2}{\kappa-1}r\right)h_n' - \frac{n^2}{4r}\left(1 - \frac{\kappa+1}{\kappa-1}r\right)h = 0$$

or, by setting

$$h_n = r^{n/2}F_n$$

the hypergeometric equation

$$r(1-r)F_n'' + \left[n + 1 - \left(n + \frac{\kappa-2}{\kappa-1}\right)r\right]F_n' + \frac{n(n+1)}{2(\kappa-1)}F_n = 0.$$

A solution regular at the origin is given in terms of the hypergeometric function by

$$h_n(r) = r^{n/2} F_n(r) = r^{n/2} {}_2F_1(\alpha_n^+, \alpha_n^-; \gamma_n; r) \tag{12.4-4}$$

where $\gamma_n = n + 1$ and

$$\alpha_n^{\pm} = \frac{1}{2(\kappa - 1)}\left[n\kappa - n - 1 \pm \left(n^2(\kappa^2 - 1) + 1\right)^{1/2}\right].$$

In the next step of the method we determine the stream function ψ_0 of the given boundary value problem for the *incompressible* flow and represent it in the form

$$\psi_0(q, \theta) = \sum_{n=0}^{\infty} \left(\tilde{A}_n q^n \cos n\theta + \tilde{B}_n q^n \sin n\theta\right). \tag{12.4-5}$$

Finally we replace $q^n \cos n\theta$ by $\psi_n^{(1)}/\delta_n$ and $q^n \sin n\theta$ by $\psi_n^{(2)}/\delta_n$ with h_n as just determined and

$$\delta_n = h_n(r_1), \qquad r_1 = \left(\frac{\kappa - 1}{\kappa + 1}\right)\frac{1}{q^{*2}}.$$

The factor δ_n is caused by the fact that

$$\lim_{q^* \to \infty} \frac{h_n(r)}{h_n(r_1)} = \lim_{q^* \to \infty} \left(\frac{r}{r_1}\right)^{n/2} = q^n.$$

This process yields an approximation to the stream function of the given compressible problem, namely,

$$\psi = \sum_{n=0}^{\infty} \left(A_n \psi_n^{(1)} + B_n \psi_n^{(2)}\right) \tag{12.4-6}$$

where $A_n = \tilde{A}_n/\delta_n$ and $B_n = \tilde{B}_n/\delta_n$.

It is interesting that Chaplygin's method of obtaining particular solutions of (12.4-1) is in principle related to the theory of analytic functions (Sec. 12.3) for finding regular solutions of the system (12.3-1). In both methods, the underlying idea is the use of the special dependence of the coefficients of the differential equations on the independent variables. Both approaches seem to work well in the construction of solutions which do not have singularities in a bounded domain in the $\theta\lambda$-plane. However, the situation changes drastically in the presence of singularities, as we shall discuss next.

12.5 BERGMAN'S INTEGRAL OPERATOR METHOD

Chaplygin's method was designed for gas jets and works well for problems of that type. However, for other problems there may be difficulties, for the following reasons. Whereas the solution of the incompressible problem satisfies both the equation and the boundary conditions, the Chaplygin solution obtained from it by that procedure of replacement, when reverted to the physical plane, need not satisfy the boundary conditions (although it satisfies the equation of the compressible problem, by construction). The reason for the discrepancy is that the streamline in the xy-plane which in the *compressible* flow corresponds to the given contour for which the *incompressible* problem has been solved will in general only approximate this contour instead of exactly coinciding with it.

Moreover, there may be singularities which restrict regions of convergence, thus causing an extension problem; and the difficulty is that the Chaplygin solution obtained from the extension of the stream function of the *incompressible* flow will generally *not* be the extension of the original Chaplygin solution. It is largely this basic difficulty which motivated Bergman's ideas.

As another point, in flow problems, one often needs solutions defined on Riemann surfaces, with singularities in some of the sheets, but regular (at the same point) in the others. Integral operators of Bergman type can handle such tasks. For examples and further discussion, see E. Lanckau in Anger [1970], p. 73.

Also by systematically choosing families of analytic functions, one gets from an integral operator complete families of particular solutions, as one needs them in connection with the application of inverse methods, for instance, for improperly posed problems, as they arise in transition problems.

Actually, Chaplygin's method is particularly suitable for constructing stream functions which are regular in the whole finite plane. It may also yield multivalued solutions, but these are special inasmuch as they have a branch point at $q = 0$ but no singularities at other points. This is a deficiency, because very simple (even incompressible) flows show that quite frequently one needs more general stream functions. For example, consider the complex potential

$$F(z) = z + \frac{1}{z}, \qquad z = x + iy.$$

The stream function is

$$\psi_0(x, y) = \operatorname{Im} F(z) = y - \frac{y}{x^2 + y^2}.$$

We see that $\psi_0(x, y) = 0$ when $y = 0$ or $x^2 + y^2 = 1$, so that ψ_0 represents an incompressible flow around a circular cylinder (whose cross section in the

xy-plane is the disk $x^2 + y^2 \leq 1$). The velocity is (cf. Kreyszig [1988], p. 922)

$$v(z) = v_1(x, y) + iv_2(x, y) = \overline{F'(z)} = 1 - \frac{1}{\bar{z}^2}.$$

Expressing z in terms of v, we have

$$F(z(v)) = (1 - \bar{v})^{-1/2} + (1 - \bar{v})^{1/2}.$$

This shows that the stream function of this flow is double-valued in the hodograph plane and has a branch point at $v = 1$.

The whole situation suggests obtaining solutions of compressible flows directly, that is, without first considering the corresponding incompressible problem. Such a direct approach is furnished by Bergman's integral operator method which we shall now consider. This method permits the construction of solutions with a great variety of singularities.

BERGMAN'S METHOD AND INTEGRAL OPERATOR

We start from the system (12.3-1). Eliminating ϕ by differentiation, we have at once

$$l'\psi_\lambda + l(\psi_{\lambda\lambda} + \psi_{\theta\theta}) = 0$$

where the prime denotes differentiation with respect to λ. By setting $\psi = l^{-1/2}\tilde{\psi}$ we obtain the more convenient equation

$$\text{(12.5-1)} \quad L\tilde{\psi} = \tilde{\psi}_{\lambda\lambda} + \tilde{\psi}_{\theta\theta} - G(\lambda)\tilde{\psi} = 0, \qquad G(\lambda) = l^{-1/2}(l^{1/2})''.$$

We now assume a solution of (12.5-1) in the form

$$\text{(12.5-2)} \qquad \tilde{\psi}(\lambda, \theta) = \sum_{n=0}^{\infty} g_n(\lambda)\tilde{h}_n(\lambda, \theta)$$

where $\tilde{h}_n$ is harmonic; convergence will be considered later. Note that this is *not* a separation of variables. Then

$$L\tilde{\psi} = \sum_{n=0}^{\infty} \left[(g_n'' - Gg_n)\tilde{h}_n + 2g_n'\tilde{h}_{n\lambda}\right].$$

Hence if

$$\text{(12.5-3)} \quad \begin{array}{ll} \text{(a)} & \tilde{h}_{n\lambda} = -\tfrac{1}{2}\tilde{h}_{n-1} \\ \text{(b)} & g_0 = 1, \qquad g_n' = g_{n-1}'' - Gg_{n-1} \end{array} \qquad n = 1, 2, \cdots$$

then, since $g_0' = 0$,

$$L\tilde{\psi} = \lim_{n\to\infty} g_{n+1}'\tilde{h}_n .$$

Assuming for the time being that the limit on the right exists and is zero, we see that (12.5-2, 3) defines a formal solution of (12.5-1).

Since (12.5-3a) is simple, for a given $\tilde{h}_0$ we can easily represent all corresponding $\tilde{h}_n$ in closed form and, at the same time, establish a relation to complex analysis. For this purpose we introduce a complex variable

$$\zeta = \lambda + i\theta$$

and complex functions f_n such that $\tilde{h}_n = \operatorname{Re} f_n$. Clearly, if the f_n satisfy

$$f_n' = -\tfrac{1}{2} f_{n-1}, \qquad\qquad ' = d/d\zeta$$

then the $\tilde{h}_n$ satisfy (12.5-3a). Starting from any analytic f_0, integrating n times, and converting the result to a single integral, we can obtain

$$f_n(\zeta) = \frac{(-1)^n}{2^n(n-1)!}\int_0^\zeta f_0(t)(\zeta - t)^{n-1}\,dt .$$

Accordingly,

$$\tilde{h}_n(\lambda,\theta) = \frac{(-1)^n}{2^n(n-1)!}\operatorname{Re}\int_0^\zeta f_0(t)(\zeta - t)^{n-1}\,dt .$$

Hence we may now use (12.5-2) to define an *integral operator B* such that

(12.5-4a)

$$\tilde{\psi}(\lambda,\theta) = \operatorname{Re} Bf_0(\lambda,\theta) = \operatorname{Re}\left[f_0(\zeta) + \int_0^\zeta k(\lambda,\theta,t) f_0(t)\,dt\right]$$

with kernel k given by

(12.5-4b) $$k(\lambda,\theta,t) = \sum_{n=1}^{\infty}\frac{(-1)^n}{2^n(n-1)!} g_n(\lambda)(\zeta - t)^{n-1}$$

and g_n satisfying (12.5-3b). Here f_0 is still arbitrary, but can be determined from the hodograph potential w_0 of the *incompressible* problem, namely, $f_0(t) = w_0'(e^t)$.

CONVERGENCE OF BERGMAN'S REPRESENTATION (12.5-2)

We shall see that the method yields convergence in the angular region

$$R: \ |\theta| < |\lambda|\sqrt{3}\,, \qquad \lambda < 0. \tag{12.5-5}$$

In this region, with every analytic function f_0 we can associate a stream function $\psi = l^{-1/2}\tilde{\psi} = l^{-1/2}\,\mathrm{Re}\,Bf_0$. Since singularities or branch points of f_0 will generally cause singularities and branch points of solutions, it follows that by choosing appropriate f_0 we can obtain stream functions of compressible flows having singularities of prescribed type and location.

Note that the kernel k of B depends only on the coefficient G, that is, on the equation of state of the flow. Hence for the usual equations of state, the coefficient functions g_n can be calculated once and for all.

Convergence of (12.5-2) and derived series can be investigated by the method of *dominants*, as in the case of operators of the first kind in Sec. 2.4. We construct functions g_n^* such that for $\lambda < 0$ and sufficiently large $n \in \mathbb{N}_0$,

$$|g_n(\lambda)| \leq g_n^*(\lambda)\,, \qquad |g_n^{(\nu)}(\lambda)| \leq g_n^{*(\nu)}(\lambda) \tag{12.5-6}$$

where $\nu \in \mathbb{N}$. This is done as follows. We assume the coefficient G to be analytic for $\lambda < 0$ and

$$G(\lambda) \ll G^*(\lambda)\,, \qquad \text{where} \qquad G^*(\lambda) = \frac{C}{(\sigma - \lambda)^2}\,, \quad C > 0; \quad \sigma < 0$$

for $\lambda < \sigma < 0$, which corresponds to subsonic flow since $q_0 = a$ in (12.3-2), so that $\lambda = 0$ when $q = a$. We now construct g_n^* from G^* in the same way as the g_n are constructed from G; that is

$$\begin{aligned} &g_0^* = 1\,, \qquad g_n^{*\prime} = g_{n-1}^{*\prime\prime} - G^* g_{n-1}^*\,, \\ &g_n^*(-\infty) = 0\,, \end{aligned} \qquad n = 1, 2, \cdots. \tag{12.5-7}$$

Then (12.5-6) holds, as follows by induction. A solution of (12.5-7) is

$$\begin{aligned} &g_n^*(\lambda) = \frac{n!}{(\sigma - \lambda)^n}\gamma_n \\ &\gamma_0 = 1\,, \qquad \gamma_{n+1} = \frac{(n+\alpha)(n+\beta)}{(n+1)^2}\gamma_n \end{aligned} \qquad n = 0, 1, \cdots \tag{12.5-8}$$

where

$$\alpha = \tfrac{1}{2} - \left(\tfrac{1}{4} + C\right)^{1/2}, \qquad \beta = \tfrac{1}{2} + \left(\tfrac{1}{4} + C\right)^{1/2}.$$

The γ_n are the coefficients of the hypergeometric series

$$ {}_2F_1(\alpha, \beta; 1; X) = \sum_{n=0}^{\infty} \gamma_n X^n . $$

Since $\alpha, \beta \neq 0, -1, -2, \cdots$, this series converges absolutely and uniformly for $|X| \leq \eta < 1$. We now consider the series

$$ k^*(\lambda, \theta, t) = \sum_{n=1}^{\infty} \frac{1}{2^n (n-1)!} g_n^*(\lambda) |\zeta - t|^{n-1} . \tag{12.5-9} $$

By (12.5-6, 7),

$$ \begin{aligned} k^*(\lambda, \theta, t) &= \frac{1}{2(\sigma - \lambda)} \sum_{n=1}^{\infty} n \gamma_n X^{n-1}, \qquad X = \frac{|\zeta - t|}{2(\sigma - \lambda)} \\ &= \frac{1}{2(\sigma - \lambda)} \frac{d}{dX} {}_2F_1(\alpha, \beta; 1; X) . \end{aligned} $$

Now, from some sufficiently large n on, each coefficient of this series is larger than the absolute value of the corresponding coefficient in (12.5-4b). Hence (12.5-4b) converges uniformly at least in the region

$$ X = \frac{|\zeta - t|}{2(\sigma - \lambda)} \leq \eta < 1, \qquad \lambda < \sigma < 0 $$

in which (12.5-9) converges and the same holds for the derived series. Since we integrate from 0 to ζ, this implies $|\zeta| < 2|\lambda|$, that is,

$$ \lambda^2 + \theta^2 < 4\lambda^2 , \qquad \lambda < 0, $$

so that the assertion involving (12.5-5) follows. This completes the proof. ■

Additional References

Bergman [1945c, 1945d, 1946a, 1946b, 1947b, 1948, 1949b, 1950b, 1952a, 1955, 1966a, 1968, 1971, 1973, 1976]

Bergman–Epstein [1947]

Bergman–Greenstone [1947]

Bers [1952, 1954a, 1954b, 1956, 1962]

Bers–Bochner–John [1954]

Cherry [1951]

Garabedian [1968]

Guderley [1962]

Kracht–Kreyszig [1986a]

Krzywoblocki [1960, 1968]

Mises [1958]

Mises–Schiffer [1948]

Nazarov [1964]

Oswatitsch [1956]

Vekua [1963]

Chapter Thirteen

Integral Operators Applied to Transonic Flow. Tricomi Equation

We continue our discussion of applications of integral operators in fluid flow, turning to ***transonic flow***. By definition, this flow simultaneously involves subsonic and supersonic speeds. In such a problem the equation for the stream function is of mixed type, namely, *elliptic* where the flow is subsonic, and *hyperbolic* where it is supersonic. The common boundary of the two types of regions is called the sonic curve. The fact that here we have to deal with an equation of mixed type causes great difficulties in studying properties of solutions.

For instance, transonic flow may arise in connection with an airfoil traveling at high subsonic speed. Then the flow is fastest on the upper side of the airfoil and may develop a supersonic region there, consisting of one or several "bubbles." See, for instance, F. Bauer et al. [1972–1975].

Similarly, if an airfoil is traveling at supersonic speed, there will be a subsonic pocket behind a detached bow wave and oblique shock waves will appear at the trailing edge, sometimes forming a fishtail pattern.

Another example of a transonic problem is a rapid flow through a nozzle of variable cross section. In this case, a ring-shaped supersonic zone may develop where the cross section is smallest.

In such problems, the ***Tricomi equation*** arises, as we shall explain in Sec. 13.1. Section 13.2 is mainly devoted to so-called ***operators of the second kind***, which seem to be very natural for equations of mixed type and other equations with singularities. In Sec. 13.3 we treat the Cauchy problem for the Tricomi equation. The special case of entire functions as initial data gives rise to an interesting discussion and is presented in Sec. 13.4. Polynomial expansions for solutions of Tricomi equations can be derived on the basis of the theory of entire functions, as is shown in Sec. 13.5. After a typical illustrative example in

Sec. 13.6, we turn to the application of ***Eichler operators*** in Sec. 13.7 and, in Sec. 13.8, to a discussion of particular solutions obtained by means of those operators. Whereas these solutions were previously derived from various sources and by different methods, they now arise from a unique source (an integral operator), along with other solutions. These solutions have been widely discussed and thoroughly investigated in the literature on fluid flow, because of their practical importance.

13.1 EQUATIONS FOR THE STREAM FUNCTION

As in the previous chapter, we consider a nonviscous and compressible flow.

For the present purpose we may start from the system [cf. (12.3-7)]

$$\phi_q = \frac{M^2 - 1}{\rho q}\psi_\theta, \qquad \phi_\theta = \frac{q}{\rho}\psi_q$$

where q and θ are the polar coordinates in the hodograph plane used before. We transform the second of these equations into the form of the second Cauchy–Riemann equation by replacing q in both equations by a new variable σ defined by $d\sigma/dq = -\rho/q$. The result is

$$\phi_\sigma = K(\sigma)\psi_\theta, \qquad \phi_\theta = -\psi_\sigma$$

where

$$K(\sigma) = \frac{1 - M^2}{\rho^2}.$$

From this we obtain another ***Chaplygin equation***, namely,

$$\psi_{\sigma\sigma} + K(\sigma)\psi_{\theta\theta} = 0. \tag{13.1-1}$$

In this equation, the coefficient K is a rather complicated function of σ, because M depends on q and the local velocity of sound. This suggests looking for simplifications. Here, we consider approximations to K near $M = 1$, which corresponds to $\sigma = 0$. The simplest approximation is that by the tangent to the curve of K at $\sigma = 0$. Hence if we impose a suitable linear transformation of the independent variable σ and denote the new variable again by σ, for simplicity, for small $|\sigma|$ we can set

$$K(\sigma) \approx \sigma.$$

Then we obtain the famous ***Tricomi equation***

$$\psi_{\sigma\sigma} + \sigma\psi_{\theta\theta} = 0. \tag{13.1-2}$$

This equation is of mixed type. Indeed, it is elliptic for $\sigma > 0$; this corresponds to subsonic flow. It is hyperbolic for $\sigma < 0$; this corresponds to supersonic flow. The sonic curve is the straight line $\sigma = 0$.

The equation is named after F. Tricomi, who was the first to formulate a properly posed problem for this equation and proved the existence and uniqueness of the solution of the problem; cf. Tricomi [1923, 1968].

In this way, Tricomi actually initiated the study of equations of mixed type. Little attention was paid to this pioneering work over the next two decades. It was only in the 1940s that the basic connection of Tricomi's results to problems in transonic flow was recognized. It was F. I. Frankl [1956], in papers he began to publish in 1945, who first drew attention to the fact that the Tricomi problem and several other mixed problems are closely related to gas flow with nearly sonic speeds. This marked the beginning of an increasing stream of publications extending Tricomi's ideas to more general equations and other problems. However, corresponding results obtained so far in this newly developing field appear to be rather fragmentary and special and, at this time, the task of developing a unified theory seems formidable. Tricomi's equation still remains a model case and the best known and most useful equation from the standpoint of most applications.

We shall exclusively consider the approximation of the Chaplygin equation (13.1-1) by the Tricomi equation, but want to mention that there are other approximations, which are more accurate but, unfortunately, also more complicated. Perhaps the best known of them is that by Tomotika and Tamada [1950] given by

$$K(\sigma) \approx \alpha(1 - e^{-2\delta\sigma}), \qquad \alpha, \delta > 0.$$

Here for $\kappa = 1.4$ (cf. Sec. 12.1) the constants α and δ have the values

$$\alpha = \left(\frac{2}{\kappa + 1}\right)^{2/(\kappa - 1)} = 0.401877 \cdots$$

$$\delta = \left(\frac{\kappa + 1}{2}\right)^{(\kappa+1)/(\kappa-1)} = 2.985984.$$

The resulting differential equation is

$$\psi_{\sigma\sigma} + \alpha(1 - e^{-2\delta\sigma})\psi_{\theta\theta} = 0.$$

It can be simplified by setting

$$\eta = e^{-\delta\sigma}.$$

Then

$$\eta^2\psi_{\eta\eta} + \eta\psi_\eta + \frac{\alpha}{\delta^2}(1 - \eta^2)\psi_{\theta\theta} = 0$$

or, if we set

$$\alpha^{-1/2}\delta\theta = \xi$$

finally

$$(1-\eta^2)\psi_{\xi\xi} + \eta^2\psi_{\eta\eta} + \eta\psi_\eta = 0. \tag{13.1-3}$$

This is called the ***Tomotika–Tamada equation***. We see that it is elliptic in the strip $|\eta| < 1$ and hyperbolic for $|\eta| > 1$. Separation of variables yields two ordinary differential equations; one of them is reducible to Bessel's equation and the other solvable in terms of sine and cosine. The result is a set of particular solutions of the form

$$\psi(\xi,\eta) = [AJ_\nu(\nu\eta) + BY_\nu(\nu\eta)]\,{\cos \atop \sin}(\nu\xi).$$

For other particular solutions which are of interest in engineering problems, see Tricomi [1968], p. 161.

13.2 INTEGRAL OPERATOR FOR THE TRICOMI EQUATION

The Tricomi equation is (cf. Sec. 13.1)

$$\psi_{\sigma\sigma} + \sigma\psi_{\theta\theta} = 0. \tag{13.2-1}$$

We transform this equation into the form (2.1-1). Using θ and

$$\gamma = \tfrac{2}{3}\sigma^{3/2}$$

as independent variables and

$$\tilde{\psi}(\gamma,\theta) = \sigma^{1/4}\psi(\sigma,\theta)$$

as a new dependent variable, we first have

$$\tilde{\psi}_{\gamma\gamma} + \tilde{\psi}_{\theta\theta} + \frac{5}{36\gamma^2}\tilde{\psi} = 0. \tag{13.2-2}$$

Observe that the coefficient of $\tilde{\psi}$ is singular precisely on the sonic line.

We mention in passing that if we transform the independent variables as before, but retain the old dependent variable, say, $\hat{\psi}(\gamma,\theta) \equiv \psi(\sigma,\theta)$, we obtain from (13.2-1)

$$\hat{\psi}_{\gamma\gamma} + \hat{\psi}_{\theta\theta} + \frac{1}{3\gamma}\hat{\psi}_\gamma = 0,$$

an equation that has been considered by Weinstein [1953]. Furthermore, noting that in the hyperbolic region the characteristics of (13.2-1) are the semicubical parabolas

$$\lambda = \theta + \tfrac{2}{3}(-\sigma)^{3/2} = const, \qquad \mu = \theta - \tfrac{2}{3}(-\sigma)^{3/2} = const$$

and introducing λ, μ as new independent variables, we obtain from (13.2-1) the ***Darboux equation***

$$\psi^*_{\lambda\mu} + \frac{1}{6(\lambda - \mu)}(\psi^*_\lambda - \psi^*_\mu) = 0.$$

We shall use (13.2-2), which seems preferable for our purpose.

COMPLEX FORM OF THE TRICOMI EQUATION

Setting

$$z_1 = \gamma + i\theta, \qquad z_2 = \gamma - i\theta$$

and writing $u(z) \equiv \tilde{\psi}(\gamma, \theta)$, we obtain from (13.2-2)

$$D_1 D_2 u + \frac{5/36}{(z_1 + z_2)^2} u = 0. \tag{13.2-3}$$

This equation is of the form (2.1-1). We now assume γ and θ to be complex. Then z_1, z_2 in $z = (z_1, z_2)$ are independent as before.

BERGMAN OPERATORS OF THE SECOND KIND FOR THE TRICOMI EQUATION

We consider the problem of obtaining an integral operator for (13.2-3) which is suitable in connection with transonic flow. The situation is as follows. If an equation (2.1-1) has holomorphic coefficients in a neighborhood of the point of reference, then an operator of the first kind can be used in most applications. However, this is no longer the case if those coefficients have singularities at the point of reference. Indeed, if for equation (13.2-3) one used an operator of the first kind, one would not obtain a representation of solutions valid everywhere in a neighborhood of the sonic line. Following Bergman [1948], in the present case we can employ so-called *operators of the second kind* (definition below). Some of these operators yield representations of solutions valid in a region which includes a portion of the sonic line in its interior, so that these operators are suitable for investigating transonic flow.

An operator of the second kind and its kernel—called a *kernel of the second kind*—can be obtained explicitly by introducing a suitable auxiliary variable Z

such that the equation (2.3-2a) corresponding to kernels $\tilde{k}$ for (13.2-3) becomes an *ordinary* differential equation. To explain the principle, let us take $j = 1$ and $\eta_1 = 0$, and choose

$$Z = \frac{t^2(z_1 - s_1)}{z_1 + z_2} = \frac{t^2(\gamma - s_1 + i\theta)}{2\gamma};$$

here (s_1, s_2) is the point of reference of the operator, as before. Setting

$$\tilde{k}_1(z, t) = \left[t^2(z_1 - s_1)\right]^{\rho} k(Z), \qquad \rho \in \mathbb{R} \tag{13.2-4}$$

we see by straightforward calculation that (2.3-2a) with $\tilde{k}_1 = k_1$ is satisfied if k is a solution of the hypergeometric equation

$$Z(1 - Z)k'' + \left[\rho + \tfrac{1}{2} - 2Z\right]k' - \tfrac{5}{36}k = 0.$$

Let $\rho \neq \frac{1}{2}(2m + 1)$, where m is an integer. Then a general solution of this equation in the disk $|Z| < 1$ is given by

$$\begin{aligned} k(Z) &= \alpha_1\, {}_2F_1\left(\tfrac{1}{6}, \tfrac{5}{6}; \rho + \tfrac{1}{2}; Z\right) \\ &\quad + \beta_1 Z^{1/2-\rho}\, {}_2F_1\left(\tfrac{2}{3} - \rho, \tfrac{4}{3} - \rho; \tfrac{3}{2} - \rho; Z\right). \end{aligned} \tag{13.2-5a}$$

Similarly, a general solution in the domain $|Z| > 1$ is

$$\begin{aligned} k(Z) &= \alpha_2 Z^{-1/6}\, {}_2F_1\left(\tfrac{1}{6}, \tfrac{2}{3} - \rho; \tfrac{1}{3}; Z^{-1}\right) \\ &\quad + \beta_2 Z^{-5/6}\, {}_2F_1\left(\tfrac{5}{6}, \tfrac{4}{3} - \rho; \tfrac{5}{3}; Z^{-1}\right). \end{aligned} \tag{13.2-5b}$$

Clearly, in order that the two formulas for k represent analytic continuations of each other, the four constants $\alpha_1, \alpha_2, \beta_1, \beta_2$ are not completely arbitrary, but it is necessary and sufficient that they satisfy two simple relations. See also Bergman–Bojanić [1962].

Note that the point (s_1, s_2) is arbitrary. Choosing suitable $s_1 \neq 0$ has the advantage that ${}_2F_1$ in (13.2-5b) becomes holomorphic at $(\gamma, \theta) = (0, 0)$ when $|t| = 1$.

The choice of ρ in (13.2-4) depends on the purpose. A value $\rho = 2 + \rho_0$ with $\rho_0 \in \mathbb{N}$ has the advantage that the Maclaurin expansions in (13.2-5a) and the Taylor expansions about $Z = \infty$ in (13.2-5b), as well as their derivatives up to the second order, are absolutely convergent, also on $|Z| = 1$.

Finally, if we take $(s_1, s_2) = (0, 0)$, θ real, and the path of integration $|t| = 1$ from -1 to 1, then $|Z| > 1$ represents a double cone with vertex at the origin which for $\theta \neq 0$ contains the sonic line $\gamma = 0$ in its interior. Its intersection with the subsonic region for (13.2-1) consists of two wedge-shaped domains in the half-plane $\gamma_1 > 0$ of the $\gamma_1\theta$-plane, where $\gamma = \gamma_1 + i\gamma_2$. For the supersonic region of (13.2-1) we obtain a similar intersection in the half-plane $\gamma_2 < 0$ of the $\gamma_2\theta$-plane.

Our result thus obtained can be formulated as follows.

13.2-1 Theorem

(a) *In* (13.2-4), *let* $\tilde{k}_1 \in C^\omega(\Omega_1 \times \Omega_2 \times \overline{B}_0)$ *and* $f \in C^\omega(\tilde{\Omega}_1)$. *Then* u *defined by*

$$u(z) = \int_{C_1} \tilde{k}_1(Z) f\left(s_1 + \tfrac{1}{2}(z_1 - s_1)\tau\right)\tau^{-1/2}\,dt$$

is a $C^\omega(\Omega_1 \times \Omega_2)$*-solution of* (13.2-3); *here* $\tilde{k}_1$ *is given by* (13.2-4, 5), $\rho \neq \frac{1}{2}(2m+1)$ *with* $\pm m \in \mathbb{N}_0$, *and* C_1 *is the upper unit semicircle from* $t = -1$ *to* $t = 1$.

(b) *A real analytic solution of the* ***Tricomi equation*** (13.2-1) *is given by the real part (or the imaginary part) of*

$$\psi(\sigma, \theta) = \sigma^{-1/4} u\left(\tfrac{2}{3}\sigma^{3/2} + i\theta, \tfrac{2}{3}\sigma^{3/2} - i\theta\right)$$

with u *defined as in part* (a). *If* f *is entire and* $s_1 = 0$, *then this representation holds in the following regions.*

(A) *In the subsonic region for* $(\sigma, \theta) \in \mathbb{R}^2$ *with* $\sigma > 0$ *and*
(i) $|\theta| < 2\sigma^{3/2}/\sqrt{3}$ *in the case of* (13.2-5a)
(ii) $|\theta| > 2\sigma^{3/2}/\sqrt{3}$ *in the case of* (13.2-5b).

(B) *In the supersonic region for* $(\sigma, \theta) \in \mathbb{R}^2$ *with* $\sigma < 0$ *and*
(i) $-2(-\sigma)^{3/2} < \theta < \frac{2}{3}(-\sigma)^{3/2}$ *in the case of* (13.2-5a)
(ii) $\theta < -2(-\sigma)^{3/2}$ *or* $\theta > \frac{2}{3}(-\sigma)^{3/2}$ *in the case of* (13.2-5b).

13.2-2 Definition

The Bergman operator T_1 defined by the kernel $\tilde{k}_1$ of the form (13.2-4, 5) is called an ***operator of the second kind*** for the Tricomi equation. (In the literature, this concept has been used almost exclusively in the case $\rho = 0$ only.)

Hence the solution obtained in Theorem 13.2-1a is generated by an operator of the second kind.

13.3 CAUCHY PROBLEMS FOR THE TRICOMI EQUATION

A frequent task is the determination of solutions of the Tricomi equation (13.2-1) in a subdomain of the subsonic region satisfying given real-analytic initial conditions on the sonic line, say,

$$\psi(0, \theta) = \sum_{n=0}^{\infty} a_n^{(1)}(\theta - \theta_0)^n \tag{13.3-1a}$$

and

$$\psi_\sigma(0, \theta) = \sum_{n=0}^{\infty} a_n^{(2)}(\theta - \theta_0)^n \tag{13.3-1b}$$

where $\psi_\sigma = \partial\psi/\partial\sigma$. To solve this problem, we split it into the two problems of determining two solutions ψ_1 and ψ_2 of (13.2-1) satisfying

$$\psi_1(0, \theta) = \sum_{n=0}^{\infty} a_n^{(1)}(\theta - \theta_0)^n, \qquad \psi_{1,\sigma}(0, \theta) = 0 \tag{13.3-2a}$$

and

$$\psi_{2,\sigma}(0, \theta) = \sum_{n=0}^{\infty} a_n^{(2)}(\theta - \theta_n)^n, \qquad \psi_2(0, \theta) = 0 \tag{13.3-2b}$$

respectively. These two problems can be solved by the operator method explained in the previous section, as follows.

APPLICATION OF THEOREM 13.2-1 TO THE CAUCHY PROBLEM (13.3-1)

We use the kernel $\tilde{k}_1$ in (13.2-4) with k given by (13.2-5b), where

(13.3-3)

(a) $\rho = \frac{1}{6}, \quad \alpha_2 = \left(\frac{3}{4}\right)^{1/6}, \quad \beta_2 = 0 \quad$ in the case of ψ_1

(b) $\rho = \frac{5}{6}, \quad \alpha_2 = 0, \quad \beta_2 = \left(\frac{3}{4}\right)^{5/6} \quad$ in the case of ψ_2.

As associated functions f_1 and f_2 we take those defined by

(13.3-4)

$$f_j(\xi) = \sum_{n=0}^{\infty} \gamma_{jn}(\xi - i\theta_0)^n, \qquad j = 1, 2,$$

$$\gamma_{jn} = (-8i)^n (n!)^2 a_n^{(j)}/[\pi(2n)!].$$

We also take $s_1 = i\theta_0$, so that

$$Z = t^2[\gamma + i(\theta - \theta_0)]/(2\gamma).$$

As in Theorem 13.2-1b, a real-analytic solution of the Tricomi equation (13.2-1) is

$$\psi = \psi_1 + \psi_2 \tag{13.3-5a}$$

with ψ_1 and ψ_2 given by the integral representations

$$\psi_j(\sigma, \theta) = \operatorname{Re} \int_{C_1} \hat{k}_{1j}(S) f_j(i\theta_0 + \xi\tau)\tau^{-1/2}\, dt\,, \qquad S = \frac{2\sigma^{3/2}}{3\xi t^2} \tag{13.3-5b}$$

where

$$\begin{aligned} &\text{(a)} \quad \hat{k}_{11}(S) = {}_2F_1\left(\tfrac{1}{6}, \tfrac{1}{2}; \tfrac{1}{3}; S\right) \\ &\text{(b)} \quad \hat{k}_{12}(S) = \sigma\, {}_2F_1\left(\tfrac{5}{6}, \tfrac{1}{2}; \tfrac{5}{3}; S\right) \end{aligned} \tag{13.3-6}$$

and

$$\xi = \tfrac{1}{3}\sigma^{3/2} + \tfrac{1}{2}i(\theta - \theta_0), \quad \sigma > 0; \qquad \tau = 1 - t^2.$$

Here C_1 is the upper unit semicircle from $t = -1$ to $t = 1$, and the f_j are given by (13.3-4).

Clearly, from these representations we also obtain the solution ψ_0 of (13.2-1) defined by

$$\psi_0(\sigma, \theta) = \operatorname{Re} \int_{C_1} \hat{k}_1(S) f(i\theta_0 + \xi\tau)\tau^{-1/2}\, dt \tag{13.3-7a}$$

where

$$\hat{k}_1 = \hat{k}_{11} + i\hat{k}_{12}\,, \qquad f = f_1 - if_2\,. \tag{13.3-7b}$$

We want to conclude that ψ_0 is also defined for $\sigma = 0$ and, second, that it satisfies the initial conditions (13.3-1). For this purpose we first note that the beta function can be represented in the form

$$B(p, q) = \int_{C_1} t^{2p-1}(1 - t^2)^{q-1}\, dt\,, \quad \operatorname{Re} p > 0, \quad \operatorname{Re} q > 0$$

and can be continued analytically to values p, q such that $p + q$ is an integer whereas p is not. For $p = -m + \frac{1}{2}$ and $q = n + \frac{1}{2}$ with $m, n \in \mathbb{N}_0$ we obtain (cf. Kratzer–Franz [1960], p. 11)

$$B\left(-m+\tfrac{1}{2}, n+\tfrac{1}{2}\right) = \frac{(-4)^m m! \pi (2n)!}{(2m)!(n-m)! 4^n n!} \qquad \text{when} \quad n \geq m$$

and 0 when $n < m$. Hence by inserting (13.3-4) and the series expansions of the hypergeometric functions (13.3-6) into (13.3-7) we arrive at

$$\psi_0(\sigma, \theta) = \operatorname{Re} \sum_{n=0}^{\infty} \sum_{m=0}^{n} \left(\alpha_{mn}^{(1)} + \alpha_{mn}^{(2)} i\sigma\right) \sigma^{3m/2} \xi^{n-m}$$

(13.3-8)
$$\alpha_{mn}^{(j)} = \frac{\Gamma_{m1j}\Gamma_{02j}(-2)^m (2n)! \pi}{\Gamma_{m2j}\Gamma_{01j} 3^m 4^n m! (n-m)! n!} (\gamma_{1n} - i\gamma_{2n})$$

$$\Gamma_{mkj} = \Gamma\left(m + k\left(\tfrac{1}{2} + \tfrac{1}{3}(-1)^j\right)\right).$$

We claim that ψ_1 and ψ_2 are solutions of the problems characterized by (13.3-2). To verify this, we need only to observe that ψ_j (j fixed) can be obtained from (13.3-7a) by taking $\hat{k}_{1j^*} = 0$ and $f_{j^*} = 0$ in (13.3-7b); here $j^* = 3 - j$, as usual.

ANALYTICITY ON THE SONIC LINE

Note that ψ_0 is generally *not* analytic in σ on the sonic line because (13.3-8) contains odd powers of $\sigma^{1/2}$. However, we shall now prove that the representation (13.3-5) is also analytic on the sonic line, so that it is analytic in both variables and the Cauchy–Kowalewski theorem guarantees uniqueness.

A proof of this analyticity can be obtained as follows.

The representation of ψ_j analogous to (13.3-8) is

(13.3-9)
$$\psi_j(\sigma, \theta) = \operatorname{Re} \sum_{n=0}^{\infty} \sum_{m=0}^{n} \tilde{\alpha}_{mn}^{(j)} \sigma^{j-1+3m/2} \xi^{n-m}$$

$$\tilde{\alpha}_{mn}^{(j)} = i^{j-1} \alpha_{mn}^{(j)}\big|_{\gamma_{j^*n}=0}.$$

If we interchange the order of summation and use the binomial expansion of ξ^{n-m}, this becomes

(13.3-10)
$$\psi_j(\sigma, \theta) = \operatorname{Re} \sum_{n=0}^{\infty} a_n^{(j)} \sigma^{j-1} (\theta - \theta_0)^n \sum_{r=0}^{n} \beta_{nr}^{(j)} (-i)^r (-X)^{r/2}$$

$$\beta_{nr}^{(j)} = \sum_{m=0}^{r} \frac{\Gamma_{m1j}\Gamma_{02j}}{\Gamma_{m2j}\Gamma_{01j}} (-2)^m \binom{r}{m}\binom{n}{r}, \qquad X = \frac{4(-\sigma)^3}{9(\theta - \theta_0)^2}$$

with Γ_{mkj} as in (13.3-8). Since $-X$ with $\sigma \geq 0$, $a_n^{(j)}$, θ, and θ_0 are all real, from (13.3-9) we have

$$\psi_j(\sigma, \theta) = \sigma^{j-1} \sum_{n=0}^{\infty} a_n^{(j)} (\theta - \theta_0)^n \sum_{k=0}^{[n/2]} \beta_{n,2k}^{(j)} X^k \tag{13.3-11}$$

with the β's defined as in (13.3-10). We now use the integral representation of the beta function

$$B\left(\frac{x}{2}, y\right) = 2\int_0^1 t^{x-1}(1-t^2)^{y-1}\, dt, \qquad x > 0, \quad y > 0,$$

the multiplication formula

$$\sqrt{\pi}\,\Gamma(2z) = 2^{2z-1}\Gamma(z)\Gamma\left(z + \tfrac{1}{2}\right)$$

(cf. Magnus et al. [1966], pp. 3, 8), and the familiar Maclaurin series of the hypergeometric function. Then we obtain from (13.3-11) the representation

$$\psi_j(\sigma, \theta) = \sigma^{j-1} \sum_{n=0}^{\infty} a_n^{(j)} (\theta - \theta_0)^n {}_2F_1\left(-\frac{n}{2}, \frac{1}{2} - \frac{n}{2}; \frac{2^j}{3}; X\right) \tag{13.3-12}$$

with X as defined before. Obviously, the series for ${}_2F_1$ terminates, the highest power being $X^{[n/2]}$. From (13.3-12) it follows that ψ_j is analytic on $\sigma = 0$. ∎

RESULTS IN THIS SECTION

We may summarize our present result as follows.

13.3-1 Theorem

(a) *For the Tricomi equation* (13.2-1) *each of the three Cauchy problems* (i) (13.3-1), (ii) (13.3-2a), *and* (iii) (13.3-2b) *involving real-analytic initial data has a unique analytic solution given by* (i) (13.3-5) *and* (ii), (iii) (13.3-5b), *respectively, where the* f_j *are defined by* (13.3-4) *and the kernels by* (13.3-6).

(b) *The integral representations in* (a) *can be converted to series representations in terms of hypergeometric functions as shown in* (13.3-12).

(c) *The Cauchy problem involving the data* (13.3-1) *also admits the solution* ψ_0 *given by* (13.3-7) *with kernels* (13.3-6) *and associated functions* (13.3-4). *This solution is not analytic in* σ *on the sonic line* ($\sigma = 0$).

It is interesting to note that (13.3-12) readily leads to series representations in terms of *Jacobi polynomials*. All we have to do is to apply the formula

$$ {}_2F_1(-m, -a-m; b+1; X) = \frac{m!}{(b+1)_m}(X-1)^m P_m^{(a,b)}\left(\frac{X+1}{X-1}\right) $$

where $(b+1)_m = (b+1)(b+2)\cdots(b+m)$, as usual [cf. Magnus et al. [1966], p. 212]. This yields at once

(13.3-13)

$$ \begin{aligned} \psi_j(\sigma,\theta) &= \sigma^{j-1}\sum_{n=0}^{\infty} A_n(\sigma,\theta) P_{[n/2]}^{(a,b)}\left(\frac{4(-\sigma)^3 + 9(\theta-\theta_0)^2}{4(-\sigma)^3 - 9(\theta-\theta_0)^2}\right) \\ A_n(\sigma,\theta) &= a_n^{(j)}\frac{[n/2]!}{(b+1)_{[n/2]}}(\theta-\theta_0)^{n-2[n/2]}\left(\frac{4}{9}(-\sigma)^3 - (\theta-\theta_0)^2\right)^{[n/2]} \end{aligned} $$

with $a = (-1)^{n+1}/2$ and $b = (-1)^j/3$.

13.4 ENTIRE CAUCHY DATA FOR THE TRICOMI EQUATION. BOREL TRANSFORM

We continue our study of the Cauchy problem for the Tricomi equation (13.2-1), but we now assume that σ and θ are *complex* and the problem is extended into space $\mathbb{C}^2$. Specifically, we assume the initial data,

(13.4-1)
$$ \psi_1(0,\theta) = \varphi_1(\tilde{\theta}), \qquad \psi_{2,\sigma}(0,\theta) = \varphi_2(\tilde{\theta}), \qquad \tilde{\theta} = \theta - \theta_0, $$

to involve two *entire* functions φ_j of exponential type $\tau_j \in [0, \infty)$.

In this case, following Sasai [1979], we can obtain further integral representations of solutions by applying the *Borel transform*, as follows.

For the special initial data $\tilde{\varphi}_j$,

$$ \tilde{\varphi}_j(\tilde{\theta}) = \tilde{\theta}^n = (\theta-\theta_0)^n, \qquad n \in \mathbb{N}_0 $$

we have from (13.3-12)

$$ \psi_{jn}(\sigma,\theta) = \sigma^{j-1}(\theta-\theta_0)^n {}_2F_1\left(-\frac{n}{2}, \frac{1}{2}-\frac{n}{2}; \frac{2^j}{3}; X\right). $$

Hence we may calculate a generating function for ψ_j:

(13.4-2)
$$\begin{aligned}\sigma^{j-1} \sum_{n=0}^{\infty} (\theta - \theta_0)^n \sum_{k=0}^{[n/2]} \frac{(-n/2)_k \left(\frac{1}{2} - n/2\right)_k}{(2^j/3)_k k!} X^k \frac{w^n}{n!} \\ = \sigma^{j-1}{}_0F_1\left(\frac{2^j}{3}; \frac{(-\sigma)^3 w^2}{9}\right) e^{(\theta-\theta_0)w}, \qquad j = 1, 2\end{aligned}$$

where ${}_0F_1$ denotes the generalized hypergeometric function of type $(0,1)$, as usual, with X defined by (13.3-10).

We now write the Maclaurin series of φ_j in the form

$$\varphi_j(\tilde{\theta}) = \sum_{n=0}^{\infty} a_n^{(j)} \tilde{\theta}^n,$$

the radius of convergence being ∞. Then the ***Borel transform*** of φ_j is

(13.4-3)
$$\Phi_j(w) = \sum_{n=0}^{\infty} \frac{n! a_n^{(j)}}{w^{n+1}}.$$

Since φ_j is entire of exponential type τ_j, it follows that the series in (13.4-3) has radius of convergence at least equal to τ_j (cf. Boas [1968], p. 73). Hence, by taking the circle $\tilde{C}_j$ of radius $\tau_j + \varepsilon$ ($\varepsilon > 0$) and center 0, applying Cauchy's integral formula, and noting (13.4-2) as well as (13.4-3), we obtain the integral representation

(13.4-4)
$$\psi_j(\sigma, \theta) = \frac{\sigma^{j-1}}{2\pi i} \int_{\tilde{C}_j} {}_0F_1\left(\frac{2^j}{3}; \frac{(-\sigma)^3 w^2}{9}\right) e^{(\theta-\theta_0)w} \Phi_j(w)\, dw.$$

Since the kernel is an entire function of σ, θ, and w, it follows that ψ_j is an entire function of σ and θ.

Our result can be formulated as follows.

13.4-1 Theorem

Let φ_j, $j = 1, 2$, be entire of exponential type τ_j. Then the entire solutions of the Cauchy problems consisting of the Tricomi equation (13.2-1) *with complex σ and θ and the data* (13.3-2a) *and* (13.3-2b), *with $\psi_1(0, \theta) = \varphi_1(\tilde{\theta})$, $\psi_{2,\sigma}(0, \theta) = \varphi_2(\tilde{\theta})$, and $\tilde{\theta} = \theta - \theta_0$, have the integral representations* (13.4-4) *with $j = 1$ and $j = 2$, respectively, where Φ_j is the Borel transform of φ_j.*

Note that for $j = 1$ and $\sigma = 0$, formula (13.4-4) reduces to the well-known *Pólya representation*, a very useful tool for investigating entire functions of exponential type (cf. Boas [1968], p. 74).

13.5 POLYNOMIAL EXPANSIONS OF SOLUTIONS OF THE TRICOMI EQUATION

In this section we shall see how one can apply results from the theory of entire functions in order to obtain useful general representations of solutions of the Tricomi equation. Specifically, we shall discuss a method of deriving polynomial expansions of ψ_1 in (13.4-4) for variable σ and constant θ. Similar representations in the case of constant σ and variable θ as well as representations for ψ_2 can be obtained in the same way.

We begin with the familiar fact that an entire function f of a single complex variable,

$$f(z) = \sum_{n=0}^{\infty} a_n z^n$$

and of finite order ρ satisfies

$$\rho = \overline{\lim_{n\to\infty}} \frac{n \log n}{\log(1/|a_n|)} \tag{13.5-1}$$

(see Boas [1968], p. 9). Furthermore, we shall make use of the following result (Boas [1968], p. 11).

13.5-1 Lemma

Let

$$\nu = \overline{\lim_{n\to\infty}} n|a_n|^{\rho/n}, \qquad 0 < \rho < \infty.$$

Then if $0 < \nu < \infty$, *the function* $f(z)$ *is of order* ρ *and type* α *if and only if* $\nu = e\alpha\rho$. *If* $\rho = 1$, *then, equivalently, by the Stirling formula,*

$$\overline{\lim_{n\to\infty}} |f^{(n)}(z)|^{1/n} = \alpha \tag{13.5-2}$$

where z is any complex number.

Let φ_1 in (13.4-1) be of order 1. Then by (13.5-1)

$$\overline{\lim_{n\to\infty}} \frac{n \log n}{\log\left(n!/|\varphi_1^{(n)}(0)|\right)} = 1.$$

Thus for all $\varepsilon > 0$ and all sufficiently large n,

$$|\varphi_1^{(n)}(0)| \leq \frac{n!}{n^{n/(1+\varepsilon)}}$$

and we can take ε so small that the right-hand sides form a monotone decreasing sequence. Then, substituting the Maclaurin series of $\varphi_1(\tilde{\theta})$ (cf. Sec. 13.4) and the hypergeometric series into (13.3-12) and interchanging the order of summation, we obtain, for an arbitrary fixed $\theta = \theta_1$,

$$\text{(13.5-3)} \quad \begin{aligned} &\text{(a)} \quad \psi_1(\sigma, \theta_1) = \sum_{\kappa=0}^{\infty} k_{3\kappa} \sigma^{3\kappa} \\ &\text{(b)} \quad k_{3\kappa} = \frac{(-1)^{\kappa}}{9^{\kappa}\kappa!\left(\frac{2}{3}\right)_{\kappa}} \sum_{n=0}^{\infty} \frac{\varphi_1^{(n+2\kappa)}(0)}{n!} (\theta_1 - \theta_0)^n . \end{aligned}$$

Also, for the preceding choice of ε,

$$\text{(13.5-4)} \qquad 1/|k_{3\kappa}| \geq 9^{\kappa}\kappa!\left(\tfrac{2}{3}\right)_{\kappa}(2\kappa)^{2\kappa/(1+\varepsilon)} e^{-|\theta_1-\theta_0|}/(2\kappa)! .$$

We now introduce the notation

$$\text{(13.5-5)} \qquad v = \psi_1|_{\theta=\theta_1} .$$

By (13.5-1, 3), the order ρ of v satisfies

$$\rho = \overline{\lim_{\kappa\to\infty}} \frac{3\kappa \log(3\kappa)}{\log(1/|k_{3\kappa}|)} \leq \frac{3}{2}(1+\varepsilon);$$

hence $\rho \leq \frac{3}{2}$. In the special case $\rho = \frac{3}{2}$ we can easily determine an upper bound for the type α_0 of v. We remember that φ_1 is of order 1. Then, by (13.5-2), the function φ_1 is of type

$$\alpha_1 = \overline{\lim_{n\to\infty}} |\varphi_1^{(n)}(0)|^{1/n} .$$

Hence for small $\varepsilon > 0$ and infinitely many n we have

$$|\varphi_1^{(n)}(0)| \leq (\alpha_1 + \varepsilon)^n .$$

Using Lemma 13.5-1, we thus obtain

$$\tfrac{3}{2}e\alpha_0 = \rho e \alpha_0 = \overline{\lim_{\kappa\to\infty}} \left(3\kappa |k_{3\kappa}|^{\rho/3\kappa}\right) \leq (\alpha_1 + \varepsilon)e$$

where the last inequality follows by using (13.5-3b) and applying the Stirling formula. Since this holds for any (sufficiently small) $\varepsilon > 0$, we conclude that

$$\alpha_0 \leq \tfrac{2}{3}\alpha_1 .$$

This yields the following result.

13.5-2 Lemma

Let φ_1 be entire of order 1 and type α_1. Then for arbitrary fixed complex θ_1, the restriction v of ψ_1 defined by (13.5-5) is of order $\frac{3}{2}$ and type not exceeding $\frac{2}{3}\alpha_1$, or of order less than $\frac{3}{2}$.

A CLASS OF ENTIRE FUNCTIONS v. APPLICATION OF NACHBIN'S THEOREM

Representation (13.4-4) now suggests characterizing those entire functions v of the variable σ^3 which can be represented in the form

$$v(\sigma^3) = \frac{1}{2\pi i}\int_C \Psi\left(\frac{(-\sigma)^3 w^2}{9}\right) V(w)\, dw$$

where $C = \{w \mid |w| = \frac{3}{2} + \varepsilon\}$, $\varepsilon > 0$, the kernel Ψ is given by

$$\Psi\left(\frac{\tilde{\sigma}^2}{9}\right) = \sum_{\kappa=0}^{\infty} \Psi_{2\kappa}\left(\frac{\tilde{\sigma}^2}{9}\right)^{\kappa} = {}_0F_1\left(\tfrac{2}{3}; \tfrac{1}{9}(-\sigma)^3 w^2\right) \tag{13.5-6}$$

$$\tilde{\sigma} = (-\sigma)^{3/2} w, \qquad \Psi_{2\kappa} = 1/\left[\left(\tfrac{2}{3}\right)_\kappa \kappa!\right]$$

and V is represented by

$$V(w) = \sum_{n=0}^{\infty} V_{2n} w^{-2n-1}.$$

Formally we have

$$v(\sigma^3) = \sum_{\kappa=0}^{\infty} v_{3\kappa}\left(\frac{(-\sigma)^3}{9}\right)^{\kappa}, \qquad v_{3\kappa} = V_{2\kappa}\Psi_{2\kappa} \tag{13.5-7}$$

$$V(w) = \sum_{n=0}^{\infty} (v_{3n}/\Psi_{2n}) w^{-2n-1}.$$

Following Boas and Buck [1958], we call a function

$$\Psi(z) = \sum_{n=0}^{\infty} \Psi_n z^n$$

a *comparison function* if $\Psi_n > 0$ and $\Psi_{n+1}/\Psi_n \downarrow 0$. We consider the class of entire functions $v(z) = \sum_{n=0}^{\infty} v_n z^n$ such that, for some α (depending on v),

$$|v(re^{i\theta})| \leq M\Psi(\alpha r), \qquad r \uparrow \infty.$$

We call the infimum of the numbers α for which this inequality holds the *Ψ-type* of v. Then *Nachbin's theorem* (cf. Boas and Buck [1958], p. 6) states that v is of Ψ-type α if and only if

$$\overline{\lim_{n\to\infty}}\, |v_n/\Psi_n|^{1/n} = \alpha .$$

Since our Ψ in (13.5-6) is a comparison function, we conclude that our function v is of Ψ-type α if and only if

$$\overline{\lim_{n\to\infty}}\, |v_{3n}/\Psi_{2n}|^{1/2n} = \alpha .$$

Now from the coefficients of the Maclaurin series of ${}_0F_1$ in (13.5-6) it follows that ${}_0F_1(\frac{2}{3}; \sigma)$ is of order $\frac{1}{2}$ and type 2. Hence ${}_0F_1(2/3; \alpha^2\sigma^3/9)$, regarded as a function of σ, is of order $3/2$ and type $2\alpha/3$. This yields

13.5-3 Lemma

If v, as a function of σ, is of order $3/2$ and type not exceeding $2\alpha/3$, or of order less than $3/2$, then v, as a function of σ^3, is of Ψ-type not exceeding α.

From Nachbin's theorem and (13.5-7) it follows that if v is of Ψ-type not exceeding α, then the series for V converges for $|w| > \alpha$. Now from Lemmas 13.5-2 and 13.5-3 we see that v is of Ψ-type not exceeding α_1, the type of φ_1. Since (13.4-2) with $j = 1$ can be written

$${}_0F_1\left(\tfrac{2}{3}; \tfrac{1}{9}(-\sigma)^3 w^2\right) = \sum_{n=0}^{\infty} \psi_{1n}(\sigma, \theta_1) w^n / [n! e^{(\theta_1 - \theta_0)w}]$$

and $\exp[(\theta_1 - \theta_0)w]$ has no zeros, we finally have the following result.

13.5-4 Theorem

Let ψ_1 be the solution of the Tricomi equation (13.2-1) *considered in Sec.* 13.4, *where φ_1 is of order* 1 *and type α_1. Then $v = \psi_1|_{\theta=\theta_1}$ (θ_1 arbitrary) has the convergent polynomial expansion*

$$v(\sigma^3) = \psi_1(\sigma, \theta_1) = \sum_{n=0}^{\infty} \frac{1}{(2n)!} M_{2n}(\psi_1(\sigma, \theta_1)) \psi_{1,2n}(\sigma, \theta_1) \tag{13.5-8a}$$

with $\psi_{1,2n}$ defined as at the beginning of Sec. 13.4,

$$M_{2n}(\psi_1(\sigma, \theta_1)) = \frac{1}{2\pi i} \int_{C_1} V(\psi_1(\sigma, \theta_1), w) w^{2n} e^{-(\theta_1 - \theta_0)w}\, dw , \tag{13.5-8b}$$

$C_1 = \{w \mid |w| = \alpha_1 + \varepsilon\}$, $\varepsilon > 0$, *and* $V(\psi_1(\sigma, \theta_1), w)$ *as defined in* (13.5-7), *taken in the case of* $v(\sigma^3) = \psi_1(\sigma, \theta_1)$.

Some further applications of the theory of entire functions to the Tricomi and related equations of mixed type can be found in Bergman [1949a], pp. 170–173.

13.6 A SPECIAL CAUCHY PROBLEM

We illustrate our results of the preceding three sections in a typical case, namely, for the solution of the Tricomi equation satisfying the initial conditions

$$\psi_1(0, \theta) = e^{\alpha_1 \theta}, \qquad \psi_{1,\sigma}(0, \theta) = 0, \qquad \alpha_1 > 0.$$

From (13.3-5b) we obtain an integral representation, which in the present case can be written in closed form:

$$\psi_1(\sigma, \theta) = \operatorname{Re} \int_{C_1} {}_2F_1\left(\frac{1}{6}, \frac{1}{2}; \frac{1}{3}; \frac{4\sigma^{3/2}}{2\sigma^{3/2} + 3i\theta} t^{-2}\right) \times$$

$$\times \sum_{n=0}^{\infty} \frac{n!}{(2n)!\pi} \left(-\frac{4}{3} i\alpha_1\right)^n (2\sigma^{3/2} + 3i\theta)^n \tau^{n-1/2}\, dt \tag{13.6-1}$$

$$= {}_0F_1\left(\frac{2}{3}; \frac{1}{9}\alpha_1^2(-\sigma)^3\right) e^{\alpha_1 \theta}.$$

This is quite remarkable and probably unexpected. Similarly, (13.3-12) yields the representation

$$\psi_1(\sigma, \theta) = \sum_{n=0}^{\infty} \frac{1}{n!} \alpha_1^n \psi_{1,n}(\sigma, \theta)$$

with the polynomials $\psi_{1,n}$ given by

$$\psi_{1,n}(\sigma, \theta) = \theta^n {}_2F_1\left(-\frac{n}{2}, \frac{1}{2} - \frac{n}{2}; \frac{2}{3}; \frac{4(-\sigma)^3}{9\theta^2}\right).$$

Furthermore, from (13.4-4) we obtain

$$\psi_1(\sigma, \theta) = \frac{1}{2\pi i} \int_{C_1} {}_0F_1\left(\tfrac{2}{3}; \tfrac{1}{9}(-\sigma)^3 w^2\right) e^{\theta w} \sum_{n=0}^{\infty} \alpha_1^n w^{-n-1}\, dw.$$

If we choose an arbitrary fixed $\theta = \theta_1$, we can apply Theorem 13.5-4. We claim that from (13.5-8) we obtain the representation

$$\psi_1(\sigma, \theta_1) = e^{\alpha_1 \theta_1} \cosh(\alpha_1 \theta_1) \sum_{n=0}^{\infty} \frac{1}{(2n)!} \alpha_1^{2n} \psi_{1,2n}(\sigma, \theta_1) \tag{13.6-2}$$

with $\psi_{1,2n}$ defined as before. Indeed, starting from the representation of ψ_1 in the last line of (13.6-1), we first have

$$v_{3\kappa} = e^{\alpha_1 \theta_1} \alpha_1^{2\kappa} / \left[\left(\tfrac{2}{3}\right)_\kappa \kappa! \right].$$

Hence, by (13.5-6) and (13.5-7), we see that

$$V(\psi_1(\sigma, \theta_1), w) = e^{\alpha_1 \theta_1} \sum_{\kappa=0}^{\infty} \alpha_1^{2\kappa} w^{-2\kappa-1}.$$

From this and (13.5-8b) it follows that

$$\begin{aligned} M_{2n}(\psi_1(\sigma, \theta_1)) &= \frac{1}{2\pi i} \int_{C_1} e^{\alpha_1 \theta_1} \sum_{\kappa=0}^{\infty} \alpha_1^{2\kappa} w^{-2\kappa-1} \sum_{m=0}^{\infty} \frac{1}{m!} (-\theta_1)^m w^{m+2n} \, dw \\ &= e^{\alpha_1 \theta_1} \sum_{\kappa=n}^{\infty} \alpha_1^{2\kappa} \frac{1}{[2(\kappa - n)]!} (-\theta_1)^{2(\kappa-n)} \\ &= e^{\alpha_1 \theta_1} \alpha_1^{2n} \cosh(\alpha_1 \theta_1). \end{aligned}$$

We now obtain (13.6-2) by (13.5-8a).

A comparison of representation (13.6-2) with (13.6-1) yields an interesting representation of ${}_0F_1$ in terms of polynomial solutions of the Tricomi equation, namely,

$${}_0F_1\left(\tfrac{2}{3}; \tfrac{1}{9}\alpha_1^2(-\sigma)^3\right) = \cosh(\alpha_1 \theta_1) \sum_{n=0}^{\infty} \frac{1}{(2n)!} \alpha_1^{2n} \psi_{1,2n}(\sigma, \theta_1).$$

Observe that the left-hand side is independent of θ_1.

13.7 EICHLER'S INTEGRAL OPERATOR FOR THE TRICOMI EQUATION

As before, we write the Tricomi equation (13.2-1) in the form (13.2-3), that is,

$$D_1 D_2 u + \frac{5/36}{(z_1 + z_2)^2} u = 0 \tag{13.7-1}$$

which is convenient for our purpose. Solutions of this equation may also be obtained and studied with the help of an integral operator introduced by

Eichler [1942, 1949]. Eichler operators have the advantage that in the case of *self-adjoint* equations, the conditions for the kernel are substantially simpler than those in the case of Bergman kernels.

EICHLER INTEGRAL OPERATORS

Eichler operators for Type I representations (Sec. 2.3) are defined as follows.

13.7-1 Definition

Suppose that equation (2.1-1), that is,

$$Lw = D_1D_2w + a_1(z)D_1w + a_2(z)D_2w + a_0(z)w = 0$$

has holomorphic coefficients on a domain $\Omega = \Omega_1 \times \Omega_2 \subset \mathbb{C}^2$. Let an integral operator E be defined on the vector space of functions $f \in C^\omega(\Omega_1)$ by

$$Ef(z) = \int_{z_0}^{z_1} \tilde{k}(z, t)f(t)\,dt - f(z_1) \tag{13.7-2}$$

where $z = (z_1, z_2)$, $\tilde{k} \in C^\omega(\Omega \times \Omega_1)$, and $z_0 \in \Omega_1$. Then E is called an ***Eichler operator*** for (2.1-1) on Ω if $LEf = 0$ on Ω.

For a self-adjoint equation (thus, $a_1 = a_2 = 0$),

$$L_0w = D_1D_2w + a_0(z)w = 0 \tag{13.7-3}$$

we readily obtain

13.7-2 Theorem

If the kernel $\tilde{k}$ *in Def.* 13.7-1 *satisfies the conditions*

$$\begin{aligned} &\text{(a)} \quad L_0\tilde{k} = 0 && \text{on } \Omega \times \Omega_1 \\ &\text{(b)} \quad D_2\tilde{k}(z, z_1) = a_0(z) && \text{on } \Omega \end{aligned} \tag{13.7-4}$$

then E is an Eichler operator for (13.7-3) *on* Ω.

Proof. This follows by substituting (13.7-2) into (13.7-3) and straightforward calculation. ■

BERGMAN–BOJANIĆ THEOREM

We next wish to establish an interesting relation between our present approach to the Tricomi equation and our previous method involving Bergman operators. For this purpose we use a result by Bergman and Bojanić [1962], p. 326, which is obtained without difficulty by direct calculation:

13.7-3 Theorem

Let V be a holomorphic function of variables γ, ξ and a parameter ρ which satisfies

$$V_{\gamma\gamma} + 2V_{\gamma\xi} + \frac{2\rho - 1}{\xi} V_\gamma + 4\tilde{a}_0(\gamma)V = 0. \tag{13.7-5}$$

Then (a) *and* (b) *hold:*

(a) *The function k_1 defined by*

$$k_1(z, t) = (t^2 z_1)^\rho V(\gamma, \xi, \rho)$$

$$\gamma = \tfrac{1}{2}(z_1 + z_2), \qquad \xi = t^2 z_1$$

satisfies the **Bergman kernel equation** (2.3-2a) *with $s = 0$ corresponding to equation* (13.7-3) *with coefficient*

$$a_0(z) = \tilde{a}_0\left(\frac{z_1 + z_2}{2}\right)$$

that is,

$$M_1 k_1 = \left\{\tau D_2 D_t - \frac{1}{t} D_2 + 2z_1 t\left[D_1 D_2 + \tilde{a}_0\left(\frac{z_1 + z_2}{2}\right)\right]\right\} k_1 = 0.$$

(b) *The function $\tilde{k}$ defined by*

$$\tilde{k}(z, t) = D_t[V(\gamma, \xi, \rho)], \qquad \xi = z_1 - t; \quad \rho = \tfrac{1}{2}$$

with γ as before and V satisfying

$$V_{\gamma\xi}(\gamma, 0, \tfrac{1}{2}) = -2\tilde{a}_0(\gamma)$$

is a kernel of an Eichler operator for (13.7-3) *with a_0 as in* (a).

APPLICATION TO THE TRICOMI EQUATION

Clearly, the Tricomi equation is of the form indicated in this theorem, with

$$a_0(z) = \tilde{a}_0(\gamma) = \frac{5/36}{(z_1 + z_2)^2} = \frac{5}{144\gamma^2}.$$

Hence if we set

$$V(\gamma, \xi, \rho) = \xi^{-\rho}\tilde{k}_1(z, t) = k(Z), \qquad Z = \xi/2\gamma$$

where $\tilde{k}_1$ is the Bergman kernel given in (13.2-4) and (13.2-5), we readily obtain solutions of (13.7-5) in the Tricomi case. Now, if in (13.2-5a) we let $\alpha_1 = 1$, $\beta_1 = 0$, and $\rho = \frac{1}{2}$, we have

$$k(Z) = {}_2F_1\left(\tfrac{1}{6}, \tfrac{5}{6}; 1; Z\right), \qquad Z = \xi/2\gamma.$$

Using

$$\frac{d}{dZ}\,{}_2F_1(a, b; c; Z) = \frac{ab}{c}\,{}_2F_1(a+1, b+1; c+1; Z)$$

and straightforward differentiation, we can verify immediately that

$$\begin{aligned}\frac{1}{2}V_{\gamma\xi}\left(\gamma, 0, \tfrac{1}{2}\right) &= -\frac{5}{144\gamma^2}\,{}_2F_1\left(\tfrac{7}{6}, \tfrac{11}{6}; 2; Z\right)\Big|_{\xi=0}\\ &= -\frac{5}{144\gamma^2} = -\tilde{a}_0(\gamma).\end{aligned}$$

Furthermore, we may now establish a connection with Legendre functions of the first kind, using (cf. Magnus et al. [1966], p. 153)

$$P_\nu^\mu(\tilde{Z}) = \frac{1}{\Gamma(1-\mu)}\left(\frac{\tilde{Z}+1}{\tilde{Z}-1}\right)^{\mu/2}{}_2F_1\left(-\nu, \nu+1; 1-\mu; \frac{1-\tilde{Z}}{2}\right).$$

Indeed, choosing $\nu = -1/6$ and $\mu = 0$, we obtain

$$\begin{gathered}D_t\,{}_2F_1\left(\tfrac{1}{6}, \tfrac{5}{6}; 1; \frac{z_1 - t}{z_1 + z_2}\right) = D_t P_{-1/6}(\tilde{Z}) = \frac{2}{z_1 + z_2}\frac{d}{d\tilde{Z}}P_{-1/6}(\tilde{Z})\\ \tilde{Z} = \frac{2t - z_1 + z_2}{z_1 + z_2}.\end{gathered}$$

This proves the following basic result.

13.7-4 Theorem

(a) *A kernel of an Eichler operator for the Tricomi equation* (13.7-1) *is given by*

$$\tilde{k}(z, t) = D_t\,{}_2F_1\left(\frac{1}{6}, \frac{5}{6}; 1; \frac{z_1 - t}{z_1 + z_2}\right) = \frac{2}{z_1 + z_2}\frac{d}{d\tilde{Z}}P_{-1/6}(\tilde{Z}) \tag{13.7-6}$$

with $\tilde{Z}$ as just defined.

(b) *An Eichler representation of solutions (according to Def.* 13.7-1 *and Theorem* 13.7-2) *for the Tricomi equation is given by*

$$u(z) = \frac{2}{z_1 + z_2} \int_{z_0}^{z_1} P'_{-1/6}(\tilde{Z}) f(t)\, dt - f(z_1) \tag{13.7-7}$$

where $' = d/d\tilde{Z}$.

Practically important applications of (13.7-7) will be considered in the next section.

Clearly, the Eichler kernel $\tilde{k}$ in (13.7-6) can also be obtained in a way similar to that of obtaining the Bergman kernel $\tilde{k}_1$ in (13.2-4, 5), that is, by introducing an auxiliary variable such that (13.7-4a) reduces to an *ordinary* differential equation, for instance, the variable $\tilde{Z}$. Then, with P defined by

$$\tilde{k}(z, t) = D_t P(\tilde{Z})$$

as a new independent variable, we obtain from (13.7-4) the Legendre equation for P with parameters $\nu = -1/6$ and $\mu = 0$, and the condition

$$P'(1) = -5/72.$$

This shows that Theorem 13.7-4 can also be obtained from Theorem 13.7-2.

13.8 FAMILIES OF SOLUTIONS OF THE TRICOMI EQUATION

The Tricomi equation plays a central role in transonic flow, as has been pointed out; see also Guderley [1962] and the references given at the end of that book. This entails great interest in various sequences of particular solutions of this equation, which were derived from various sources and by different procedures. These solutions were obtained by Carrier and Ehlers, Chaplygin, Darboux, Falkovich, Germain and Bader, Guderley, and Tomotika and Tamada, some of whose works are listed and discussed in the book by Guderley [1962]. It is most remarkable that our operator approach permits us to derive all these solutions (and others) from a unique source, namely, from representation (13.7-7). This illustrates again the unifying power of the theory of integral operators defined on spaces of complex analytic functions.

APPLICATION OF THEOREM 13.7-4

In (13.7-7) we first choose the path from ∞ to z_1 represented by

$$t = \tilde{Z}\gamma + i\theta = \tfrac{1}{2}\tilde{Z}(z_1 + z_2) + \tfrac{1}{2}(z_1 - z_2), \qquad \tilde{Z} \in [1, \infty).$$

Since $P'_{-1/6}(\tilde{Z}) = O(\tilde{Z}^{-7/6})$, we require that on the path,

$$(13.8\text{-}1) \qquad |f(t)| < C|t|^{1/6-\varepsilon}, \qquad C > 0; \quad \varepsilon > 0.$$

We now recall from Sec. 13.2 that u in (13.7-1) and ψ in the original form (13.2-1) of the Tricomi equation are related by

$$(13.8\text{-}2) \qquad \psi(\sigma,\theta) = \sigma^{-1/4} u\left(\tfrac{2}{3}\sigma^{3/2} + i\theta, \tfrac{2}{3}\sigma^{3/2} - i\theta\right).$$

Integrating by parts and reversing the sense of integration, from Theorem 13.7-4b we thus obtain the following result (cf. also Lanckau [1962]):

13.8-1 Theorem

Let

$$(13.8\text{-}3) \qquad \psi(\sigma,\theta) = \sigma^{-1/4} \int_1^{\infty} P_{-1/6}(\tilde{Z}) \frac{\partial}{\partial \tilde{Z}} f\left(\tfrac{2}{3}\sigma^{3/2}\tilde{Z} + i\theta\right) d\tilde{Z}$$

where $P_{-1/6}$ is the Legendre function of the first kind, and f is assumed to be holomorphic on a domain Ω_1 containing the path of integration and satisfying (13.8-1) *on the path. Then ψ is a solution of the Tricomi equation* (13.2-1). *This representation is valid in the intersection of the subsonic region* ($\sigma > 0$) *and the domain in the $\sigma\theta$-plane which corresponds to Ω_1.*

FAMILIES OF SOLUTIONS

From this theorem we can now derive sets of particular solutions of the Tricomi equation by choosing various classes of functions f. Let us show this for some cases of practical interest. We first choose

$$(13.8\text{-}4) \qquad f_\nu(t) = c_0 t^{(4\nu+1)/6}, \quad c_0 \neq 0; \quad \nu < 0; \quad \nu \neq -\tfrac{1}{4}.$$

We now set

$$\xi = -\theta^2/\gamma^2, \qquad \gamma = \tfrac{2}{3}\sigma^{3/2}.$$

Then Theorem 13.8-1 yields

$$\psi_\nu(\sigma,\theta) = \sigma^\nu h_\nu(\xi)$$

$$(13.8\text{-}5) \qquad h_\nu(\xi) = c_{1\nu} \int_1^{\infty} P_{-1/6}(\tilde{Z})\left(\tilde{Z} + \sqrt{\xi}\right)^{(4\nu-5)/6} d\tilde{Z}$$

$$c_{1\nu} = (4\nu + 1)\tfrac{1}{6}\left(\tfrac{2}{3}\right)^{(4\nu+1)/6} c_0.$$

h_ν is complex valued and such that $h_{\nu 1} = \operatorname{Re} h_\nu$ and $h_{\nu 2} = \operatorname{Im} h_\nu$ constitute a basis of solutions of the hypergeometric equation

$$\xi(1-\xi)h'' + \left(\frac{1}{2} + \frac{4-2\nu}{3}\xi\right)h' - \frac{\nu(\nu-1)}{9}h = 0.$$

These ψ_ν in (13.8-5) are *Guderley's solutions*; cf. Guderley [1947, 1962]; see also Carrier and Ehlers [1948]. Note that we have obtained an *integral representation* of these solutions. This again illustrates the fact that the method of integral operators can produce new results in the *theory of special functions*.

In the case $\nu = -\frac{1}{4}$, the formulas (13.8-3) and (13.8-4) would only generate the trivial solution. But this gap can be closed if we choose

$$f_{-1/4}(t) = c_0 \log t, \qquad c_0 \neq 0. \tag{13.8-6}$$

Then we obtain from (13.8-3), a representation similar to (13.8-5), namely

$$\begin{aligned} \psi_{-1/4}(\sigma,\theta) &= \sigma^{-1/4} h_{-1/4}(\xi) \\ h_{-1/4}(\xi) &= c_0 \int_1^\infty P_{-1/6}(\tilde{Z})(\tilde{Z} + \sqrt{\xi})^{-1}\, d\tilde{Z} \\ &= c_0 2\pi P_{-1/6}(\sqrt{\xi}). \end{aligned} \tag{13.8-7}$$

It is known that Guderley's solutions can be transformed into *Darboux's solutions*. Hence the latter may also be obtained directly from Theorem 13.8-1. Similarly, the *Tomotika–Tamada solutions* are obtained by slight changes in (13.8-4) and (13.8-6):

$$f_\nu(t) = c_0(t-\theta_0)^{(4\nu+1)/6}, \qquad \nu < 0; \quad \nu \neq -\tfrac{1}{4},$$

$$f_{-1/4}(t) = c_0 \log(t - \theta_0)$$

where $\theta_0 = const.$

Finally, Chaplygin's solutions result from the functions

$$f_m(t) = c_0 \exp\left(-\frac{m\pi}{\theta_0}t\right), \qquad \operatorname{Re}\frac{mt}{\theta_0} \geq 0. \tag{13.8-8}$$

It is interesting that for this choice of f, the integral (13.8-3) represents essentially a modified Bessel function

$$K_\nu(z) = \frac{\pi}{2\sin\nu\pi}\left[e^{i\nu\pi/2}J_{-\nu}(iz) - e^{-i\nu\pi/2}J_\nu(iz)\right]$$

of order $\nu = \frac{1}{3}$. Indeed, we have (cf. Magnus et al. [1966], p. 85)

$$\int_1^\infty e^{-zt}P_{\nu-1/2}(t)\,dt = 2^{1/2}(\pi z)^{-1/2}K_\nu(z), \qquad \operatorname{Re} z > 0.$$

Hence (13.8-3) with $f = f_m$ given by (13.8-8) yields the result

$$\psi_m(\sigma, \theta) = c_2 \sigma^{1/2} \exp\left(-i\frac{m\pi}{\theta_0}\theta\right) K_{1/3}\left(\frac{2m\pi}{3\theta_0}\sigma^{3/2}\right) \tag{13.8-9}$$

where $c_2 = -2c_0(m/3\theta_0)^{1/2}$. These Chaplygin solutions have the important property that they vanish at infinity, as is needed in various applications.

It is quite remarkable that we have obtained all those sequences of solutions from the same source—our integral representation—and that all of them correspond to very simple functions f, which one would probably choose on the outset if one started experimenting with (13.7-7) and (13.8-3).

We mention that, by taking another path of integration, we can also represent the *solutions of Falkovich* and *of Germain and Bader* by means of (13.7-7). As can be seen from Lanckau [1962], for this purpose we can set

$$z_1^{(0)} = -\gamma_0 + i\theta_0 = -\gamma_0 < 0$$

and take as the path of integration in (13.7-7) an arc of the circle through the points $-\gamma_0$, γ_0, and $z_1 = \gamma + i\theta$ $(\theta \neq 0)$ in the $\gamma\theta$-plane ($\gamma = \frac{2}{3}\sigma^{3/2}$ for $\sigma > 0$), namely, the arc from $-\gamma_0$ to z_1 not passing through γ_0. Then (13.7-7) gives those solutions if we set

$$f_m(t) = \left(\frac{t + \gamma_0}{t - \gamma_0}\right)^m, \qquad m \neq 0$$

and, since for $m = 0$ this would only generate the trivial solution,

$$f_0(t) = \log \frac{t + \gamma_0}{t - \gamma_0}.$$

To gain full appreciation of the present results, we should recall that in transonic flow, almost all specific results have been obtained by the so-called *indirect method* or *inverse method*, that is, by searching large classes of special solutions for members that may correspond to practically relevant flows. On the one hand, this accounts for the importance of special solutions and explains why they have been investigated in great detail. On the other hand it shows that *integral operators are particularly suitable for the indirect method*, for the simple reason that, by taking a sufficiently large class of analytic functions we may readily obtain a large supply of special solutions having similar properties.

We finally want to mention that the indirect method and its variants are also useful in elasticity theory; cf. Neményi [1951]. For further general remarks on the method in fluid flow as well as on relevant work by K. Morawetz and others, see F. Bauer et al. [1972–1975]. For results and conjectures in the mathematical theory of subsonic and transonic gas flows, see also

Bers [1954a] (this paper includes many references) and Bers [1958].

From our previous discussion it is clear that sequences of particular solutions of the Tricomi equation obtained from Bergman or Eichler operators can be used in the approximate solution of boundary and initial value problems (cf. Bergman [1946b]). See also Guderley–Yoshihara [1950]. For a survey of other approximative methods (finite differences, finite elements) for the Tricomi and other mixed type equations, see Aziz and Leventhal [1976]; cf. also Trangenstein [1975]. For further aspects of a general theory of these equations we refer to Bitsadze [1964].

In the present chapter we have discussed some basic ideas and techniques of the integral operator approach to the Tricomi equation. This seems an area of particularly promising further research in connection with analytical and numerical applications to new problems (particularly those involving more complicated domains), the study of further relations to the theory of special functions and, last but not least, the discovery and investigation of new integral operators.

Additional References

Aleksandrov–Kucherov [1970]
Aziz–Leventhal [1976]
Aziz–Schneider [1985]
Bauer–Garabedian–Korn [1972–1975]
Bergman [1946b, 1948, 1949a, 1952a, 1966a, 1970a]
Bergman–Bojanić [1962]
Bers [1954a, 1958]
Bitsadze [1964]
Carrier–Ehlers [1948]
Courant–Friedrichs [1976]
Eichler [1942, 1949]
Falkovich [1946, 1947]
Frankl [1956]
Garabedian [1968]
Garabedian–Korn [1976]
Guderley [1956, 1962]
Guderley–Yoshihara [1950]
Kracht–Kreyszig [1986b]
Krzywoblocki [1960, 1968]
Lanckau [1958–1959, 1962, 1963]
Mackie–Pack [1955]
Manwell [1971, 1979]
Neményi [1951]
Rosenthal [1970, 1974a, 1974b]
Sasai [1979]
Schneider [1977]
Schubert–Schleiff [1969]
Sedov [1965]
Smirnov [1978]
Sorokina [1966]
Stark [1964, 1966]
Tomotika–Tamada [1950]
Trangenstein [1975]
Tricomi [1923, 1968]
Weinstein [1953]
Wolfersdorf [1965]

Bibliography

Agostinelli, C. [1937]: Risoluzione per un campo circolare o sferico di un problema più generale di quello di Dirichlet. *Atti Accad. Sci. Torino Cl. Sci. Fis. Mat. Natur.* **72**, 317–328.

Aleksandrov, V. M., and V. A. Kucherov [1970]: On the method of orthogonal polynomials in plane mixed problems of elasticity theory. *Prikl. Mat. Mekh.* **34**, 643–652.

Andersson, K. G. [1971]: Propagation of analyticity of solutions of partial differential equations with constant coefficients. *Ark. Mat.* **8**, 277–302.

Anger, G. (Ed.) [1970]: *Elliptische Differentialgleichungen* (Kolloquium, Berlin, 1969), vol. I. Berlin: Akademie-Verlag.

Aupetit, B. (Ed.) [1980]: *Complex Approximation* (Proceedings, Quebec, Canada, 1978). Boston, MA: Birkhäuser.

Avila, G. S. S., and R. P. Gilbert [1967]: On the analytic properties of solutions of the equation $\Delta u + xu_x + yu_y + c(r)u = 0$. *Duke Math. J.* **34**, 353–362.

Aziz, A. K., and S. H. Leventhal [1976]: Numerical solution of linear partial differential equations of elliptic-hyperbolic type. In B. Hubbard (Ed.): *Numerical Solution of Partial Differential Equations*, vol. III, pp. 55–88. New York: Academic Press.

Aziz, A. K., and M. Schneider [1985]: The existence of generalized solutions for a class of quasilinear equations of mixed type. *J. Math. Anal. Appl.* **107**, 425–445.

Azzam, A., and E. Kreyszig [1980]: Construction of kernels of integral operators for linear partial differential equations. *Applicable Anal.* **10**, 261–273.

Azzam, A., and E. Kreyszig [1981]: On Bergman operators of exponential type. *Ann. Polon. Math.* **39**, 37–48.

Balk, M. B. [1963]: The Picard theorem for entire bi-analytic functions. *Dokl. Akad. Nauk SSSR* **152**, 1282–1285. (In Russian.)

Balk, M. B. [1965]: The big Picard theorem for entire bi-analytic functions. *Uspekhi Mat. Nauk* **20**, 159–165. (In Russian.)

Balk, M. B. [1966]: On the values taken by entire polyanalytic functions. *Dokl. Akad. Nauk SSSR* **167**, 12–15. (In Russian.)

Balk, M. B. [1983]: Polyanalytic functions. In E. Lanckau and W. Tutschke [1983], pp. 68–84.

Balk, M. B., and M. F. Zuev [1970]: On polyanalytic functions. *Uspekhi Mat. Nauk* **25**, 203–226.

Bauer, F., P. Garabedian, and D. Korn [1972–1975]: *Supercritical Wing Sections*, vols. I and II. New York: Springer.

Bauer, K. W. [1965]: Über eine der Differentialgleichung $(1 \pm z\bar{z})^2 w_{z\bar{z}} \pm n(n+1)w = 0$ zugeordnete Funktionentheorie. *Bonn. Math. Schr.* **23**, iii + 98 pp.

Bauer, K. W. [1966a]: Über die Lösungen der elliptischen Differentialgleichung $(1 \pm z\bar{z})^2 w_{z\bar{z}} + \lambda w = 0$. I, II. *J. Reine Angew. Math.* **221**, 48–84, 176–196.

Bauer, K. W. [1966b]: Über eine partielle Differentialgleichung 2. Ordnung mit zwei unabhängigen komplexen Variablen. *Monatsh. Math.* **70**, 385–418.

Bauer, K. W. [1967]: Über eine Klasse verallgemeinerter Cauchy–Riemannscher Differentialgleichungen. *Math. Z.* **100**, 17–28.

Bauer, K. W. [1968]: Zur Verallgemeinerung des Poissonschen Satzes. *Ann. Acad. Sci. Fenn. Ser. A I Math.* **437**, 28 pp.

Bauer, K. W. [1971]: Über Differentialgleichungen der Form $F(z, \bar{z}) w_{z\bar{z}} - n(n+1)w = 0$. *Monatsh. Math.* **75**, 1–13.

Bauer, K. W. [1973]: Differentialoperatoren bei partiellen Differentialgleichungen. In S. Ruscheweyh [1973b], pp. 7–17.

Bauer, K. W. [1974]: Differentialoperatoren bei einer Klasse verallgemeinerter Tricomi-Gleichungen. *Z. Angew. Math. Mech.* **54**, 715–721.

Bauer, K. W. [1975]: Eine verallgemeinerte Darboux–Gleichung. I, II. *Monatsh. Math.* **80**, 1–11, 265–276.

Bauer, K. W. [1976a]: Polynomoperatoren bei Differentialgleichungen der Form $w_{z\bar{z}} + Aw_{\bar{z}} + Bw = 0$. *J. Reine Angew. Math.* **283/284**, 364–369.

Bauer, K. W. [1976b]: Zur Darstellung pseudoanalytischer Funktionen. In V. E. Meister, N. Weck, and W. L. Wendland [1976], pp. 101–111.

Bauer, K. W. [1978]: On a differential equation in the theory of pseudo-holomorphic functions. *J. Math. Soc. Japan* **30**, 457–461.

Bauer, K. W. [1981]: Zur Bestimmung nicht-einfacher Riemann-Funktionen. *Ber. Math.-Statist. Sekt. Forschungszentrum Graz* **173**, 23 pp.

Bauer, K. W. [1982a]: Differentialoperatoren bei adjungierten Differentialgleichungen. *Ber. Math.-Statist. Sekt. Forschungszentrum Graz* **183**, 29 pp.

Bauer, K. W. [1982b]: On a class of Riemann functions. *Applicable Anal.* **13**, 109–126.

Bauer, K. W. [1984]: Riemannfunktionen und Differentialoperatoren. *Z. Anal. Anwendungen* **3**, 7–17.

Bauer, K. W., and H. Florian [1976]: Bergman-Operatoren mit Polynomerzeugenden. In R. P. Gilbert and R. J. Weinacht [1976], pp. 85–93.

Bauer, K. W., and G. Jank [1971]: Differentialoperatoren bei einer inhomogenen elliptischen Differentialgleichung. *Rend. Istit. Mat. Univ. Trieste* **3**, 140–168.

Bauer, K. W., and E. Peschl [1966]: Ein allgemeiner Entwicklungssatz für die Lösungen der Differentialgleichung $(1 + \varepsilon z\bar{z})^2 w_{z\bar{z}} + \varepsilon n(n+1)w = 0$ in der Nähe isolierter Singularitäten. *Bayer. Akad. Wiss. Math.-Natur. Kl. Sitzungsber.* 1965 II, 113–146.

Bauer, K. W., and E. Peschl [1967]: Eindeutige Lösungen einer partiellen Differentialgleichung mit mehrdeutigen Erzeugenden. *Arch. Math. (Basel)* **18**, 285–289.

Bauer, K. W., and S. Ruscheweyh [1980]: *Differential Operators for Partial Differential Equations and Function Theoretic Applications*. New York: Springer.

Begehr, H. [1976]: Das Schwarzsche Lemma und verwandte Sätze für pseudoanalytische Funktionen. In R. P. Gilbert and R. J. Weinacht [1976], pp. 11–21.

Begehr, H. [1983]: Value distribution theory. In E. Lanckau and W. Tutschke [1983], pp. 59–67.

Begehr, H., and R. P. Gilbert [1988]: *Constructive Methods for Elliptic Equations*. New York: Springer.

Behnke, H., and F. Sommer [1972]: *Theorie der analytischen Funktionen einer komplexen Veränderlichen*, 3rd ed. reprint. Berlin: Springer.

Behnke, H., and P. Thullen [1970]: *Theorie der Funktionen mehrerer komplexer Veränderlichen*, 2nd ed. New York: Springer.

Beltrami, E. [1911]: Sulle teoria delle funzioni potenzali simmetriche. *Opere Matematice* (Milano) **3**, 349–377.

Bengel, G. [1984]: Convergence of formal solutions of singular partial differential equations. Conference on Linear Partial and Pseudo-Differential Operators, Torino, 1982. *Rend. Sem. Mat. Univ. Politec. Torino*, 1983, Special Issue, 41–50.

Berenstein, C. A., and B. A. Taylor [1979]: A new look at interpolation theory for entire functions of one variable. *Adv. in Math.* **33**, 109–143.

Berglez, P. [1979]: Lineare Differentialoperatoren bei einem Paar adjungierter Differentialgleichungen. *Ber. Math.-Statist. Sekt. Forschungszentrum Graz* **118**, ii + 80 pp.

Berglez, P. [1980]: Über eine partielle Differentialgleichung, die bei der Behandlung gewisser inhomogener Differentialgleichungen von Bedeutung ist. *Ber. Math.-Statist. Sekt. Forschungszentrum Graz* **122**, i + 18 pp.

Berglez, P. [1982]: Characterization of certain differential operators in the representation of pseudo-analytic functions. *Rend. Istit. Mat. Univ. Trieste* **14**, 27–31.

Berglez, P. [1983]: On the representation of the Riemann function by differential operators. *Glas. Mat. Ser. III* **18**(38), 295–303.

Bergman, S. [1930]: Über Kurvenintegrale von Funktionen zweier komplexen Veränderlichen, die die Differentialgleichung $\Delta V + V = 0$ befriedigen. *Math. Z.* **32**, 386–406.

Bergman, S. [1933]: Herstellung von vollständigen Funktionssystemen in zwei komplexen Veränderlichen. *Sitzungsber. Berlin. Math. Ges.* **32**, 76–77.

Bergman, S. [1936]: Über ein Verfahren zur Konstruktion der Näherungslösungen der Gleichung $\Delta u + \tau^2 u = 0$. *Prikl. Mat. Mekh.* **3**, 97–107.

Bergman, S. [1937a]: Sur un lien entre la théorie des équations aux dérivées partielles elliptiques et celle des fonctions d'une variable complexe. *C. R. Acad. Sci. Paris* **205**, 1198–1200, 1360–1362.

Bergman, S. [1937b]: Zur Theorie der Funktionen, die eine lineare partielle Differentialgleichung befriedigen. *Mat. Sb. (N.S.)* **2**(44), 1169–1198.

Bergman, S. [1940]: The approximation of functions satisfying a linear partial differential equation. *Duke Math. J.* **6**, 537–561.

Bergman, S. [1942]: The Hodograph Method in the Theory of Compressible Fluids. Supplement to R. von Mises and K. O. Friedrichs: *Fluid Dynamics*. Providence, R.I.: Brown University.

Bergman, S. [1943a]: A formula for the stream function of certain flows. *Proc. Nat. Acad. Sci. U.S.A.* **29**, 276–281.

Bergman, S. [1943b]: Linear operators in the theory of partial differential equations. *Trans. Amer. Math. Soc.* **53**, 130–155.

Bergman, S. [1944a]: Solutions of linear partial differential equations of the fourth order. *Duke Math. J.* **11**, 617–649.

Bergman, S. [1944b]: The determination of some properties of a function satisfying a partial differential equation from its series development. *Bull. Amer. Math. Soc.* **50**, 535–546.

Bergman, S. [1945a]: A class of nonlinear partial differential equations and their properties. *Bull. Amer. Math. Soc.* **51**, 545–554.

Bergman, S. [1945b]: Certain classes of analytic functions of two real variables and their properties. *Trans. Amer. Math. Soc.* **57**, 299–331.

Bergman, S. [1945c]: *On Two-Dimensional Flows of Compressible Fluids*. NACA Tech. Note 972. Washington: National Advisory Committee for Aeronautics.

Bergman, S. [1945d]: *Graphical and Analytic Methods for the Determination of a Flow of a Compressible Fluid around an Obstacle*. NACA Tech. Note 973. Washington: National Advisory Committee for Aeronautics.

Bergman, S. [1946a]: *Methods for the Determination and Computation of Flow Patterns of a Compressible Fluid*. NACA Tech. Note 1018. Washington: National Advisory Committee for Aeronautics.

Bergman, S. [1946b]: *On Supersonic and Partially Supersonic Flows*. NACA Tech. Note 1096. Washington: National Advisory Committee for Aeronautics.

Bergman, S. [1946c]: A class of harmonic functions in three variables and their properties. *Trans. Amer. Math. Soc.* **59**, 216–247.

Bergman, S. [1947a]: Functions satisfying certain partial differential equations of elliptic type and their representation. *Duke Math. J.* **14**, 349–366.

Bergman, S. [1947b]: Two-dimensional subsonic flows of a compressible fluid and their singularities. *Trans. Amer. Math. Soc.* **62**, 452–498.

Bergman, S. [1948]: Two-dimensional transonic flow patterns. *Amer. J. Math.* **70**, 856–891.

Bergman, S. [1949a]: An initial value problem for a class of equations of mixed type. *Bull. Amer. Math. Soc.* **55**, 165–174.

Bergman, S. [1949b]: Operator methods in the theory of compressible fluids. *Proc. Symposia Appl. Math.* **1**, 19–40. New York: American Mathematical Society.

Bergman, S. [1950a]: On solutions with algebraic character of linear partial differential equations. *Trans. Amer. Math. Soc.* **68**, 461–507.

Bergman, S. [1950b]: Determination of axially symmetric flow patterns of a compressible fluid. *J. Math. Phys.* **29**, 133–145.

Bergman, S. [1952a]: On solutions of linear partial differential equations of mixed type. *Amer. J. Math.* **74**, 444–474.

Bergman, S. [1952b]: Operatorenmethoden in der Gasdynamik. *Z. Angew. Math. Mech.* **32**, 33–45.

Bergman, S. [1952c]: The coefficient problem in the theory of linear partial differential equations. *Trans. Amer. Math. Soc.* **73**, 1–34.

Bergman, S. [1954a]: On zero and pole surfaces of functions of two complex variables. *Trans. Amer. Math. Soc.* **77**, 413–454.

Bergman, S. [1954b]: Essential singularities of solutions of a class of linear partial differential equations in three variables. *J. Rational Mech. Anal.* **3**, 539–560.

Bergman, S. [1955]: On representation of stream functions of subsonic and supersonic flows of compressible fluids. *J. Rational Mech. Anal.* **4**, 883–905.

Bergman, S. [1957]: On singularities of solutions of certain differential equations in three variables. *Trans. Amer. Math. Soc.* **85**, 462–488.

Bergman, S. [1958]: Properties of solutions of certain differential equations in three variables. *J. Math. Mech.* **7**, 87–101.

Bergman, S. [1961]: Some properties of a harmonic function of three variables given by its series development. *Arch. Rational Mech. Anal.* **8**, 207–222.

Bergman, S. [1962]: Value distribution of meromorphic functions of two complex variables. *Bull. Soc. Sci. Lett. Łódź* **13**, 1–15.

Bergman, S. [1963]: On the coefficient problem in the theory of a system of linear partial differential equations. *J. Analyse Math.* **11**, 249–274.

Bergman, S. [1966a]: Applications of integral operators to singular differential equations and to computations of compressible fluid flows. In *Numerical Solution of Partial Differential Equations* (Proc. Sympos., Univ. Maryland, 1965), pp. 257–287. New York: Academic Press.

Bergman, S. [1966b]: Approximation of harmonic functions of three variables by harmonic polynomials. *Duke Math. J.* **33**, 379–387.

Bergman, S. [1968]: Application of the kernel function for the computation of flows of compressible fluids. *Quart. Appl. Math.* **26**, 301–310.

Bergman, S. [1970a]: On an initial value problem in the theory of two-dimensional transonic flow patterns. *Pacific J. Math.* **32**, 29–46.

Bergman, S. [1970b]: *The Kernel Function and Conformal Mapping*, 2nd ed. Providence, R.I.: American Mathematical Society.

Bergman, S. [1971]: *Integral Operators in the Theory of Linear Partial Differential Equations*, 3rd ed. New York: Springer.

Bergman, S. [1973]: On stream functions of compressible fluid flows. In S. Ruscheweyh [1973b], pp. 19–27.

Bergman, S. [1976]: On the mathematical theory of flow patterns of compressible fluids. In V. E. Meister, N. Weck, and W. L. Wendland [1976], pp. 1–9.

Bergman, S., and R. Bojanić [1961]: On some relations between integral operators. *Scripta Math.* **25**, 317–321.

Bergman, S., and R. Bojanić [1962]: Application of integral operators to the theory of partial differential equations with singular coefficients. *Arch. Rational Mech. Anal.* **10**, 323–340.

Bergman, S., and B. Epstein [1947]: Determination of a compressible fluid flow past an oval-shaped obstacle. *J. Math. Phys.* **26**, 195–222.

Bergman, S., and L. Greenstone [1947]: Numerical determination by use of special computational devices of an integral operator in the theory of compressible fluids. *J. Math Phys.* **26**, 1–9.

Bergman, S., and J. G. Herriot [1962]: *Calculation of Particular Solutions of Linear Partial Differential Equations by the Method of Integral Operators*. Tech. Report. Stanford, CA.: Applied Mathematics and Statistics Laboratories, Stanford University.

Bergman, S., and J. G. Herriot [1965]: Numerical solution of boundary-value problems by the method of integral operators. *Numer. Math.* **7**, 42–65.

Bergman, S., and M. Schiffer [1948]: Kernel functions in the theory of partial differential equations of elliptic type. *Duke Math. J.* **15**, 535–566.

Bergman, S., and M. Schiffer [1951]: A majorant method for non-linear partial differential equations. *Proc. Nat. Acad. Sci. U.S.A.* **37**, 744–749.

Bergman, S., and M. Schiffer [1953]: *Kernel Functions and Elliptic Differential Equations in Mathematical Physics*. New York: Academic Press.

Bers, L. [1950]: Partial differential equations and generalized analytic functions. *Proc. Nat. Acad. Sci. U.S.A.* **36**, 130–136.

Bers, L. [1952]: *Theory of Pseudo-Analytic Functions*. Mimeographed Notes. New York: Institute of Mathematical Science, New York University.

Bers, L. [1954a]: Results and conjectures in the mathematical theory of subsonic and transonic gas flows. *Comm. Pure Appl. Math.* **7**, 79–104.

Bers, L. [1954b]: Existence and uniqueness of a subsonic flow past a given profile. *Comm. Pure Appl. Math.* **7**, 441–504.

Bers, L. [1956]: An outline of the theory of pseudoanalytic functions. *Bull. Amer. Math. Soc.* **62**, 291–331.

Bers, L. [1958]: *Mathematical Aspects of Subsonic and Transonic Gas Dynamics*. New York: Wiley.

Bers, L. [1962]: Function theoretic aspects of the theory of elliptic partial differential equations. In R. Courant and D. Hilbert [1962], vol. II, pp. 374–406.

Bers, L., S. Bochner, and F. John [1954]: *Contributions to the Theory of Partial Differential Equations*. Princeton, N.J.: Princeton University Press.

Bers, L., and A. Gelbart [1943]: On a class of differential equations in mechanics of continua. *Quart. Appl. Math.* **1**, 168–188.

Bers, L., and A. Gelbart [1944]: On a class of functions defined by partial differential equations. *Trans. Amer. Math. Soc.* **56**, 67–93.

Bhatnagar, S. C., and R. P. Gilbert [1977]: Bergman type operators for pseudoparabolic equations in several space variables. *Math. Nachr.* **76**, 61–68.

Bianchi, L. [1910]: *Vorlesungen über Differentialgeometrie*, 2nd ed. Leipzig: Teubner.

Bieberbach, L. [1921]: Neuere Untersuchungen über Funktionen von komplexen Variablen. *Encyklopädie Math. Wiss.*, vol. II, Analysis 3.1, Part II C 4. Leipzig: Teubner.

Bieberbach, L. [1955]: *Analytische Fortsetzung*. Berlin: Springer.

Bieberbach, L. [1968]: *Lehrbuch der Funktionentheorie*. 2 vols., 2nd ed. New York: Johnson Reprint.

Bitsadze, A. V. [1964]: *Equations of the Mixed Type*. New York: Macmillan.

Bitsadze, A. V. [1968]: *Boundary Value Problems for Second Order Elliptic Equations*. New York: Wiley.

Bizadse, W. A. (Bitsadze, A. V.) [1957]: *Zum Problem der Gleichungen vom gemischten Typus*. Berlin: Deutscher Verlag der Wissenschaften.

Blohina, G. N. [1970]: Theorems of Phragmén-Lindelöf type for linear elliptic equations of second order. *Math. USSR-Sb.* **11**, 467–490.

Boas, R. P., Jr. [1968]: *Entire Functions*, 2nd ed. New York: Academic Press.

Boas, R. P., Jr., and R. C. Buck [1958]: *Polynomial Expansions of Analytic Functions*. Berlin: Springer.

Böhmer, K. [1970]: Eine verallgemeinerte Fuchs'sche Theorie. *Manuscripta Math.* **3**, 343–356.

Borel, É. [1894]: Sur une application d'un théorème de M. Hadamard. *Bull. Sci. Math. (2)* **18**, 22–25.

Bosch, W., and P. Krajkiewicz [1970]: The big Picard theorem for polyanalytic functions. *Proc. Amer. Math. Soc.* **26**, 145–150.

Brackx, F. [1976]: On (k)-monogenic functions of a quaternion variable. In R. P. Gilbert and R. J. Weinacht [1976], pp. 22–44.

Brackx, F., R. Delanghe, and F. Sommen [1982]: *Clifford Analysis*. London: Pitman.

Bragg, L. R. [1969]: Hypergeometric operator series and related partial differential equations. *Trans. Amer. Math. Soc.* **143**, 319–336.

Bragg, L. R., and J. W. Dettman [1983]: Analogous function theories for the heat, wave, and Laplace equations. *Rocky Mountain J. Math.* **13**, 191–214.

Bragg, L. R., and J. W. Dettman [1984]: Analogous function theories for certain singular partial differential equations. *Rocky Mountain J. Math.* **14**, 383–396.

Brown, R. M. [1970]: The Geometry of the Singularities of Harmonic Functions. Thesis, Indiana University, 94 pp.

Čanak, M. [1978]: Operator vom Schwarzschen Typus für eine Klasse der p-analytischen Funktionen. *Math. Balkanica* **8**, 49–55.

Carleman, T. [1933]: Sur les systèmes linéaires aux dérivées partielles du premier ordre à deux variables. *C. R. Acad. Sci. Paris* **197**, 471–474.

Carrier, G. F., and F. E. Ehlers [1948]: On some singular solutions of the Tricomi equation. *Quart. Appl. Math.* **6**, 331–334.

Carroll, R. W. [1979]: *Transmutation and Operator Differential Equations*. Amsterdam: North-Holland.

Carroll, R. W. [1982]: *Transmutation, Scattering Theory and Special Functions*. Amsterdam: North-Holland.

Carroll, R. W. [1984a]: Some remarks on the Bergman–Gilbert integral operator. *J. Math. Pures Appl.* **63**, 1–14.

Carroll, R. W. [1984b]: Some topics in transmutation. In I. W. Knowles and R. T. Lewis (Eds.): *Differential Equations* (Proc. Conf., Birmingham, AL., 1983), pp. 87–104. Amsterdam: North-Holland.

Case, B. A. [1985]: Function theory for a Beltrami algebra. *Internat. J. Math. Math. Sci.* **8**, 247–256.

Chang, P. H. [1983]: A certain recursion formula of Euler–Poisson–Darboux equations. *Global Analysis—Analysis on Manifolds* (dedicated to M. Morse), pp. 51–72. Leipzig: Teubner.

Chang, Y. F., and D. Colton [1977]: The numerical solution of parabolic partial differential equations by the method of integral operators. *Internat. J. Comput. Math. Sect. B* **6**, 229–239.

Chaplygin, S. A. [1904]: On gas jets. *Scient. Ann. Moscow Univ. Phys.-Math. Sect.* **21**, 1–121. (In Russian.)

Chaundy, T. W. [1935, 1936, 1939]: Hypergeometric partial differential equations. I, II, III. *Quart. J. Math. Oxford Ser.* **6**, 288–303; **7**, 306–315; **10**, 219–240.

Chaundy, T. W. [1938]: Linear partial differential equations. I. *Quart. J. Math. Oxford Ser.* **9**, 234–240.

Chernoff, H. [1946]: Complex solutions of partial differential equations. *Amer. J. Math.* **68**, 455–478.

Cherry, T. M. [1951]: Relation between Begman's and Chaplygin's methods of solving the hodograph equation. *Quart. Appl. Math.* **9**, 92–94.

Cholewinski, F. M., and D. T. Haimo [1968]: Classical analysis and the generalized heat equation. *SIAM Rev.* **10**, 67–80.

Cohn, H. [1947]: The Riemann function for $\partial^2 u/\partial x\, \partial y + H(x+y)u = 0$. *Duke Math. J.* **14**, 297–304.

Cohn, H. [1970]: Note on when a Riemann function can be obtained from first stage iteration. In *Studies and Essays* (Presented to Yu-why Chen on his 60th birthday), pp. 49–52. Taipei: Mathematics Research Center, National Taiwan University.

Cohn, H. [1973]: A functionally stable iteration process for hyperbolic equations. *J. Math. Phys. Sci.* **7**, 341–347.

Colton, D. [1969]: A contribution to the analytic theory of partial differential equations. *J. Differential Equations* **5**, 117–135.

Colton, D. [1972]: Bergman operators for elliptic equations in four independent variables. *SIAM J. Math. Anal.* **3**, 401–412.

Colton, D. [1975a]: Complete families of solutions for parabolic equations with analytic coefficients. *SIAM J. Math. Anal.* **6**, 937–947.

Colton, D. [1975b]: Runge's theorem for parabolic equations in two space variables. *Proc. Roy. Soc. Edinburgh Sect. A* **73**, 307–315.

Colton, D. [1976a]: Walsh's theorem for the heat equation. In W. N. Everitt and B. D. Sleeman (Eds.): *Ordinary and Partial Differential Equations* (Proc. 4th Conf., Dundee, Scotland, 1976), pp. 54–60. New York: Springer.

Colton, D. [1976b]: The approximation of solutions to initial boundary value problems for parabolic equations in one space variable. *Quart. Appl. Math.* **33**, 377–386.

Colton, D. L. [1976c]: *Solution of Boundary Value Problems by the Method of Integral Operators*. London: Pitman.

Colton, D. L. [1976d]: *Partial Differential Equations in the Complex Domain*. London: Pitman.

Colton, D. L. [1980]: *Analytic Theory of Partial Differential Equations*. London: Pitman.

Colton, D., and R. P. Gilbert [1968]: Singularities of solutions to elliptic partial differential equations with analytic coefficients. *Quart. J. Math. Oxford (2)* **19**, 391–396.

Colton, D., and R. P. Gilbert [1970]: Nonlinear, analytic partial differential equations with generalized Goursat data. *Duke Math. J.* **37**, 367–376.

Colton, D. L., and R. P. Gilbert [1971]: New results on the approximation of solutions to partial differential equations. In P. F. Hsieh and A. W. J. Stoddart [1971], pp. 207–212, 213–220.

Colton, D. L., and R. P. Gilbert (Eds.) [1974]: *Constructive and Computational Methods for Differential and Integral Equations* (Sympos., Indiana Univ., 1974). New York: Springer.

Colton, D., and R. Kress [1983]: *Integral Equation Methods in Scattering Theory*. New York: Wiley.

Colton, D., and W. Watzlawek [1977]: Complete families of solutions to the heat equation and generalized heat equation in R^n. *J. Differential Equations* **25**, 96–107.

Conway, J. B. [1973]: *Functions of One Complex Variable*. New York: Springer.

Copson, E. T. [1958]: On the Riemann–Green function. *Arch. Rational Mech. Anal.* **1**, 324–348.

Copson, E. T. [1970]: On Hadamard's elementary solution. *Proc. Roy. Soc. Edinburgh Sect. A* **69**, 19–27.

Copson, E. T. [1971]: Hadamard's elementary solution and Frobenius's method. *SIAM Rev.* **13**, 222–230.

Courant, R., and K. O. Friedrichs [1976]: *Supersonic Flow and Shock Waves*, reprint. New York: Springer.

Courant, R., and D. Hilbert [1962]: *Methods of Mathematical Physics*, vol. II. New York: Wiley.

Daggit, E. [1970]: The use of infinitesimal transformations in predicting the form of the Riemann(–Green) function. *J. Math. Anal. Appl.* **29**, 91–108.

Danilyuk, Ī. Ī. [1975]: *Nonregular Boundary Value Problems in the Plane*. Moscow: Nauka. (In Russian.)

Darboux, G. [1972]: *Leçons sur la théorie génèrale des surfaces*, vol. II. 3rd ed. New York: Chelsea.

Davis, R. M. [1956]: On a regular Cauchy problem for the Euler–Poisson–Darboux equation. *Ann. Mat. Pura Appl. (4)* **42**, 205–226.

Delanghe, R. [1969]: On the solutions of $\Box u = 0$. *Bull. Soc. Roy. Sci. Liège* **38**, 605–609.

Delanghe, R. [1970]: On regular-analytic functions with values in a Clifford algebra. *Math. Ann.* **185**, 91–111.

Delsarte, J. [1938]: Sur certains transformations fonctionnelles relatives aux équations linéaires aux dérivées partielles du second ordre. *C. R. Acad. Sci. Paris* **206**, 1780–1782.

Delsarte, J., and J. Lions [1957]: Transmutations d'opérateurs différentiels dans le domaine complexe. *Comment. Math. Helv.* **32**, 113–128.

Dettman, J. W. [1970]: The wave, Laplace and heat equations and related transforms. *Glasgow Math. J.* **11**, 117–125.

Diaz, J. B., and G. S. S. Luford [1955]: On two methods of generating solutions of linear partial differential equations by means of definite integrals. *Quart. Appl. Math.* **12** (1954/55), 422–427.

Diaz, J. B., and G. S. S. Ludford [1956]: On a theorem of Le Roux. *Canad. J. Math.* **8**, 82–85.

Diaz, J. B., and G. S. S. Ludford [1957]: On the integration methods of Bergman and Le Roux. *Quart. Appl. Math.* **14** (1956/57), 428–432.

Dienes, P. [1957]: *The Taylor Series*, 1st ed., reprint. New York: Dover.

Dieudonné, J. [1978]: L'évolution récente de la théorie des équations linéaires aux dérivées partielles. *Actas del V Congreso de la Agrupación de Matemáticos de Expresión Latina* (Madrid, 1978), pp. 104–113.

Dimiev, S. [1983]: Fonctions presque holomorphes. *Complex Analysis* (Warsaw, 1979), pp. 61–75. Banach Center Publication 11, PWN, Warsaw.

Dinghas, A. [1961]: *Vorlesungen über Funktionentheorie*. Berlin: Springer.

Dont, M. [1973]: Sets of removable singularities of an equation. *Acta Univ. Carolin.—Math. Phys.* **14**, 23–30.

Douglis, A. [1953]: A function-theoretic approach to elliptic systems of equations in two variables. *Comm. Pure Appl. Math.* **6**, 259–289.

Douglis, A., and L. Nirenberg [1955]: Interior estimates for elliptic systems of partial differential equations. *Comm. Pure Appl. Math.* **8**, 503–538.

Du, X. H. [1981]: A note on Olevski's formula for selfadjoint hyperbolic equations of the second order. *Kexue Tongbao (English)* **26**, 869–873.

Duff, G. F. D., and D. Naylor [1966]: *Differential Equations of Applied Mathematics*. New York: Wiley.

Ďurikovič, V. [1968]: On the uniqueness of solutions and the convergence of successive approximations in the Darboux problem for certain differential equations of the type $u_{xy} = f(x, y, u, u_x, u_y)$. *Arch. Math. (Brno)* **4**, 223–235.

Ecker, K., and H. Florian [1967]: Exponentialoperatoren bei Differengleichungen mit n Variablen. *Inst. Angew. Math. Tech. Univ. Graz Ber.* **67-5**, iii + 13 pp.

Ecker, K., and H. Florian [1969]: Über Integraloperatoren bei Differentialgleichungen mit n Variablen. *Glas. Mat.* **4**(24), 79–87.

Edenhofer, J. [1973]: Analytische Funktionen auf Algebren. Thesis, Technical University of München, iii + 81 pp.

Edenhofer, J. [1976]: A solution of the biharmonic Dirichlet problem by means of hypercomplex analytic functions. In V. E. Meister, N. Weck, and W. L. Wendland [1976], pp. 192–202.

Eichhorn, W. [1961]: Funktionentheorie in Algebren über dem reellen Zahlkörper und ihre Anwendung auf partielle Differentialgleichungen. Thesis, University of Würzburg, 117 pp.

Eichhorn, W. [1969]: Eine aus Fragen der Verallgemeinerung der Funktionentheorie erwachsende Charakterisierung der Algebra der komplexen Zahlen. *Jahresber. Deutsch. Math.-Verein.* **71**, 123–137.

Eichler, M. [1942]: Allgemeine Integration linearer partieller Differentialgleichungen von elliptischem Typ bei zwei Grundvariablen. *Abh. Math. Sem. Univ. Hamburg* **15**, 179–210.

Eichler, M. [1949]: On the differential equation $u_{xx} + u_{yy} + M(x)u = 0$. *Trans. Amer. Math. Soc.* **65**, 259–278.

Eichler, M. [1950]: Eine Modifikation der Riemannschen Integrationsmethode bei partiellen Differentialgleichungen vom hyperbolischen Typ. *Math. Z.* **53**, 1–10.

Eisenstat, S. C. [1974]: On the rate of convergence of the Bergman–Vekua method for the numerical solution of elliptic boundary value problems. *SIAM J. Numer. Anal.* **11**, 654–680.

Eke, B. G. [1970]: Value distribution of regular functions of two complex variables. *Ann. Polon. Math.* **23**, 151–158.

Erdélyi, A., W. Magnus, F. Oberhettinger, and F. G. Tricomi (Eds.) [1953–1955]: *Higher Transcendental Functions*, vols. I–III. New York: McGraw-Hill.

Falkovich, S. V. [1946]: On the theory of the Laval nozzle. *Prikl. Mat. Mekh.* **10**, 503–512. (In Russian.)

Falkovich, S. V. [1947]: A class of Laval nozzles. *Prikl. Mat. Mekh.* **11**, 223–230. (In Russian.)

Fichera, G. [1979]: The problem of the completeness of systems of particular solutions of partial differential equations. In: R. Ansorge, K. Glashoff, and B. Werner (Eds.): *Numerical Mathematics* (Sympos., Inst. Appl. Math., Univ. Hamburg, 1979), pp. 25–41. Boston: Birkhäuser.

Florian, H. [1965]: Normale Integraloperatoren. *Monatsh. Math.* **69**, 18–29.

Florian, H., and R. Heersink [1974]: Über eine partielle Differentialgleichung mit $p + 2$ Variablen und deren Zusammenhang mit den allgemeinen Kugelfunktionen. *Manuscripta Math.* **12**, 339–349.

Florian, H., and R. Heersink [1979]: Ein verallgemeinertes Riemann–Hilbertsches Randwertproblem. *Applicable Anal.* **8** (1978/79), 277–284.

Florian, H., and G. Jank [1971]: Polynomerzeugende bei einer Klasse von Differentialgleichungen mit zwei unabhängigen Variablen. *Monatsh. Math.* **75**, 31–37.

Florian, H., and J. Püngel [1979]: Riemann-Funktionen als Lösungen gewöhnlicher Differentialgleichungen. *Ber. Math.-Statist. Sekt. Forschungszentrum Graz* **106**, 47 pp.

Florian, H., and J. Püngel [1981]: Riemann-Funktionen für Gleichungen vom Typ $u_{z\zeta} + h(a + bz + c\zeta + dz\zeta)u = 0$. *Ber. Math.-Statist. Sekt. Forschungszentrum Graz* **176**, 31 pp.

Florian, H., J. Püngel, and W. Tutschke [1985]: Complex methods for linear elliptic equations in formally hyperbolic canonical form. *Ber. Math.-Statist. Sekt. Forschungszentrum Graz* **258**, 28 pp.

Florian, H., J. Püngel, and H. Wallner [1983a]: Über die Riemannfunktion für Gleichungen höherer Ordnung. *Ber. Math.-Statist. Sekt. Forschungszentrum Graz* **203**, 30 pp.

Florian, H., J. Püngel, and H. Wallner [1983b]: Darstellungen von Riemannfunktionen für $\partial^n w/\partial z_1 \partial z_2 \cdots \partial z_n + C(z_1, z_2, \cdots, z_n)w = 0$. *Ber. Math.-Statist. Sekt. Forschungszentrum Graz* **204**, 29 pp.

Frank, G., and W. Hennekemper [1981]: Einige Ergebnisse über die Werteverteilung meromorpher Funktionen und ihrer Ableitungen. *Resultate Math.* **4**, 39–54.

Frankl, F. I. [1956]: Subsonic flow about a profile with a supersonic zone. *Prikl. Mat. Mekh.* **20**, 196–202.

Friedlander, F. G., and A. E. Heins [1969]: On the representation theorems of Poisson, Riemann and Volterra for the Euler–Poisson–Darboux equation. *Arch. Rational Mech. Anal.* **33**, 219–230.

Friedman, A. [1964]: *Partial Differential Equations of Parabolic Type*. Englewood Cliffs, N.J.: Prentice-Hall.

Fryant, A. J. [1977]: Growth and complete sequences of generalized axisymmetric potentials. *J. Approximation Theory* **19**, 361–370.

Fryant, A. J. [1979]: Growth and complete sequences of generalized bi-axially symmetric potentials. *J. Differential Equations* **31**, 155–164.

Fueter, R. [1934]: Die Funktionentheorie der Differentialgleichungen $\Delta u = 0$ und $\Delta\,\Delta u = 0$ mit vier reellen Variablen. *Comment. Math. Helv.* **7**, 307–330.

Fueter, R. [1948]: Über die Funktionentheorie in einer hyperkomplexen Algebra. *Elem. Math.* **3**, 89–94.

Gaier, D. [1980]: *Vorlesungen über Approximation im Komplexen*. Basel: Birkhäuser.

Gaier, D. [1984]: Approximation im Komplexen. *Jahresber. Deutsch. Math.-Verein.* **86**, 151–159.

Gakhov, F. D. [1966]: *Boundary Value Problems*. Oxford: Pergamon Press.

Ganin, M. P. [1957]: Dirichlet's problem for the equation $\Delta U + 4n(n+1)(1+x^2+y^2)^{-2}U = 0$. *Uspekhi Mat. Nauk* **12**, 205–209. (In Russian.)

Garabedian, P. R. [1954]: Application of analytic continuation of boundary value problems. *J. Rational Mech. Anal.* **3**, 383–393.

Garabedian, P. R. [1968]: Analytic methods for the numerical computation of fluid flows. In R. P. Gilbert and R. G. Newton [1968], pp. 169–184.

Garabedian, P., and D. Korn [1976]: A systematic method for computer design of supercritical airfoils in cascade. *Comm. Pure Appl. Math.* **29**, 369–382.

Geddes, R. L., and A. G. Mackie [1977]: Riemann functions for self-adjoint equations. *Applicable Anal.* **7**, 43–47.

Germain, P., and R. Bader [1953]: Solutions élémentaires de certaines équations aux dérivées partielles du type mixte. *Bull. Soc. Math. France* **81**, 145–174.

Gilbert, R. P. [1960a]: On the singularities of generalized axially symmetric potentials. *Arch. Rational Mech. Anal.* **6**, 171–176.

Gilbert, R. P. [1960b]: Singularities of solutions of the wave equation. *J. Reine Angew. Math.* **205**, 75–81.

Gilbert, R. P. [1960c]: Singularities of three-dimensional harmonic functions. *Pacific J. Math.* **10**, 1243–1255.

Gilbert, R. P. [1964]: On generalized axially symmetric potentials whose associates are distributions. *Scripta Math.* **27**, 245–256.

Gilbert, R. P. [1969a]: A method of ascent for solving boundary value problems. *Bull. Amer. Math. Soc.* **75**, 1286–1289.

Gilbert, R. P. [1969b]: *Function Theoretic Methods in Partial Differential Equations*. New York: Academic Press.

Gilbert, R. P. [1970a]: The construction of solutions for boundary value problems by function theoretic methods. *SIAM J. Math. Anal.* **1**, 96–114.

Gilbert, R. P. [1970b]: Integral operator methods for approximating solutions of Dirichlet problems. In L. Collatz, G. Meinardus, H. Unger, and H. Werner (Eds.): *Iterationsverfahren, Numerische Mathematik, Approximationstheorie* (Vortragsauszüge Tagungen Oberwolfach, 1968–1969), pp. 129–146. Basel: Birkhäuser.

Gilbert, R. P. [1973]: Pseudohyperanalytic function theory. In S. Ruscheweyh [1973b], pp. 53–63.

Gilbert, R. P. [1974]: *Constructive Methods for Elliptic Equations*. New York: Springer.

Gilbert, R. P. (Ed.) [1982]: *Plane Ellipticity and Related Problems* (Proc. Special Session, 87th Meet. Amer. Math. Soc., San Francisco, 1981). Providence, R.I.: American Mathematical Society.

Gilbert, R. P. [1983]: Recent results and development in generalized hyperanalytic function theory. In E. Lanckau and W. Tutschke [1983], pp. 174–184.

Gilbert, R. P., and K. E. Atkinson [1970]. Integral operator methods for approximating solutions of Dirichlet problems. *Sér. Internat. Anal. Numér. Suisse* **15**, 129–146.

Gilbert, R. P., and J. L. Buchanan [1983]: *First Order Elliptic Systems. A Function Theoretic Approach*. New York: Academic Press.

Gilbert, R. P., and D. L. Colton [1971]: On the numerical treatment of partial differential equations by function theoretic methods. In B. Hubbard (Ed.): *Numerical Solutions of Partial Differential Equations II*, pp. 273–326. New York: Academic Press.

Gilbert, R. P., and J. Conlan [1975]: Iterative schemes for almost linear elliptic systems with nonlinear initial data. *J. Reine Angew. Math.* **277**, 193–206.

Gilbert, R. P., and G. Hile [1974]: Generalized hypercomplex function theory. *Trans. Amer. Math. Soc.* **195**, 1–29.

Gilbert, R. P., and H. C. Howard [1965]: On solutions of the generalized bi-axially symmetric Helmholtz equation generated by integral operators. *J. Reine Angew. Math.* **218**, 109–120.

Gilbert, R. P., and H. C. Howard [1967]: Role of the integral operator method in the theory of potential scattering. *J. Math. Phys.* **8**, 41–148.

Gilbert, R. P., H. C. Howard, and S. Aks [1965]: Singularities of analytic functions having integral representations, with a remark about the elastic unitary integral. *J. Math. Phys.* **6**, 1157–1162.

Gilbert, R. P., and G. S. Hsiao [1976]: Constructive function theoretic methods for higher order pseudoparabolic equations. In V. E. Meister, N. Weck, and W. L. Wendland [1976], pp. 51–67.

Gilbert, R. P., R. B. Knight, and C. Y. Lo [1969]: Nonlinear, analytic partial differential equations with generalized Goursat data. II. *Bul. Inst. Politehn. Iaşi (N.S.)* **14**(18) (1968), 39–44.

Gilbert, R. P., and D. K. Kukral [1974]: Function theoretic methods for elliptic equations with more than four variables. *Rend. Sem. Mat. Univ. Politec. Torino* **31** (1971/73), 275–285.

Gilbert, R. P., and D. Kukral [1975]: A function theoretic method for $\Delta_3^2 u + Q(x)u = 0$. *Math. Nachr.* **67**, 199–212.

Gilbert, R. P., and C. Y. Lo [1971]: On the approximation of solutions of elliptic partial differential equations in two and three dimensions. *SIAM J. Math. Anal.* **2**, 17–30.

Gilbert, R. P., and R. G. Newton (Eds.) [1968]: *Analytic Methods in Mathematical Physics* (Conf. Indiana Univ., Bloomington, 1968). New York: Gordon and Breach.

Gilbert, R. P., and M. Schneider [1976]: Generalized meta and pseudo parabolic equations in the plane. University of Delaware Tech. Report G-2. i + 29 pp.

Gilbert, R. P., and L. Wei [1983]: Algorithms for generalized Cauchy kernels. *Complex Variables Theory Appl.* **2**, 103–124.

Gilbert, R. P., and R. J. Weinacht (Eds.) [1976]: *Function Theoretic Methods in Differential Equations*. London: Pitman.

Gilbert, R. P., and W. L. Wendland [1974–1975]: Analytic, generalized, hyperanalytic function theory and an application to elasticity. *Proc. Roy. Soc. Edinburgh Sect. A* **73**, 317–331.

Goldschmidt, B. [1980]: *Verallgemeinerte analytische Vektoren im R^n*. Habilitationsschrift, Martin-Luther-University, Halle-Wittenberg.

Goluzin, G. M. [1969]: *Geometric Theory of Functions of a Complex Variable*. Providence, R.I.: American Mathematical Society.

Goman, O. G. [1984]: Representation of p-analytic functions by analytic ones. *Ukrainian Math. J.* **36**, 93–97.

Görtler, H. [1939]: Zum Übergang von Unterschall- zu Überschallgeschwindigkeiten in Düsen. *Z. Angew. Math. Mech.* **19**, 325–337.

Görtler, H. [1940]: Gasströmungen mit Übergang von Unterschall- zu Überschallgeschwindigkeiten. *Z. Angew. Math. Mech.* **20**, 254–262.

Grabarska, H. [1983]: Integral operators for systems of parabolic equations. *Demonstratio Math.* **16**, 787–796.

Grauert, H., and K. Fritzsche [1976]: *Several Complex Variables*. New York: Springer.

Gronau, D. [1976]: Logarithmenfreie Singularitäten von linearen partiellen Differentialgleichungen. *Ber. Math.-Statist. Sekt. Forschungszentrum Graz* **51**, ii + 17 pp.

Gronau, D. [1981]: Meromorphic solutions and solutions by differential operators of linear partial differential equations with constant coefficients. In *Qualitative Theory of Differential Equations* (Szeged, 1979), vols. I and II, pp. 307–323. Amsterdam: North-Holland.

Gronau, D. [1985]: Existence and uniqueness theorems for partial differential equations in complex domains. *Ber. Math.-Statist. Sekt. Forschungszentrum Graz* **246**, 21 pp.

Gronau, D., and L. Reich [1975]: Über lokale regulär-singuläre Lösungen einer Klasse linearer partieller Differentialgleichungen zweiter Ordnung im Komplexen. *Österreich. Akad. Wiss. Math.-Natur. Kl. Sitzungsber. II* **183**, 321–333.

Guderley, K. G. [1947]: *Singularities at the Sonic Velocity*. Tech. Rep. F-TR-1171-ND. Wright Field, Dayton, OH.: Headquarters Air Materiel Command.

Guderley, G. [1956]: On the development of solutions of Tricomi's differential equation in the vicinity of the origin. *J. Rational Mech. Anal.* **5**, 747–790.

Guderley, K. G. [1962]: *The Theory of Transonic Flow*. Oxford: Pergamon Press.

Guderley, K. G., and H. Yoshihara [1950]: The flow over a wedge profile at Mach number 1. *J. Aeronaut. Sci.* **17**, 723–735.

Gunning, R. C., and H. Rossi [1965]: *Analytic Functions of Several Complex Variables*. Englewood Cliffs, N.J.: Prentice-Hall.

Gutknecht, M. H . [1983]: On complex rational approximation. I, II. In *Computational Aspects of Complex Analysis* (Proc. Conf. Braunlage, 1982), pp. 79–132. Dordrecht: Reidel.

Haack, W., and W. Wendland [1969]: *Vorlesungen über partielle und Pfaffsche Differentialgleichungen*. Basel: Birkhäuser.

Habetha, K. [1967]: Über die Werteverteilung pseudoanalytischer Funktionen. *Ann. Acad. Sci. Fenn. Ser. A I* **406**, 20 pp.

Habetha, K. [1973]: Über lineare elliptische Differentialgleichungssysteme mit analytischen Koeffizienten. In S. Ruscheweyh [1973b], pp. 65–89.

Habetha, K. [1976]: Eine Bemerkung zur Funktionentheorie in Algebren. In V. E. Meister, N. Weck, and W. L. Wendland [1976], pp. 502–509.

Habetha, K. [1983]: Function theory in algebras. In E. Lanckau and W. Tutschke [1983], pp. 225–237.

Hadamard, J. [1952]: *Lectures on Cauchy's Problem in Linear Partial Differential Equations*, reprint. New York: Dover.

Haimo, D. T. [1973]: Widder temperature representations. *J. Math. Anal. Appl.* **41**, 170–178.

Harke, O. [1981]: Untersuchungen zum Picardschen Satz für verallgemeinerte analytische Funktionen. *Wiss. Z. Martin-Luther-Univ. Halle-Wittenberg Math.-Natur. Reihe* **30**, 29–37.

Hayman, W. K. [1964]: *Meromorphic Functions*. Oxford: Oxford University Press.

Heersink, R. [1972]: Spezielle Operatoren zur Lösung partieller Differentialgleichungen. *Ber. Math.-Statist. Sekt. Forschungszentrum Graz* **72-2**, iii + 61 pp.

Heersink, R. [1976a]: Über Lösungsdarstellungen und funktionentheoretische Methoden bei elliptischen Differerntialgleichungen. *Ber. Math.-Statist. Sekt. Forschungszentrum Graz* **67**, i + 79 pp.

Heersink, R. [1976b]: Characterization of certain differential operators in the solution of linear partial differential equations. *Glasgow Math. J.* **17**, 83–88.

Heersink, R. [1983]: Zur Charakterisierung spezieller Lösungsdarstellungen für elliptische Gleichungen. *Österreich. Akad. Wiss. Math.–Natur. Kl. Sitzungsber. II* **192**, 267–293.

Heersink, R. [1985]: Über das Randverhalten von gewissen verallgemeinerten holomorphen Funktionen. *Complex Variables Theory Appl.* **4**, 119–136.

Heins, A. E. [1965]: Axially symmetric boundary value problems. *Bull. Amer. Math. Soc.* **71**, 787–808.

Heins, A. E., and R. C. MacCamy [1963]: Integral representations of axially symmetric potential functions. *Arch. Rational Mech. Anal.* **13**, 371–385.

Heins, M. [1968]: *Complex Function Theory*. New York: Academic Press.

Henrici, P. [1952]: Bergmans Integraloperator erster Art und Riemannsche Funktion. *Z. Angew. Math. Phys.* **3**, 228–232.

Henrici, P. [1953]: Zur Funktionentheorie der Wellengleichung. *Comment. Math. Helv.* **27**, 235–293.

Henrici, P. [1957]: A survey of I. N. Vekua's theory of elliptic partial differential equations with analytic coefficients. *Z. Angew. Math. Phys.* **8**, 169–203.

Henrici, P. [1960]: Complete systems of solutions for a class of singular elliptic partial differential equations. In R. E . Langer, (Ed.): *Boundary Problems in Differential Equations*, pp. 19–34. Madison, WI.: Univ. Wisconsin Press.

Henrici, P. [1974–1986]: *Applied and Computational Complex Analysis*, vols. I–III. New York: Wiley.

Hilb, E., and O. Szász [1922]: Allgemeine Reihenentwicklungen. *Encyklopädie Math. Wiss*. II C 11, pp. 1229–1276. Leipzig: Teubner.

Hill, C. D. [1968]: A method for the construction of reflection laws for a parabolic equation. *Trans. Amer. Math. Soc.* **133**, 357–372.

Hille, E. [1959–1962]: *Analytic Function Theory*, 2 vols. Boston: Ginn.

Hobson, E. W. [1931]: *The Theory of Spherical and Ellipsoidal Harmonics*. London: Cambridge University Press.

Hoefer, E. T. [1974]: Properties of Solutions of Linear Partial Differential Equations Given by Bergman Integral Operators. Ph.D. Thesis. SUNY at Buffalo. iv + 105pp.

Holland, A. S. B. [1973]: *Introduction to the Theory of Entire Functions*. New York: Academic Press.

Hörmander, L. [1983–1985]: *The Analysis of Linear Partial Differential Operators*, vols. I–IV. New York: Springer.

Horváth, J. [1963]: A generalization of the Cauchy–Riemann equations. In *Contributions to Differential Equations*, vol. I, pp. 39–58. New York: Interscience/Wiley.

Hsieh, P. F., and A. W. J. Stoddart (Eds.) [1971]: *Analytic Theory of Differential Equations* (Proc. Conf., Kalamazoo, 1970). New York: Springer.

Hua, L. K. [1963]: *Harmonic Analysis of Functions of Several Complex Variables in the Classical Domains*. Providence, R.I.: American Mathematical Society.

Ince, E. L. [1956]: *Ordinary Differential Equations*. New York: Dover.

Ingersoll, B. M. [1948]: The regularity domains of solutions of linear partial differential equations in terms of the series development of the solution. *Duke Math. J.* **15**, 1045–1056.

Jank, G. [1971]: *Funktionentheoretische Methoden bei partiellen Differentialgleichungen*. Habilitationsschrift, Technische Hochschule Graz, ii + 60 pp.

Jank, G. [1973]: Integral- und Differentialoperatoren bei einer Differentialgleichung mit mehreren komplexen Veränderlichen. *Period. Math. Hungar.* **3**, 305–312.

Jank, G. [1979]: Lösungen der Euler–Darboux–Gleichung, die über die reelle Achse fortsetzbar sind. *Math. Nachr.* **91**, 375–387.

Jank, G., and S. Ruscheweyh [1973]: Eine Bemerkung zur Darstellung gewisser pseudoanalytischer Funktionen. *Ber. Ges. Math. Datenverarb. Bonn* **75**, 17–19.

Jank, G., and S. Ruscheweyh [1975]: Funktionenfamilien mit einem Maximumprinzip und elliptische Differentialgleichungen. II. *Monatsh. Math.* **79**, 103–113.

Jank, G., and L. Volkmann [1985]: *Einführung in die Theorie der ganzen und meromorphen Funktionen mit Anwendungen auf Differentialgleichungen*. Basel: Birkhäuser.

Jank, G., and K.-J. Wirths [1976]: Generalized maximum principles in certain classes of pseudoanalytic functions. In R. P. Gilbert and R. J. Weinacht [1976], pp. 63–67.

Kaplan, S. [1979]: Formal and convergent power series solutions of singular partial differential equations. *Trans. Amer. Math. Soc.* **256**, 163–183.

Kapoor, G. P., and A. Nautiyal [1984]: On the approximation of generalized axisymmetric potentials. *Glas. Mat. Ser. III* **19**(39), 123–133.

Kaucher, E. W., and W. L. Miranker [1984]: *Self-Validating Numerics for Function Space Problems. Computation with Guarantees for Differential and Integral Equations.* New York: Academic Press.

Kemnitz, H. [1985]: Polynomial expansions for solutions of the system $D^k_{X_1}U(X_1, \cdots, X_r) = D_{X_k}U(X_1, \cdots, X_r)$. *SIAM J. Math. Anal.* **16**, 379–391.

Kleinman, R. E. [1961]: Integral Representations of Solutions of the Helmholtz Equation with Application to Diffraction by a Strip. Thesis, Technische Hochschule Delft. v + 123 pp.

Klink, H.-K. [1984]: Über komplexe Methoden in der Kontrolltheorie: Randgesteuerte Probleme. *Math. Nachr.* **118**, 201–208.

Kneis, G. [1970a]: Eine kanonische Gestalt für indefinite quadratische Differentialformen und globale Darstellungen für negativgekrümmte Flächen im R^3. Thesis, University of Halle-Wittenberg.

Kneis, G. [1970b]: Über eine Darstellung für negativ gekrümmte Flächen durch die Lösungen einer verallgemeinerten Laplaceschen Differentialgleichung. *Math. Nachr.* **47**, 40–45.

Korevaar, J. [1980]: Polynomial and rational approximation in the complex domain. In *Aspects of Contemporary Complex Analysis* (Proc. Conf. Durham, 1979), pp. 251–292. New York: Academic Press.

Kracht, M. [1970]: Integraloperatoren für die Helmholtzgleichung. *Z. Angew. Math. Mech.* **50**, 389–396.

Kracht, M. [1974a]: Zur Existenz und Charakterisierung von Bergman-Operatoren. I, II. *J. Reine Angew. Math.* **265**, 202–220; **266**, 140–158.

Kracht, M. [1974b]: *Über Bergman-Operatoren für lineare partielle Differentialgleichungen zweiter Ordnung.* Habilitationsschrift, University of Düsseldorf. iii + 111 pp.

Kracht, M. [1976]: Spezielle Bergman-Operatoren nebst Anwendungen. In V. E. Meister, N. Weck, and W. L. Wendland [1976], pp. 327–341.

Kracht, M., and E. Kreyszig [1969]: Bergman-Operatoren mit Polynomen als Erzeugenden. *Manuscripta Math.* **1**, 369–376.

Kracht, M., and E. Kreyszig [1970]: Zur Darstellung von Lösungen der Gleichung $\Delta\Psi + c(1 + x^2 + y^2)^{-2}\Psi = 0$. *Z. Angew. Math. Mech.* **50**, 375–380.

Kracht, M., and E. Kreyszig [1975]: Zur Konstruktion gewisser Integraloperatoren für partielle Differentialgleichungen. I, II. *Manuscripta Math.* **17**, 79–103 and 171–186.

Kracht, M., and E. Kreyszig [1979]: Constructive methods for certain Bergman operators. *Applicable Anal.* **8** (1978/79), 233–252.

Kracht, M., and E. Kreyszig [1986a]: The Tricomi equation and transition problems. In J. M. Rassias (Ed.): *Mixed Type Equations.* Leipzig: Teubner.

Kracht, M., and E. Kreyszig [1986b]: The integral operator approach to the Tricomi equation. In J. M. Rassias (Ed.): *Mixed Type Equations.* Leipzig: Teubner.

Kracht, M., and E. Kreyszig [1988]: On a method of generating the complex Riemann(–Vekua) function. In preparation.

Kracht, M., E. Kreyszig, and G. Schröder [1981]: Definitionsprinzipien für Operatoren Bergmanscher Art und einige Anwendendungen. *Elem. Math.* **36**, 25–36.

Kracht, M., E. Kreyszig, and G. Schröder [1982]: Integral operators in the function theoretic method for partial differential equations. *SIAM Rev.* **24**, 52–71.

Kracht, M., and G. Schröder [1973]: Bergmansche Polynom-Erzeugende erster Art. *Manuscripta Math.* **9**, 333–355.

Kraft, R. [1969]: Riemann functions for linear systems in two independent variables. *J. Math. Anal. Appl.* **25**, 74–79.

Krajkiewicz, P. [1973]: The Picard theorem for multianalytic functions. *Pacific J. Math.* **48**, 423–439.

Kratzer, A., and W. Franz [1960]: *Transzendente Funktionen*. Leipzig: Akad. Verlagsges. Geest und Portig.

Kreyszig, E. [1955]: On a class of partial differential equations. *J. Rational Mech. Anal.* **4**, 907–923.

Kreyszig, E. [1956]: On certain partial differential equations and their singularities. *J. Rational Mech. Anal.* **5**, 805–820.

Kreyszig, E. [1957a]: Relations between properties of solutions of partial differential equations and the coefficients of their power series development. *J. Rational Mech. Anal.* **6**, 361–381.

Kreyszig, E. [1957b]: On coefficient problems of solutions of partial differential equations of the fourth order. *J. Rational Mech. Anal.* **6**, 811–822.

Kreyszig, E. [1958a]: On some relations between partial and ordinary differential equations. *Canad. J. Math.* **10**, 183–190.

Kreyszig, E. [1958b]: Coefficient problems in systems of partial differential equations. *Arch. Rational Mech. Anal.* **1**, 283–294.

Kreyszig, E. [1958c]: On singularities of solutions of partial differential equations in three variables. *Arch. Rational Mech. Anal.* **2**, 151–159.

Kreyszig, E. [1960]: Zur Behandlung elliptischer partieller Differentialgleichungen mit funktionentheoretischen Methoden. *Z. Angew. Math. Mech.* **40**, 334–342.

Kreyszig, E. [1962]: Über den Aufbau der Theorie elliptischer partieller Differentialgleichungen auf funktionentheoretischer Grundlage. *Bul. Inst. Politehn. Iaşi (N. S.)* **8**(12), 19–46.

Kreyszig, E. [1963]: Kanonische Integraloperatoren zur Erzeugung harmonischer Funktionen von vier Veränderlichen. *Arch. Math. (Basel)* **14**, 193–203.

Kreyszig, E. [1964]: *Differential Geometry*, revised ed. Toronto: Univ. Toronto Press.

Kreyszig, E. [1968]: Über zwei Klassen Bergmanscher Operatoren. *Math. Machr.* **37**, 197–202.

Kreyszig, E. [1970]: Bergman-Operatoren der Klasse P. *Monatsh. Math.* **74**, 437–444.

Kreyszig, E. [1971]: On Bergman operators for partial differential equations in two variables. *Pacific J. Math.* **36**, 201–208.

Kreyszig, E. [1972]: Zur Theorie der Exponentialoperatoren und der Bergman-Operatoren erster Art. *Bul. Inst. Politehn. Iaşi Sect. I* **18**(22), 127–137.

Kreyszig, E. [1973]: Gewöhnliche Differentialgleichungen für Erzeugende gewisser Bergman-Operatoren. *J. Reine Angew. Math.* **262/263**, 74–81.

Kreyszig, E. [1975]: Representations of solutions of certain partial differential equations related to Liouville's equation. *Abh. Math. Sem. Univ. Hamburg* **44**, 32–44.

Kreyszig, E. [1978]: *Introductory Functional Analysis With Applications*. New York: Wiley.

Kreyszig, E. [1980]: Function theoretic integral operator methods for partial differential equations. *Canad. Math. Bull.* **23**, 127–135.

Kreyszig, E. [1982]: On the construction of a class of Bergman kernels for partial differential equations. *Abh. Math. Sem. Univ. Hamburg* **52**, 120–132.

Kreyszig, E. [1988]: *Advanced Engineering Mathematics*, 6th ed. New York: Wiley.

Kriszten, A. [1952]: Hyperkomplexe und pseudo-analytische Funktionen. *Comment. Math. Helv.* **26**, 6–35.

Krzywoblocki, M. Z. von [1960]: *Bergman's Linear Integral Operator Method in the Theory of Compressible Fluid Flow*. Wien: Springer.

Krzywoblocki, M. Z. von [1963]: Association of partial differential equations with ordinary ones and its applications. *International Symposium on Nonlinear Differential Equations and Nonlinear Mechanics*, pp. 231–240.

Krzywoblocki, M. Z. von [1968]: Bergman's and Gilbert's operators in elasticity, electromagnetism, fluid dynamics, wave mechanics. In R. P. Gilbert and R. G. Newton [1968], pp. 207–247.

Kühn, E. [1974]: Über die Funktionentheorie und das Ähnlichkeitsprinzip einer Klasse elliptischer Differentialgleichungssysteme in der Ebene. Thesis, University of Dortmund, iii + 51 pp.

Kühnau, R. [1973]: Über drehstreckungssymmetrische Potentiale. *Math. Nachr.* **56**, 201–205.

Kukral, D. K. [1973]: On a Bergman–Whittaker type operator in five or more variables. *Proc. Amer. Math. Soc.* **39**, 122–124.

Kürcz, A. [1978]: Zur Lösung parabolischer Gleichungen und Systeme mittels Integraloperatoren. *Wiss. Z. Tech. Hochsch. Karl-Marx-Stadt* **20**, 499–509.

Lammel, E. [1950–1951]: Über eine zur Differentialgleichung $(a_0\, \partial^n/\partial x^n + a_1\, \partial^n/\partial x^{n-1}\, \partial y + \cdots + \partial^n/\partial y^n)U(x, y) = 0$ gehörige Funktionentheorie, I. *Math. Ann.* **122**, 109–126.

Lanckau, E. [1958–1959]: Eine Anwendung der Bergmanschen Operatorenmethode auf Profilströmungen im Unterschall. *Wiss. Z. Tech. Hochsch. Dresden* **8**, 200–207.

Lanckau, E. [1962]: Eine einheitliche Darstellung der Lösungen der Tricomischen Gleichung. *Z. Angew. Math. Mech.* **42**, 180–186.

Lanckau, E. [1963]: Über eine lineare elliptische Differentialgleichung zweiter Ordnung mit einem singulären Koeffizienten. *Wiss. Z. Martin-Luther-Univ. Halle-Wittenberg* **12**, 51–60.

Lanckau, E. [1967]: Randwerttreue Bergmansche Integraloperatoren. *Wiss. Z. Tech. Hochsch. Ilmenau* **13**, 377–379.

Lanckau, E. [1969]: Zur Anwendung Bergmanscher Integraloperatoren. In: *Beiträge Anal. Angew. Math.* Wiss. Beiträge 1968/69 (M 1), University of Halle, pp. 115–121.

Lanckau, E. [1970]: Konstruktive Methoden zur Lösung von elliptischen Differentialgleichungen mittels Bergman-Operatoren. In G. Anger [1970], pp. 67–76.

Lanckau, E. [1971]: *Zur Lösung gewisser partieller Differentialgleichungen mittels parameterabhängiger Bergman-Operatoren*. Leipzig: Barth.

Lanckau, E. [1974a]: Integralgleichungen und Randwertprobleme für partielle Differentialgleichungen von elliptischem Typ. *Math. Nachr.* **62**, 233–239.

Lanckau, E. [1974b]: Parameterabhängige Bergman-Operatoren. *Wiss. Z. Tech. Hochsch. Karl-Marx-Stadt* **16**, 361–365.

Lanckau, E. [1975]: Integral equations and boundary value problems for elliptic partial differential equations. *Acta Univ. Carolin.—Math. Phys.* **15** (1974), 81–83.

Lanckau, E. [1978]: Bergmansche Integraloperatoren für dreidimensionale Gleichungen und Gleichungssysteme. *Beiträge Anal.* **12**, 99–112.

Lanckau, E. [1979a]: Bergmansche Integraloperatoren für dreidimensionale Gleichungen. II. *Beiträge Anal.* **13**, 83–87.

Lanckau, E. [1979b]: Die Riemannfunktion selbstadjungierter Gleichungen. *Wiss. Z. Tech. Hochsch. Karl-Marx-Stadt* **21**, 535–540.

Lanckau, E. [1980a]: Solving of linear partial differential equations by using complex integral operators. In W. Tutschke [1980], Part 2, pp. 146–158.

Lanckau, E. [1980b]: General Vekua operators. In J. Ławrynowicz (Ed.): *Analytic Functions* (Proc. Conf., Kozubnik, Poland, 1979), pp 301–311. New York: Springer.

Lanckau, E. [1981]: Zur Behandlung pseudoparabolischer Differentialgleichungen mit funktionentheoretischen Methoden. *Beiträge Anal.* **16**, 87–96.

Lanckau, E. [1983a]: On the representation of Bergman–Vekua operators for three-dimensional equations. *Z. Anal. Anwendungen* **2**, 1–10.

Lanckau, E. [1983b]: Bergmansche Integraloperatoren für instationäre Prozesse in der Ebene. *Z. Anal. Anwendungen* **2**, 309–320.

Lanckau, E. [1986]: Bergman-Vekua operators and "Generalized Axially Symmetric Potential Theory." *Z. Anal. Anwendungen* **5**, 147–156.

Lanckau, E., and W. Tutschke (Eds.) [1983]: *Complex Analysis—Methods, Trends, and Applications*. Berlin: Akademie-Verlag.

Landau, E., and D. Gaier [1986]: *Darstellung und Begründung einiger neuerer Ergebnisse der Funktionentheorie*. 3rd ed. Berlin: Springer.

Lawrentjew, M. A., and B. W. Schabat [1967]: *Methoden der komplexen Funktionentheorie*. Berlin: Deutscher Verlag der Wissenschaften.

Leis, R. [1986]: *Initial Boundary Value Problems in Mathematical Physics*. New York: Wiley.

Lelong, P., and L. Gruman [1986]: *Entire Functions of Several Complex Variables*. New York: Springer.

Lewy, H. [1959]: On the reflection laws of second order differential equations in two independent variables. *Bull. Amer. Math. Soc.* **65**, 37–58.

Lieb, I. [1970]: Die Cauchy–Riemannschen Differentialgleichungen auf streng pseudokonvexen Gebieten. Beschränkte Lösungen. *Math. Ann.* **190**, 6–44.

Lions, J. [1956]: Opérateurs de Delsarte et problèmes mixtes. *Bull. Soc. Math. France* **84**, 9–95.

Lions, J. [1959]: Opérateurs de transmutation singuliers et équations d'Euler–Poisson–Darboux généralisées. *Rend. Sem. Mat. Fis. Milano* **28**, 3–16.

Lo, C. Y. [1980]: Expansions in terms of generalized Helmholtz polynomials and associated functions. *Applicable Anal.* **10**, 197–211.

Ludford, G. S. S. [1953]: Riemann's method of integration: Its extensions with an application. *Collect. Math.* **6**, 293–323.

Luecking, D. H., and L. A. Rubel [1984]: *Complex Analysis. A Functional Analysis Approach*. New York: Springer.

Mackie, A. G. [1954]: One-dimensional unsteady motion of a gas initially at rest and the dam-break problem. *Proc. Cambridge Phil. Soc.* **50**, 131–138.

Mackie, A. G. [1955]: Contour integral solutions of a class of differential equations. *J. Rational Mech. Anal.* **4**, 733–750.

Mackie, A. G. [1965]: Green's functions and Riemann's method. *Proc. Edinburgh Math. Soc. (2)* **14**, 293–302.

Mackie, A. G., and D. C. Pack [1955]: Transonic flow past finite wedges. *J. Rational Mech. Anal.* **4**, 177–199.

Magnus, W., F. Oberhettinger, and R. P. Soni [1966]: *Formulas and Theorems for the Special Functions of Mathematical Physics*, 3rd ed. New York: Springer.

Malonek, H. [1980]: Über die Gültigkeit von Sätzen vom Phragmén-Lindelöfschen Typ bei reellen partiellen Differentialgleichungen und ihre Herleitung durch komplexe Methoden. Thesis, Martin-Luther-Univ. Halle-Wittenberg, 179 pp.

Malonek, H., W. Willner, and T. Woithe [1981]: Herleitung qualitativer Eigenschaften von Lösungen reeller partieller Differentialgleichungen mit dem Hauptteil $\Delta^n u$ mittels komplexer Methoden. *Beiträge Anal.* **15** (1980), 159–169.

Manwell, A. R. [1971]: *The Hodograph Equations. An Introduction to the Mathematical Theory of Plane Transonic Flow*. Edinburgh: Oliver and Boyd.

Manwell, A. R. [1979]: *The Tricomi Equation with Applications to the Theory of Plane Transonic Flow*. London: Pitman.

Marden, M. [1971]: Value distribution of harmonic polynomials in several variables. *Trans. Amer. Math. Soc.* **159**, 137–154.

Marzuq, M. M. H. [1984]: Bergman spaces for the solutions of linear partial differential equations. *Rev. Colombiana Mat.* **18**, 137–143.

Masuda, K. [1974]: On Schwarz's lemma for $\Delta u + c(x)u = 0$. *Proc. Japan Acad.* **50**, 555–560.

McCoy, P. A. [1978]: Extremal properties of real biaxially symmetric potentials. *Pacific J. Math.* **74**, 381–389.

McCoy, P. A. [1979a]: Polynomial approximation and growth of generalized axisymmetric potentials. *Canad. J. Math.* **31**, 49–59.

McCoy, P. A. [1979b]: Polynomial approximation of generalized biaxisymmetric potentials. *J. Approx. Theory* **25**, 153–168.

McCoy, P. A. [1979c]: Analytical properties of generalized biaxially symmetric potentials. *Applicable Anal.* **8**, 201–209.

McCoy, P. A. [1980]: Singularities of solutions to linear second order elliptic partial differential equations with analytic coefficients by approximation methods. *Pacific J. Math.* **91**, 397–406.

McCoy, P. A. [1982]: Best approximation of solutions to a class of elliptic partial differential equations. *Houston J. Math.* **8**, 517–523.

McCoy, P. A., and J. D'Archangelo [1976]: Value distribution of biaxially symmetric harmonic polynomials. *Canad. J. Math.* **28**, 769–773.

Meister, E. [1983a]: *Integraltransformationen mit Anwendungen auf Probleme der mathematischen Physik*. Frankfurt/Main: Lang.

Meister, E. [1983b]: *Randwertaufgaben der Funktionentheorie*. Stuttgart: Teubner.

Meister, V. E., N. Weck, and W. L. Wendland (Eds.) [1976]: *Function Theoretic Methods for Partial Differential Equations* (Proc. Internat. Sympos., Darmstadt, Germany, 1976). New York: Springer.

Mel'nik, V. I. [1969]: Singular points of a power series. *Mat. Zametki* **6**, 267–276. (In Russian.)

Michlin, S. G., and S. Prößdorf [1980]: *Singuläre Integraloperatoren*. Berlin: Akademie-Verlag.

Millar, R. F. [1970]: The location of singularities of two-dimensional harmonic functions. I. *SIAM J. Math. Anal.* **1**, 333–344.

Millar, R. F. [1971]: Singularities of solutions to linear second order analytic elliptic equations in two independent variables. I. *Applicable Anal.* **1**, 101–121.

Millar, R. F. [1973]: Singularities of solutions to linear, second-order, analytic elliptic equations in two independent variables. II. *Applicable Anal.* **2**, 301–320.

Millar, R. F. [1976]: The singularities of solutions to analytic boundary value problems. In V. E. Meister, N. Weck, and W. L. Wendland [1976], pp. 73–87.

Millar, R. F. [1980]: The analytic continuation of solutions to elliptic boundary value problems in two independent variables. *J. Math. Anal. Appl.* **76**, 498–515.

Millar, R. F. [1983]: The analytic continuation of solutions of the generalized axially symmetric Helmholtz equation. *Arch. Rational Mech. Anal.* **81**, 349–372.

Miranda, C. [1970]: *Partial Differential Equations of Elliptic Type*, 2nd ed. New York: Springer.

Mises, R. von [1958]: *Mathematical Theory of Compressible Fluid Flow*. New York. Academic Press.

Mises, R. von, and M. Schiffer [1948]: On Bergman's integration method in two-dimensional compressible fluid flow. *Adv. Appl. Mech.* **1**, 249–285.

Mitchell, J. [1946]: Some properties of solutions of partial differential equations given by their series development. *Duke Math. J.* **13**, 87–104.

Mitchell, J. [1959a]: Integral theorems for solutions of linear partial differential equations in three variables. Mimeographed notes, 30 pp.

Mitchell, J. [1959b]: Representation theorems for solutions of linear partial differential equations in three variables. *Arch. Rational Mech. Anal.* **3**, 439–459.

Mitchell, J. [1973]: Approximation to the solutions of linear partial differential equations given by Bergman integral operators. In S. Ruscheweyh [1973b], pp. 97–107.

Miyake, M. [1981]: Global and local Goursat problems in a class of holomorphic or partially holomorphic functions. *J. Differential Equations* **39**, 445–463.

Morrey, C. B., Jr., and L. Nirenberg [1957]: On the analyticity of the solutions of linear elliptic systems of partial differential equations. *Comm. Pure Appl. Math.* **10**, 271–290.

Mülthei, H. N., and H. Neunzert [1969]: Untersuchung pseudoparabolischer Differentialgleichungen mit Hilfe einer verallgemeinerten Riemannschen Integrationsmethode. *Math. Z.* **111**, 257–266.

Mülthei, H. N., and H. Neunzert [1970]: Pseudoparabolische Differentialgleichungen mit charakteristischen Vorgaben im komplexen Gebiet. *Math. Z.* **113**, 24–32.

Musiałek, J. [1970]: Local properties of certain solutions of differential equations of elliptic type. *Comment. Math. Prace Mat.* **13**, 171–192.

Naas, J., and W. Tutschke [1968]: Fortsetzung reeller Kurven zu isotropen Kurven und Anfangswertprobleme für verallgemeinerte analytische Funktionen. *Monatsber. Deutsch. Akad. Wiss. Berlin* **10**, 244–249.

Naas, J., and W. Tutschke [1976]: Komplexe Analysis und ihr Allgemeinheitsgrad. *Math Nachr.* **75**, 127–131.

Nautiyal, A. [1983]: On the growth of entire solutions of generalized axially symmetric Helmholtz equation. *Indian J. Pure Appl. Math.* **14**, 718–721.

Nazarov, G. I. [1964]: Bergman functions in the theory of flow of a compressible fluid. *Tomsk. Gos. Univ. Učen. Zap.* **49**, 13 pp. (In Russian.)

Nef, W. [1942–1944]: Die Funktionentheorie der partiellen Differentialgleichungen zweiter Ordnung. *Bull. Soc. Fribourgeoise Sci. Natur.* **37**, 348–375.

Nehari, Z. [1967]: Some function-theoretic aspects of linear second order differential equations. *J. Analyse Math.* **18**, 259–276.

Neményi, P. F. [1951]: Recent developments in inverse and semi-inverse methods in the mechanics of continua. *Advances Appl. Mech.* **2**, 123–151.

Netanyahu, E. [1954]: On the singularities of solutions of differential equations of the elliptic type. *J. Rational Mech. Anal.* **3**, 755–761.

Netto, E. [1927]: *Lehrbuch der Combinatorik*, 2nd ed., reprint. New York: Chelsea.

Nevanlinna, R. [1929]: *Le théorème de Picard-Borel et la théorie des fonctions méromorphes*. Paris: Gauthier-Villars.

Nielsen, K. [1944a]: On Bergman operators for linear partial differential equations. *Bull. Amer. Math. Soc.* **50**, 195–201.

Nielsen, K. L. [1944b]: Some properties of functions satisfying partial differential equations of elliptic type. *Duke Math. J.* **11**, 121–137.

Nielsen, K., and B. Ramsay [1943]: On particular solutions of linear partial differential equations. *Bull. Amer. Math. Soc.* **49**, 156–162.

Oberhettinger, F. [1974]: *Tables of Mellin Transforms*. New York: Springer.

Olevskij, M. N. [1952]: On the Riemann function for the differential equations $\partial^2 u/\partial x^2 - \partial^2 u/\partial t^2 + [\rho_1(x) + \rho_2(t)]u = 0$. *Dokl. Akad. Nauk SSSR* **87**, 337–340. (In Russian.)

Oswatitsch, K. [1956]: *Gas Dynamics*. New York: Academic Press.

Papadakis, J. S., and D. H. Wood [1977]: An addition formula for Riemann functions. *J. Differential Equations* **24**, 397–411.

Payne, L. E. [1975]: *Improperly Posed Problems in Partial Differential Equations*. Philadelphia: SIAM.

Peschl, E. [1955]: Les invariants différentiels non holomorphes et leur rôle dans la théorie des fonctions. *Rend. Sem. Mat. Messina* **1**, 100–108.

Peschl, E. [1963]: Über die Verwendung von Differentialinvarianten bei gewissen Funktionenfamilien und die Übertragung einer darauf gegründeten Methode auf partielle Differentialgleichungen vom elliptischen Typus. *Ann. Acad. Sci. Fenn. Ser. A I Math.* **336/6**, 22 pp.

Peschl, E. [1968]: *Funktionentheorie I*, revised. Mannheim: Bibliographisches Institut.

Peschl, E. [1983]: Die konforme Abbildung von Kreisbogenpolygongebieten einer besonderen Klasse, bei denen sich das Problem der akzessorischen Parameter lösen läßt. *Ber. Math.-Statist. Sekt. Forschungszentrum Graz* **202**, 1–8.

Peschl, E., and L. Reich [1971]: Eine Linearisierung kontrahierender biholomorpher Abbildungen und damit zusammenhängender analytischer Differentialgleichungssysteme. *Monatsh. Math.* **75**, 153–162.

Peters, J. M. H. [1984]: Some traditional approaches to the solution of Helmholtz's equation in spherical coordinates. *Internat. J. Math. Ed. Sci. Tech.* **15**, 335–344.

Petrov, V. A. [1967]: Bianalytic polynomials with exceptional values in the sense of Picard. *Smolensk. Gos. Ped. Inst. Učen. Zap.* **18**, 92–96. (In Russian.)

Položiĭ, G. M. [1960]: (p, q)-analytic functions of a complex variable and some of their applications. *Issledovaniya po Sovremennym Problemam Teorii Funkciĭ Kompleksnogo Peremennogo*, pp. 483–515. Moscow: Gosudarstvenniĭ Izdat. Fiz.-Mat. Lit. (In Russian.)

Položiĭ, G. M. [1965a]: Applications of p-analytic functions and (p, q)-analytic functions. In *Applied Theory of Functions in Continuum Mechanics* (Proc. Internat. Sympos., Tbilisi, 1963), vol. I, pp. 309–326. Moscow: Nauka. (In Russian.)

Položiĭ, G. M. [1965b]: *Generalization of the Theory of Analytic Functions of a Complex Variable. p-Analytic and* (p, q)*-Analytic Functions and Some of Their Applications.* Kiev: Izdat. Kiev University.

Pólya, G., and G. Szegö [1972–1976]: *Problems and Theorems in Analysis*. 2 vols. New York: Springer.

Priwalow, I. I. [1956]: *Randeigenschaften analytischer Funktionen*, 2nd ed. Berlin: Deutscher Verlag der Wissenschaften.

Püngel, J. [1978]: Lineare Abbildungen zwischen Lösungsmengen partieller Differentialgleichungen im Komplexen. *Ber. Math.-Statist. Sekt. Forschungszentrum Graz* **91**, 81 pp.

Püngel, J. [1980]: Zur Darstellung von Riemannfunktionen durch Integro-Differentialoperatoren. *Österreich. Akad. Wiss. Math.-Natur. Kl. Sitzungsber. II* **189**, 55–66.

Püngel, J. [1981]: Riemannfunktionen mit speziellen Potenzreihenentwicklungen. *Applicable Anal.* **11**, 199–210.

Püngel, J. [1982]: Riemann functions for generalized Euler equations. I: General results. *Bull. Appl. Math.* **100–109**, 77–93.

Püngel, J. [1984]: Lösungsdarstellungen für partielle Differentialgleichungen in Differentialringen. (Habilitationsschrift, Tech. Univ. Graz) *Ber. Math.-Statist. Sekt. Forschungszentrum Graz* **214**, iv + 102 pp.

Rassias, J. M. (Ed.) [1986]: *Mixed Type Equations*. Leipzig: Teubner.

Reddy, A. R. [1972–1974]: Best polynomial approximation of certain entire functions. *J. Approximation Theory* **5** (1972), 97–112; **12** (1974), 199–200.

Reich, L. [1972]: Über multiplikative und algebraisch verzweigte Lösungen der Differentialgleichung $(1 + \varepsilon z\bar{z})^2 w_{z\bar{z}} + \varepsilon n(n+1)w = 0$. *Ber. Ges. Math. Datenverarb. Bonn* **57**, 13–28.

Reinartz, E. [1974]: *Eine konstruktive Methode zur Lösung elliptischer Randwertprobleme*. GMD-Ber. 94. Bonn: Gesellschaft für Mathematik und Datenverarbeitung.

Renelt, H. [1985]: Über Integraltransformationen, die analytische Funktionen in Lösungen elliptischer Differentialgleichungssysteme überführen. *Ann. Polon. Math.* **45**, 1–9.

Riemann, B. [1860]: Über die Fortpflanzung ebener Luftwellen von endlicher Schwingungsweite. *Abh. Königl. Ges. Wiss. Göttingen* **8**. (Reprinted in: *Collected Works of B. Riemann*, pp. 156–175. New York: Dover, 1953.)

Roelcke, W. [1956]: Über die Wellengleichung bei Grenzkreisgruppen erster Art. *Sitzungsber. Heidelberger Akad. Wiss., Math.-Natur. Kl.* 1953/55, 4. Abh., pp. 161–267. Heidelberg: Springer.

Ronkin, L. I. [1974]: *Introduction to the Theory of Entire Functions of Several Variables*. Providence, R. I.: American Mathematical Society.

Rosenthal, P. [1970]: On the Bergman integral operator for an elliptic partial differential equation with a singular coefficient. *Pacific J. Math.* **35**, 493–497.

Rosenthal, P. [1974a]: On the location of the singularities of the function generated by the Bergman operator of the second kind. *Proc. Amer. Math. Soc.* **44**, 157–162.

Rosenthal, P. [1974b]: A note on the singular points of the function generated by the Bergman operator of the second kind. *Proc. Amer. Math. Soc.* **44**, 163–166.

Rudin, W. [1974]: *Real and Complex Analysis*, 2nd ed. New York: McGraw-Hill.

Ruscheweyh, S. [1969]: Gewisse Klassen verallgemeinerter analytischer Funktionen. *Bonn. Math. Schr.* **39**, iv + 79 pp.

Ruscheweyh, S. [1970]: *Operatoren bei partiellen Differentialgleichungen. II.* (Steiermärk. Math. Sympos., Grottenhof-Hardt, 1970) Graz: Institute für Angewandte Mathematik, Universität und Technische Hochschule, 11 pp.

Ruscheweyh, S. [1972a]: Über Funktionen mit einem Maximumprinzip und eine Klasse von Differentialoperatoren in der Funktionentheorie. Habilitationsschrift, Bonn, ii + 80 pp.

Ruscheweyh, S. [1972b]: Über den Rand des Einheitskreises fortsetzbare Lösungen der Differentialgleichung von Peschl und Bauer. *Ber. Ges. Math. Datenverarb. Bonn* **57**, 29–36.

Ruscheweyh, S. [1973a]: Eine Verallgemeinerung der Leibnizschen Produktregel. *Math. Nachr.* **58**, 241–245.

Ruscheweyh, S. (Ed.) [1973b]: Ergebnisse einer Tagung über "Funktionentheoretische Methoden bei partiellen Differentialgleichungen," Bonn, 19. 21. Oktober 1972. *Ber. Ges. Math. Datenverarb. Bonn* **77**, 179 pp.

Ruscheweyh, S. [1974a]: Geometrische Eigenschaften der Lösungen der Differentialgleichung $(1 - z\bar{z})^2 w_{z\bar{z}} - n(n+1)w = 0$. *J. Reine Angew. Math.* **270**, 143–157.

Ruscheweyh, S. [1974b]: Funktionenfamilien mit einem Maximumprinzip und elliptische Differentialgleichungen. I. *Monatsh. Math.* **78**, 246–255.

Ruscheweyh, S. [1976]: Hardy spaces of λ-harmonic functions. In R. P. Gilbert and R. J. Weinacht [1976], pp. 68–84.

Ruscheweyh, S. [1982]: *Convolutions in Geometric Function Theory*. (Séminaire de Mathématiques Supérieures 83). Montréal: Les Presses de l'Université de Montréal.

Ruscheweyh, S., and K.-J. Wirths [1976]: Riemann's mapping theorem for n-analytic functions. *Math. Z.* **149**, 287–297.

Ruscheweyh, S., and K.-J. Wirths [1985]: On a radius of convexity problem with application to the Peschl–Bauer equation. *Complex Variables Theory Appl.* **4**, 173–179.

Rusev, P. [1983]: Expansion of analytic functions in series of classical orthogonal polynomials. *Complex Analysis* (Warsaw, 1979), pp. 287–298. Banach Center Publication 11, PWN, Warsaw.

Saff, E. B. [1971]: Regions of meromorphy determined by the degree of best rational approximation. *Proc. Amer. Math. Soc.* **29**, 30–38.

Sario, L., and K. Noshiro [1966]: *Value Distribution Theory*. New York: Van Nostrand.

Sasai, T. [1979]: The Cauchy problem for Tricomi's equation and the relations between generating functions and entire functions. *J. Math. Anal. Appl.* **69**, 563–570.

Sastry, M. S. K. [1971]: Meta-analytic functions. *Trans. Amer. Math. Soc.* **157**, 399–415.

Scheffers, G. [1893–1894]: Verallgemeinerung der Grundlage der gewöhnlichen complexen Funktionen. I, II. *Ber. Verh. Sächs. Ges. Wiss. Leipzig, Math.-Phys. Kl.* **45**, 828–848; **46**, 120–134.

Schep, A. R. [1981]: Compactness properties of an operator which imply that it is an integral operator. *Trans. Amer. Math. Soc.* **265**, 111–119.

Schneider, M. [1977]: *Introduction to Partial Differential Equations of Mixed Type*. Newark, DE.: Institute of Mathematical Sciences, University of Delaware, i + 55 pp.

Schubert, H., and M. Schleiff [1969]: Über zwei Randwertprobleme des inhomogenen Systems der Cauchy-Riemannschen Differentialgleichungen mit einer Anwendung auf ein Problem der stationären schallnahen Strömung. *Z. Angew. Math. Mech.* **49**, 621–630.

Schwarz, H. A. [1890]: *Gesammelte Abhandlungen*, vol. I. Berlin: Springer.

Scott, E. J. [1973]: Determination of the Riemann function. *Amer. Math. Monthly* **80**, 906–909.

Sedov, L. I. [1965]: *Two-Dimensional Problems in Hydrodynamics and Aerodynamics*. New York: Interscience.

Sewell, W. E. [1935]: Degree of approximation by polynomials to continuous functions. *Bull. Amer. Math. Soc.* **41**, 111–117.

Sewell, W. E. [1942]: *Degree of Approximation by Polynomials in the Complex Domain*. Princeton, N.J.: Princeton University Press.

Shapiro, V. L. [1970]: Removable sets for pointwise solutions of the generalized Cauchy–Riemann equations. *Ann. of Math. (2)* **92**, 82–101.

Singh, J. P. [1969]: On the order and type of entire functions of several complex variables. *Riv. Mat. Univ. Parma (2)* **10**, 111–121.

Smirnov, M. M. [1978]: *Equations of Mixed Type*. Providence, RI: American Mathematical Society.

Smirnov, V. I., and N. A. Lebedev [1968]: *Functions of a Complex Variable. Constructive Theory*. Cambridge, MA: M.I.T. Press.

Sneddon, I. N. [1966]: *Mixed Boundary Value Problems in Potential Theory*. Amsterdam: North-Holland.

Snyder, H. H. [1968]: A *Hypercomplex Function Theory Associated with Laplace's Equation*. Berlin: Deutscher Verlag der Wissenschaften.

Sorokina, N. G. [1966]: On the strong solvability of the Tricomi problem. *Ukrain. Mat. Zh.* **18**, 65–77.

Spindelböck, K., and L. Reich [1975]: Algebraisch verzweigte Lösungen einer linearen partiellen Differentialgleichung zweiter Ordnung. *Österreich. Akad. Wiss. Math.-Natur. Kl. Sitzungsber. II* **183**, 423–443.

Sreedharan, V. P. [1965]: Function-theoretic solutions of certain boundary-value problems. *J. Math. Mech.* **14**, 211–230.

Stark, J. M. [1964]: *Transonic Flow Patterns Generated by Bergman's Integral Operator*. Stanford, CA.: Department of Mathematics, Stanford University. ii + 92 pp.

Stark, J. M. [1966]: Application of Bergman's integral operators to transonic flows. *Internat. J. Non-Linear Mech.* **1**, 17–34.

Stecher, M. [1975]: Integral operators and the non-characteristic Cauchy problem for parabolic equations. *SIAM J. Math. Anal.* **6**, 796–811.

Strelic, Š., and I. V. Kiseljus [1968]: Holomorphic solutions of partial differential equations. *Litovsk. Mat. Sb.* **8**, 811–825. (In Russian.)

Suschowk, D. K. [1962]: On a class of singular solutions of the Euler–Poisson–Darboux equation. *Arch. Rational Mech. Anal.* **11**, 50–61.

Takasu, T. [1956]: Some classes of function-theoretical solutions of the Laplacian and the wave equation. *Yokohama Math. J.* **4**, 65–79.

Temliakoff, A. [1935]: Zum Wachstumsproblem der harmonischen Funktionen des dreidimensionalen Raumes. *Mat. Sb.* **42**, 707–718.

Titchmarsh, E. C. [1948]: *Introduction to the Theory of Fourier Integrals*, 2nd ed. London: Oxford University Press.

Tollmien, W. [1937]: Zum Übergang von Unterschall- in Überschallströmungen. *Z. Angew. Math. Mech.* **17**, 117–136.

Tomantschger, K. W. [1983a]: Bergman kernels of first kind for $W_{z\zeta} - (s + z\zeta)W = 0$, $s \in Z$. *Bull. Appl. Math.* **30**, no. 194, 16 pp.

Tomantschger, K. W. [1983b]: Integral operators for $W_{z\zeta} - z^{M_1-1}\zeta^{M_2-1}W = 0$, $M_\nu \in Z$. *Bull. Appl. Math.* **30**, no. 198, 15 pp.

Tomantschger, K. W. [1985]: Bergman kernels of the first kind for a class of self-adjoint partial differential equations. *Bull. Appl. Math.* **39**, no. 330, 133–138.

Tomotika, S., and K. Tamada [1950]: Studies on two-dimensional transonic flows of compressible fluid. I, II. *Quart. Appl. Math.* **7**, 381–397; **8**, 127–136.

Tonti, N. E., and D. H. Trahan [1970]: Analytic functions whose real parts are bounded below. *Math. Z.* **115**, 252–258.

Trangenstein, J. A. [1975]: A Finite Element Method for the Tricomi Problem in the Elliptic Region. Thesis. Ithaca, N.Y.: Cornell University.

Tricomi, F. G. [1923]: Sulle equazioni lineari alle derivate parziali di 2° ordine di tipo misto. *Atti Accad. Naz. Lincei Rend. Cl. Sci. Fis. Mat. Natur. (5)* **14**, 133–247.

Tricomi, F. G. [1968]: *Repertorium der Theorie der Differentialgleichungen*. Berlin: Springer.

Tutschke, W. [1970]: Pseudoholomorphe Funktionen einer und mehrerer komplexer Variabler im Zusammenhang mit der Theorie vollständiger Differentialgleichungssysteme. In G. Anger [1970], pp. 165–173.

Tutschke, W. [1977]: *Partielle komplexe Differentialgleichungen in einer und in mehreren komplexen Variablen*. Berlin: Deutscher Verlag der Wissenschaften.

Tutschke, W. (Ed.) [1980]: *Komplexe Analysis und ihre Anwendung auf partielle Differentialgleichungen*. Parts 1–3. Wiss. Beiträge 1980/41 (M 18). Halle: Martin-Luther-Universität Halle-Wittenberg.

Tutschke, W. [1983]: *Partielle Differentialgleichungen. Klassische, funktionalanalytische und komplexe Methoden*. Leipzig: Teubner.

Tutschke, W., and C. Withalm [1983]: The Cauchy–Kovalevska theorem for pseudoholomorphic functions in the sense of L. Bers. *Complex Variables Theory Appl.* **1**, 389–393.

Vekua, I. N. [1937]: Sur la représentation générale des solutions des équations aux dérivées partielles du second ordre. *Dokl. Akad. Nauk SSSR (N.S.)* **17**, 295–299.

Vekua, I. N. [1963]: *Verallgemeinerte analytische Funktionen*. Berlin: Akademie-Verlag.

Vekua, I. N. [1967]: *New Methods for Solving Elliptic Equations*. New York: Wiley and Amsterdam: North-Holland.

Volkmer, H. [1979]: Integralrelationen mit variablen Grenzen für spezielle Funktionen der mathematischen Physik. Thesis, University of Konstanz, 86 pp.

Vol'man, V. I., and Yu. A. Pampu [1977]: Riemann's functions for the equation $\Delta u_m + (1/r)\,\partial u_m/\partial r + (1 - m^2/r^2)u_m = 0$. *Differential Equations* **13**, 1192–1194.

Wahl, W. von [1970]: Klassische Lösungen nichtlinearer gedämpfter Wellengleichungen im Großen. *Manuscripta Math.* **3**, 7–33.

Wahlberg, C. [1977]: Riemann's function for a Klein–Gordon equation with a nonconstant coefficient. *J. Phys. A* **10**, 867–878.

Wallner, H. [1980]: Riemann-Funktionen als Lösungen partieller Differentialgleichungssysteme. *Ber. Math.-Statist. Sekt. Forschungszentrum Graz* **136**, 28 pp.

Wallner, H. [1981a]: Riemann-Funktionen mit mehreren Hilfsveränderlichen. *Ber. Math.-Statist. Sekt. Forschungszentrum Graz* **174**, 28 pp.

Wallner, H. [1981b]: A note on a paper by Azzam and Kreyszig concerning the construction of integral operators. *Applicable Anal.* **12**, 265–266.

Wallner, H. [1983]: Associated differential operators and their Bergman kernels. *Complex Variables Theory Appl.* **1**, 395–403.

Wallner, H. [1985]: Bemerkung zu einer Arbeit von E. Kreyszig. *Abh. Math. Sem. Univ. Hamburg* **55**, 1–2.

Walsh, J. L. [1965]: The convergence of sequences of rational functions of best approximation with some free poles. In: *Approximation of Functions*. Proc. Sympos. General Motors Res. Lab., 1964, pp. 1–16. Amsterdam: Elsevier.

Walsh, J. L. [1969]: *Interpolation and Approximation by Rational Functions in the Complex Domain*, 5th ed. Providence, R.I.: American Mathematical Society.

Warnecke, G. [1968]: Über die Darstellung von Lösungen der partiellen Differentialgleichung $(1 + \delta z\bar{z})^2 w_{z\bar{z}} = \delta - \varepsilon e^{2w}$. *Bonn. Math. Schr.* **34**, iii + 75 pp.

Warnecke, G. [1970]: Über einige Probleme bei einer nichtlinearen Differentialgleichung zweiter Ordnung im Komplexen. *J. Reine Angew. Math.* **239/240**, 353–362.

Watzlawek, W. [1969]: Zur Lösungsdarstellung bei gewissen linearen partiellen Differentialgleichungen zweiter Ordnung. *Monatsh. Math.* **73**, 461–472.

Watzlawek, W. [1971a]: Zur Konstruktion von Bergman-Operatoren. *Z. Angew. Math. Mech.* **51**, 617–620.

Watzlawek, W. [1971b]: Über lineare partielle Differentialglelichungen zweiter Ordnung mit Fundamentalsystemen. *J. Reine Angew. Math.* **247**, 69–74.

Watzlawek, W. [1971c]: Über Zusammenhänge zwischen Fundamentalsystemen, Riemann-Funktion und Bergman-Operatoren. *J. Reine Angew. Math.* **251**, 200–211.

Watzlawek, W. [1972]: Über lineare partielle Differentialgleichungen zweiter Ordnung mit Bergman-Operatoren der Klasse P. *Monatsh. Math.* **76**, 356–369.

Watzlawek, W. [1973]: Hyperbolische und parabolische Differetialgleichungen der Klasse P. In S. Ruscheweyh [1973b], pp. 147–179.

Watzlawek, W. [1983a]: Parameter shifting and complete families of solutions to a singular parabolic equation. *Partial Differential Equations*, pp. 405–412. Warsaw: Banach Center Publication 10.

Watzlawek, W. [1983b]: Wärmepolynome-Modell für besondere Lösungssysteme bei linearen partiellen Differentialgleichungen. *Ber. Math.-Statist. Sekt. Forschungszentrum Graz* **211**, 34 pp.

Watzlawek, W. [1985]: Zur Verallgemeinerung von Wärmepolynomen. *Monatsh. Math.* **100**, 67–76.

Weinstein, A. [1953]: Generalized axially symmetric potential theory. *Bull. Amer. Math. Soc.* **59**, 20–38.

Wen, G. C. [1984]: Function theoretic methods for nonlinear elliptic complex equations. *Bull. Math. Soc. Sci. Math. R. S. Roumanie (N.S.)* **28**(76), 87–90.

Wendland, W. L. [1979]: *Elliptic Systems in the Plane*. London: Pitman.

White, A. [1961]: Singularities of harmonic functions of three variables generated by Whittaker–Bergman operators. *Ann. Polon. Math.* **10**, 81–100.

White, A. M. [1962]: Singularities of a harmonic function of three variables given by its series development. MRC Tech. Summ. Rep. 312, Mathematics Research Center, University of Wisconsin, 28 pp.

Whittaker, E. T., and G. N. Watson [1963]: *A Course of Modern Analysis*, 4th ed., reprint. Cambridge: Cambridge University Press.

Widder, D. V. [1966]: Some analogies from classical analysis in the theory of heat conduction. *Arch. Rational Mech. Anal.* **21**, 108–119.

Widder, D. V. [1975]: *The Heat Equation*. New York: Academic Press.

Wildenhain, G. [1968]: *Potentialtheorie linearer elliptischer Differentialgleichungen beliebiger Ordnung*. Berlin: Akademie-Verlag.

Withalm, C. [1972]: Über einen algebraischen Zugang zu den pseudoanalytischen Funktionen und eine Erweiterung des Ähnlichkeitsprinzips. I, II. *Ber. Math.-Statist. Sekt. Forschungszentrum Graz* **72-1**, i + 58 pp.; **72-3**, i + 24 pp.

Withalm, C. [1985]: Über einen Integraloperator für pseudoholomorphe Funktionen modulo eines Haupterzeugendensystems. *Complex Variables Theory Appl.* **4**, 155–161.

Wittich, H. [1968]: *Neuere Untersuchungen über eindeutige analytische Funktionen*. Berlin: Springer.

Wittich, H. [1982]: Anwendungen der Werteverteilungslehre auf gewöhnliche Differentialgleichungen. *Ann. Acad. Sci. Fenn. Ser. A I Math.* **7**, 89–97.

Wolfersdorf, L. [1965]: Abelsche Integralgleichungen und Randwertprobleme für die verallgemeinerte Tricomi-Gleichung. *Math. Nachr.* **29**, 161–178.

Wolfersdorf, L. von [1970]: Ein Kopplungsproblem für verallgemeinerte analytische Funktionen. *Math. Nachr.* **45**, 243–261.

Wood, D. H. [1976]: Simple Riemann functions. *Bull. Amer. Math. Soc.* **82**, 737–739.

Wu, H. [1969]: An *n*-dimensional extension of Picard's theorem. *Bull. Amer. Math. Soc.* **75**, 1357–1361.

Young, E. C. [1970]: A characteristic initial value problem for the Euler–Poisson–Darboux equation. *Ann. Mat. Pura Appl. (4)* **85**, 365–367.

Yu, C.-L. [1974]: Integral representation, analytic continuation and the reflection principle under the complementing boundary condition for higher order elliptic equations in the plane. *SIAM J. Math. Anal.* **5**, 209–223.

Zemanian, A. H. [1968]: *Generalized Integral Transformations*. New York: Wiley.

Symbol Index

($j = 1, 2$, $j^* = 3 - j$; $\mathbb{N}, \mathbb{N}_0, \mathbb{Z}, \mathbb{R}, \mathbb{R}^n, \mathbb{C}, \mathbb{C}^n$, as usual)

Domains:

B_0, 9
$B_j(s_j, r_j)$, 9
$\dot{B}_1(z_1^{(0)}, \rho)$, 250
G_j, 21
$N(C_j, \delta)$, 14
Ω, 9
Ω_0, 14
Ω_j, 9
$\tilde{\Omega}_j$, 14
Ω_1^*, 36, 125
$\hat{\Omega}_1$, 125
Ω_3, 36
$\check{\Omega}_3$, 37

Functions:

b_j, 9
c_j, 9
e_j, 9
e_{jr}, 69
e_j^R, 137
f_j, 14
$\tilde{f}_j$, 51
h_j, 26
h_{jr}, 70
H_{jm}, 26
k_j, 15
$\tilde{k}_j$, 14
$\tilde{k}_1^{\times}$, 225
R, 128
$\tilde{R}$, 136, 168
$\hat{R}$, 168
R_j, 127
η_j, 9
φ_j, 31

Special Functions of Mathematical Physics:

${}_0F_1$, 334
${}_0F_2$, 43
${}_1F_1$, 42
${}_2F_1$, 23
F_3, 169
F_B, 190
F_B^+, 188
J_ν, 13
K_ν, 346
$P_m^{(a,b)}$, 333
P_n, 168
Q_n, 258
B, 13
Γ, 331
Ξ_2, 170
Φ_3, 41

Classes of Functions:

$\mathscr{B}_n$, 296
$C^0(\Omega_0 \times N(C_j, \delta))$, 15
$C^\omega(\Omega_j)$, 9
$C^\omega(\tilde{\Omega}_j)$, 14
$C^\omega(\Omega)$, 9
F, 37
$\mathscr{H}_n(\Omega_1)$, 300
$K(\tilde{k}_1)$, 224
$K(\tilde{k}_1^{\times})$, 225
$V(\Omega_j)$, 53
$V(\tilde{\Omega}_j)$, 14
$V(\Omega)$, 53
$\tilde{V}(\Omega)$, 53

Solutions:

u, 110
$\hat{u}$, 147
u_j, 9
w, 9, 16
W, 11
w_0, 93
w_j, 15
w_{jr}, 69
$\tilde{w}$, 53, 59, 167
$\tilde{w}_j$, 51, 53
$\hat{w}$, 10, 167
$\check{w}$, 77, 234
ψ, 309, 323
ψ_j, 329

Integral Operators:

E, 341
J, 160
J_j, 127
$\underset{\sim}{P}_1$, 35
$\underset{\sim}{P}_2$, 35
T, 110
T_j, 14
$T_1^{\times}$, 225

Integro-Differential Operators:

S_j, 29
$S_{j\mu}$, 28

Differential Operators:

D, 85
$\hat{D}$, 85
D_j, 9
$\check{D}_j$, 77
D_t, 9
D_x, 35
$\tilde{D}_p$, 87
$\overline{D}_1$, 257
D_{jr}, 69
$D_j^{(r)}$, 69
$\tilde{D}_{p,m\mu}$, 87
$\tilde{E}$, 37
$\tilde{\tilde{E}}$, 37
L, 9
$\tilde{L}$, 136
$\hat{L}$, 10, 125
$\check{L}$, 65, 77
L^*, 98, 126
L_j, 9
$L^{(r)}$, 68
$\tilde{L}^{(n)}$, 102
$L^{(r)*}$, 99
$\check{L}^{(r)}$, 77
L_ζ, 135
M_j, 15
M_{jm}, 59
$\tilde{P}_n$, 159
$\tilde{\tilde{P}}_n$, 159
$\tilde{\tilde{P}}_n^*$, 161
$\tilde{Q}_n$, 159
S, 36
$\tilde{S}$, 36
$\tilde{T}_j$, 51, 53
$\overline{\tilde{T}}_1$, 274
$\tilde{T}_{1(\kappa)}$, 269
$\tilde{T}_{1(\omega)}$, 266
$\tilde{\tilde{T}}_p$, 88
δ_r, 171
Δ, Δ_2, 10, 147
Δ_n, 151
ϑ_{jm}, 73

Classes of Differential Operators:

P_{jn}, 47
P_{jn}^0, 47
FP_{jn}, 47
FP_{jn}^0, 47
$(FP_{jn}^0)^*$, 64

Author Index

A

Agostinelli, C., 257, 349
Aks, S., 208, 361
Aleksandrov, V. M., 348, 349
Andersson, K. G., 349
Anger, G., 317, 349, 368, 376
Ansorge, R., 359
Atkinson, K. E., 3, 222, 361
Aupetit, B., 222, 349
Avila, G. S. S., 208, 349
Aziz, A. K., 348, 349
Azzam, A., 118, 119, 120, 123, 349

B

Bader R., 344, 360
Balk, M. B., 223, 250, 255, 298, 299, 304, 305, 349, 350
Bauer, F., 3, 322, 347, 348, 350
Bauer, K. W., 52, 62, 63, 67, 68, 81, 86, 88, 97, 100, 107, 141, 188, 194, 195, 196, 255, 257, 258, 260, 299, 304, 305, 350, 351
Begehr, H., 223, 255, 305, 351
Behnke, H., 17, 210, 214, 252, 269, 291, 351
Beke, E., 202
Beltrami, E., 312, 351
Bengel, G., 351
Berenstein, C. A., 351
Berglez, P., 107, 188, 196, 351
Bergman, S., 2, 3, 12, 15, 21, 44, 109, 113, 123, 124, 154, 165, 182, 199, 208, 209, 221, 222, 252, 255, 306, 321, 326, 327, 339, 341, 348, 351, 352, 353, 354
Bers, L., 4, 298, 305, 312, 314, 321, 348, 354
Bhatnagar, S. C., 44, 354
Bianchi, L., 257, 354
Bieberbach, L., 197, 201, 354
Bitsadze, A. V., 348, 354, 355
Blohina, G. N., 305, 355
Boas, R. P., Jr., 334, 335, 337, 338, 355
Bochner, S., 321, 354
Böhmer, K., 123, 355
Bojanić, R., 327, 341, 353
Borel, É., 201, 223, 355
Bosch, W., 250, 255, 355
Brackx, F., 298, 305, 355
Bragg, L. R., 259, 305, 355
Brown, R. M., 208, 355
Buchanan, J. L., 4, 298, 305, 361
Buck, R. C., 337, 338, 355

C

Čanak, M., 355
Carleman, T., 312, 355
Carrier, G. F., 344, 346, 348, 355
Carroll, R. W., 125, 147, 158, 159, 161, 164, 165, 355
Case, B. A., 305, 355
Chang, P. H., 356
Chang, Y. F., 356
Chaplygin, S. A., 344, 356
Chaundy, T. W., 171, 172, 189, 190, 196, 356
Chernoff, H., 223, 236, 255, 356
Cherry, T. M., 321, 356
Cholewinski, F. M., 356
Cohn, H., 166, 180, 181, 182, 183, 196, 356
Collatz, L., 361
Colton, D. L., 3, 4, 7, 8, 35, 36, 44, 165, 208, 222, 356, 357, 361
Conlan, J., 361
Conway, J. B., 210, 252, 269, 291, 357
Copson, E. T., 166, 171, 173, 196, 357
Courant, R., 312, 314, 348, 354, 357

D

Daggit, E., 180, 182, 183, 184, 196, 357
Danilyuk, Ī. Ī., 357
Darboux, G., 68, 107, 257, 344, 357
D'Archangelo, J., 255, 370
Davis, R. M., 357
Delanghe, R., 298, 305, 355, 357
Delsarte, J., 159, 165, 357
Dettman, J. W., 259, 305, 355, 357
Diaz, J. B., 124, 127, 165, 358
Dienes, P., 197, 201, 203, 205, 358
Dieudonné, J., 358
Dimiev, S., 358
Dinghas, A., 223, 358
Dirichlet, P. G. L., 1
Dont, M., 208, 358
Douglis, A., 298, 358
Du, X. H., 166, 180, 196, 358
Duff, G. F. D., 358
Ďurikovič, V., 222, 358

E

Ecker, K., 123, 358
Edenhofer, J., 298, 305, 358
Ehlers, F. E., 344, 346, 348, 355
Eichhorn, W., 298, 305, 358
Eichler, M., 2, 196, 341, 348, 358, 359
Eisenstat, S. C., 165, 222, 359
Eke, B. G., 359
Epstein, B., 321, 353
Erdélyi, A., 41, 153, 170, 174, 271, 359
Everitt, W. N., 356

F

Falkovich, S. V., 344, 348, 359
Fichera, G., 222, 359
Florian, H., 52, 62, 67, 107, 122, 123, 165, 182, 185, 188, 192, 193, 196, 350, 358, 359
Frank, G., 255, 359
Frankl, F. I., 324, 348, 360
Franz, W., 331, 366
Friedlander, F. G., 165, 196, 360
Friedman, A., 35, 360
Friedrichs, K. O., 348, 352, 357
Fritzsche, K., 362
Frobenius, G., 80
Fryant, A. J., 255, 360
Fueter, R., 298, 305, 360

G

Gaier, D., 6, 209, 222, 360, 368
Gakhov, F. D., 360
Ganin, M. P., 257, 360
Garabedian, P. R., 165, 321, 348, 350, 360
Geddes, R. L., 182, 184, 196, 360
Gelbart, A., 314, 354
Germain, P., 344, 360
Gilbert, R. P., 3, 4, 7, 44, 125, 147, 151, 152, 154, 158, 165, 208, 222, 258, 298, 305, 349, 350, 351, 354, 355, 357, 360, 361, 362, 365, 367, 374
Glashoff, K., 359
Goldschmidt, B., 362
Goluzin, G. M., 287, 288, 293, 295, 305, 362
Goman, O. G., 305, 362
Görtler, H., 362
Grabarska, H., 362
Grauert, H., 362
Greenstone, L., 321, 354
Gronau, D., 208, 305, 362
Gruman, L., 368
Guderley, K. G., 321, 344, 346, 348, 362
Gunning, R. C., 362
Gutknecht, M. H., 209, 363

H

Haack, W., 363
Habetha, K., 255, 298, 305, 363
Hadamard, J., 170, 196, 223, 363
Haimo, D. T., 356, 363
Harke, O., 255, 363
Hayman, W. K., 223, 234, 255, 363
Heersink, R., 44, 52, 62, 107, 359, 363
Heins, A. E., 44, 165, 196, 360, 363
Heins, M., 363
Hennekemper, W., 255, 359
Henrici, P., 3, 44, 126, 165, 169, 170, 196, 270, 305, 363, 364
Hersh, R., 159
Herriot, J. G., 3, 354
Hilb, E., 364
Hilbert, D., 312, 314, 354, 357
Hile, G., 361
Hill, C. D., 364
Hille, E., 197, 214, 219, 285, 291, 295, 364
Hobson, E. W., 364
Hoefer, E. T., 209, 214, 221, 222, 364
Holland, A. S. B., 364
Hörmander, L., 364

Horváth, J., 305, 364
Howard, H. C., 208, 361
Hsiao, G. S., 361
Hsieh, P. F., 357, 364
Hua, L. K., 364
Hubbard, B., 349, 361

I

Ince, E. L., 116, 364
Ingersoll, B. M., 208, 364

J

Jank, G., 44, 52, 62, 67, 107, 223, 255, 305, 350, 359, 364
John, F., 321, 354

K

Kaplan, S., 364
Kapoor, G. P., 365
Kaucher, E. W., 222, 365
Kemnitz, H., 365
Kiseljus, I. V., 375
Kleinman, R. E., 365
Klink, H.-K., 365
Kneis, G., 365
Knight, R. B., 361
Knowles, I. W., 355
Korevaar, J., 209, 365
Korn, D., 348, 350, 360
Kracht, M., 44, 52, 58, 62, 64, 66, 67, 72, 107, 124, 165, 196, 257, 305, 321, 348, 365, 366
Kraft, R., 196, 366
Krajkiewicz, P., 223, 246, 250, 251, 252, 255, 298, 299, 304, 305, 355, 366
Kratzer, A., 331, 366
Kress, R., 357
Kreyszig, E., 44, 46, 52, 58, 62, 64, 66, 67, 72, 107, 116, 118, 119, 120, 122, 123, 165, 196, 208, 257, 305, 308, 318, 321, 348, 349, 365, 366, 367
Kriszten, A., 298, 305, 367
Krzywoblocki, M. Z. von, 306, 321, 348, 367
Kucherov, V. A., 348, 349
Kühn, E., 298, 305, 367
Kühnau, R., 367
Kukral, D. K., 361, 367
Kürcz, A., 44, 367

L

Lammel, E., 298, 305, 367
Lanckau, E., 4, 8, 36, 44, 165, 188, 189, 190, 196, 250, 255, 317, 345, 347, 348, 349, 351, 361, 363, 367, 368
Landau, E., 368
Langer, R. E., 364
Lawrentjew, M. A., 268, 368
Ławrynowicz, J., 368
Lebedev, N. A., 375
Leis, R., 368
Lelong, P., 368
Leventhal, S. H., 348, 349
Lewis, R. T., 355
Lewy, H., 165, 368
Lieb, I., 305, 368
Lions, J., 159, 165, 357, 368
Lo, C. Y., 62, 222, 361, 369
Ludford, G. S. S., 124, 127, 165, 168, 196, 358, 369
Luecking, D. H., 369

M

MacCamy, R. C., 44, 363
Mackie, A. G., 165, 171, 172, 182, 184, 196, 348, 360, 369
Magnus, W., 42, 123, 143, 175, 297, 332, 333, 343, 346, 359, 369
Malonek, H., 305, 369
Manwell, A. R., 348, 369
Marden, M., 369
Marzuq, M. M. H., 27, 44, 369
Masuda, K., 305, 369
McCoy, P. A., 208, 222, 255, 369, 370
Meinardus, G., 361
Meister, V. E., 44, 350, 353, 358, 361, 363, 365, 370
Mel'nik, V. I., 370
Michlin, S. G., 370
Millar, R. F., 165, 208, 370
Miranda, C., 313, 370
Miranker, W. L., 222, 365
Mises, R. von, 306, 321, 352, 370
Mitchell, J., 208, 209, 214, 222, 370
Miyake, M., 370
Morawetz, C. S., 347
Morrey, C. B., Jr., 370
Mülthei, H. N., 165, 371
Musiałek, J., 371

N

Naas, J., 371
Nautiyal, A., 255, 365, 371
Naylor, D., 358
Nazarov, G. I., 306, 321, 371
Nef, W., 298, 305, 371
Nehari, Z., 371
Neményi, P. F., 347, 348, 371
Netanyahu, E., 208, 371
Netto, E., 104, 279, 371
Neumann, C., 1
Neunzert, H., 165, 371
Nevanlinna, R., 223, 371
Newton, R. G., 44, 360, 362, 367
Nielsen, K., 114, 123, 371
Nirenberg, L., 358, 371
Noshiro, K., 374

O

Oberhettinger, F., 162, 359, 369, 371
Olevskij, M. N., 166, 176, 196, 371
Oswatitsch, K., 321, 371

P

Pack, D. C., 348, 369
Pampu, Y. A., 186, 196, 376
Papadakis, J. S., 176, 196, 371
Payne, L. E., 3, 371
Peschl, E., 52, 62, 63, 107, 257, 305, 350, 371, 372
Peters, J. M. H., 372
Petrov, V. A., 372
Picard, É., 223, 312, 313
Poincaré, H., 1, 223
Položiĭ, G. M., 4, 298, 305, 372
Pólya, G., 201, 202, 372
Priwalow, I. I., 372
Prößdorf, S., 370
Püngel, J., 107, 165, 182, 185, 188, 189, 191, 192, 193, 196, 359, 372

R

Ramsay, B., 114, 371
Rassias, J. M., 365, 366, 372
Reddy, A. R., 372
Reich, L., 305, 362, 372, 373, 375
Reinartz, E., 209, 222, 373
Renelt, H., 373
Riemann, B., 1, 167, 168, 373
Roelcke, W., 258, 305, 373
Ronkin, L. I., 373
Rosenthal, P., 208, 209, 221, 348, 373
Rossi, H., 362
Rubel, L. A., 369
Rudin, W., 212, 373
Ruscheweyh, S., 52, 62, 63, 107, 255, 257, 258, 260, 269, 270, 290, 293, 305, 350, 351, 353, 361, 363, 364, 370, 373, 374, 377
Rusev, P., 374

S

Saff, E. B., 222, 374
Sario, L., 374
Sasai, T., 333, 348, 374
Sastry, M. S. K., 374
Schabat, B. W., 268, 368
Scheffers, G., 298, 374
Schep, A. R., 374
Schiffer, M., 306, 321, 354, 370
Schleiff, M., 348, 374
Schneider, M., 348, 349, 362, 374
Schröder, G., 44, 52, 62, 67, 107, 305, 366
Schubert, H., 348, 374
Schwarz, H. A., 1, 257, 374
Scott, E. J., 185, 196, 374
Sedov, L. I., 348, 374
Sewell, W. E., 209, 216, 374
Shapiro, V. L., 374
Singh, J. P., 374
Sleeman, B. D., 356
Smirnov, M. M., 375
Smirvov, V. I., 348, 375
Sneddon, I. N., 375
Snyder, H. H., 298, 305, 375
Sommen, F., 305, 355
Sommer, F., 17, 210, 214, 252, 269, 291, 351
Soni, R. P., 369
Sorokina, N. G., 348, 375
Spindelböck, K., 305, 375
Sreedharan, V. P., 165, 375
Stark, J. M., 348, 375
Stecher, M., 44, 375
Stoddart, A. W. J., 357, 364
Strelic, Š., 375
Suschowk, D. K., 208, 375
Szász, O., 364
Szegö, G., 201, 202, 372

T

Takasu, T., 375
Tamada, K., 324, 344, 348, 375
Taylor, B. A., 351
Temliakoff, A., 375
Thullen, P., 351
Thyssen, M., 159
Titchmarsh, E. C., 175, 375
Tollmien, W., 375
Tomantschger, K. W., 44, 375
Tomotika, S., 324, 344, 348, 375
Tonti, N. E., 255, 375
Trahan, D. H., 255, 375
Trangenstein, J. A., 348, 375
Tricomi, F. G., 68, 107, 324, 325, 348, 359, 376
Tutschke, W., 4, 44, 165, 250, 255, 298, 305, 349, 351, 359, 361, 363, 368, 371, 376

U

Unger, H., 361

V

Vekua, I. N., 2, 3, 4, 7, 124, 126, 128, 165, 196, 298, 305, 312, 321, 376
Volkmann, L., 223, 255, 364
Volkmer, H., 165, 196, 376
Vol'man, V. I., 186, 196, 376

W

Wahl, W. von, 376
Wahlberg, C., 185, 196, 376
Wallner, H., 44, 119, 122, 182, 188, 193, 196, 359, 376
Walsh, J. L., 6, 209, 214, 215, 222, 376
Warnecke, G., 377
Watson, G. N., 269, 272, 377
Watzlawek, W., 44, 52, 62, 67, 107, 124, 165, 357, 377
Weck, N., 44, 350, 353, 358, 361, 363, 365, 370
Wei, L., 165, 222, 362
Weinacht, R. J., 44, 350, 351, 355, 362, 365, 374
Weinstein, A., 326, 348, 377
Wen, G. C., 377
Wendland, W. L., 4, 44, 222, 305, 350, 353, 358, 361, 362, 363, 365, 370, 377
Werner, B., 359
Werner, H., 361
White, A. M., 208, 377
Whittaker, E. T., 269, 272, 377
Widder, D. V., 35, 377
Wildenhain, G., 377
Willner, W., 369
Wirths, K.-J., 260, 305, 364, 374
Withalm, C., 305, 376, 377, 378
Wittich, H., 197, 223, 255, 378
Woithe, T., 369
Wolfersdorf, L. von, 348, 378
Wood, D. H., 176, 182, 183, 184, 196, 371, 378
Wu, H., 378

Y

Yoshihara, H., 348, 362
Young, E. C., 378
Yu, C.-L., 165, 378

Z

Zemanian, A. H., 162, 378
Zuev, M. F., 350

Subject Index

A

Adjoint equation, 98, 126
Adjoint operator, 98, 99
Approximation of solutions, 209–222
Approximation theorems and generalizations for solutions
  Mergelyan, 212–213
  Runge, 210–211
  Sewell, 216–217
  Walsh, 214–215
Associated function. (*See also* Goursat problem; Inversion problem.)
  with respect to the operator T_j, 15
  with respect to the operator $\tilde{T}_j$, 51
  of a solution w_j, 15
  of a solution $\tilde{w}_j$, 51
  standard, 117
  uniqueness, 34
Associated functions
  in terms of Goursat data, 32, 58
  relations between f_j and $\tilde{f}_j$, 51, 59, 60

B

Bauer differential operator, 52, 64, 261. (*See also* Class P operator, Polynomial kernel.)
Bauer–Peschl equation, 30, 52, 56
  differential operator $\tilde{T}_j$ for, 56
  kernel of the first kind for, 30
  Riemann function for, 142
Bauer–Peschl equation with $\lambda = -1$, 256–305. (*See also* Class $\mathscr{B}_n$; Function theory of class $\mathscr{H}_n(\Omega_1)$.)
  Bauer differential operator for, 261
    multiplication theorem, 269–270
  behavior of solutions near an isolated singularity, 292
Bauer–Peschl equation with $\lambda = -1$ (*cont.*)
  Bergman integral-free form of solutions, 262
    as multianalytic function, 298
  exceptional value of a solution, 290, 292
  function theory of special solutions, 298–305
  Goursat problem for, 261
    representations of solution, 262–265
  Great Picard theorem, analog for solutions, 254, 290
  Hadamard's three circles theorem, analog for solutions, 295
  maximum modulus theorem for solutions, 273
  Poisson's formula for disks, generalization of, 267, 269
  polynomial kernels of the first kind for, 260
  Privalov's uniqueness theorem, analog for solutions, 288
  real-valued solution, 274
  relations to other equations, 257–260
  representations of solutions, 260–272
  Riemann function for, 261
  Riemann's theorem on removable singularities, analog for solutions, 293
  Schottky's theorem, analog for solutions, 293
  Schwarz's lemma, analog for solutions, 288
  solutions constant on circles, 276, 282
  solutions with equal imaginary parts, 275
  solutions with equal real parts, 274–275
  T_1-associated function f_1, 262, 264
  $\tilde{T}_1$-associated function f_1, 263–265
Beke's theorem, 201–202
Bergman–Bojanić theorem, 342
Bergman–Gilbert kernel, extended, 163–164

Bergman–Gilbert operator, 147–164. (*See also* Elliptic equations, radial case; Gilbert's *G*-function; Gilbert's representation of solutions.)
  definition, 154
  relation to Carroll's theory of transmutations, 158–164
Bergman kernel $\tilde{k}_j$, 15. (*See also* Kernel.)
Bergman metric, 182
Bergman operator
  definition, 15
  first kind, 20
  general theory, 8–44
  historical root, 12
  point of reference, 15
  representation of solutions by, 15–16
  second kind, 328
Bergman representation theorem, generalized 15–16
Bergman space, 28
Bernoulli's law, 311
Bessel function
  of the first kind, J_ν, 13, 118, 158, 164, 168, 181, 183, 186, 192, 225, 232
  modified, K_ν, 346, 347
Beta function, 13, 149, 155, 332
  analytic continuation, 330–331
Borel criterion for rational functions, 201
Borel transform, 334
Boundary value problem, 158, 257, 314

C

Cauchy inequalities, 23, 219
Cauchy problem
  applied in determining the Riemann function, 168
  for the Tricomi equation, 329–340
Cauchy-Riemann equations, 299, 309
  generalizations, 299, 312, 323
Cauchy's integral theorem, 16, 303, 304
Chaplygin equations, 312, 323
  Tomotika–Tamada approximation, 324
Characteristic initial value problem, *see* Goursat problem
Chaundy's equation of generalized hypergeometric type, 171
Class $\mathscr{B}_n$, 296, 297
Class *F*, 37
Class $\mathscr{H}_n(\Omega_1)$, 300–304
Class $K(\tilde{k}_1)$, 224
Classes P_{jn}, P_{jn}^0, FP_{jn}, FP_{jn}^0, 47
Class *P* kernel, *see* Polynomial kernel
Class *P* operator, 45–107
  definition, 46
  existence theory, 63–107
  inversion problem, 57–62
Class *P* operators and corresponding partial differential equations
  general construction principles, 81–82, 92–93
  special cases, 63–68
Class *P* operator of the first kind, 46
  inversion problem, 57–58
Closed-form kernel (except class *P* kernel), 108–123
  special classes, 110, 119, 120
Coefficient problem for solutions, 197–206
Colton operators, 35, 41–43
Combined class *P* and exponential kernels, 118–120
Comparison function, 337
Complex potential, 309
Compressible fluid flow, 306–321
  Bergman's method, 317–321
  Bernoulli's law, 311
  Chaplygin's equation, 312, 323
  Chaplygin's method, 314–316
  complex potential, 309
  equation of continuity, 307–308
  equation of state, 307
  hodograph method, 312
  hodograph plane, 311
  Mach number, 311
  Method of pseudoanalytic functions, 312–314
  stream function, 309, 323–325
  velocity potential, 308–309
Construction principles for class *P* operators and corresponding equations, 81–82, 92–93, 99–102
Continuity equation, 307–308
Curvature constant, 182

D

Darboux equation, 326
Darboux type equation, 171, 189
Differential operator T_j, 51, 53–62. (*See also* Bauer differential operator, Class *P* operator.)
  definition, 51, 53
  relation to polynomial kernel, 51, 55
  solutions by means of, 53, 57
Dirichlet's formula, 33, 132, 133, 145, 146

Distribution of a-points, 230–233
Duhamel product, 43

E

Eichler operator, 341–344. (*See also* Tricomi equation.)
  criterion for self-adjoint equations, 341
  definition, 341
  generation by means of the Bergman–Bojanić theorem, 342
  representation of solutions, 341, 343, 344
Elementary solution
  generated from elementary associated function, 211
  in the sense of Hadamard, 170
Elliptic equation, 10, 125
Elliptic equation, radial case
  in two variables, 147, 149–150
  in n variables, 151–152
Entire functions
  of exponential type, 333, 334
  of finite order, 335
  of special form, existence of, 202
Equation of state, 307
Euler equation, 281
Euler–Poisson–Darboux equation, 259
Existence and construction of polynomial kernels, 63–107
Existence and representation theorems for
  exponential kernels, 111
  kernels of the first kind, 21–22
  kernels of the second kind, 327–328
  polynomial kernels for Type I representations, 72–73
  polynomial kernels for Type II representations, 78, 92–93
Exponential operators and kernels, 109–118
  definition, 110
  for equations with constant coefficients, 113, 117
  existence, 111
  ordinary differential equation for solutions, 114
  representation, 111
Extended Bergman–Gilbert kernel, 163–164

F

Flow. (*See also* Compressible fluid flow.)
  compressible, 306
  incompressible, 306
Flow (*cont.*)
  irrotational, 308
  nonviscous, 307
  sinks of, 308
  sources of, 308
  steady, 308
  subsonic, 313, 322
  supersonic, 313, 322
  transonic, 322
  two-dimensional, 309
Formal adjoint operator, *see* Adjoint operator
Fourier cosine transform, 168
Fourier integral method, 175–176
Frobenius irreducibility, 80
  and minimality of polynomial kernels, 81
Fuchs–Frobenius theory, 110
Function theoretical methods, 2
Function theory of class $\mathscr{H}_n(\Omega_1)$, 300–304. (*See also* Bauer–Peschl equation with $\lambda = -1$.)
  analytic continuation, 302
  Bauer derivative, 300
  Cauchy–Riemann equations, generalized (higher-order), 299
  Cauchy's theorem, generalized, 303, 304
  contour integral, generalized, 303, 304
  definition of $\mathscr{H}_n(\Omega_1)$, 300
  expansion theorem, 301
  function element, regular, 302
  identity theorem, analog, 301
  monodromy theorem, analog, 302–303
  nonisolatedness of a-points, 302
  primitive, generalized, 303, 304
Fundamental domain, 126
Fundamental solution, 170
  and Riemann function, 170–171

G

GASPT equation, 44, 258, 259
Generalized
  analytic function, 4, 298
  Cauchy–Riemann equations, 299, 312, 323
  constant, 301
  Hadamard's theorem on poles, 206
  holomorphic function of class $\mathscr{H}_n(\Omega_1)$, 300–304
  hyperanalytic function, 4, 298
  Leibniz formula, 270
  power, 300
Generating operator $\tilde{E}$, 37, 40–42

Gilbert's G-function. (*See also* Bergman–Gilbert operator.)
for Bauer–Peschl equation, 157
definition, 154
for Helmholtz equation, 158
independence of dimension, 154
integral transform of, 156
relation to Riemann function, 155
Gilbert's representations of solutions, 152, 154, 157, 158
Goursat problem, 12, 13, 31, 32, 57–59, 143, 175, 185, 261–264
Great Picard theorem
analog for solutions of the Bauer–Peschl equation, 290
analog for solutions by means of class P operators of the first kind, 254
for multianalytic functions, 252

H

Hadamard's elementary solution, 170
Hadamard's theorem on poles, 206
generalized, 206
Hadamard's three circles theorem, 295
analog for solutions of the Bauer–Peschl equation, 295
Hankel determinant $H_r^{(p)}(g)$, 200, 201
Hardy space of λ-harmonic functions, 260
Heat equation, 259
Helmholtz equation, 12, 29, 138, 169, 224
kernel of the first kind for, 30
n-dimensional, 157, 163
extended Bergman-Gilbert kernel for, 164
Gilbert's G-function for, 158
Riemann function for, 139
solution as series in terms of Bessel functions, 13
Hodograph method, 312
Hurwitz's theorem, 290
Hyperanalytic function, 298
generalized, 4, 298
Hypergeometric differential equation, 142, 315, 327, 346
Hypergeometric function
${}_2F_1$, 23, 123, 142, 156, 157, 168, 171, 174, 175, 181, 183, 185, 187, 316, 321, 327, 330, 332, 339, 343
Hypergeometric function (*cont.*)
generalized
type (0, 1), ${}_0F_1$, 334, 338, 339, 340
type (0, 2), ${}_0F_2$, 43
type (1, 1), ${}_1F_1$, 42, 50, 139, 185
of two variables
F_3, 169, 170
Ξ_2 (confluent), 170, 187
Φ_3, of Horn–Birkeland type, 41
of three variables
F_B^+ (confluent), 188, 189
of four variables
F_B, of Lauricella type, 190
Hypergeometric series, multiple, 172

I

Ill-posed problems, 3
Indirect (inverse) method, 3, 347
Integral equation for the Riemann function, 128
Integral-free form of solutions, 51
uniqueness, 57, 59, 93, 97, 262, 264
Integral operator
Bergman, *see* Bergman operator, Class P operator, Exponential operator, Operator T_j
Bergman–Gilbert, 147–161
Colton, 35, 41–43
Eichler, 341–344
Lanckau, 37
Le Roux, 125–128
Riemann–Vekua, generalized, 44. (*See also* Riemann function.)
Integral relation for first-kind kernels in Type II representations, 138
Integro-differential equations for class P_{jn}, 64
Inversion problem (*See also* Goursat problem.)
for class P operators, 57–62
for class P operators of the first kind, 57–58
for differential operators $\tilde{T}_j$, 58–60
for operators of the first kind, 31–34
Iteration
Picard, 166, 180, 187
functionally stable, 182

J

Jacobi polynomials $P_m^{(a,b)}$, 333

K

Kernel, Bergman kernel $\tilde{k}_j$, 15
Bergman–Gilbert, extended, 163
of class P, 46
of closed form, 108–123
exponential, 110
of the first kind, 20
determination from the Riemann function, 141
existence, 21–22
recursive system for, 21, 28, 47
representation, 21–22, 28–29
uniqueness of its even part, 29
polynomial, 46
rational, 120–123
of the second kind, 328
for the Tricomi equation, 326, 328
Kernel equation, 8, 15
recursive system, 21, 28, 47
Kummer's function ${}_1F_1$, 42, 50, 139, 185

L

Lagrange–Bürmann formula, 270
Laguerre function, 183
Lanckau's equation, 36–44
generating operator $\tilde{E}$, 37
associated class F, 37
representation of solutions, 37–38
for special cases, 36, 39–43
Lanckau's integral operator, 37
Laplace
equation, 103, 269, 299, 309
invariants h_{jr}, 70, 77
operator of hyperbolic geometry, 257
second integral for Legendre polynomials, 269
transform, 162, 176, 185
one-sided, 176
two-sided, 162
Laplacian, n-dimensional, Δ_n, 151
Leau-Faber theorem, 202–203
Legendre differential equation, 280, 344
Legendre function
of the first kind, P_n, 168, 171, 258, 269, 271, 272, 280, 295, 343, 345
of the second kind, Q_n, 258, 271, 272, 280, 295
recursion formula of Hermite and Schläfli, 272
representation, 271, 272
Legendre polynomial, 142, 172, 259
Laplace's second integral for, 269
Le Roux kernel R_j, 127
Le Roux operator J_j, 125–128
criterion for, 127
definition, 127
solution of Goursat problems by means of, 128
Liouville equation, 101
Liouville's formula, 27, 42
Little Picard theorem, 246
generalized for solutions, 247
$\log^+$, 233

M

Mach number, 311
Maximal convergence, 214, 215
Mellin transform, 161
generalized, 162
Method
of ascent, 147
of dominants (majorants), 23, 129, 132, 320
indirect, 3, 347
of Laplace invariants, 68
Mixed problems, 324
Multianalytic function, 250–252, 298
definition, 250
exceptional value of, 252
Great Picard theorem for, 252
isolated essential singularity of, 251

N

Nachbin's theorem, 338
n-Analytic function, 259–260
Nevanlinna's
characteristic function, 234
First Fundamental Theorem, 234
generalized for solutions, 234
Second Fundamental Theorem, modified, 236–237
generalized for solutions, 244
n-Harmonic functions, 259, 260
Nonisolatedness of a-points, 302
Nontangential path, 288
Number of a-points of a function, 231
generalized, 231–233

O

Operator T_j. (*See also* Bergman operator.)
 of class P, *see* Class P operator
 exponential, 109–118
 of the first kind, 20–34
  existence and representation theorem, 21–22
  Bergman space preservation property, 27–28
  inversion problem, 31–34
  representation of solutions by means of, 15–16
 of the second kind, 221, 328
Operator $T_1^{\times}$, 225
Order of growth
 of a function, 226–228
 of a solution, 229–230
Ordinary differential equation
 Bessel, 325
 Euler, 281
 hypergeometric, 142, 315, 327, 346
 Legendre, 280, 344
 for solutions of partial differential equations, 114–117, 201–202
 for Tricomi equation kernels, 327

P

Parabolic equation, 34–36, 40, 42, 257
 analyticity of solutions, 35
Partial differential equations
 Bauer–Peschl, *see* Bauer–Peschl equation
 Chaplygin, 312, 323
 Chaundy, 171
 with constant coefficients, 109
 Darboux, 326
 Darboux type, 171, 172, 189
 of Eisenstein series, 258
 elliptic, 10, 39, 125, 147, 149–152
 Euler–Poisson–Darboux, 259
 of generalized axially symmetric potential theory (GASPT), 44, 258, 259
 of generalized hypergeometric type, 171
 heat, 259
 Helmholtz, *see* Helmholtz equation
 hyperbolic, 39, 167, 257
 Lanckau, 36–44
 Laplace, 103, 269, 299, 309
Partial differential equations (*cont.*)
 Liouville, 101
 mixed type, *see* Tomotika—Tamada equation, Tricomi equation
 parabolic, 34–36, 40, 257
 pseudoparabolic, 42
 real elliptic, 10, 125
 real hyperbolic, 167
 reduced, 9
 Tomotika–Tamada, 325
 Tricomi, *see* Tricomi equation
 ultrahyperbolic, 257
 wave, 3, 257, 259
 Weinstein, 325–326
Picard iteration, 166, 180, 187
Picard–Krajkiewicz theorem, 252
 analog for solutions, 254
Picard theorem (on analyticity of solutions), 313
Picard theorem (on exceptional values), 246, 247, 252, 254, 290
Point of reference, 15
Poisson formula for harmonic functions, 269
Pole-like singularities, 200–207
Polyanalytic function, 250
Pólya representation, 334
Polynomial kernel
 definition, 46
 degree, 46
 existence for Type I representations, 64, 68–77
 existence for Type II representations, 77–107
 of the first kind, 46, 48, 50
 generation of higher degree, 48
 minimality, 46
  and Frobenius irreducibility of associated equation, 81
 recursion for its coefficients, 47
 for special equations, 49, 50, 64–67, 76, 87, 96–99, 120
 transformation of, 76
Power
 formal, 313
 generalized, 300
Privalov's uniqueness theorem, 288
 for solutions of the Bauer–Peschl equation, 288
Proximity function, 233
 analog of, 233
Pseudoanalytic function, 4, 298, 313
Pseudoparabolic equation of third order, 42

R

Rational kernels, 120–123
Real
  -analytic solution, 12, 148
  elliptic equation, 10, 125
  hyperbolic equation, 167
  solution, 11, 148
    singularities of, 200, 203, 206
Reduced equation, 9
  reducing factor e_j, 9, 137
Reduction of Type II representation to Type I, 31, 276
Representation of solutions, 15–16, 21–22, 26, 32, 35, 37–38, 51, 57–60, 72–73, 92–93, 111, 143–144, 149–152, 155, 260–269, 298, 328, 332–334, 341, 343–345
Riemann function, 128–147, 166–196
  defining Goursat problem, 134
  defining Volterra-type equation, 128
  determination from the Bergman kernel of the first kind, 137–138, 140
  equivalent definitions, 134
  existence, 133
  nonsimple, 185–187
  relation to Green's function, 171
  simple (in the sense of Wood), 183–184
  symmetry, 135
  transformation behavior, 136–137
  uniqueness, 133
Riemann functions
  addition formula for, 176–179
  multiplication formula for, 180
Riemann functions, methods for explicit determination and results by:
  Bauer, 194–195
  Chaundy, 171–172
  Chaundy–Mackie, 172–175
  Cohn, 180–182
  Copson, 173
  Daggit, 183
  Du, 180
  Florian–Püngel, 192–193
  Hadamard, 170–171
  Henrici, 169–170
  Kracht–Kreyszig, 137–140, 195–196
  Lanckau, 188
  Olevskij, 176–180
  Püngel, 189–191
  Riemann, 167–170
  Scott, 185
  Vol'man–Pampu, 186–187
Riemann functions, methods for explicit determination and results by (*cont.*):
  Wahlberg, 185–186
  Wallner, 193–194
  Wood, 183–184
Riemann functions for special equations, 139, 142, 166–196
Riemann mapping theorem for n-analytic functions, 260
Riemann–Vekua function, *see* Riemann function
Riemann–Vekua operator, generalized, 44
Riemann–Vekua representation, 143
Rouché's theorem, application of, 232, 248

S

Schottky's theorem, 293
  analog for solutions of the Bauer–Peschl equation, 293
Schwarz's formula, 268
Schwarz's lemma, 285
  generalized, 285
Self-adjoint equation, existence of class P operators, 50, 98–107
Singularities of real solutions, 200, 203, 206
Singularities of solutions
  dependence on associated function, 207
  location, 203, 206, 207
  number, 200, 203, 206
  type, 200, 201, 203, 205, 207, 208
Solution, real, 11
  real-analytic, 12
Sonic curve (sonic line), 328, 329
Standard associated function, 117
Stream function, 309, 323–325
Subsonic flow, 313, 322
Supersonic flow, 313, 322
Surface harmonic, 153

T

Time-dependent processes in the plane, *see* Lanckau's equation
T_j-associated function f_j, 15
  relation to $\tilde{f}_j$, 51, 59–60
$\tilde{T}_j$-associated function $\tilde{f}_j$, 51, 53
  relation to f_j, 51, 59–60
Tomotika–Tamada equation, 325
Translation principle, 2, 6
Transmutation, 159, 161
Transonic flow, 322–348

Tricomi equation
  application of the Bergman–Bojanić theorem, 342–343
  behavior of solutions on the sonic line, 331–332
  Bergman operator of the second kind for, 328
  Bergman representation theorem for solutions of, 328
    applied to a Cauchy problem, 332
  Cauchy problem for, 329–340
  derivation of the equation, 323
    complex form, 326
    related equations, 325, 326
  Eichler operator for, 343
  Eichler representation theorem for solutions of, 344
    Lanckau's form of, 345
  entire Cauchy data for solutions of, 333–340
  integral operator approach, 322–348
  particular solutions by:
    Carrier–Ehlers, 346
    Chaplygin, 346–347
    Darboux, 346
    Falkovich, 347
    Germain–Bader, 347
    Guderley, 345–346
    Tomotika–Tamada, 346
  polynomial expansion of solutions of, 338
Type I representation of solutions, 19
Type II representation of solutions, 19

U

Ultrahyperbolic equation, 257
Uniqueness of associated function, 32, 57–60, 262–265

V

Value distribution theory of solutions, 223–255
Velocity potential 308–309
Volterra integral equation, 128, 131

W

Wave equation, 3, 257
Weinstein equation, 325–326
Wirtinger operator, 10

Z

Zero solution, general representation by means of differential operators, 93, 264